IEE ELECTROMAGNETIC WAVES SERIES 41

Series Editors: Professor P. J. B. Clarricoats
Professor J. R. Wait
Professor E. V. Jull

Approximate boundary conditions in electromagnetics

Other volumes in this series:

Approximate boundary conditions in electromagnetics

T. B. A. Senior and
J. L. Volakis

The Institution of Electrical Engineers

Published by: The Institution of Electrical Engineers, London,
United Kingdom

© 1995: The Institution of Electrical Engineers

The Institution of Electrical Engineers,
Michael Faraday House,
Six Hills Way, Stevenage,
Herts. SG1 2AY, United Kingdom

British Library Cataloguing in Publication Data

A CIP catalogue record for this book
is available from the British Library

ISBN 0 85296 849 3

Printed in England by Bookcraft, Bath

Contents

Preface

Non-metallic materials and composites are now commonplace in modern vehicle construction and are crucial to the operation of many microwave devices. It is no longer sufficient to confine attention to metallic surfaces, and the need to compute scattering and other electromagnetic phenomena in the presence of material structures has led to the investigation of new simulation and solution techniques. In most cases it is necessary to introduce some type of simplification to make the problem tractable and to achieve a solution simple enough to use. Approximate boundary conditions have proved effective for simplifying calculations dealing with material surfaces whose solution would be difficult or impractical without them. They have been used extensively to generate analytical solutions to canonical problems and are easily incorporated into numerical codes. In fact, approximate boundary conditions of a special type are essential in finite element and finite difference codes where they are used to confine the computational domain to the immediate vicinity of the scattering or radiating structure.

This book provides the first comprehensive treatment of a variety of approximate boundary conditions applied to electromagnetics. Methods for deriving them are discussed and analytical solutions are developed for a number of canonical problems involving material interfaces, coatings, layered structures and perturbed surfaces. These analytical solutions are important in the development of high-frequency electromagnetic scattering codes and should therefore be of interest to practicing engineers as well as graduate students concerned with high-frequency diffraction by impedance structures. The chapters are written in a pedagogical manner with the background material necessary to make the book suitable for graduate students who have completed a standard first course in electromagnetic theory. An extensive list of references is appended to each chapter for further reading from the original sources, but a substantial amount of the material is new, and has not appeared in the literature before. Throughout the book, many numerical calculations are given which should be invaluable to prospective users in demonstrating the accuracy afforded by the approximate boundary conditions. Overall, the book is intended to provide an up-to-date and comprehensive coverage of the genealogy and application of standard and higher order impedance boundary conditions, without overlooking such delicate points as solution uniqueness and accuracy.

A key feature of the book is the treatment of higher order boundary conditions (referred to in the book as generalised impedance boundary conditions). In contrast to the standard (or first order) condition, these have additional degrees of freedom which allow improved simulation of more complex material properties in scattering and propagation problems. They are of particular interest at the present time, and one of their important applications is the simulation of non-reflecting surfaces. The last chapter of the book is, indeed, devoted to the development of this class of non-reflecting (or absorbing) boundary condition and should be of interest to students and scientists concerned with finite methods. This chapter surveys traditional and more recent two- and three-dimensional absorbing boundary conditions and compares them with the generic forms developed in the earlier part of the book.

We are grateful to Professor J. R. Wait, Series Editor of the IEE Electromagnetic Waves Series, and to the IEE for giving us this opportunity to collect together results which were previously scattered in a variety of technical journals, and to develop those results necessary to fill the gaps. Although a labour, it has also been a labour of love that has given us a great deal of personal satisfaction, and we hope that the end product will prove useful to the reader.

We are indebted to our graduate students, past and present, who have contributed to this work, and we recognise in particular the contributions of Drs. M. I. Herman, M. A. Ricoy, H. H. Syed and A. Chatterjee, as well as those of T. Özdemir, S. R. LeGault and S. Bindiganavale. Last but not least, we acknowledge the skill of Mr. R. Carnes in converting a hand-written manuscript into camera-ready copy.

T. B. A. Senior   November 1994
J. L. Volakis

Chapter 1

Introduction

Approximate boundary conditions provide an approximate relationship between the electric and magnetic field on a chosen surface. In general, their purpose is to simplify the analytical or numerical solution of wave problems involving complex structures. They are important in all disciplines, e.g. acoustics, hydrodynamics and electromagnetics, where boundary conditions are involved, and are becoming more so as we seek to model more complicated situations. Some versions have also been around for a long time and, although the classical condition $\mathbf{E}_{\text{tan}} = 0$ at the surface of a metal is often regarded as exact, it is in fact an approximation for all metals even at microwave frequencies.

In electromagnetics (to which we shall devote our attention) approximate boundary conditions are now widely used in scattering, propagation and waveguide analyses to simulate the material and geometric properties of surfaces. Take, for example, a finite body immersed in a homogeneous medium and illuminated by an electromagnetic field. Knowing the material properties of the body, it is possible, in principle, to find the scattered field exterior to the body, by taking into account the behaviour of the fields within the body. However, the task would be greatly simplified if the properties could be simulated via a boundary condition involving only the external fields imposed at the outer surface, thereby converting a two (or more) media problem into a single medium one. This is the objective of an approximate boundary condition. Although it is generally convenient for the surface of the actual body to coincide with the surface where the approximate boundary condition is enforced, this is not necessary. The only requirement is that in the region of interest the field obtained using the postulated condition approximates the exact field to an adequate degree of accuracy. If the exact solution of the original problem were known, we could always construct *a posteriori* an equivalent boundary condition which, if imposed at the surface, would precisely reproduce the fields throughout the region exterior to the body. Unfortunately, the condition would almost certainly be unique to that particular situation and, to be useful, an approximate boundary condition has to be applicable to a class of excitations and body configurations over and beyond the specific one for which it was derived.

Typically, an approximate boundary condition is derived by considering the solution of a canonical problem which is similar to the original, but more complicated. For example, if a homogeneous lossy dielectric body is replaced by an impenetrable surface which coincides with the outer boundary of the dielectric, this new surface can be chosen to satisfy the boundary condition extracted from the exact solution of a circular dielectric cylinder of the same composition. Alternatively, a coated body can be simulated by a surface which satisfies the boundary condition obtained from the exact solution of the corresponding coated circular cylinder. For greater simplicity, though, one could choose to simulate a curvilinear coated surface by a boundary condition derived from the reflection properties of a planar coated surface. As expected, the simulation based on the planar coating is less accurate, but because of its simplicity, it can be implemented in a greater variety of situations. An example of this is the physical optics approximation that continues to be widely used in spite of its lack of rigour.

As the title implies, the book's goal is to familiarise engineers and scientists with approximate boundary conditions and their applications in electromagnetics. We discuss both the development of approximate boundary conditions and their application to electromagnetic problems of practical interest. We begin with the derivation of the most simple and widely used boundary conditions and then proceed to their more complicated forms. The accuracy of the different approximate boundary conditions is discussed in connection with their application to scattering by specific configurations such as metal-backed dielectric coatings, homogeneous single-layer structures and multi-layer configurations. Nearly half of the book is devoted to the development of analytical solutions for scattering by a variety of material discontinuities which are important in diffraction theory, and in all cases numerical calculations are presented. Most of these are presented in the open literature for the first time, and should help in understanding the results and their implications. The last chapter of the book is devoted to a special class of approximate boundary conditions which are now widely used in finite methods for the numerical solution of the scalar and vector wave equations. This particular chapter should be of interest to readers in computational electromagnetics, and shows that the usual absorbing boundary conditions employed in numerical solutions are a special class of approximate boundary conditions. Because of this connection, the methods discussed in the book are also suitable for generating improved absorbing boundary conditions for numerical solutions of the wave equation.

Having given a brief overview of the book, we next discuss the contents of each chapter in sufficient detail to guide readers in their study and use of the book.

The simplest approximate boundary conditions are the (standard) impedance conditions applicable at the surface of a lossy dielectric, and the related transition conditions which model a thin dielectric layer as a current sheet. Although these have been available for many years (the transition conditions were first proposed 80 years ago), it was not until World War II that the impedance boundary condition came into use in connection with ground wave propagation over the Earth. Since then it has become an essential tool for modeling an opaque surface for which a local surface impedance can be identified. Many numerical codes have been written making use of these conditions, and analytical solutions to canonical impedance problems have been developed for use in high frequency techniques such as the geometrical and physical theories of diffraction. As a result of their simplicity, ease of use and successful application, improved or higher order versions of the impedance boundary conditions are now being considered for electromagnetic applications. These higher order impedance boundary conditions, often referred to as generalised impedance boundary conditions (GIBCs), permit the simulation of more complicated material and composite surfaces with greater accuracy. Special types are the so-called absorbing boundary conditions (ABCs) which are required in the implementation of finite element and finite difference numerical solutions of the wave equation. Interest in generalised impedance and absorbing boundary conditions has increased substantially over the past decade because of the popularity of composites and the development of computational tools for radar scattering, antenna, microwave circuit and propagation problems. To a certain degree, this motivated the writing of this book.

Depending on the form in which the boundary condition is expressed, it may involve (at most) a first derivative of the field, and for this reason it is referred to as a first order condition. In Chapter 2 the derivation of the first order (standard) impedance boundary and transition conditions is described. The close connection between them is established and various equivalent mathematical forms of the boundary conditions are given. Also, their accuracy is examined in modelling planar metal-backed coatings, dielectric layers and cylinders, and recommendations are provided regarding the limits of their validity. We also present a generalisation of the standard impedance and transition conditions for modelling anisotropic surfaces, and discuss their application to rough and corrugated surfaces, including corrugated waveguides. The final section of the chapter deals with the uniqueness of boundary value problems involving the standard impedance boundary conditions, leading to restrictions on the impedance values for modelling passive structures.

One of the most successful applications of the standard impedance boundary condition deals with problems in diffraction theory and Chapter 3 is completely devoted to diffraction by half plane or sheet junctions satisfying the standard impedance or transition conditions. Explicit diffraction coefficient expressions are given for impedance, resistive and conductive half planes as well as for material junctions simulated using transition conditions. Both polarisations are treated for two- and three-dimensional excitations, and efforts have been made to present the results in a compact form while avoiding cumbersome symbolic notations. Throughout the book, the dual integral equation method, rather than the Wiener-Hopf approach, is used for solving the boundary value problems involving half planes and junctions. The dual integral equation method is discussed in detail at the beginning of the chapter, and it is shown that this method, although completely equivalent to the Wiener-Hopf approach, is much simpler to use. The first application discussed in Chapter 3 is plane wave diffraction by a resistive half plane, since its solution can be used to generate the corresponding ones for conductive and impedance sheets by exploiting duality and the equivalence of the associated conditions. The solution for diffraction by a resistive half plane is carried out in sufficient detail to be followed by readers not previously exposed to this type of method. The relationship of the integrand poles to the geometrical optics fields is discussed, and uniform solutions in the context of the uniform geometrical and asymptotic theories of diffraction are presented. Many calculated data are also included for the uniform and non-uniform diffraction coefficients showing the effect of the impedance, resistivity or conductivity on the echowidth of the half plane. The latter part of the chapter deals with impedance and sheet junctions excited by a plane wave at skew incidence, i.e. incidence not in a plane perpendicular to the junction. Although rather involved because of the presence of two independent current density components, the skew incidence solution procedure is similar to that in the two-dimensional case. The chapter closes with generalisations to curved edges, and extensions of the analysis methods to sheet junctions satisfying non-uniform boundary conditions.

Chapter 4 is concerned with the problem of plane wave diffraction by impedance wedges illuminated at normal and skew incidence. This is an important problem because the diffraction coefficients are essential in formulating high frequency codes for impedance structures. The first part of the chapter is devoted to the solution at normal incidence and is essentially an expanded version of Maliuzhinets' work. Here, particular attention is given to the pedagogical development of the solution, and the presentation of the diffraction coefficients and associated special functions in a form suitable for numerical implementation. A substantial portion of the chapter deals with the skew incidence solution since, to date, an exact skew incidence solution for arbitrary wedge angles has eluded scientists. However, an approximate solution is possible and is presented here for the first time, thereby opening the way for constructing high frequency solutions for complex scatterers based on the geometrical or physical theories of diffraction. The latter theory is developed for impedance structures and several numerical calculations are presented for simple bodies that demonstrate the accuracy of the approximate skew incidence dyadic diffraction coefficient. Also, as in Chapter 3, many additional calculations are given showing the behaviour of the diffraction coefficient for isolated impedance wedges illuminated at normal and skew incidence.

To improve the accuracy of the standard impedance boundary or transition conditions, one approach is to consider conditions which involve higher derivatives of the field at the surface. This is similar to retaining more terms in a Taylor or Maclaurin series expansion and, in fact, a class of GIBCs can be derived on the basis of such an expansion. Chapter 5 is devoted to the derivation and application of generalised impedance boundary and transition conditions which contain up to second order derivatives of the field. They are consequently referred to as second order generalised impedance boundary and transition conditions, respectively, and the first part of Chapter 5 presents various equivalent mathematical forms, followed by their application in modelling dielectric coatings, thin layers and wire meshes. Explicit forms of the conditions are given for these situations, and it is shown how their accuracy can be improved by considering alternative expressions dictated by the material properties. The latter part of the chapter is concerned with the use of the conditions in diffraction analyses. It is shown that explicit solutions for the coated half plane and wedge (as well as layer and coating junctions) are possible when simulated by second order GIBCs. These solutions are carried out by the dual integral equation or Maliuzhinets method presented in Chapters 3 and 4, respectively. In the case of the second order boundary conditions, however, additional care is required to ensure the uniqueness of the solution and this is a difficulty encountered with all higher order boundary and transition conditions beyond the first. This chapter clarifies in a concise manner the solution uniqueness issues, and develops the supplemental conditions which must be enforced in any analytical or numerical implementation of the higher order impedance or transitions conditions.

Third (and higher) order conditions have still more degrees of freedom and, in principle at least, provide even better simulations. For most practical applications it is sufficient to restrict attention to even order conditions, and their general form is discussed in Chapter 6. Uniqueness proofs are developed and we show how to construct surfaces which are complementary in the sense of Babinet's principle. Then, in Chapter 7, we illustrate the application of GIBCs to curved surfaces. Explicit second order GIBCs are derived for homogeneous and coated circular cylinders, and a procedure is presented for generating GIBCs of any order. Numerical calculations are also included which demonstrate the improved accuracy of the second order GIBC over the standard impedance boundary conditions (SIBCs). The latter part of Chapter 7 is devoted to the simulation of non-circular coated cylinders. Specific second order GIBC expressions are derived, and these are used to develop high frequency solutions in the context of the geometrical theory of diffraction. Creeping and surface wave diffraction coefficients are derived in terms of Fock-type functions, and calculations are included which demonstrate the accuracy of the simulation.

In all of the above, an approximate boundary condition was required to simulate an actual surface, and the parameters were chosen to achieve this end. However, there is another purpose for an approximate boundary condition, and this is to *create* a boundary which does not perturb a field incident upon it—in effect, to simulate a surface which is actually not there. The resulting conditions can be regarded as GIBCs for non-reflecting surfaces, and are generally referred to as absorbing boundary conditions (ABCs). They are of growing importance in numerical work where they are used to terminate the computational domain in a finite element or finite difference solution of the wave equation, and are discussed in Chapter 8. The various methods that have been proposed for the construction of ABCs are summarised and the resulting ABCs are cast in a common form for comparison purposes. Both two- and three-dimensional ABCs are derived in a consistent manner, and particular attention is given to conformal boundary conditions, i.e. conditions which can be applied at a doubly curved surface. We also establish the connection between GIBCs and ABCs and show how the procedure described in Appendix A for the derivation of GIBCs, applicable at the curved surface of a dielectric body, can be used to develop a general class of ABCs.

Chapter 2

First order conditions

2.1 Basic equations

We consider only time harmonic electromagnetic fields with a time dependence specified by the factor $e^{j\omega t}$ which is omitted. In a stationary, linear, isotropic, homogeneous medium which is free of sources, the field is described by Maxwell's equations

$$\nabla \times \mathbf{E} = -jkZ\mathbf{H} \qquad \nabla \times \mathbf{H} = jkY\mathbf{E} \qquad (2.1)$$

where $\mathbf{E}$ and $\mathbf{H}$ are the complex phasors representing the electric and magnetic fields respectively, $Z = 1/Y = \sqrt{\mu/\epsilon}$ is the intrinsic impedance of the medium, and $k = \omega\sqrt{\epsilon\mu}$ is the propagation constant or wave number. The permittivity and permeability of the medium are ϵ and μ respectively, and these may be complex, incorporating the effect of losses. In the particular case of a vacuum, $Z = 120\pi$ ohm, and we distinguish the corresponding parameters by the suffix "o". SI units are used throughout.

A consequence of (2.1) is

$$\nabla \cdot \mathbf{E} = \nabla \cdot \mathbf{H} = 0 \qquad (2.2)$$

and by eliminating $\mathbf{H}$ or $\mathbf{E}$ from (2.1) it follows that

$$(\nabla^2 + k^2)\mathbf{E} = 0 \qquad (\nabla^2 + k^2)\mathbf{H} = 0 \qquad (2.3)$$

We observe that all of the above equations are invariant under the transformation

$$\mathbf{E} \to \mathbf{H}, \quad \mathbf{H} \to -\mathbf{E}, \quad \epsilon \leftrightarrow \mu, \quad Z \leftrightarrow Y$$

and this is referred to as the *duality* of Maxwell's equations.

At an interface between two media, the properties of the overall medium change discontinuously, and this may result in discontinuities in some components of the field. The transition conditions (generally referred to as boundary conditions) relating the fields on the two sides of the interface can be deduced

7

from the integral form of Maxwell's equations using standard procedures (see, for example, TAI (1994)). The conditions are

$$[\hat{n} \times \mathbf{E}]_-^+ = -\mathbf{J}_{m(s)} \qquad [\hat{n} \cdot \mu\mathbf{H}]_-^+ = \rho_{m(s)} \qquad (2.4)$$

$$[\hat{n} \times \mathbf{H}]_-^+ = \mathbf{J}_{e(s)} \qquad [\hat{n} \cdot \epsilon\mathbf{E}]_-^+ = \rho_{e(s)} \qquad (2.5)$$

where $[\ ]_-^+$ denotes the discontinuity between the upper $(+)$ and lower $(-)$ sides of the interface, $\hat{n}$ is a unit vector normal to the interface directed into the upper medium, and $\mathbf{J}_{e(s)}$ and $\rho_{e(s)}$ are the surface electric current and charge distributions respectively. The suffix "m" denotes the corresponding magnetic quantities, introduced for convenience.

If the lower medium is a perfect electric conductor (pec material), all fields there are zero by definition and $\mathbf{J}_{m(s)} = \rho_{m(s)} = 0$. Then

$$\hat{n} \times \mathbf{E}^+ = 0 \qquad (2.6)$$
$$\hat{n} \cdot \mathbf{H}^+ = 0 \qquad (2.7)$$

from (2.4), and these two conditions are equivalent. In practice, it is usually more convenient to employ (2.6), and this is the standard pec boundary condition. Alternatively, if the lower medium is a perfect magnetic conductor (pmc material), all fields there are again zero and $\mathbf{J}_{e(s)} = \rho_{e(s)} = 0$. Then

$$\hat{n} \times \mathbf{H}^+ = 0 \qquad (2.8)$$

or equivalently

$$\hat{n} \cdot \mathbf{E}^+ = 0 \qquad (2.9)$$

and (2.8) is the standard pmc boundary condition. Admittedly, both cases are idealisations and, though a pmc material is even more difficult to achieve than a pec one, it is a useful concept for mathematical purposes. If neither medium is a pec or pmc, the interface cannot support a surface current, and $\mathbf{J}_{e(s)} = \mathbf{J}_{m(s)} = 0$. The resulting "boundary" conditions are

$$[\hat{n} \times \mathbf{E}]_-^+ = 0 \qquad [\hat{n} \times \mathbf{H}]_-^+ = 0 \qquad (2.10)$$

showing the continuity of the tangential components of the electric and magnetic fields across the interface. Moreover, if either medium has non-zero (but finite) loss, $\rho_{e(s)} = \rho_{m(s)} = 0$ and hence

$$[\hat{n} \cdot \mu\mathbf{H}]_-^+ = 0 \qquad [\hat{n} \cdot \epsilon\mathbf{E}]_-^+ = 0 \qquad (2.11)$$

which are equivalent to (2.10).

The above conditions all involve the unit vector normal to the interface $\hat{n}$ but, at a surface singularity such as a vertex or edge, the Gaussian curvature is infinite and there is no unique normal. To ensure a unique solution to the boundary value problem it is therefore necessary to invoke an additional boundary condition in the form of an edge condition. This requires that the energy contained in any finite volume about the surface singularity is finite (MEIXNER, 1949) and restricts the maximum allowed singularity of any field component. For example, in the case of a wedge-type singularity of interior angle 2Ω, it can be shown (VAN BLADEL, 1985) that the components of the electric and magnetic fields parallel to the edge behave like ρ^ν and those perpendicular to the edge like $\rho^{\nu-1}$ as $\rho \to 0$, where ρ is the perpendicular distance from the edge and $\nu = \pi/(2\pi - 2\Omega)$.

In many instances we will be concerned with scattering in an infinitely extended region, and this demands that we impose a pseudo boundary condition at infinity. If the scatterer and all primary sources are located within a finite distance from a fixed origin $r = 0$, the fields must satisfy the Silver-Müller radiation condition

$$\lim_{r\to\infty}[\mathbf{r} \times (\nabla\times) - jkr]\mathbf{E} = 0$$

$$\lim_{r\to\infty}[\mathbf{r} \times (\nabla\times) - jkr]\mathbf{H} = 0$$

(2.12)

uniformly in $\hat{r}$. For plane wave incidence it is necessary to separate the incident and scattered fields, and only the scattered fields are required to satisfy (2.12). In the special case of a two-dimensional problem where the scatterer and the incident field are both independent of a Cartesian coordinate z, the radiation condition (2.12) must be replaced by

$$\lim_{\rho\to\infty} \rho^{1/2}\left(\frac{\partial}{\partial\rho} + jk\right) E_z = 0$$

$$\lim_{\rho\to\infty} \rho^{1/2}\left(\frac{\partial}{\partial\rho} + jk\right) H_z = 0$$

(2.13)

and for a plane wave at broadside incidence, (2.13) is applicable to the scattered field components only.

2.2   Classical boundary conditions

For scalar fields such as those in acoustics or heat conduction, the classical boundary conditions are the Dirichlet and Neumann ones, and this is natural in view of the Helmholtz representation of a scalar field (see, for example, GOODMAN (1968)). In acoustics the boundary condition on the velocity potential U at a soft or yielding surface S is the Dirichlet condition

$$U = 0$$

(2.14)

corresponding to zero excess pressure on S. At a hard or rigid surface the normal component of the fluid velocity is zero, giving rise to the Neumann condition

$$\frac{\partial U}{\partial n} = 0 \qquad (2.15)$$

In practice, however, the surface may yield a little under the influence of the pressure, and a boundary condition which simulates this to some degree is

$$\frac{\partial U}{\partial n} - jk_0 \frac{Z_0}{\eta} U = 0 \qquad (2.16)$$

where η is the specific impedance of the surface material and Z_0 is the intrinsic impedance of the surrounding medium, and (2.16) is often employed as a mathematical condition without further justification. Like (2.14) and (2.15), it is a true boundary condition applied to the velocity potential exterior to the scatterer, and requires no consideration of the interior potential.

In heat conduction, with U representing the temperature, (2.14) and (2.15) are the isothermal and adiabatic boundary conditions respectively, and (2.16) can be derived as an approximation to the Stefan-Boltzmann radiation law (SOMMERFELD, 1949). The parameter η/Z_0 is then a negative purely imaginary quantity.

There are many similarities to the boundary conditions appropriate in electromagnetics, and the similarities are not limited to two-dimensional problems which are obviously scalar ones for the z components of the field. For example, at a pec material, (2.6) and (2.2) imply

$$E_t = 0 \qquad \frac{\partial E_n}{\partial n} = 0 \qquad (2.17)$$

where $\hat{t}$ is a tangential vector, and likewise, from (2.7) and (2.2),

$$H_n = 0 \qquad \frac{\partial H_t}{\partial n} = 0 \qquad (2.18)$$

Thus, to the extent that we can work with individual components of the field, the boundary conditions at a pec material are equivalent to a Dirichlet one for E_t and H_n and a Neumann one for E_n and H_t. In the dual case of a pmc material, the scalar equivalents of (2.8) and (2.9) are Dirichlet boundary conditions for E_n and H_t and Neumann boundary conditions for E_t and H_n.

The development of an electromagnetic equivalent of (2.16) is attributable to scientists at the P. N. Lebedev Institute of Physics in Russia. In the early years of World War II, RYTOV (1940) considered the problem of a highly conducting body and, at the suggestion of M. A. Leontovich, sought a boundary condition which would be an improvement over the condition of (2.6) or (2.7) for infinite conductivity. For a homogeneous body whose boundary is a coordinate surface in a system of orthogonal curvilinear coordinates, Rytov showed that the tangential components of the electric field at the boundary could be expressed as power series in the (small) penetration depth d, with coefficients which involve the tangential magnetic field components and their tangential derivatives. He also examined the case when the conductivity is a function of position as well as the simpler case of a planar boundary, and derived some of the restrictions on the validity of the results. A more general version of Rytov's analysis is summarised in Appendix A.

Using arguments similar to those of Rytov, GRÜNBERG (1942) obtained directly a boundary condition applicable at the planar surface of a homogeneous lossy half space $y < 0$. For a field incident from above, Grünberg showed that to the first order in d

$$\frac{\partial E_y}{\partial y} - jk_0 \frac{Z}{Z_0} E_y = 0 \qquad (2.19)$$

where Z is the intrinsic impedance of the lower medium and, in a study of coastal refraction, used (2.19) to model the land with the sea treated as a perfect conductor. For a rectilinear coastline, the field on the surface of the land was cast as a Wiener-Hopf integral equation, but the solution was obtained only as a perturbation of that for an infinite perfectly conducting surface.

As part of a comprehensive study of ground wave propagation, FEINBERG (1944) examined the local behaviour of the field at the curved surface of an inhomogeneous lossy non-metallic body, and developed an expression for $\partial E_n/\partial n$ in terms of the material properties and other components and derivatives of the field. He also considered the case of a surface uniformly and isotropically inhomogeneous with a (small scale) roughness, and used the resulting boundary condition to determine (FEINBERG, 1945) the field of a dipole above a plane surface simulating an imperfect ground. Finally, the problem of coastal refraction was addressed (FEINBERG, 1946).

In his first paper FEINBERG (1944) noted that Leontovich had developed a new form for the boundary condition at a highly conducting surface that is "clearer in its fundamentals and more simple mathematically". As given by LEONTOVICH (1948) the new form was

$$\hat{n} \times \mathbf{E} = Z\hat{n} \times (\hat{n} \times \mathbf{H}) \qquad (2.20)$$

where $\hat{n}$ is the outward unit vector normal to the surface, but in a footnote Leontovich indicated that the boundary condition was originally proposed by SCHUKIN (1940). Indeed, for a planar earth, (2.20) in component form *was* derived by Schukin using a method almost identical to that later employed by GRÜNBERG (1942). In spite of this, (2.20) and its equivalent forms are often referred to as the Leontovich boundary condition, and in his 1948 paper Leontovich used Rytov's results to establish bounds on the validity of the condition.

The first application of (2.20) to a scattering (as opposed to propagation) problem was by FOCK (1946) who developed an asymptotic approximation to the field close to the surface of a large homogeneous conducting body when illuminated by a plane wave. JONES AND PIDDUCK (1950) examined the two-dimensional problem of a metallic wedge for plane wave illumination and determined the first order correction to the field scattered by a perfectly conducting wedge, and SENIOR (1952) considered the analogous problem of a metallic half plane. The half plane models a semi-infinite slab whose thickness τ is such that

$$d \ll \tau \ll \lambda_0$$

where d is the penetration depth, and it was verified that the boundary condition (2.20) is applicable everywhere provided the edge is assumed to be semi-cylindrical. The Wiener-Hopf integral equations were solved to produce a closed form expression for the scattered field, and this is believed to be the first use of (2.20) in a mathematically exact analysis.

2.3   First order impedance conditions

Equations (2.19) and (2.20) are examples of a first order impedance boundary condition, sometimes referred to as an SIBC (Standard Impedance Boundary Condition). To see how they arise, we will first summarise the analysis in SENIOR (1960a) for a lossy homogeneous material occupying the half space $y < 0$ where x, y, z are Cartesian coordinates (see Fig. 2–1). The permittivity

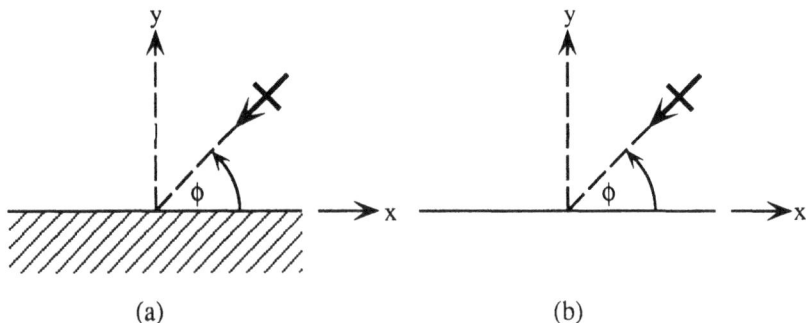

(a)   (b)

Figure 2–1: *Dielectric interface and impedance surface*

and permeability of the material are ϵ and μ respectively, and the complex refractive index is $N = \sqrt{\epsilon\mu/\epsilon_0\mu_0}$. The region $y > 0$ is free space containing the sources of the electromagnetic field, and we seek a boundary condition imposed at the surface $y = 0+$.

Inside the material, the field (distinguished by a prime) is such that

$$\left(\frac{\partial^2}{\partial x^2} + \frac{\partial^2}{\partial y^2} + \frac{\partial^2}{\partial z^2} + N^2 k_0^2 \right) E_y' = 0 \tag{2.21}$$

If $|N| \gg 1$, E'_y varies rapidly in the y direction, leading to a large value of $\partial^2 E'_y / \partial y^2$, and by comparison the other two derivatives in (2.21) are negligible. Hence, to the leading order in $|N|$,

$$\left(\frac{\partial^2}{\partial y^2} + N^2 k_0^2 \right) E'_y = 0$$

and the solution that satisfies the radiation condition at $y = -\infty$ is

$$E'_y(x, y, z) = E'_y(x, 0-, z) \, e^{jNk_0 y}$$

giving

$$\left. \frac{\partial E'_y}{\partial y} \right|_{y=0-} = jNk_0 \, E'_y(x, 0-, z) \qquad (2.22)$$

At the interface $y = 0$ the exact boundary conditions are

$$E_y = \frac{\epsilon}{\epsilon_0} E'_y$$

and

$$\frac{\partial E_y}{\partial y} = \frac{\partial E'_y}{\partial y}$$

as indicated by (2.11) and (2.2) respectively. It follows that

$$\frac{\partial E_y}{\partial y} = jNk_0 E'_y = jNk_0 \frac{\epsilon_0}{\epsilon} E_y$$

which can be written as

$$\frac{\partial E_y}{\partial y} - jk_0 \frac{\eta}{Z_0} E_y = 0 \qquad (2.23)$$

where

$$\eta = Z \qquad (2.24)$$

Similarly

$$\frac{\partial H_y}{\partial y} - jk_0 \frac{Z_0}{\eta} H_y = 0 \qquad (2.25)$$

from duality, and the field derivatives identify these as first order impedance boundary conditions.

For many applications an alternative but equivalent form is more convenient. Using (2.1) and (2.2) the condition (2.23) can be written as

$$-\frac{\partial E_x}{\partial x} - \frac{\partial E_z}{\partial z} + \eta \left(\frac{\partial H_z}{\partial x} - \frac{\partial H_x}{\partial z} \right) = 0$$

so that

$$\frac{\partial}{\partial x} (E_x - \eta H_z) = -\frac{\partial}{\partial z} (E_z + \eta H_x)$$

Similarly, from (2.25)

$$\frac{\partial}{\partial z}\left(E_x - \eta H_z\right) = \frac{\partial}{\partial x}\left(E_z + \eta H_x\right)$$

and therefore

$$\left(\frac{\partial^2}{\partial x^2} + \frac{\partial^2}{\partial z^2}\right)\Phi = 0$$

where $\Phi = E_x - \eta H_z$ or $E_z + \eta H_x$. The only solution not exponentially large at $x = \pm\infty$ or $z = \pm\infty$ is $\Phi = 0$, implying

$$E_x = \eta H_z, \qquad E_z = -\eta H_x \qquad (2.26)$$

These conditions can be expressed in vector form as

$$\hat{y} \times \mathbf{E} = \eta\,\hat{y} \times (\hat{y} \times \mathbf{H}) \qquad (2.27)$$

where η is given in (2.24), and this is the Leontovich boundary condition for a planar surface. In the special case of a pec material, $\eta = 0$, and the condition reduces to (2.6), but it is not necessary for η to be small. Indeed,

$$\frac{\eta}{Z_0} = \frac{1}{N}\frac{\mu}{\mu_0} = N\frac{\epsilon_0}{\epsilon} \qquad (2.28)$$

and for a magnetic material, $|\eta|$ can exceed Z_0 without violating the requirement that $|N| \gg 1$.

As evident from the above derivation, (2.27) can be obtained from (2.23) and (2.25) by a process of tangential integration, but this is only true if η has the same value in (2.23) as it does in (2.25). Conversely, (2.23) and (2.25) can be obtained from (2.27) by tangential differentiation, and this highlights the fact that (2.23) and (2.25) are more singular than (2.27).

For an inhomogeneous material with functions of position ϵ and μ, a similar analysis is possible. If, for example, $\epsilon = \epsilon(x, z)$ with μ constant and

$$|k_0^{-1}\nabla \ln Z| \ll 1 \qquad (2.29)$$

the boundary conditions analogous to (2.23) and (2.25) are (SENIOR, 1960a)

$$\frac{\partial E_y}{\partial y} - jk_0\frac{\eta}{Z_0}E_y = -\frac{1}{\eta}\left(E_x\frac{\partial\eta}{\partial x} + E_z\frac{\partial\eta}{\partial z}\right) \qquad (2.30)$$

$$\frac{\partial H_y}{\partial y} - jk_0\frac{Z_0}{\eta}H_y = \frac{Y_0}{\eta}\left(E_x\frac{\partial\eta}{\partial z} - E_z\frac{\partial\eta}{\partial x}\right) \qquad (2.31)$$

with $\eta = Z$ as before but, in spite of the added terms, (2.27) is unchanged. In fact, to the leading order in $|N|$, (2.27) is unaffected by any variation of ϵ and/or μ either laterally or in depth provided these are slow on the scale of the local wavelength in the material. This is an important feature of (2.27).

It explains, for example, its ability to model progressive changes in material properties if η is given its local value at each point of the surface. It also shows that in the case of small random and statistically uniform variations in properties, the average fields are determined by the average value of Z and not by the average values of ϵ and μ individually (SENIOR, 1960a). This is in agreement with the conclusion reached by FEINBERG (1944) using a somewhat different argument.

Physically at least, (2.27) is very reasonable. Since $\eta = Z$ is simply the impedance looking into the material half space, (2.27) is equivalent to an impedance match, and this suggests an extension to other types of surface. Thus, for an anisotropic material

$$\hat{y} \times \mathbf{E} = \overline{\overline{\eta}} \cdot \hat{y} \times (\hat{y} \times \mathbf{H}) \tag{2.32}$$

with (say)

$$\overline{\overline{\eta}} = (\eta_{11}\hat{x} + \eta_{12}\hat{z})\hat{x} + (\eta_{21}\hat{x} + \eta_{22}\hat{z})\hat{z} \tag{2.33}$$

and for a curved surface

$$\hat{n} \times \mathbf{E} = \overline{\overline{\eta}} \cdot \hat{n} \times (\hat{n} \times \mathbf{H}) \tag{2.34}$$

where $\hat{n}$ is the outward unit vector normal and $\overline{\overline{\eta}}$ is specified by a local impedance match. An equivalent representation of (2.34) is

$$\hat{n} \times (\hat{n} \times \mathbf{E}) = -\Delta \overline{\overline{\eta}}^{-1} \cdot \hat{n} \times \mathbf{H} \tag{2.35}$$

where $\Delta = \det \overline{\overline{\eta}}$, and in the particular case of an isotropic material,

$$\hat{n} \times \mathbf{E} = \eta \, \hat{n} \times (\hat{n} \times \mathbf{H}) \tag{2.36}$$

or

$$\hat{n} \times (\hat{n} \times \mathbf{E}) = -\eta \, \hat{n} \times \mathbf{H} \tag{2.37}$$

For a planar surface $y = 0$ with constant surface impedance η, the boundary conditions (2.27) and (2.23)–(2.25) are equivalent, but this is not necessarily true in other coordinate systems, and it is not permissible simply to replace $\hat{y}$ by $\hat{n}$. Thus, for the surface $\phi = $ constant in cylindrical polar coordinates, (2.27) implies

$$\frac{1}{\rho}\frac{\partial E_\phi}{\partial \phi} - jk_0 \frac{\eta}{Z_0} E_\phi = -\frac{1}{\rho}\frac{\partial}{\partial \rho}(\rho \eta) H_z$$

$$\frac{1}{\rho}\frac{\partial H_\phi}{\partial \phi} - jk_0 \frac{Z_0}{\eta} H_\phi = \frac{1}{\rho}\frac{\partial}{\partial \rho}\left(\frac{\rho}{\eta}\right) E_z$$

(SENIOR, 1978a) which are no longer scalar boundary conditions involving only the normal components. Moreover, for a planar surface with a tensor surface impedance or a curved surface with any impedance, there are no scalar counterparts to (2.32) or (2.34). Thus, (2.34) is a more generally applicable form of the impedance boundary condition, and although there are a few special cases for which (2.23) and (2.25) are more convenient, these are the exceptions.

A surface subject to an impedance boundary condition supports electric and magnetic currents

$$\mathbf{K_e} = \hat{n} \times \mathbf{H}$$

and

$$\mathbf{K_m} = -\hat{n} \times \mathbf{E}$$

respectively, and these are evidently related. Thus, from (2.36)

$$\mathbf{K_m} = -\eta\,\hat{n} \times \mathbf{K_e} \tag{2.38}$$

and similarly

$$\mathbf{K_e} = \frac{1}{\eta}\,\hat{n} \times \mathbf{K_m} \tag{2.39}$$

so that in the solution of a boundary value problem it is sufficient to determine just one. The duality implicit in the condition is also a mathematically useful feature (SENIOR, 1962), and if the scattered field for a field $(\mathbf{E^i}, \mathbf{H^i}) = (\mathbf{F}, Y_0\mathbf{G})$ incident on the body is $(\mathbf{E^s}, \mathbf{H^s}) = \{\,\mathbf{f}(\eta), Y_0\,\mathbf{g}(\eta)\,\}$, the scattered field for the field $(\mathbf{E^i}, \mathbf{H^i}) = (-Z_0\mathbf{G}, \mathbf{F})$ is $(\mathbf{E^s}, \mathbf{H^s}) = \{\,-Z_0\,\mathbf{g}(1/\eta), \mathbf{f}(1/\eta)\,\}$.

For a lossy homogeneous body for which $\eta = Z$, the conditions under which (2.36) can be justified have been discussed by RYTOV (1940), LEONTOVICH (1948) and SENIOR (1960a). Some of these are evident from the results in Appendix A, while others are obvious from physical considerations. In summary the requirements are

(i) On the surface the external field is slowly varying on a scale of λ_0.

(ii) Any variation of the material properties is slow on a scale of the local wavelength in the material, i.e.

$$|k_0^{-1} N^{-1} \nabla \ln Z| \ll 1 \tag{2.40}$$

(iii) At each point of the material

$$|N| \gg 1 \tag{2.41}$$

(iv) and

$$|\mathrm{Im}.\ N| k_0 \rho \gg 1 \tag{2.42}$$

where ρ is the smallest radius of curvature or dimension of the body.

In some circumstances it is possible to replace (iv) by

(v)

$$|N|k_0\rho \gg 1 \qquad (2.43)$$

but such cases are exceptional. We can also relax (ii) in favour of the weaker condition

(vi)

$$|k_0^{-1}\nabla \ln Z| \ll 1 \qquad (2.44)$$

if all variations of the material properties are parallel to the surface, or by using a modified value of η that takes into account the variation of $\ln Z$ perpendicular to the surface.

These restrictions are quite severe, but experience has shown that the impedance boundary condition often provides good accuracy in circumstances where it is difficult to justify its use. One example is the edge of a wedge or half plane where (iv) is certainly violated, but results obtained with (2.36) applied right up to the edge have proved to be in good agreement with experimental (and other) data. By ascribing a curvature to the edge it may be possible to satisfy (iv), but condition (i) is still violated.

2.4   First order transition conditions

There are two pairs of transition conditions which are closely related to an SIBC. To see how these come about, consider first a thin layer consisting of a highly conducting non-magnetic dielectric. If the conductivity is σ, the layer will support an electric current whose volume density is

$$\mathbf{J}_e = \sigma \mathbf{E}' \qquad (2.45)$$

where $\mathbf{E}'$ is the tangential electric field within the material, and if the layer thickness τ is such that $\tau \ll \lambda_0$, we can replace the layer by an equivalent sheet current

$$\mathbf{J}_{e(s)} = \tau \mathbf{J}_e$$

Thus, from (2.45),

$$\mathbf{E}' = (\sigma\tau)^{-1}\mathbf{J}_{e(s)} \qquad (2.46)$$

and, since the tangential electric field is continuous across the surface of the layer, (2.46) can be expressed in terms of the external field as

$$\mathbf{E}_{tan} = (\sigma\tau)^{-1}\mathbf{J}_{e(s)} \qquad (2.47)$$

that is

$$\hat{n} \times (\hat{n} \times \mathbf{E}) = -R_e\mathbf{J}_{e(s)} \qquad (2.48)$$

where

$$R_e = (\sigma\tau)^{-1} \qquad (2.49)$$

is the resistivity of the sheet. For a block of material with lateral dimensions a, b and thickness τ, the resistance across the block is consistent with (2.49) only if $a = b$. Thus, the units of R_e are ohms per square, implying that R_e is the resistance (Ω) measured between the edges of a square sheet.

Let us now consider the more general case of a thin layer of lossy, non-magnetic ($\mu = \mu_0$) material with complex permittivity ϵ. If the layer is immersed in free space, the volume equivalence principle (HARRINGTON, 1961) allows us to replace the layer by the equivalent polarisation current

$$\mathbf{J}_e = jk_0 Y_0 \left(\frac{\epsilon}{\epsilon_0} - 1 \right) \mathbf{E}'$$

On the assumption that $k_0 \tau \ll 1$, the component of $\mathbf{J}_e$ normal to the layer can be neglected (HARRINGTON AND MAUTZ, 1975), and the tangential components replaced by the sheet currents

$$\mathbf{J}_{e(s)} = \tau \mathbf{J}_e$$

as before. For the external field components we then have

$$\mathbf{E}_{\text{tan}} = R_e \mathbf{J}_{e(s)} \tag{2.50}$$

where

$$R_e = -\frac{jZ_0}{k_0 \tau \left(\frac{\epsilon}{\epsilon_0} - 1 \right)} \tag{2.51}$$

with $\mathbf{E}_{\text{tan}}$ continuous across the sheet. We observe that if $\epsilon = \epsilon' - j\frac{\sigma}{\omega}$, (2.51) reduces to (2.49) as σ increases.

In vector form the transition conditions are

$$\hat{n} \times \mathbf{E} = R_e \hat{n} \times \mathbf{J}_{e(s)} \tag{2.52}$$

with

$$\mathbf{J}_{e(s)} = [\hat{n} \times \mathbf{H}]_-^+ \tag{2.53}$$

and

$$[\hat{n} \times \mathbf{E}]_-^+ = 0 \tag{2.54}$$

(see Fig. 2–2) where $\hat{n}$ is the outward unit vector normal to the upper side of the sheet and $[\ ]_-^+$ denotes the discontinuity between the upper $(+)$ and lower $(-)$ sides. The conditions define a *resistive* sheet whose properties are entirely specified by the single measurable quantity R_e (Ω/square). As evident from (2.52) and (2.54) the sheet supports only an electric current whose strength is proportional to the common value of the tangential electric field at the surface. When $R_e = 0$ the sheet is perfectly conducting and therefore opaque, but in general it is partially transparent. When $R_e = \infty$ it ceases to exist. The idea originated (BATEMAN, 1915) with Levi-Cività, and in recent years the sheets have found many fruitful applications. Although an infinitesimally thin sheet is obviously an idealisation, sheets are readily available with thickness no more than about 0.1 mm presenting an almost constant resistance as high as 2000 Ω/square over a wide range of frequencies. By changing the amount and type of loading employed, it is even possible to fabricate sheets with a specified variation of R_e over the surface.

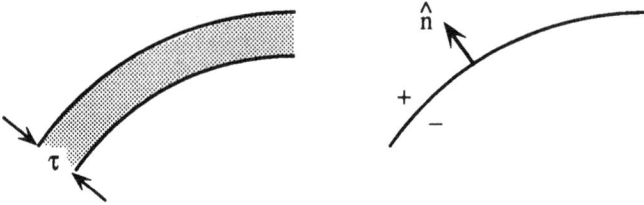

Figure 2-2: *A thin layer and its sheet simulation*

The electromagnetic dual of an (electrically) resistive sheet is a (magnetically) conductive one simulating a thin layer of lossy material with $\epsilon = \epsilon_0$. The corresponding conditions are the duals of (2.51)–(2.54), namely

$$\hat{n} \times \mathbf{H} = R_m \hat{n} \times \mathbf{J}_{m(s)} \tag{2.55}$$

with

$$\mathbf{J}_{m(s)} = -[\hat{n} \times \mathbf{E}]^+_- \tag{2.56}$$

and

$$[\hat{n} \times \mathbf{H}]^+_- = 0 \tag{2.57}$$

where

$$R_{\mathrm{m}} = -\frac{jY_0}{k_0\tau \left(\dfrac{\mu}{\mu_0} - 1\right)} \tag{2.58}$$

is the conductivity whose units are S/square. The sheet supports only a magnetic current and, although obviously not as easy to achieve in practice, it proves to be important for many purposes. When $R_{\mathrm{m}} = 0$ the sheet is a pmc surface (or ideal ferrite) and when $R_{\mathrm{m}} = \infty$ it ceases to exist.

In the special case of planar sheets, the conditions can be expressed in terms of the normal field components and their normal derivatives. The procedure for doing so is the same as that needed to deduce (2.23) and (2.25) from (2.27) and thus, for a resistive sheet in the plane $y = 0$

$$\begin{aligned}
\left(\frac{2R_e}{Z_0} - \frac{1}{jk_0}\frac{\partial}{\partial y}\right) E_y^+ - \left(\frac{2R_e}{Z_0} + \frac{1}{jk_0}\frac{\partial}{\partial y}\right) E_y^- &= 0 \\[2mm]
\left(\frac{Z_0}{2R_e} - \frac{1}{jk_0}\frac{\partial}{\partial y}\right) H_y^+ + \left(\frac{Z_0}{2R_e} + \frac{1}{jk_0}\frac{\partial}{\partial y}\right) H_y^- &= 0
\end{aligned} \tag{2.59}$$

with

$$\left[\frac{\partial E_y}{\partial y}\right]_-^+ = 0, \qquad [H_y]_-^+ = 0 \tag{2.60}$$

Similarly, for a conductive sheet

$$\left(\frac{Y_0}{2R_m} - \frac{1}{jk_0}\frac{\partial}{\partial y}\right) E_y^+ + \left(\frac{Y_0}{2R_m} + \frac{1}{jk_0}\frac{\partial}{\partial y}\right) E_y^- = 0$$

$$\left(\frac{2R_m}{Y_0} - \frac{1}{jk_0}\frac{\partial}{\partial y}\right) H_y^+ - \left(\frac{2R_m}{Y_0} + \frac{1}{jk_0}\frac{\partial}{\partial y}\right) H_y^- = 0 \tag{2.61}$$

with

$$[E_y]_-^+ = 0, \qquad \left[\frac{\partial H_y}{\partial y}\right]_-^+ = 0 \tag{2.62}$$

and these constitute alternative forms for the transition conditions.

A more general situation is that in which the resistivity and conductivity of the sheets are tensors, and the transition conditions are then

$$\hat{n} \times \left(\mathbf{E}^+ + \mathbf{E}^-\right) = 2\overline{\overline{R}}_e \cdot \hat{n} \times [\hat{n} \times \mathbf{H}]_-^+ \tag{2.63}$$

with

$$[\hat{n} \times \mathbf{E}]_-^+ = 0 \tag{2.64}$$

for the resistive sheet, and

$$\hat{n} \times \left(\mathbf{H}^+ + \mathbf{H}^-\right) = -2\overline{\overline{R}}_m \cdot \hat{n} \times [\hat{n} \times \mathbf{E}]_-^+ \tag{2.65}$$

with

$$[\hat{n} \times \mathbf{H}]_-^+ = 0 \tag{2.66}$$

for the conductive one. Equations (2.63) and (2.65) can be written alternatively as

$$[\hat{n} \times \mathbf{H}]_-^+ = -\frac{1}{2\Delta_e}\overline{\overline{R}}_e^T \cdot \hat{n} \times \left\{\hat{n} \times \left(\mathbf{E}^+ + \mathbf{E}^-\right)\right\} \tag{2.67}$$

$$[\hat{n} \times \mathbf{E}]_-^+ = \frac{1}{2\Delta_m}\overline{\overline{R}}_m^T \cdot \hat{n} \times \left\{\hat{n} \times \left(\mathbf{H}^+ + \mathbf{H}^-\right)\right\} \tag{2.68}$$

where $\overline{\overline{R}}_{e,m}^T$ is the transpose of $\overline{\overline{R}}_{e,m}$ and $\Delta_{e,m} = \det \overline{\overline{R}}_{e,m}$. For some man-made structures such as a carbon-epoxy panel where the fibres have a preferred orientation, a tensor resistive sheet may provide an appropriate simulation.

To model a layer whose permittivity and permeability both differ from their free space values, a logical approach is to consider a combination sheet (SENIOR, 1985) consisting of coincident resistive and conductive sheets. For a combination sheet the transition conditions are (2.63) or (2.67) and (2.65) or (2.68). Such a sheet supports electric and magnetic currents and these are, in general, coupled in as much as each is affected by the presence of the other. The exception is a planar sheet when the resistive and conductive parts scatter independently of each other (SENIOR, 1985). It follows that for any planar configuration it is sufficient to restrict the development of analytical and/or numerical procedures to the simple case of a resistive sheet. By application of the duality principle the solution for the corresponding conductive sheet can be deduced, and the solution for the combination sheet is then the sum of the two. Thus, for a planar structure, we can achieve the added generality of a combination sheet without any increase in complexity.

By addition and subtraction of (2.63) and (2.68) we obtain

$$\hat{n} \times \mathbf{E}^{\pm} = \left(\overline{\overline{R}}_e \pm \frac{1}{4\Delta_m}\overline{\overline{R}}_m^T\right) \cdot \hat{n} \times \left(\hat{n} \times \mathbf{H}^+\right)$$
$$- \left(\overline{\overline{R}}_e \mp \frac{1}{4\Delta_m}\overline{\overline{R}}_m^T\right) \cdot \hat{n} \times \left(\hat{n} \times \mathbf{H}^-\right) \qquad (2.69)$$

showing that in general a combination sheet is partially transparent, but if

$$\overline{\overline{R}}_e = \frac{1}{4\Delta_m}\overline{\overline{R}}_m^T \qquad (2.70)$$

implying $4\Delta_m = (4\Delta_e)^{-1}$, it becomes opaque. The transition conditions then reduce to boundary conditions and (2.69) is simply

$$\hat{n} \times \mathbf{E}^{\pm} = \pm\overline{\overline{\eta}} \cdot \hat{n} \times \left(\hat{n} \times \mathbf{H}^{\pm}\right) \qquad (2.71)$$

with

$$\overline{\overline{\eta}} = 2\overline{\overline{R}}_e = \frac{1}{2\Delta_m}\overline{\overline{R}}_m^T \qquad (2.72)$$

We recognise (2.71) as the first order impedance boundary condition (2.34), showing that a combination sheet includes as a special case an impedance sheet subject to the same SIBC on both sides. When the resistivity and conductivity are scalars, (2.72) implies

$$R_e = \frac{\eta}{2}, \qquad R_m = \frac{1}{2\eta} \qquad (2.73)$$

and therefore

$$\eta = \left(\frac{R_e}{R_m}\right)^{1/2}$$

From (2.51) and (2.58)

$$\eta = \left(\frac{\mu - \mu_0}{\epsilon - \epsilon_0}\right)^{1/2}$$

which reduces to (2.24) for a homogeneous lossy material whose thickness exceeds the penetration depth.

The connection between the impedance boundary condition (2.34) and the transition conditions (2.63)–(2.66) can now be expressed as follows. Consider an infinitely thin layer subject to the same SIBC on both sides. The layer is, of course, opaque, and supports electric and magnetic currents. By addition and subtraction of the boundary conditions we obtain

$$\hat{n} \times \left(\mathbf{E}^+ + \mathbf{E}^-\right) = \bar{\bar{\eta}} \cdot \hat{n} \times [\hat{n} \times \mathbf{H}]_-^+ \tag{2.74}$$

$$[\hat{n} \times \mathbf{E}]_-^+ = \bar{\bar{\eta}} \cdot \hat{n} \times \left\{\hat{n} \times \left(\mathbf{H}^+ + \mathbf{H}^-\right)\right\} \tag{2.75}$$

If the magnetic current is suppressed (for a planar layer this can be achieved by confining attention to electric fields whose tangential components are symmetrical about the layer), the result is a resistive sheet with resistivity

$$\bar{\bar{R}}_e = \frac{1}{2}\bar{\bar{\eta}} \tag{2.76}$$

satisfying the transition condition (2.74). Similarly, by suppressing the electric current, we obtain a conductive sheet whose conductivity is

$$\bar{\bar{R}}_{\mathrm{m}} = \frac{1}{2\Delta_\eta}\bar{\bar{\eta}}^T \tag{2.77}$$

where $\Delta_\eta = \det \bar{\bar{\eta}}$, satisfying the transition condition (2.75). By analogy with the SIBC, we shall refer to (2.63) and (2.65) as sheet transition conditions or STCs.

2.5   Accuracy

As noted earlier, one of the first applications of an impedance boundary condition was to simulate the land in a study of ground wave propagation, and a summary of more recent applications has been given by KING AND WAIT (1976). The accuracy depends on how well the SIBC models the electromagnetic properties of the land and, in the case of a planar surface, one way to obtain a measurement is by comparing the plane wave reflection coefficients.

Consider a half space $y < 0$ composed of a homogeneous dielectric having permittivity ϵ and permeability μ, with free space above. For an incident plane wave polarised perpendicular to the plane of incidence with

$$\mathbf{E}^i = \hat{z}e^{jk_0(x\cos\phi + y\sin\phi)} \tag{2.78}$$

as shown in Fig. 2–1(a), the reflection coefficient based on the component E_z or H_y is (STRATTON, 1941)

$$R_\perp = -\frac{\sqrt{1 - N^{-2}\cos^2\phi} - \dfrac{1}{N}\dfrac{\mu}{\mu_0}\sin\phi}{\sqrt{1 - N^{-2}\cos^2\phi} + \dfrac{1}{N}\dfrac{\mu}{\mu_0}\sin\phi} \tag{2.79}$$

where $N = \sqrt{\epsilon\mu/\epsilon_0\mu_0}$ is the complex refractive index of the material, and if $|N| \gg 1$

$$R_\perp = -\frac{1 - \dfrac{1}{N}\dfrac{\mu}{\mu_0}\sin\phi}{1 + \dfrac{1}{N}\dfrac{\mu}{\mu_0}\sin\phi} + O\left(|N|^{-2}\right) \tag{2.80}$$

By comparison, the boundary condition (2.27) or, equivalently, (2.23) implies

$$R_\perp = -\frac{1 - \eta Y_0 \sin\phi}{1 + \eta Y_0 \sin\phi} \tag{2.81}$$

and (2.80) and (2.81) agree if

$$\eta = \frac{1}{N}\frac{\mu}{\mu_0} Z_0 = Z \tag{2.82}$$

as indicated in (2.28). Similarly, for a plane wave

$$\mathbf{H}^i = \hat{z}e^{jk_0(x\cos\phi + y\sin\phi)} \tag{2.83}$$

polarised in the plane of incidence, the reflection coefficient based on H_z or E_y is (STRATTON, 1941)

$$R_\| = -\frac{\sqrt{1 - N^{-2}\cos^2\phi} - \dfrac{1}{N}\dfrac{\epsilon}{\epsilon_0}\sin\phi}{\sqrt{1 - N^{-2}\cos^2\phi} + \dfrac{1}{N}\dfrac{\epsilon}{\epsilon_0}\sin\phi} \tag{2.84}$$

i.e.

$$R_\| = -\frac{1 - \dfrac{1}{N}\dfrac{\epsilon}{\epsilon_0}\sin\phi}{1 + \dfrac{1}{N}\dfrac{\epsilon}{\epsilon_0}\sin\phi} + O\left(|N|^{-2}\right) \tag{2.85}$$

By comparison, the boundary condition (2.27) or (2.23) implies

$$R_\| = -\frac{1 - \dfrac{Z_0}{\eta}\sin\phi}{1 + \dfrac{Z_0}{\eta}\sin\phi} \tag{2.86}$$

and these agree if

$$\eta = N\frac{\epsilon_0}{\epsilon} Z_0 = Z \tag{2.87}$$

as before. For normal incidence ($\phi = \pi/2$), (2.81) and (2.86) are exact for all $|N|$, but for any given $|N| \gg 1$, their accuracy decreases with decreasing ϕ, and is lowest at grazing incidence ($\phi = 0$).

The behaviour is illustrated in Fig. 2–3 where, for a non-magnetic ($\mu = \mu_0$)

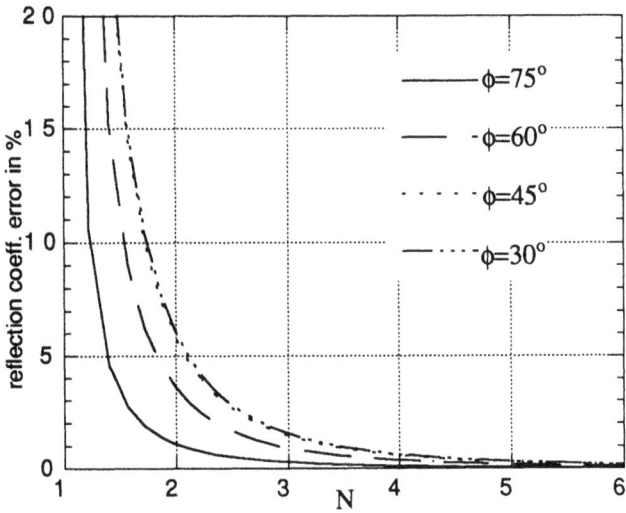

(a)

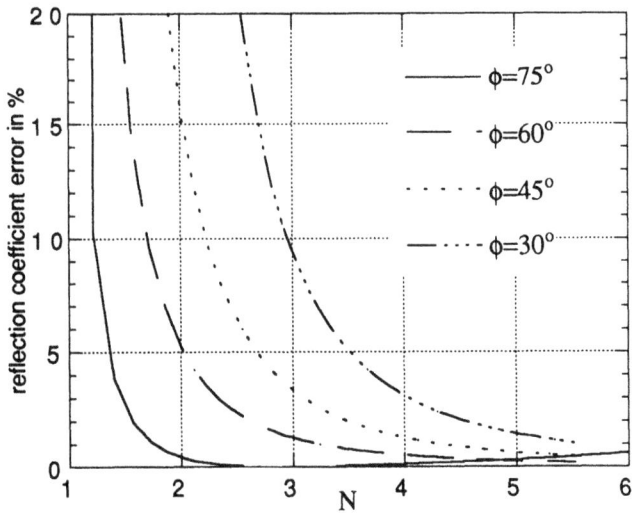

(b)

Figure 2–3: *Percentage error in the reflection coefficient for real ϵ and $\mu = \mu_0$: (a) perpendicular polarisation; (b) parallel polarisation*

lossless dielectric, the percentage errors in the magnitudes of the reflection coefficients are plotted as functions of N for four different angles of incidence. Although the graphs are superficially similar, there are significant differences between the two polarisations. For perpendicular (or E) polarisation the error is a maximum for $\phi \approx 45$ degrees, and is less than about 2 percent for all ϕ if $N \gtrsim 3$. For parallel (or H) polarisation, however, the error increases with decreasing ϕ. This is attributable to the Brewster angle ϕ_B which decreases with increasing N and is not precisely simulated by the SIBC. At $\phi = \phi_B$ the error is infinite and, as a result, the error far exceeds 2 percent for all $N \lesssim 8$. The accuracy is no better if N is complex (see Fig. 2–4) or for a magnetic material, and we remark that if $\mu \neq \mu_0$ both polarisations possess a Brewster angle.

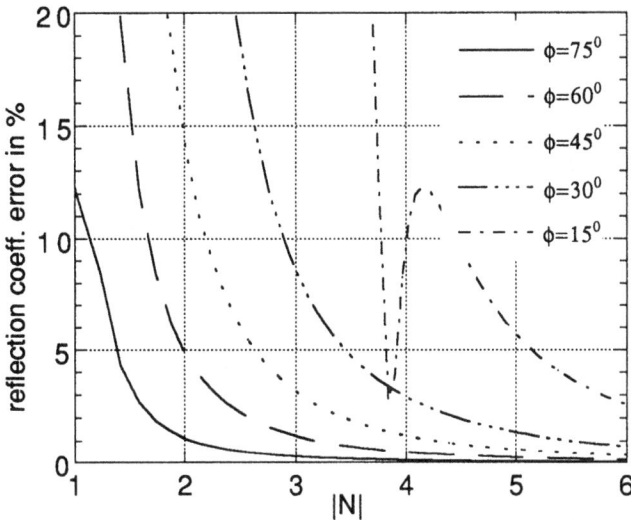

Figure 2–4: *Percentage error in the reflection coefficient for* $\arg \epsilon = 10°$ *and* $\mu = \mu_0$: *parallel polarisation*

One way to accurately reproduce the Brewster angle is to use a second order boundary condition (SENIOR AND VOLAKIS, 1987), but there is another approach. Recognising that for parallel polarisation

$$\phi_B = \sin^{-1} \sqrt{\frac{N^2 - 1}{(\epsilon/\epsilon_0)^2 - 1}} \tag{2.88}$$

the surface impedance η in (2.86) is replaced by

$$\eta = Z_0 \sin \phi_B \tag{2.89}$$

which reduces to (2.87) for sufficiently large ϵ/ϵ_0. With the impedance modified in this manner, the SIBC has the correct Brewster angle for all ϵ and μ, and the percentage error for parallel polarisation is shown in Fig. 2–5(b). The error now decreases with decreasing ϕ. It is less than 2 percent for all ϕ if $N \gtrsim 4$, and this is true also for magnetic and/or lossy materials. By using the impedance (2.89) in (2.81), we do not significantly increase the error for

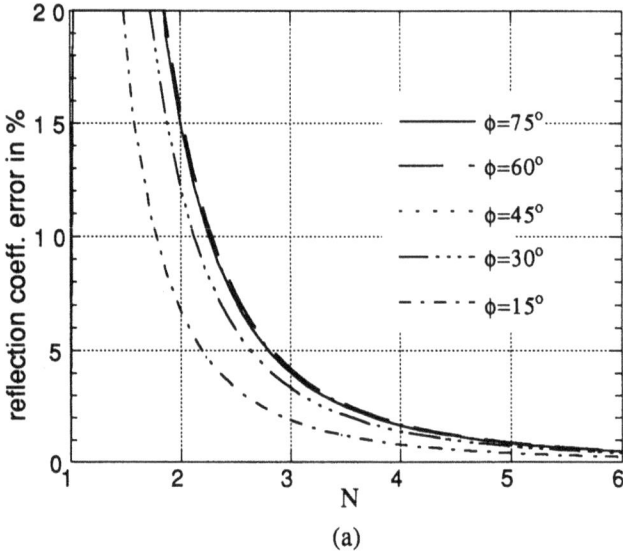

(a)

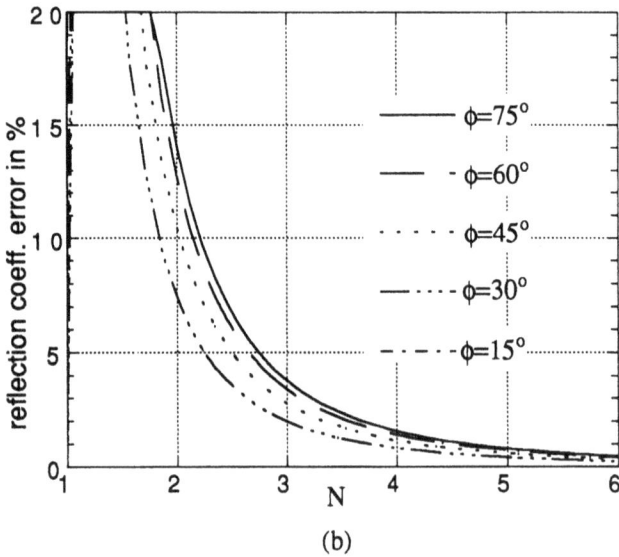

(b)

Figure 2–5: *Percentage error in the reflection coefficient for real ϵ and $\mu = \mu_0$ with $\eta = Z_0 \sin \phi_B$: (a) perpendicular polarisation; (b) parallel polarisation*

perpendicular polarisation (see Fig. 2–5(a)), but we do sacrifice the duality of the expression for η.

One of the most common applications of an SIBC is for simulating a thin dielectric coating applied to a metal surface, and this boundary condition continues to be an important tool for studying the effect of radar absorbing materials. For a homogeneous dielectric layer of thickness t on top of a pec material in the plane $y = -t$ as shown in Fig. 2–6(a), the reflection coefficient

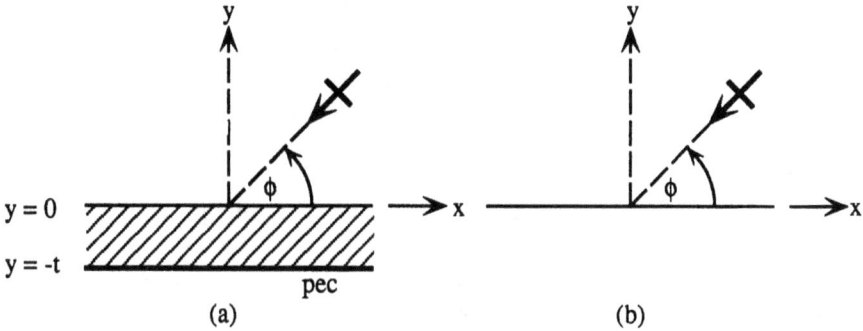

Figure 2–6: *Metal-backed dielectric layer and impedance surface*

for the incident plane wave (2.78) is (BHATTACHARYYA AND SENGUPTA, 1991)

$$R_\perp = -\frac{\sqrt{1 - N^{-2}\cos^2\phi} - j\frac{1}{N}\frac{\mu}{\mu_0}\sin\phi\tan\left(Nk_0t\sqrt{1 - N^{-2}\cos^2\phi}\right)}{\sqrt{1 - N^{-2}\cos^2\phi} + j\frac{1}{N}\frac{\mu}{\mu_0}\sin\phi\tan\left(Nk_0t\sqrt{1 - N^{-2}\cos^2\phi}\right)} \quad (2.90)$$

and if $|N| \gg 1$, $R_\perp$ can be approximated as

$$R_\perp = -\frac{1 - j\frac{1}{N}\frac{\mu}{\mu_0}\sin\phi\tan(Nk_0t)}{1 + j\frac{1}{N}\frac{\mu}{\mu_0}\sin\phi\tan(Nk_0t)} \quad (2.91)$$

Comparison with (2.81) shows agreement if the surface impedance is

$$\eta = jZ\tan(Nk_0t) \quad (2.92)$$

where Z is the intrinsic impedance of the material. We observe that as Im. $N \to \infty$, $\tan(Nk_0t) \to -j$ so that $\eta \to Z$ corresponding to the impedance of a half space, independent of the layer thickness. Indeed, the limit is achieved if $t \gg d$ where d is the penetration depth in the material.

Similarly, for the incident plane wave (2.83) polarised in the plane of incidence

$$R_\parallel = -\frac{\sqrt{1 - N^{-2}\cos^2\phi} + j\frac{1}{N}\frac{\epsilon}{\epsilon_0}\sin\phi\cot\left(Nk_0t\sqrt{1 - N^{-2}\cos^2\phi}\right)}{\sqrt{1 - N^{-2}\cos^2\phi} - j\frac{1}{N}\frac{\epsilon}{\epsilon_0}\sin\phi\cot\left(Nk_0t\sqrt{1 - N^{-2}\cos^2\phi}\right)} \quad (2.93)$$

(BHATTACHARYYA AND SENGUPTA, 1991), which can be approximated as

$$R_\parallel = -\frac{1 + j\dfrac{1}{N}\dfrac{\epsilon}{\epsilon_0}\sin\phi\cot(Nk_0 t)}{1 - j\dfrac{1}{N}\dfrac{\epsilon}{\epsilon_0}\sin\phi\cot(Nk_0 t)} \tag{2.94}$$

and this is identical to (2.86) if η has the value given in (2.92). In spite of the fact that (2.90) and (2.93) are not the electromagnetic duals of each other (because of the pec backing), the approximate reflection coefficients (2.91) and (2.94) do satisfy duality as evidenced by the common value of η.

For very thin coatings with $|N|k_0 t \lesssim 0.2$, we can replace $\tan x$ by x, and (2.90) and (2.91) then agree precisely, as do (2.93) and (2.94) if $|N| \gg 1$. More generally, for lossless coatings the exact and approximate reflection coefficients all have unit magnitudes and the only errors are in the phase. These errors are zero at normal incidence but can be significant at oblique angles even for relatively thin coatings. This is illustrated in Fig. 2–7 where the phase error is plotted as a function of N for a coating of thickness $t = 0.05\lambda_0$ composed of a lossless material with $\mu = \mu_0$. The peaks in the error curves for perpendicular polarisation near $N = 5$ are due to the fact that $\tan Nk_0 t$ is infinite for this value of N. For thicker coatings the errors are larger, and if ϵ and/or μ are complex there are also amplitude errors, but the phase errors tend to be smaller than they are for lossless coatings.

The above calculations have been based on the location of the impedance surface at the upper surface of the layer as shown in Fig. 2–6. This is not always appropriate and, in the case of a metal plate coated on both sides, the logical location is at the plate. A displacement of the impedance surface away from the upper surface of the layer can produce a substantial increase in the phase error and, as we shall see, one way to avoid this is to use a higher order boundary condition.

In the case of a curved surface, the accuracy of an SIBC can be judged by examining the known solutions for scattering by a dielectric sphere or cylinder. Consider a sphere of radius a composed of a homogeneous dielectric immersed in free space. For the incident field

$$\mathbf{E}^i = \hat{x}e^{jk_0 z}$$

representing a linearly polarised plane wave propagating in the direction of the negative z axis, the exact solution can be expressed as an infinite sum of spherical vector wave functions (STRATTON, 1941). The coefficients in the expansion are

$$a_n = \frac{\psi_n(k_0 a) + j\Gamma_1\,\psi_n'(k_0 a)}{\chi_n(k_0 a) + j\Gamma_1\,\chi_n'(k_0 a)}$$

$$b_n = \frac{\psi_n'(k_0 a) - j\Gamma_2\,\psi_n(k_0 a)}{\chi_n'(k_0 a) - j\Gamma_2\,\chi_n(k_0 a)} \tag{2.95}$$

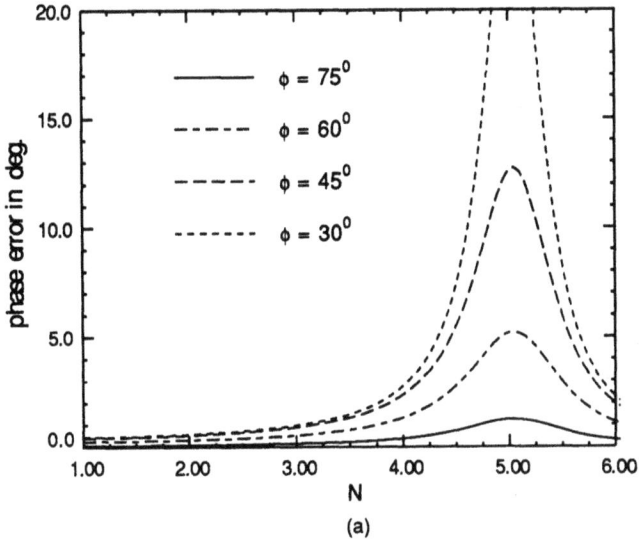

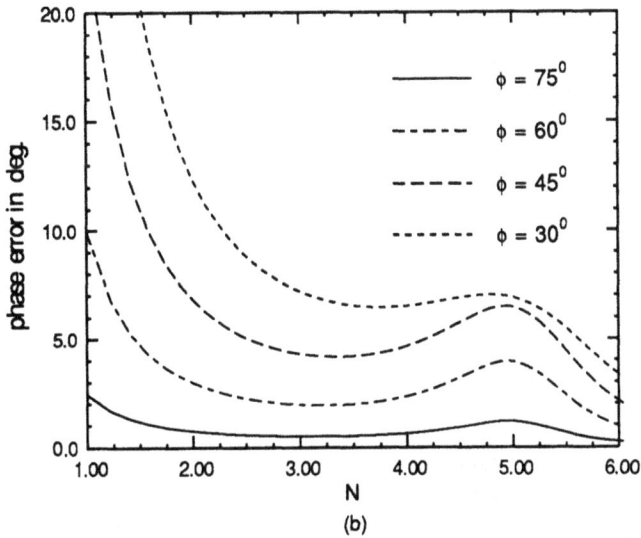

Figure 2-7: *Phase error in the reflection coefficient for a lossless layer with* $\mu = \mu_0$ *and* $t = 0.05\lambda_0$: *(a) perpendicular polarisation; (b) parallel polarisation*

with

$$\Gamma_1 = j \frac{\mu}{N\mu_0} \frac{\psi_n(Nk_0a)}{\psi_n'(Nk_0a)}, \qquad \Gamma_2 = -j \frac{\mu}{N\mu_0} \frac{\psi_n'(Nk_0a)}{\psi_n(Nk_0a)}$$

and

$$\psi_n(x) = x\,j_n(x), \qquad \chi_n(x) = x\,h_n^{(2)}(x)$$

where $j_n(x)$ is the spherical Bessel function, $h_n^{(2)}(x)$ is the spherical Hankel function of the second kind, and the prime denotes differentiation with respect to the entire argument. The parameters Γ_1 and Γ_2 are modal impedances, but as Im. $N \to \infty$

$$\frac{\psi_n(Nk_0a)}{\psi_n'(Nk_0a)} \to -j$$

and therefore $\Gamma_1, \Gamma_2 \to Z$, the same for all modes.

Alternatively, for a sphere subject to the SIBC (2.37) with $\hat{n} = \hat{r}$, the coefficients are as shown in (2.95) with

$$\Gamma_1 = \Gamma_2 = \eta \tag{2.96}$$

and, if $\eta = Z$ as specified in (2.82) and (2.87), agreement is obtained for large values of Im. N.

Based on detailed examinations of computed data, WANG (1987) has developed a criterion for the applicability of the SIBC. He notes, in particular, the importance of having (2.42) satisfied, as opposed to the weaker condition (2.43), and this is illustrated in Fig. 2–8(a) where the exact and approximate bistatic scattering cross-sections are plotted for spheres having $k_0a = 10$ with $N = 10$, $\mu = \mu_0$ and $N = 20$, $\mu = 2\mu_0$. Since Z is the same for both, the solution obtained using the SIBC is also the same, whereas the exact cross-sections show substantial differences. In contrast, as shown in Fig. 2–8(b), the agreement is excellent if some loss is introduced into the dielectric. From many computations of this type, Wang concluded that necessary conditions for the application of an SIBC are

$$|N| \gtrsim 10 \qquad \text{and} \qquad (\text{Im. } N)k_0\rho \gtrsim 2.3 \tag{2.97}$$

where ρ is the smallest radius of curvature (cf. (2.42)), and verified these for non-spherical surfaces using moment method data for prolate spheroids.

The exact eigenfunction solutions for dielectric-coated metal spheres and cylinders can also be used to check the accuracy of an SIBC with the surface impedance given in (2.92). It turns out that for lossless coatings of any appreciable thickness the errors can be substantial, and this is illustrated in Fig. 2–9 where the total field for a coated cylinder is computed at a distance of $0.05\lambda_0$ above the surface. The errors associated with the SIBC are as large as 20 dB, but most of these can be eliminated (SENIOR AND VOLAKIS, 1991) by using a boundary condition of higher order than the first. This topic is discussed in Section 7.2.

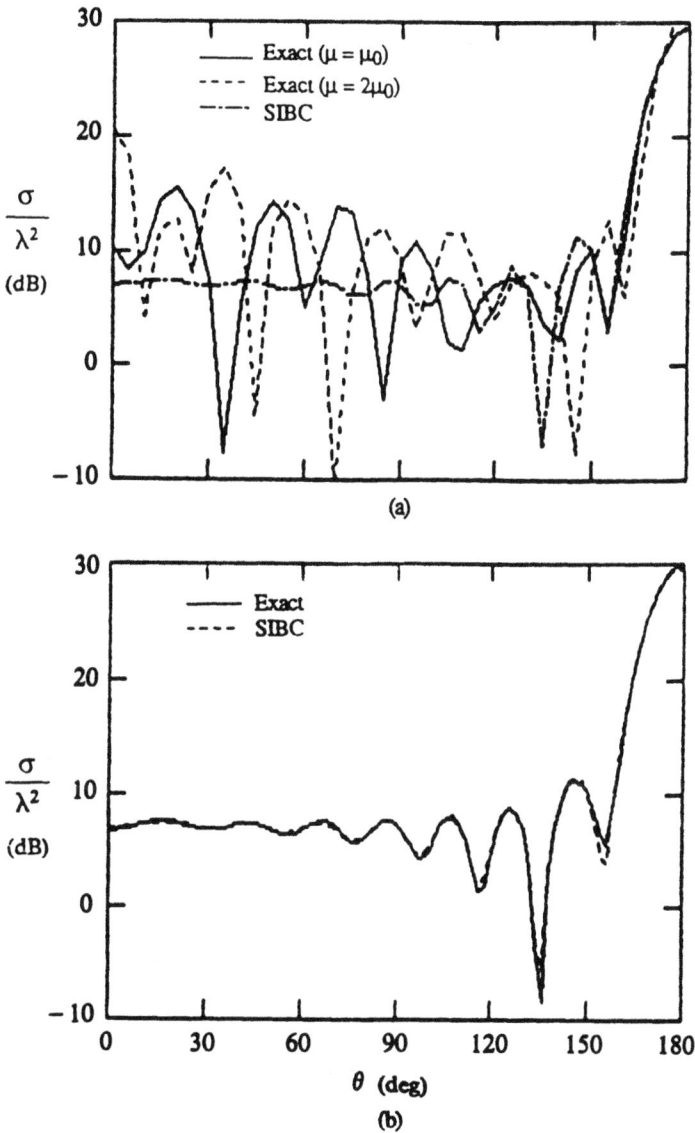

Figure 2–8: *Bistatic scattering cross sections of a sphere having $k_0a = 10$:*
(a) $N = 10$, $\mu = \mu_0$ and $N = 20$, $\mu = 2\mu_0$; (b) $N = 9.9(1 - j0.1)$, $\mu = \mu_0$
and $N = 19.8(1 - j0.1)$, $\mu = 2\mu_0$ (WANG, 1987; copyright © IEEE)

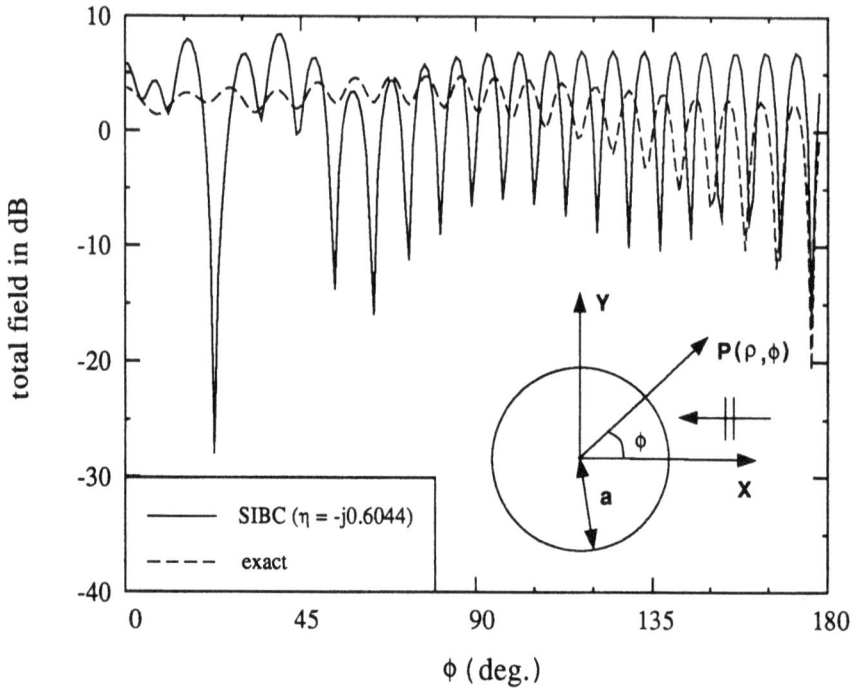

Figure 2–9: *Bistatic scattering pattern of a perfectly conducting circular cylinder of radius $a = 2.93\lambda_0$ coated with a layer $0.07\lambda_0$ thick having $\epsilon = 4\epsilon_0$, $\mu = \mu_0$ for parallel (or H) polarisation. The field was computed at a distance of $0.05\lambda_0$ above the surface of the coating: exact solution (- - - -), SIBC (———) with $\eta = -j0.6044$*

2.6   Surface perturbations

2.6.1   Displacement

An SIBC can also be used to simulate a small displacement or disturbance of the surface. A simple example is the displacement of an ideal pec or pmc surface from $y = 0$ to $y = -t$ where $k_0|t| \ll 1$. If, for the incident plane wave (2.78), the reflection coefficients in the two cases are $R_\perp$ and $R'_\perp$ respectively, then

$$R'_\perp = e^{-2jk_0 t \sin \phi} R_\perp \qquad (2.98)$$

which can be approximated as

$$R'_\perp = \left\{ \frac{1 - jk_0 t \sin \phi}{1 + jk_0 t \sin \phi} + O\left[(k_0 t)^2\right] \right\} R_\perp$$

and to the first order in $k_0 t$ the effect can be modelled using an SIBC with

$$\eta = jZ_0 k_0 t \tag{2.99}$$

applied at $y = -t$. The impedance is inductive (capacitive) if the new surface is located behind (in front of) the original surface, and the accuracy of the simulation can be judged from Fig. 2–10 where the phase is plotted as a

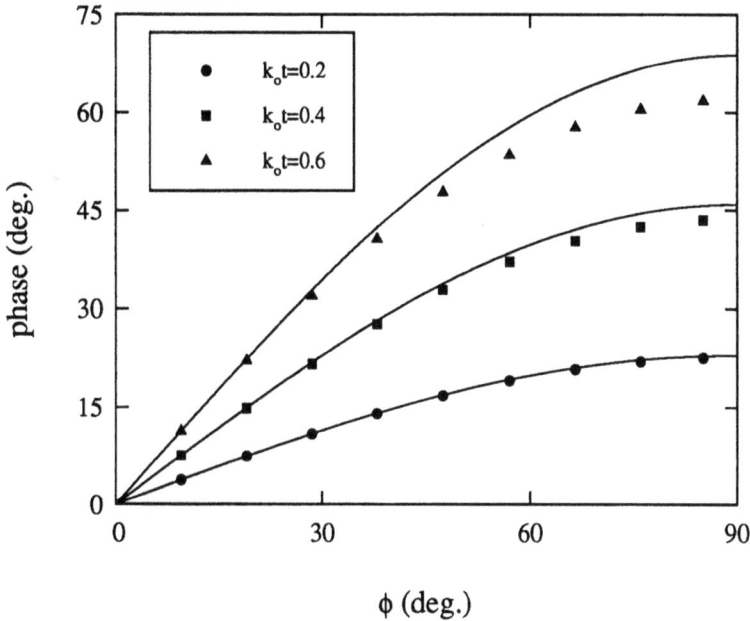

Figure 2–10: *Exact phase (——) compared with the approximate phase*
($\diamond, \nabla, \Delta$) implied by the SIBC

function of ϕ for several $k_0 t$. We observe that for all ϕ the error is less than 5 degrees if $|t| \lesssim 0.08\lambda_0$, but if $t > 0$ it is clear that the resulting field is only meaningful in $y \geq 0$.

2.6.2   Uniform roughness

Another type of perturbation is a uniform isotropic roughness, and this has been discussed by FEINBERG (1944) and SENIOR (1960b). Consider a pec surface whose height is

$$y = \xi(x, z) \tag{2.100}$$

where the rms values of ξ and the slopes $\partial\xi/\partial x$ (= ξ_x) and $\partial\xi/\partial z$ (= ξ_z) are assumed small. From the boundary condition at the perfectly conducting surface we have

$$E_x = -\xi_x E_y \qquad E_z = -\xi_z E_y \tag{2.101}$$

and by expansion in a Taylor series

$$E_x(x,\xi,z) = E_x(x,0,z) + \xi\frac{\partial}{\partial y}E_x(x,0,z) + \frac{1}{2}\xi^2\frac{\partial^2}{\partial y^2}E_x(x,0,z) + \cdots$$

with similar expressions for the other components of the electric field. Since (2.101) shows that E_x is of the first order in small quantities (denoted collectively by δ), it follows that

$$E_x(x,\xi,z) = E_x + \xi\frac{\partial E_x}{\partial y} + O(\delta^3)$$

where the field quantities on the right are evaluated at $y = 0$. Likewise

$$E_y(x,\xi,z) = E_y + \xi\frac{\partial E_y}{\partial y} + O(\delta^2)$$

$$E_z(x,\xi,z) = E_z + \xi\frac{\partial E_z}{\partial y} + O(\delta^3)$$

and hence, from (2.101),

$$E_x = -\xi_x E_y - \xi\frac{\partial E_x}{\partial y} + O(\delta^3)$$

$$\tag{2.102}$$

$$E_z = -\xi_z E_y - \xi\frac{\partial E_z}{\partial y} + O(\delta^3)$$

at $y = 0$. These can be written alternatively as

$$\frac{\partial E_y}{\partial y} = \frac{\partial}{\partial x}\left(\xi_x E_y + \xi\frac{\partial E_x}{\partial y}\right) + \frac{\partial}{\partial z}\left(\xi_z E_y + \xi\frac{\partial E_z}{\partial y}\right) + O(\delta^3) \tag{2.103}$$

(FEINBERG, 1944) and

$$\hat{y} \times \{\mathbf{E} + \nabla(\xi E_y)\} = -jZ_0k_0\xi\,\hat{y} \times (\hat{y} \times \mathbf{H}) + O(\delta^3) \tag{2.104}$$

Apart from the term involving E_y, (2.104) is an SIBC with impedance $-jZ_0k_0\xi$ applied at the surface $y = 0$. For a statistically defined surface, however, only the average value of $\xi = \xi(x,z)$ has meaning. To order δ the field components for a smooth surface can be inserted into the right hand sides of (2.102) and (2.103) and, if the surface is such that $\bar{\xi} = 0$ (implying $\bar{\xi}_x = \bar{\xi}_z = 0$) where the bar denotes the average, the boundary condition satisfied by the average field is simply $\hat{y} \times \mathbf{E} = 0$, which is the same as for a smooth pec surface.

To display the roughness effect it is necessary to obtain expressions for E_y, $\partial E_x/\partial y$ and $\partial E_z/\partial y$ which are accurate to $O(\delta)$ and can be substituted into the right hand sides of (2.102) and (2.103) to make explicit the terms $O(\delta^2)$. We can do this by using integral expressions for the field components in $y \geq 0$. Thus

$$E_y(x,0,z) = E^i_y(x,0,z) - \frac{1}{2\pi} \iint_{-\infty}^{\infty} \frac{\partial E_y}{\partial y'} \phi \, dx' \, dy'$$

where $\phi = \rho^{-1} e^{-jk_0\rho}$ with

$$\rho = \left\{ (x-x')^2 + (z-z')^2 \right\}^{1/2}$$

and by inserting the expression (2.103) for $\partial E_y/\partial y'$ and using Maxwell's equation, we find

$$
\begin{aligned}
E_y(x,0,z) &= E^i_y(x,0,z) - \frac{1}{2\pi} \iint_{-\infty}^{\infty} \left\{ jZ_0k_0(\xi_{x'} H_z - \xi_{z'} H_x) \right. \\
&\left. + \left(k_0^2 + \frac{\partial^2}{\partial x'^2} + \frac{\partial^2}{\partial z'^2} \right)(\xi' E_y) \right\} \phi \, dx' \, dz' + O(\delta^3)
\end{aligned}
$$

Also

$$
\begin{aligned}
\frac{\partial}{\partial y} E_x(x,0,z) &= \frac{\partial}{\partial y} E^i_x(x,0,z) + \frac{1}{2\pi} \iint_{-\infty}^{\infty} \left(k_0^2 + \frac{\partial^2}{\partial x'^2} + \frac{\partial^2}{\partial z'^2} \right) \\
&\cdot \left\{ jZ_0k_0\xi' H_z - \frac{\partial}{\partial x'}(\xi' E_y) \right\} \phi \, dx' \, dz' + O(\delta^3)
\end{aligned}
$$

and hence, from (2.102),

$$E_x(x,0,z) = -\xi_x E^i_y(x,0,z) - \xi \frac{\partial}{\partial y} E^i_x(x,0,z) + \frac{1}{2\pi} \iint_{-\infty}^{\infty} P_x \phi \, dx' \, dz' + O(\delta^3)$$

(2.105)

where

$$
\begin{aligned}
P_x &= jZ_0k_0 \left\{ \xi_x\xi_{x'} H_z - \xi_x\xi_{z'} H_x + \left(k_0^2 + \frac{\partial^2}{\partial x'^2} + \frac{\partial^2}{\partial z'^2} \right)(\xi\xi' H_z) \right\} \\
&+ \left(k_0^2 + \frac{\partial^2}{\partial x'^2} + \frac{\partial^2}{\partial z'^2} \right) \left(\frac{\partial}{\partial x} + \frac{\partial}{\partial x'} \right)(\xi\xi' E_y)
\end{aligned}
$$

(2.106)

(SENIOR, 1960b). Similarly

$$E_z(x,0,z) = -\xi_z E^i_y(x,0,z) - \xi \frac{\partial}{\partial y} E^i_z(x,0,z) + \frac{1}{2\pi} \iint_{-\infty}^{\infty} P_z \phi \, dx' \, dz' + O(\delta^3)$$

(2.107)

where

$$
\begin{aligned}
P_z &= -jZ_0k_0 \left\{ \xi_z\xi_{z'} H_x - \xi_z\xi_{x'} H_z + \left(k_0^2 + \frac{\partial^2}{\partial x'^2} + \frac{\partial^2}{\partial z'^2} \right)(\xi\xi' H_x) \right\} \\
&+ \left(k_0^2 + \frac{\partial^2}{\partial x'^2} + \frac{\partial^2}{\partial z'^2} \right) \left(\frac{\partial}{\partial z} + \frac{\partial}{\partial z'} \right)(\xi\xi' E_y)
\end{aligned}
$$

(2.108)

The next task is to consider the ensemble average. On the assumption that $y = 0$ is the mean surface, $\bar{\xi} = \bar{\xi}_x = \bar{\xi}_z = 0$ and, for a uniform isotropic roughness,

$$\overline{\xi(x,z)\,\xi(x',z')} = \xi_0^2 F(\rho) \tag{2.109}$$

where ξ_0 is the standard deviation and $F(\rho)$ is the correlation function. To the second order in δ the field components H_x, H_z and E_y can be replaced by the components for a smooth surface and thereby excluded from the averaging process. If

$$H_x(x',0,z') = H_x(x,0,z)\,e^{-j\{k_x(x'-x)+k_z(z'-z)\}}$$

with similar expressions for the other components, the boundary condition on the average field is as shown in (2.32) where the components of the tensor surface impedance $\bar{\bar{\eta}}$ are

$$\eta_{11} = jZ_0 k_0 \xi_0^2 \left\{ \frac{1}{2}\frac{k_x^2 - k_z^2}{k_0^2} A + B \right\} \tag{2.110}$$

$$\eta_{12} = \eta_{21} = jZ_0 k_0 \xi_0^2 \frac{k_x k_z}{k_0^2} A \tag{2.111}$$

$$\eta_{22} = jZ_0 k_0 \xi_0^2 \left\{ -\frac{1}{2}\frac{k_x^2 - k_z^2}{k_0^2} A + B \right\} \tag{2.112}$$

with

$$A = \int_0^\infty \left[\left(\frac{\partial^2}{\partial \rho^2} + \frac{1}{\rho}\frac{\partial}{\partial \rho} + k_0^2 \right) \{F\,J_0(\tau\rho)\} \right.$$
$$\left. - \left(\frac{k_0}{\tau} \right) \left(\frac{\partial^2 F}{\partial \rho^2} - \frac{1}{\rho}\frac{\partial F}{\partial \rho} \right) J_2(\tau\rho) \right] e^{-jk_0\rho}\,d\rho \tag{2.113}$$

$$B = \int_0^\infty \left[\left(1 - \frac{1}{2}\frac{\tau^2}{k_0^2} \right) \left(\frac{\partial^2}{\partial \rho^2} + \frac{1}{\rho}\frac{\partial}{\partial \rho} + k_0^2 \right) \{F\,J_0(\tau\rho)\} \right.$$
$$\left. - \frac{1}{2} \left(\frac{\partial^2 F}{\partial \rho^2} + \frac{1}{\rho}\frac{\partial F}{\partial \rho} \right) J_0(\tau\rho) \right] e^{-jk_0\rho}\,d\rho \tag{2.114}$$

Here

$$\tau = (k_x^2 + k_z^2)^{1/2}$$

and J_n is the Bessel function of order n. The surface impedance is a function of k_x and k_z and, hence, the incident field direction. Particular results for normal incidence ($k_x = k_z = 0$ implying $\tau = 0$) and grazing incidence ($\tau = k_0$ with $k_x = 0$ or $k_z = 0$) have been derived by FEINBERG (1944) and SENIOR (1960b), but a more widely applicable boundary condition can be obtained by averaging over all angles of incidence. We then have

$$\eta_{12} = \eta_{21} = 0 \tag{2.115}$$

with

$$\eta_{11} = \eta_{22} = jZ_0 k_0 \xi_0^2 C \tag{2.116}$$

where

$$C = \frac{1}{4} \int_0^\infty \left[\left(\frac{\partial^2}{\partial \rho^2} + \frac{1}{\rho} \frac{\partial}{\partial \rho} + 2k_0^2 \right) \left\{ F J_0 \left(\frac{k_0 \rho}{\sqrt{2}} \right) \right\} \right.$$
$$\left. - \frac{k_0}{\sqrt{2}} \frac{\partial F}{\partial \rho} J_1 \left(\frac{k_0 \rho}{\sqrt{2}} \right) \right] e^{-jk_0\rho} \, d\rho \qquad (2.117)$$

For a roughness whose scale ℓ is such that $k_0\ell \ll 1$,

$$C \simeq \frac{1}{4} \int_0^\infty \frac{1}{\rho} \frac{\partial F}{\partial \rho} e^{-jk_0\rho} \, d\rho \qquad (2.118)$$

and if

$$F(\rho) = e^{-4\rho^2/\ell^2} \qquad (2.119)$$

the integral can be evaluated to give

$$C = -\frac{\sqrt{\pi}}{2\ell} \qquad (2.120)$$

For larger values of $k_0\ell$ (but still not large compared with unity), the approximations made in going from (2.117) to (2.118) are no longer valid but, if (2.119) is inserted into (2.117), a quasi-analytical evaluation is still possible. It is found that

$$\eta_{11} = \eta_{22} = Z_0 \left(\tfrac{1}{2} k_0 \xi_0 \right)^2 (L - jM) \qquad (2.121)$$

(SENIOR, 1960b) and L and M are plotted as functions of $k_0\ell$ in Fig. 2–11. If $k_0\ell \ll 1$

$$M \simeq \frac{2\sqrt{\pi}}{k_0\ell} \qquad L \simeq 0$$

in accordance with (2.116) and (2.120) but, as $k_0\ell$ increases, L becomes more significant, and L and M are equal for $k_0\ell \simeq 2.1$.

A purely imaginary impedance is equivalent to the displacement of an ideal surface parallel to itself (see Section 2.6.1). In the present case, the fact that the imaginary part is always negative is a consequence of the random nature of the irregularities and the chosen location of the simulating surface. With this located at the mean surface, the ensemble averaging eliminates the first order effect, leaving only a negative second order one. Because of the roughness, it appears that the pec surface is displaced *outwards*, and for a convex body this is expected to *increase* the scattering. As $k_0\ell$ increases, however, the impedance changes from purely reactive to part resistive and part reactive, and the resistivity has an opposing effect on the scattering. Which effect dominates depends on the scattering body, and measured data for rough spheres (HIATT ET AL., 1960) confirm this behaviour.

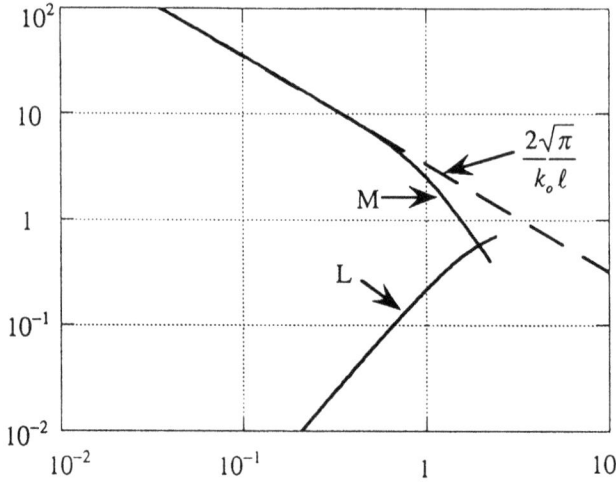

Figure 2-11: *L and M (see (2.121)) as functions of $k_0\ell$*

2.6.3   Corrugations

A periodic roughness in the form of corrugations is important in connection with waveguides and (flared) horns where the anisotropic surface impedance is used to suppress (or at least attenuate) unwanted higher order modes. Examples are the corrugated pec structures shown in Figs. 2–12(a) and (b) where the grooves are parallel to the z axis and are filled with a material having relative permittivity ϵ_r and relative permeability μ_r. It is assumed that $k_0T \ll 1$, implying $k_0w \ll 1$ *a fortiori*, and we seek an average surface impedance

$$\bar{\bar{\eta}} = \eta_{11}\hat{x}\hat{x} + \eta_{22}\hat{z}\hat{z} \qquad (2.122)$$

at $y = 0$ to simulate the effect of the corrugations.

If the polarisation is such that $\mathbf{H} = \hat{z}H_z$, only the lowest order mode can propagate in a groove and, because this is a TEM mode, we can determine the effective surface impedance presented by a single groove from the input impedance Z_{1n} of a transmission line. For the geometry in Fig. 2–12(a)

$$\frac{E_x}{H_z} = \frac{1}{w}Z_{1n}$$

where

$$Z_{1n} = jZw \tan(Nk_0d)$$

and Z and N are the intrinsic impedance and refractive index, respectively, of the material in the groove. Since $\eta = 0$ on the metal lying in the plane $y = 0$, it follows that

$$\eta_{22} = jZ\frac{w}{T}\tan(Nk_0d) \qquad (2.123)$$

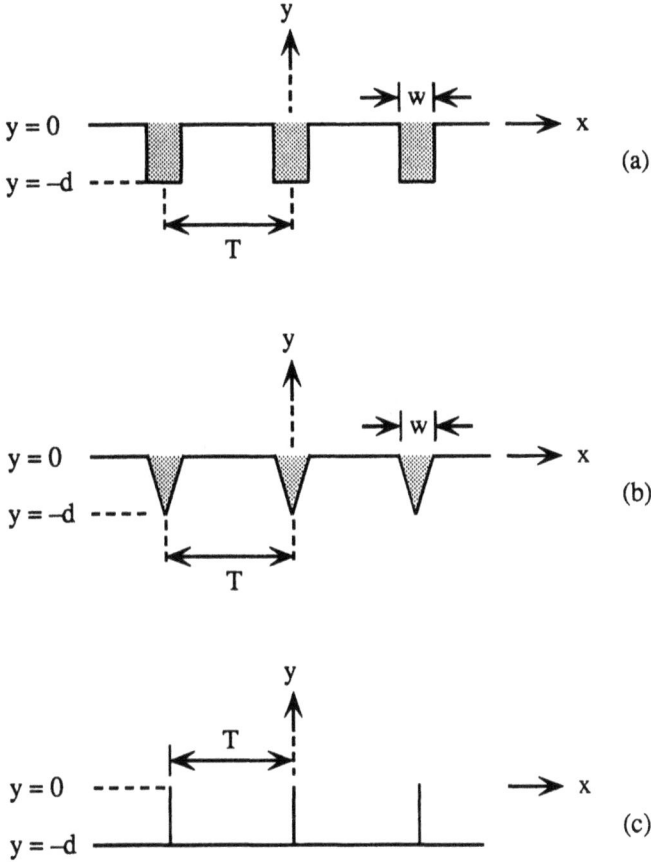

Figure 2–12: *Examples of corrugated surfaces*

Similarly, for the V-shaped grooves in Fig. 2–12(b),

$$\eta_{22} = jZ\frac{w}{T}\frac{J_1(Nk_0d)}{J_0(Nk_0d)} \tag{2.124}$$

(SENIOR ET AL., 1990) where J_n is the Bessel function of order n. As expected, $\eta_{11} = 0$ if $w = 0$ or $d = 0$ since then the grooves do not exist.

For E polarisation such that $\mathbf{E} = \hat{z}E_z$, all of the modes are evanescent, but if we again assume that the first order mode dominates, simple formulas for the surface impedance can be found. In a parallel plate waveguide of width w the propagation constant of this mode is $-jk_0p$ where

$$p = \left\{\left(\frac{\pi}{k_0w}\right)^2 - N^2\right\}^{1/2}$$

and, since E_z/H_x is independent of position across the guide, a transmission line analogy can be made. Thus, for a single rectangular groove

$$Z_{1n} = jZ_0\mu_r \frac{w}{p} \tanh(k_0 pd)$$

giving

$$\eta_{11} = -jZ_0\mu_r \frac{w}{pT} \tanh(k_0 pd) \tag{2.125}$$

and for the V-shaped grooves

$$\eta_{11} = -jZ_0\mu_r \frac{w}{pT} \frac{I_1(k_0 pd)}{I_0(k_0 pd)} \tag{2.126}$$

(SENIOR ET AL. 1990), where I_n is the modified Bessel function of order n. In most practical situations $|\eta_{22}|$ is so small that we can take $\eta_{22} = 0$.

Another type of periodic roughness is an array of equally spaced perfectly conducting strips protruding from a plane as shown in Fig. 2–12(c). The effective surface impedance can be obtained from the solution for a plane wave incident on an infinite stack of perfectly conducting half planes and, if $k_0 T \ll 1$ and $k_0 d \ll 1$, it is found that (WEINSTEIN, 1969)

$$\eta_{22} = jk_0 Z_0 (d - \alpha) \qquad \eta_{11} = 0 \tag{2.127}$$

with

$$\alpha = -\frac{T}{\pi} \ln \left\{ \frac{1}{2} \left(1 + e^{-2\pi d/T} \right) \right\} \tag{2.128}$$

where the boundary condition is now imposed at $y = -\alpha$. To transfer the condition to the surface $y = 0$, a higher order boundary condition would be required. If the perfectly conducting surface $y = -d$ is not present, leaving only the parallel strips, the structure is partially transparent and can be simulated using a resistive sheet. For this and other partially transparent periodic surfaces such as a grating composed of perfectly conducting wires or narrow strips in the plane $y = 0$, the equivalent resistivities are given by WEINSTEIN (1969) and MARCUVITZ (1986).

When the impedances in (2.122)–(2.128) are used to simulate the corrugated wall of a waveguide, it is necessary to assign Z (or Z_0) and N their values appropriate to the particular waveguide mode. DYBDAL ET AL. (1971) determined the modes that can be supported by a rectangular guide with corrugated walls, and showed that, if two adjacent walls are corrugated, the resulting anisotropic surface impedances must satisfy a consistency condition for any modes to exist. Although their analysis has been criticised by MCISAAC (1974) and NARASIMHAN AND RAO (1974), the condition is easily derived by considering a right angled wedge whose open quadrant is $x, y \geq 0$, $-\infty < z < \infty$, illuminated by a plane wave whose propagation vector is

$$\hat{k} = k_x \hat{x} + k_y \hat{y} + k_z \hat{z}$$

If the surfaces $y = 0$, $x \geq 0$ and $x = 0$, $y \geq 0$ are subject to an SIBC with

$$\overline{\overline{\eta}} = \eta_1 \hat{x}\hat{x} + \eta_2 \hat{z}\hat{z}$$

and

$$\overline{\overline{\eta}} = \eta_3 \hat{y}\hat{y} + \eta_4 \hat{z}\hat{z}$$

respectively, the image method provides a solution consisting of four plane waves if and only if

$$k_x k_y k_z \Gamma = 0 \qquad (2.129)$$

where

$$\Gamma = \eta_1 \eta_4 + \eta_2 \eta_3 - \eta_1 \eta_3 \qquad (2.130)$$

(SENIOR, 1978b). The equation $\Gamma = 0$ is the consistency condition developed by DYBDAL ET AL. (1971) and, if (2.129) is not satisfied, the solution of the boundary value problem does not have a discrete angular spectrum. It can be shown that (2.129) is equivalent to the requirement that $\mathbf{E} \cdot \mathbf{H} = 0$ at the vertex.

2.7   Uniqueness

If a boundary condition is to be useful, it must ensure a unique solution of the boundary value problem, and for the tensor SIBC (2.34) we now seek the restrictions on the tensor elements in order that it does. The analysis is a simple extension of the standard uniqueness proof (see, for example, HARRINGTON (1961)) for electromagnetic fields, and requires that the *net* real power flow out of the surface into the surrounding medium is never positive. This is appropriate for passive surfaces which are not themselves a source of power.

Consider a volume V bounded externally by a sphere S_∞ of large radius and internally by the surface S at which the boundary condition is imposed (see Fig. 2–13). If $(\mathbf{E}, \mathbf{H})$ is a field generated by the current sources $\mathbf{J}_e$ and $\mathbf{J}_m$ in V,

$$\nabla \times \mathbf{E} = -j\omega\mu\mathbf{H} - \mathbf{J}_m, \qquad \nabla \times \mathbf{H} = j\omega\epsilon\mathbf{E} + \mathbf{J}_e$$

where ϵ and μ are the permittivity and permeability, respectively, of the medium comprising V and, when the divergence theorem is applied to $\mathbf{E} \times \mathbf{H}^*$, we obtain

$$\left(\iint_S + \iint_{S_\infty} \right) \mathbf{E} \times \mathbf{H}^* \cdot \hat{n}\, dS = j\omega \iiint_V \left(\mu|\mathbf{H}|^2 - \epsilon^*|\mathbf{E}|^2 \right) dV$$
$$+ \iiint_V \left(\mathbf{E} \cdot \mathbf{J}_e^* + \mathbf{H}^* \cdot \mathbf{J}_m \right) dV$$

where the asterisk denotes the complex conjugate. The integral over S_∞ vanishes due to the radiation condition. Hence, if $(\mathbf{E}^{(1)}, \mathbf{H}^{(1)})$ and $(\mathbf{E}^{(2)}, \mathbf{H}^{(2)})$ are two fields generated by the same sources, the difference field

$$\partial\mathbf{E} = \mathbf{E}^{(1)} - \mathbf{E}^{(2)} \qquad \partial\mathbf{H} = \mathbf{H}^{(1)} - \mathbf{H}^{(2)} \qquad (2.131)$$

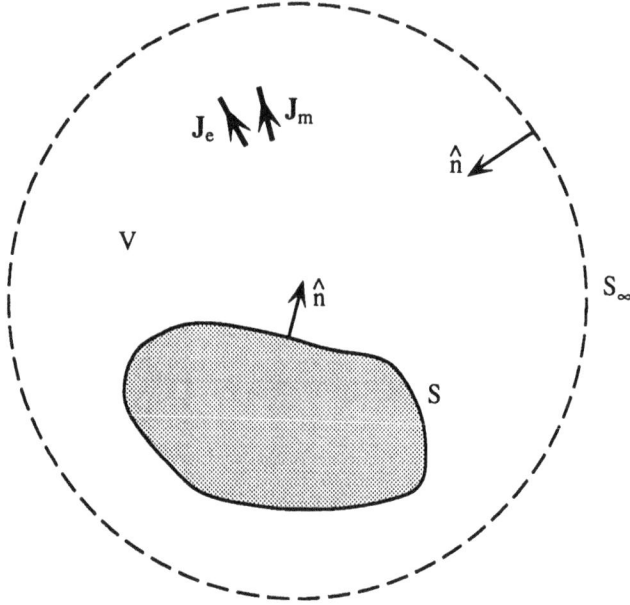

Figure 2–13: *Geometry for the uniqueness proof*

satisfies

$$\iint_S \partial \mathbf{H}^* \cdot (\hat{n} \times \partial \mathbf{E}) \, dS = j\omega \iiint_V \left(\mu |\partial \mathbf{H}|^2 - \epsilon^* |\partial \mathbf{E}|^2 \right) \, dV$$

Let

$$\epsilon = \epsilon' - j\epsilon'' \qquad \mu = \mu' - j\mu''$$

and assume that the medium has some loss so that ϵ'' and/or $\mu'' > 0$ at every point of V. Then

$$\text{Re.} \iint_S \partial \mathbf{H}^* \cdot (\hat{n} \times \partial \mathbf{E}) \, dS = \omega \iiint_V \left(\mu'' |\partial \mathbf{H}|^2 + \epsilon'' |\partial \mathbf{E}|^2 \right) \, dV \qquad (2.132)$$

and this is the basis for the proof.

If $\hat{n} \times \mathbf{E} = 0$ on S (it is sufficient if it has a specified value there), $\hat{n} \times \partial \mathbf{E} = 0$ and the left hand side of (2.132) vanishes. Hence, $\partial \mathbf{E} = 0$ and/or $\partial \mathbf{H} = 0$ at every point of V and the solution is unique; similarly if $\hat{n} \times \mathbf{H} = 0$ on S. For the scalar impedance boundary condition (2.36), the integrand of the surface integral is

$$\eta \partial \mathbf{H}^* \cdot \hat{n} \times (\hat{n} \times \partial \mathbf{H}) = -\eta |\hat{n} \times \partial \mathbf{H}|^2$$

and if

$$\text{Re.} \, \eta \geq 0 \qquad (2.133)$$

at every point of S, the left hand side of (2.132) is ≤ 0. The only possible solution of (2.132) is then $\partial \mathbf{H} = 0$ and/or $\partial \mathbf{E} = 0$ throughout V and on S. The condition (2.133) is both necessary and sufficient for uniqueness in the case of a passive surface. Finally, for the tensor impedance boundary condition (2.34), we introduce the local tangent vectors $\hat{s}$ and $\hat{t}$, such that $\hat{s}$, $\hat{n}$, $\hat{t}$ form a right handed system of orthogonal unit vectors, and write

$$\overline{\overline{\eta}} = (\eta_{11}\hat{s} + \eta_{12}\hat{t})\hat{s} + (\eta_{21}\hat{s} + \eta_{22}\hat{t})\hat{t} \tag{2.134}$$

Then

$$\partial\mathbf{H} \cdot (\hat{n} \times \partial\mathbf{E}) = -\eta_{11}|\partial H_s|^2 - \eta_{22}|\partial H_t|^2 - \eta_{12}\,\partial H_t\,\partial H_s^* - \eta_{21}\,\partial H_s\,\partial H_t^*$$

and by the same argument as before the solution is unique if

$$\text{Re. } \eta_{11}, \text{ Re. } \eta_{22} \geq 0 \tag{2.135}$$

and

$$\eta_{21} = -\eta_{12}^* \tag{2.136}$$

We note that uniqueness still holds if the tensor elements η_{ij} are functions of position, including the case when one or more is discontinuous on S. Since we have assumed that the medium surrounding the scatterer is lossy, it is necessary to regard a lossless medium as a limiting case of a lossy medium.

There is a similar proof for a scalar impedance boundary condition of the form

$$\frac{\partial U}{\partial n} = \alpha U \tag{2.137}$$

where α may be a function of position on S and U is a solution of the scalar wave equation. Suppose $U^{(1)}$ and $U^{(2)}$ are two solutions generated by the same sources in V and let

$$W = U^{(1)} - U^{(2)} \tag{2.138}$$

Then W satisfies the boundary condition (2.137) on S and the homogeneous scalar wave equation

$$(\nabla^2 + k_0^2)W = 0$$

throughout V. Application of the divergence theorem to $W\nabla W^*$ gives

$$\left(\iint_S + \iint_{S_\infty}\right) W^*\frac{\partial W}{\partial n}\,dS = k_0^2 \iiint_V |W|^2\,dV - \iiint_V |\nabla W|^2\,dV$$

(SENIOR, 1993) and the integral over S_∞ vanishes by virtue of the radiation condition. Hence, if k_0 has a small negative imaginary part corresponding to a slight loss in V,

$$\text{Im. }\iint_S W^*\frac{\partial W}{\partial n}\,dS = \text{Im. } k_0^2 \iiint_V |W|^2\,dV \tag{2.139}$$

and, when the boundary condition is inserted, we obtain

$$\text{Im. }\iint_S \alpha|W|^2\,dS = \text{Im. } k_0^2 \iiint_V |W|^2\,dV \tag{2.140}$$

Since the right hand side is never positive, and is zero only if $W = 0$ throughout V, it follows that if

$$\text{Im.}\ \alpha \geq 0 \tag{2.141}$$

at all points of S, $W = 0$ in V and on S, showing that the solution is unique. This is a necessary and sufficient condition for uniqueness.

In the special case of a planar surface $y = 0$ with $U = E_y$ and $\alpha = jk_0\eta/Z_0$ (see (2.23)) or $U = H_y$ with $\alpha = jk_0Z_0/\eta$ (see (2.25)), the restriction on η is as shown in (2.133).

We have so far considered only passive surfaces. For most practical purposes, these are the ones of interest, but it should be noted that the α defined by (2.141) are merely a subset of those for which the solution of the boundary value problem is unique. This is easily demonstrated for a planar surface subject to the boundary condition (2.137). The restriction (2.141) then ensures that $|R| \leq 1$ for all real angles of incidence, where R is the reflection coefficient, but the solution is still unique if $|R|$ is bounded for these same angles, i.e. if $\alpha \neq -ia$ where $0 < a \leq 1$. For all other values of α, the boundary condition leads to a well-posed mathematical problem, with Im. $\alpha < 0$ corresponding to an active surface and Im. $\alpha \geq 0$ to a passive one.

References

Bateman, H. (1915), *Electrical and Optical Wave Motion*, Cambridge University Press, London, p. 19.

Bhattacharyya, A. K. and Sengupta, D. L. (1991), *Radar Cross Section Analysis and Control*, Artech House, Boston, p. 180.

Dybdal, R. B., Peters, L., Jr. and Peake, W. H. (1971), "Rectangular waveguides with impedance walls", *IRE Trans. Microwave Theory Tech.*, **MTT-19**, pp. 2–9. See also comments by P. R. McIsaac and M. S. Narasimhan and V. V. Rao (1974), *IRE Trans. Microwave Theory Tech.*, **MTT-22**, pp. 972–973.

Feinberg, E. (1944), "On the propagation of radio waves along an imperfect surface", *J. Phys. USSR*, **8**, pp. 317–330.

Feinberg, E. (1945), "On the propagation of radio waves along an imperfect surface", *J. Phys. USSR*, **9**, pp. 1–6.

Feinberg, E. (1946), "On the propagation of radio waves along an imperfect surface", *J. Phys. USSR*, **10**, pp. 410–418.

Fock, V. (1946), "The field of a plane wave near the surface of a conducting body", *J. Phys. USSR*, **10**, pp. 399-409.

Goodman, J. W. (1968), *Introduction to Fourier Optics*, McGraw-Hill Book Co., New York, pp. 35–37.

Grünberg, G. A. (1942), "Theory of the coastal refraction of electromagnetic waves", *J. Phys. USSR*, **6**, pp. 185–209.

Harrington, R. F. (1961), *Time-Harmonic Electromagnetic Fields*, McGraw-Hill Book Co., New York, pp. 100-102, 126.

Harrington, R. F. and Mautz, J. R. (1975), "An impedance sheet approximation for thin dielectric shells", *IEEE Trans. Antennas Propagat.*, **AP-23**, pp. 531–534.

Hiatt, R. E., Senior, T. B. A. and Weston, V. H. (1960), "A study of surface roughness and its effect on the backscattering cross section of spheres", *Proc. IRE*, **48**, pp. 2008–2016.

Jones, D. S. and Pidduck, F. B. (1950), "Diffraction by a metal wedge at large angles", *Quart. J. Math. Oxford*, 1, pp. 229–237.

King, R. J. and Wait, J. R. (1976), "Electromagnetic groundwave propagation theory and experiment", in *Symposia Mathematica*, Vol. 17, Academic Press, New York, pp. 107–208.

Leontovich, M. A. (1948), *Investigations on Radiowave Propagation, Part II*, Printing House of the Academy of Sciences, Moscow, pp. 5–12.

Marcuvitz, N. (1986), *Waveguide Handbook* (2nd edition), Peter Peregrinus Ltd., Institution of Electrical Engineers, London, pp. 280 et seq.

Meixner, J. (1949), "Die Kantenbedingung in der Theorie der Beugung elektromagnetischer Wellen an Vollkommen Leitenden Ebenen Schirmen", *Ann. Physik*, **441**, pp. 2–9.

Rytov, S. M. (1940), "Calcul du skin-effet par la méthode des perturbations", *J. Phys. USSR*, **2**, pp. 233–242. The paper was republished in Russian in *Zhur. Eksp. i Teoret. Fiz.*, **10**, pp. 180–189 (1940). An English translation of the latter has been made by V. Kerdemelidis and K. M. Mitzner, Northrop Navair, Hawthorne CA 90250.

Schukin, A. N. (1940), *Propagation of Radio Waves*, Publishing House Svyazizdat, Moscow, pp. 48–51 (in Russian). The authors are indebted to Professor P. Ya. Ufimtsev for providing a translation of the relevant pages.

Senior, T. B. A. (1952), "Diffraction by a semi-infinite metallic sheet", *Proc. Roy. Soc. London, Ser. A*, **213**, pp. 436–458.

Senior, T. B. A. (1960a), "Impedance boundary conditions for imperfectly conducting surfaces", *Appl. Sci. Res. B*, **8**, pp. 418–436.

Senior, T. B. A. (1960b), "Impedance boundary conditions for statistically rough surfaces", *Appl. Sci. Res. B*, **8**, pp. 437–462.

Senior, T. B. A. (1962), "A note on impedance boundary conditions", *Can. J. Phys.*, **40**, pp. 663–665.

Senior, T. B. A. (1978a), "Some problems involving imperfect half planes", in *Electromagnetic Scattering* (ed. P. L. E. Uslenghi), Academic Press, New York, pp. 185–219.

Senior, T. B. A. (1978b), "Skew incidence on a right-angled impedance wedge", *Radio Sci.*, **13**, pp. 639–647.

Senior, T. B. A. (1985), "Combined resistive and conductive sheets", *IEEE Trans. Antennas Propagat.*, **AP-33**, pp. 577–579.

Senior, T. B. A. (1993), "Generalized boundary and transition conditions and the uniqueness of solution", University of Michigan Radiation Laboratory Report RL–891, Ann Arbor MI.

Senior, T. B. A. and Volakis, J. L. (1987), "Sheet simulation of a thin dielectric layer", *Radio Sci.*, **22**, pp. 1261–1272.

Senior, T. B. A., Sarabandi, K. and Natzke, J. R. (1990), "Scattering by a narrow gap", *IEEE Trans. Antennas Propagat.*, **AP-38**, pp. 1102–1110.

Senior, T. B. A. and Volakis, J. L. (1991), "Generalized impedance boundary conditions in scattering", *Proc. IEEE*, **79**, pp. 1413–1420.

Sommerfeld, A. (1949), *Partial Differential Equations in Physics,* Academic Press Inc., New York, pp. 63–64.

Stratton, J. A. (1941), *Electromagnetic Theory*, McGraw-Hill Book Co., New York, pp. 492–494, 564–565.

Tai, C.-T. (1994), *Dyadic Green's Functions in Electromagnetic Theory*, IEEE Press, Piscataway NJ, pp. 42–47.

Van Bladel, J. (1985), *Electromagnetic Fields,* Hemisphere Pub. Co., New York, pp. 385–387.

Wang, D.-S. (1987), "Limits and validity of the impedance boundary condition on penetrable surfaces", *IEEE Trans. Antennas Propagat.*, **AP-35**, pp. 453–457.

Weinstein, L. A. (1969), *The Theory of Diffraction and the Factorization Method* (translation from the Russian by Petr Beckmann), The Golem Press, Boulder CO, pp. 298–302.

Chapter 3

Application to planar structures

3.1   Introduction

The first order conditions have proved useful for simulating the properties of imperfectly conducting and coated geometries in propagation, scattering and antenna studies. Analytical and numerical solutions have been developed for a variety of configurations, and in the next two chapters we concentrate on the application of the impedance boundary and transition conditions to problems amenable to analytical solution. The coated half-plane and junction, the material half-plane and junction, and the coated wedge are particular examples of these. In this chapter we consider the analysis of half-planes and junctions that are modelled using the equivalent impedance or sheet geometries illustrated in Fig. 3–1. The accuracy of the SIBC in simulating dielectric coatings was discussed in Chapter 2, and it was pointed out that the SIBC provides a reasonable simulation for thin (particularly lossy) coatings. When in doubt, the exact reflection coefficients can be compared with those of the SIBC surface to evaluate the validity of the approximation. However, such a comparison does not account for simulation errors caused by an abrupt termination of the coating. A numerical analysis indeed demonstrates that the SIBC approximation, although sufficiently accurate far from the edge, is substantially in error within a distance of about $\lambda_0/10$ from the edge. This is illustrated in Figs. 3–2 and 3–3 for a coated right-angled wedge and a coated half-plane. The coating thickness in these figures is $0.06\lambda_0$ and the incidence angle is $135°$ from the top face. The impedance calculations were carried out using a hybrid finite element–boundary integral code that is, in principle, exact. Nevertheless, in spite of the inaccurate specification of the surface impedance close to the edge, it has been shown numerically (SANCER, personal communication) that the SIBC provides reasonably accurate results for the field away from the surface (particularly in the far field) whenever its accuracy is acceptable for the infinite layer. An exception to this is the case of near edge-on incidence ($\phi \approx 180°$) for coatings more than about 0.05λ thick.

The classical technique for the analysis of impedance or resistive/conductive half-planes and junctions is the Wiener-Hopf method (WIENER AND HOPF,

47

original geometry

equivalent impedance or
resistive sheet geometry

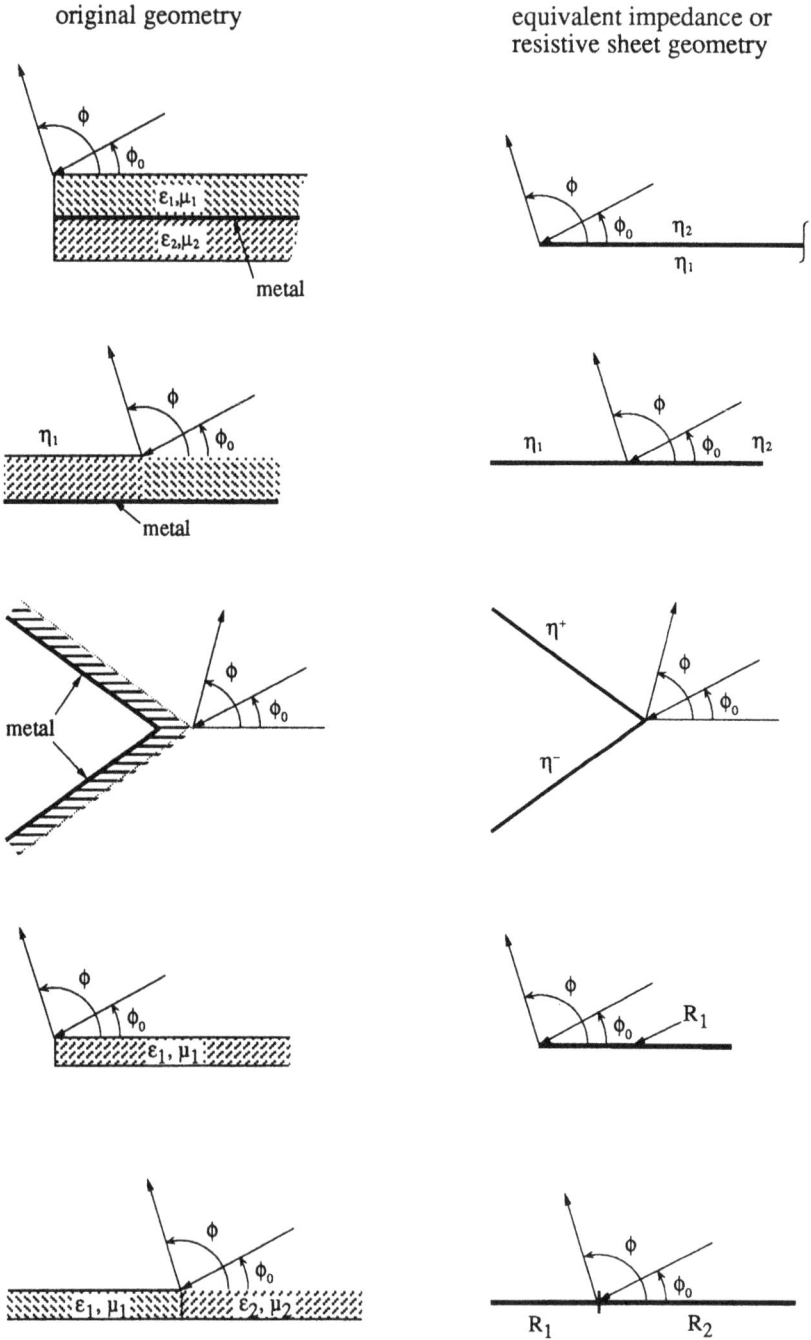

Figure 3-1: *Geometries amenable to analytical solution*

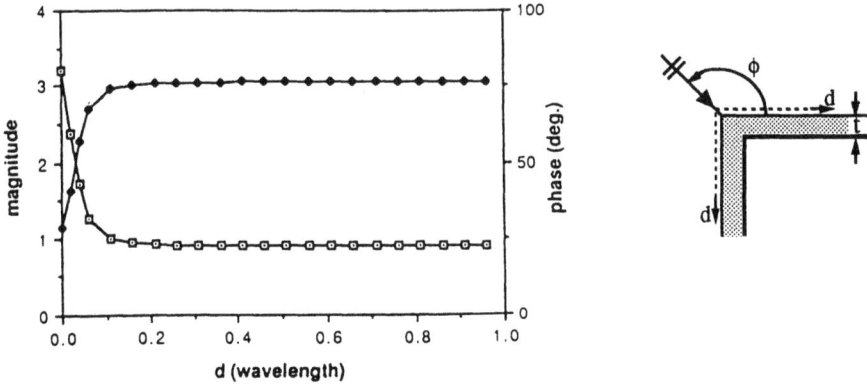

Figure 3–2: *Magnitude (□□□) and phase (◇◇◇) of the actual surface impedance for a coated right-angled wedge with* $t = 0.062\lambda_0$, $\epsilon_r = 2(1-j)$, $\mu_r = 2$ *and* $\phi = 135°$

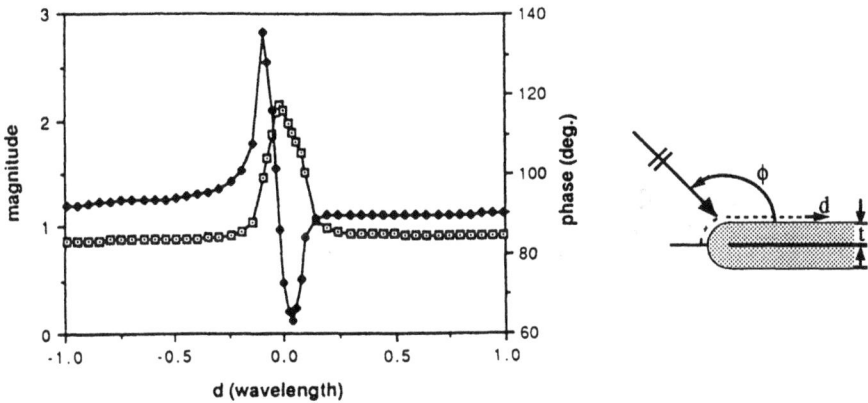

Figure 3–3: *Magnitude (□□□) and phase (♦♦♦) of the actual surface impedance for a coated half plane with* $t = 0.062\lambda_0$, $\epsilon_r = 2$, $\mu_r = 2$ *and* $\phi = 135°$

1931). An alternative but equivalent approach is the dual integral equation method (CLEMMOW, 1951) and, since this is a more concise method, we will use it consistently throughout the text for the analysis of half-plane structures. In the next section we first describe the dual integral equation approach and compare it with the Wiener-Hopf method. The rest of the chapter presents the uniform and non-uniform diffraction coefficients for resistive, conductive and impedance half-planes, as well as junctions formed by these semi-infinite structures.

3.2   Wiener-Hopf versus dual integral equation methods

The analysis of half-plane diffraction problems is generally carried out using the Wiener-Hopf method. The first application of the method to a perfectly conducting half-plane was by COPSON (1946) and, independently, by CARLSON AND HEINS (1947), but an appreciation of the method's scope is generally attributed to Schwinger (see remarks by CARLSON AND HEINS (1947) and MILES (1949)). SENIOR (1952) was the first to apply the Wiener-Hopf method to the impedance half-plane problem. He later (SENIOR, 1959) extended the analysis to skew incidence, and then repeated it (SENIOR, 1975b) starting with the boundary conditions (2.23)–(2.25) on the normal components.

The Wiener-Hopf method provides the analytical solution of the integral equation that results from the application of the boundary conditions. This is accomplished by taking the Fourier transform of the integral equation, thereby reducing it to a functional equation which is in turn solved by invoking the properties of functions that are analytic (free of poles and branch cuts) in the upper and lower halves of the complex plane of the transform variable. For the problems at hand, the method can become tedious because of having to switch between the spatial and transform quantities when enforcing the boundary conditions. Also, additional steps arise because the unknown is the current density on the half plane, whereas the boundary conditions involve the field quantities. To avoid these complications, CLEMMOW (1951) proposed an alternative approach that works directly with the spectra of the field quantities without having to return to the spatial expressions until the analysis is complete. Clemmow's approach, referred to as the dual integral equation method, is mathematically equivalent to the Wiener-Hopf method, but requires fewer steps to arrive at the solution. For this reason we will use it to develop the half-plane and junction solutions.

A key feature of the dual integral equation method is the *a priori* introduction of angular spectrum representations, such as (BOOKER AND CLEMMOW, 1950)

$$E_z(\rho, \phi) = \int_C P(\cos \alpha)\, e^{-jk_0 \rho \cos(\phi \mp \alpha)}\, d\alpha \qquad y \gtrless 0 \qquad (3.1)$$

$$H_x(\rho, \phi) = -\frac{Y_0}{jk_0} \frac{\partial E_z}{\partial y}$$

$$= \pm Y_0 \int_C \sin \alpha \, P(\cos \alpha)\, e^{-jk_0 \rho \cos(\phi \mp \alpha)}\, d\alpha \qquad y \gtrless 0 \quad (3.2)$$

These are valid for the fields due to a z-directed electric current distribution in the $y = 0$ plane as illustrated in Fig. 3–4. The integration contour C is defined in Fig. 3–5(a), (ρ, ϕ) denote the cylindrical coordinates of the field observation point and $P(\cos \alpha)$ is referred to as the angular spectrum of the field.

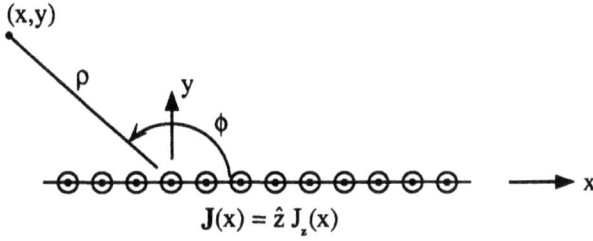

Figure 3–4: *Illustration of a z-directed current sheet in the $y = 0$ plane*

To prove the validity of (3.1) and (3.2), we begin by defining the Fourier transform pair

$$\tilde{J}_z(k_x) \;=\; \mathcal{F}\{J_z(x)\} = \int_{-\infty}^{\infty} J_z(x)\, e^{jk_x x}\, dx \tag{3.3}$$

$$J_z(x) \;=\; \mathcal{F}^{-1}\{\tilde{J}_z(k_x)\} = \frac{1}{2\pi} \int_{-\infty}^{\infty} \tilde{J}_z(k_x)\, e^{-jk_x x}\, dk_x \tag{3.4}$$

where, in this case, J_z represents the electric current density. The field radiated by $J_z(x)$ is given by

$$E_z(x,y) = -\frac{k_0 Z_0}{4} \int_{-\infty}^{\infty} J_z(x')\, H_0^{(2)}\left(k_0\sqrt{(x-x')^2 + y^2}\right) dx' \tag{3.5}$$

where $H_0^{(2)}$ is the zeroth order Hankel function of the second kind.

Taking the transform of both sides of (3.5) and invoking the convolution

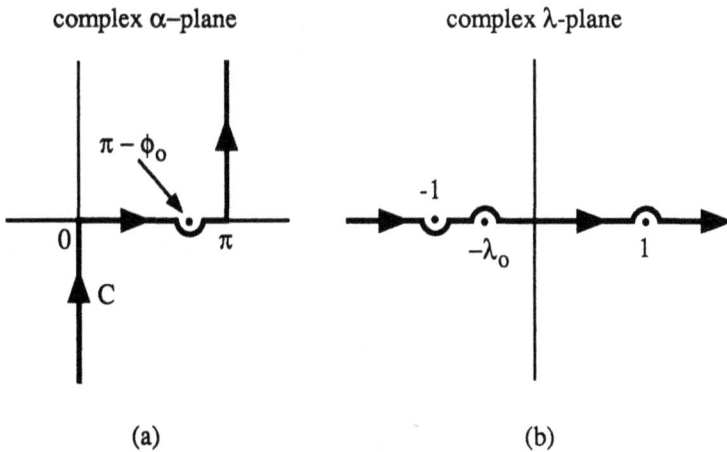

Figure 3–5: *Contours in the complex α- and λ-planes*

theorem yields

$$\tilde{E}_z(k_x, y) = -\frac{k_0 Z_0}{2k_y}\, \tilde{J}_z(k_x)\, e^{-jk_y|y|} \tag{3.6}$$

where

$$k_y = \begin{cases} \sqrt{k_0^2 - k_x^2} & k_x < k_0 \\ -j\sqrt{k_x^2 - k_0^2} & k_x > k_0 \end{cases}$$

and we have used the result $\mathcal{F}\{H_0^{(2)}(k_0|x|)\} = 2/k_y$. Next, from the inverse transform of (3.6),

$$E_z(x, y) = -\frac{k_0 Z_0}{4\pi} \int_{-\infty}^{\infty} \frac{\tilde{J}_z(k_x)}{k_y} e^{-jk_y|y|} e^{-jk_x x}\, dk_x \tag{3.7}$$

and on making the change of variable

$$k_x = k_0 \lambda = k_0 \cos\alpha, \qquad k_y = k_0\sqrt{1 - \lambda^2} = k_0 \sin\alpha \tag{3.8}$$

we get

$$E_z(x, y) = -\frac{k_0 Z_0}{4\pi} \int_C \tilde{J}_z(k_0 \cos\alpha)\, e^{-jk_0(x\cos\alpha + |y|\sin\alpha)}\, d\alpha \tag{3.9}$$

where C is the contour shown in Fig. 3–5. Finally, in terms of the cylindrical coordinates (ρ, ϕ) where $x = \rho\cos\phi$, $y = \rho\sin\phi$, we have

$$E_z(\rho, \phi) = -\frac{k_0 Z_0}{4\pi} \int_C \tilde{J}_z(k_0 \cos\alpha)\, e^{-jk_0\rho\cos(\phi\mp\alpha)}\, d\alpha \tag{3.10}$$

When this is compared with (3.1), we see that the angular spectrum $P(\cos\alpha)$ is proportional to the angular spectrum of the current density. Specifically,

$$P(\cos\alpha) = -\frac{k_0 Z_0}{4\pi}\, \tilde{J}_z(k_0 \cos\alpha) \tag{3.11}$$

and consequently (3.1) and (3.2) are nothing more than the usual spectral representations of the scattered field (3.5) in terms of the angular variable α. Moreover, (3.2) implies

$$\tilde{H}_x(k_x, y = 0^\pm) = \mp\tfrac{1}{2}\, \tilde{J}_z(k_x)\, e^{-jk_0\lambda|y|}, \qquad y \gtrless 0$$

or

$$\tilde{H}_x(k_x, y = 0^+) - \tilde{H}_x(k_x, y = 0^-) = -\tilde{J}_z(k_x) \tag{3.12}$$

which follows immediately from (2.5).

3.3   Resistive half-plane (E-polarisation)

As mentioned in Section 2.4, the solution to a problem involving the resistive sheet boundary conditions (2.59) can be readily generalised to configurations of the same shape that instead satisfy the conductive sheet or impedance boundary conditions (see (2.61) and (2.34)). Alternatively the E-polarisation solutions for resistive and conductive half-planes can be used to construct all other solutions involving resistive, conductive and impedance geometries of the same shape. This is the approach followed here, and we first consider the diffraction of an E-polarised plane wave by a resistive half-plane. The half-plane occupies the right half $(x > 0)$ of the $y = 0$ plane as illustrated in Fig. 3–6.

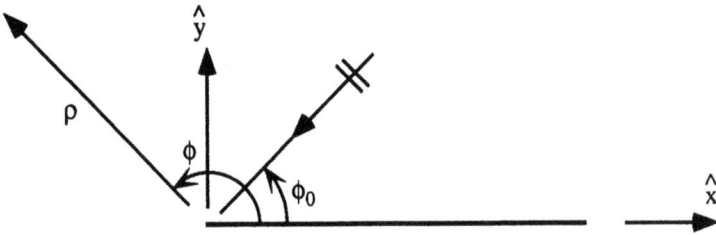

Figure 3–6: *Half-plane geometry and the coordinate system*

3.3.1   Dual integral equation solution

We assume that the plane wave

$$\mathbf{E}^{\mathrm{i}} = \hat{z}\, e^{jk_0(x\cos\phi_0 + y\sin\phi_0)} = \hat{z}\, e^{jk_0\rho\cos(\phi-\phi_0)} \tag{3.13}$$

impinges on the resistive half-plane shown in Fig. 3–6. From Maxwell's equations, the corresponding magnetic field is given by

$$\begin{aligned}
\mathbf{H}^{\mathrm{i}} &= -\frac{jY_0}{k_0}\,\hat{z} \times \nabla E_z^{\mathrm{i}} = Y_0(-\hat{\rho} \times \hat{z})\, E_z^{\mathrm{i}} \\
&= Y_0(-\hat{x}\sin\phi_0 + \hat{y}\cos\phi_0)\, E_z^{\mathrm{i}}
\end{aligned}$$

and from (2.52) and (2.53), the boundary condition satisfied by the half-plane is

$$E_z = -R_{\mathrm{e}}[H_x]_-^+ \tag{3.14}$$

for $y = 0$, $x > 0$. In this, R_{e} denotes the resistivity in Ω per square and, as usual, $[\ ]_-^+$ implies the discontinuity in H_x between the $y = 0^+$ and $y = 0^-$ sides of the half-plane. It goes without saying that the H_x field is continuous on the extension $(x < 0)$ of the $y = 0$ plane and, in addition, E_z is continuous everywhere, since only electric currents are supported by the resistive half-plane.

In accordance with the dual integral equation method, we begin our analysis by introducing the angular spectrum representations

$$E_z^s(x, y \gtrless 0) = \int_C P_e(\cos \alpha) \, e^{-jk_0 \rho \cos(\phi \mp \alpha)} \, d\alpha \qquad (3.15)$$

$$H_x^s(x, y \gtrless 0) = \pm Y_0 \int_C \sin \alpha \, P_e(\cos \alpha) \, e^{-jk_0 \rho \cos(\phi \mp \alpha)} \, d\alpha \qquad (3.16)$$

(see (3.1) and (3.2)) where

$$\mathbf{E}^s = \mathbf{E} - \mathbf{E}^i \qquad (3.17)$$

$$\mathbf{H}^s = \mathbf{H} - \mathbf{H}^i \qquad (3.18)$$

are the scattered fields generated by the electric currents induced on the resistive half-plane. The unknown spectral function $P_e(\cos \alpha)$ is proportional to the spectrum of the electric current density, as indicated in (3.11), and our goal is to determine $P_e(\cos \alpha)$ via the dual integral equation approach. Having found $P_e(\cos \alpha)$, we can then evaluate (3.15) using the steepest descent method.

The first step is to construct integral equations on the basis of the boundary conditions and the representations (3.15)–(3.16). To this end, we introduce the transformation (3.8), allowing us to rewrite the fields on the plane $y = 0$ as

$$E_z^s(x, 0) = \int_{-\infty}^{\infty} \frac{P_e(\lambda)}{\sqrt{1 - \lambda^2}} e^{-jk_0 x \lambda} \, d\lambda \qquad (3.19)$$

$$H_x^s(x, 0^\pm) = \pm Y_0 \int_{-\infty}^{\infty} P_e(\lambda) \, e^{-jk_0 x \lambda} \, d\lambda \qquad (3.20)$$

and we note that the limit $\lambda \to -\infty$ corresponds to $\alpha \to \pi + j\infty$. Since H_x is continuous across the half-plane $x < 0$, and given that $H_x^s(x, 0^+) = -H_x^s(x, 0^-)$, it is then necessary that $H_x^s(x, 0^\pm) = 0$ for $x < 0$, implying the integral equation

$$\int_{-\infty}^{\infty} P_e(\lambda) \, e^{-jk_0 x \lambda} \, d\lambda = 0, \qquad x < 0 \qquad (3.21)$$

Also, from (3.14), (3.17) and (3.18), it follows that

$$E_z^s(x, 0) + E_z^i(x, 0) = -R_e \left[H_x^s(x, 0^+) - H_x^i(x, 0-) \right], \qquad x > 0 \qquad (3.22)$$

and by using (3.19), (3.20) and (3.13), we obtain

$$\int_{-\infty}^{\infty} \left(\frac{1}{\sqrt{1 - \lambda^2}} + \frac{2R_e}{Z_0} \right) P_e(\lambda) \, e^{-jk_0 x \lambda} \, d\lambda = -e^{jk_0 x \cos \phi_0}, \qquad x > 0 \qquad (3.23)$$

The pair (3.21) and (3.23) form a set of (dual) integral equations that can be solved for $P_e(\lambda)$ using Cauchy's residue theorem. Concentrating first on (3.21) we observe that since $x < 0$ the path of integration can be closed by an infinite semicircular contour in the upper half λ-plane without altering the value of the integral. From Cauchy's residue theorem, it then follows that the right hand side of (3.21) is zero only if $P_e(\lambda)$ is analytic in the upper half λ-plane. We can therefore set it equal to an "upper" half-plane function, namely:

$$P_e(\lambda) = U(\lambda) \qquad (3.24)$$

and when this is substituted into (3.23) we have

$$\int_{-\infty}^{\infty} \left(\frac{1}{\sqrt{1-\lambda^2}} + \frac{2R_e}{Z_0} \right) U(\lambda) e^{-jk_0 x \lambda} \, d\lambda = -e^{jk_0 x \cos\phi_0}, \qquad x > 0 \qquad (3.25)$$

To solve (3.25) we again close the path of integration by an infinite semicircular contour, but this time in the lower half-plane (since $x > 0$). To recover the right hand side of (3.25) the integrand must have the form

$$\left(\frac{1}{\sqrt{1-\lambda^2}} + \frac{2R_e}{Z_0} \right) U(\lambda) = \frac{1}{2\pi j} \frac{L_1(\lambda)}{L_1(-\lambda_0)} \frac{1}{\lambda + \lambda_0} + L_2(\lambda) \qquad (3.26)$$

as required by Cauchy's residue theorem, where $L_{1,2}(\lambda)$ are unknown "lower" functions (i.e. functions analytic in the lower half λ-plane) and $\lambda_0 = \cos\phi_0$. Equation (3.26) is sufficient to determine $U(\lambda)$, and the standard procedure is to decompose all functions appearing in that equation into functions that are analytic as well as free of zeros in either the upper or lower half-plane. This procedure is commonly referred to as the factorisation or splitting of a complex function as

$$F(\lambda) = F_+(\lambda) \, F_-(\lambda)$$

where $F_+(\lambda)$ is the "upper split" function and $F_-(\lambda)$ is the "lower split" function. They must be distinguished from the aforementioned "upper" and "lower" functions in that the split functions are also free of zeros in their respective regions of analyticity. For example, the function $\sqrt{1-\lambda^2}$ can be decomposed as

$$\sqrt{1-\lambda^2} = \sqrt{1+\lambda}\sqrt{1-\lambda}$$

where $\sqrt{1+\lambda}$ is recognised to be free of branch points in the lower half λ-plane (see the indentation of the integration path in the λ-plane, shown in Fig. 3–5), and correspondingly the function $\sqrt{1-\lambda}$ is analytic in the upper half-plane. The factorisation of a function into a product (or sum) of upper (+) and lower (−) split functions is possible only if the function is itself free of singularities and zeros on the path of integration. This is an important step in the Wiener-Hopf or dual integral equation method, and can be a difficult task if analytical results are desired. However, direct integral expressions can be employed (NOBLE, 1958; MITTRA AND LEE, 1971; KOBAYASHI, 1990) that can be evaluated numerically. This is referred to as numerical splitting. Appendix B.1 presents a rather efficient numerical procedure (RICOY AND VOLAKIS, 1989) that has proven very successful in evaluating the pertinent integral expressions.

Returning to (3.26), we observe that $[(1/\sqrt{1-\lambda^2})+2R_e]$ is the only function that has yet to be factorised as a product of upper and lower split functions. It can be written as

$$\left(\frac{1}{\sqrt{1-\lambda^2}} + \frac{2R_e}{Z_0}\right)^{-1} = K_+(\bar{\eta},\lambda)\,K_-(\bar{\eta},\lambda) \tag{3.27}$$

where $\bar{\eta}$ is equal to twice the normalised resistivity, namely:

$$\bar{\eta} = \frac{2R_e}{Z_0} \tag{3.28}$$

$K_+(\bar{\eta},\lambda)$ is the upper split function and $K_-(\bar{\eta},\lambda)$ is the lower split function. Also, on the basis of the factorisation process (see Appendix B.1),

$$K_+(\bar{\eta},-\lambda) = K_-(\bar{\eta},\lambda) \tag{3.29}$$

Explicit expressions for the split functions $K_\pm(\bar{\eta},\lambda)$ were first given by SENIOR (1952) who later rewrote them in terms of the more convenient MALIUZHINETS (1958) half-plane function. From SENIOR (1975a) we have

$$K_+(\bar{\eta},\cos\alpha) = \frac{4}{\sqrt{\bar{\eta}}}\sin\frac{\alpha}{2}\left\{\frac{\psi_\pi\left(\frac{3\pi}{2}-\alpha-\theta\right)\psi_\pi\left(\frac{\pi}{2}-\alpha+\theta\right)}{\left(\psi_\pi\left(\frac{\pi}{2}\right)\right)^2}\right\}^2$$
$$\cdot\left\{\left[1+\sqrt{2}\cos\left(\frac{\frac{\pi}{2}-\alpha+\theta}{2}\right)\right]\left[1+\sqrt{2}\cos\left(\frac{\frac{3\pi}{2}-\alpha-\theta}{2}\right)\right]\right\}^{-1} \tag{3.30}$$

where

$$\sin\theta = \frac{1}{\bar{\eta}} \tag{3.31}$$

and $\psi_\pi(\alpha)$ is the Maliuzhinets half-plane function given by

$$\psi_\pi(\alpha) = \exp\left\{-\frac{1}{8\pi}\int_0^\alpha\left(\frac{\pi\sin u - 2\sqrt{2}\pi\sin\frac{u}{2}+2u}{\cos u}\right)du\right\} \tag{3.32}$$

in which α may be complex. From (3.27) it is seen that

$$K_+(\bar{\eta}\to\infty,\lambda) = \frac{1}{\sqrt{\bar{\eta}}} \tag{3.33}$$

and

$$K_+(\bar{\eta}=0,\lambda) = \sqrt{1-\lambda} = \sqrt{2}\sin\frac{\alpha}{2} \tag{3.34}$$

A very accurate non-integral approximation of $\psi_\pi(\alpha)$ is (VOLAKIS AND SENIOR, 1985; see also OSIPOV, 1990)

$$\psi_\pi(\alpha) \approx \begin{cases} 1 - 0.0139\alpha^2 & \text{Im. } \alpha < 4.2 \\ 1.05302\sqrt{\cos\frac{1}{4}(\alpha-j\ln 2)}\exp\left\{\frac{j\alpha}{2\pi}e^{j\alpha}\right\} & \text{Im. } \alpha > 4.2 \end{cases} \tag{3.35}$$

provided Re. $\alpha < \pi/2$. Otherwise the identities

$$\psi_\pi(\alpha) = \left\{\psi_\pi\left(\frac{\pi}{2}\right)\right\}^{2} \frac{\cos\left(\frac{\alpha}{4} - \frac{\pi}{8}\right)}{\psi_\pi(\alpha - \pi)}$$

$$\psi_\pi(-\alpha) = \psi_\pi(\alpha) \tag{3.36}$$

$$\psi_\pi(\alpha^*) = \psi_\pi^*(z)$$

(asterisk denotes the complex conjugate) must be employed as many times as necessary until the argument is suitable for use in (3.35). We remark that the approximation (3.35) has a maximum amplitude error of less than one percent. Typical plots of $K_+(\bar{\eta}, \cos\alpha)$ as a function of α (real values only) for different values of $\bar{\eta}$ are shown in Fig. 3–7. Surprisingly, in spite of its complicated functional form, the split function $K_+(\bar{\eta}, \cos\alpha)$ has a very simple graphical representation. FORTRAN subroutines for the evaluation of the $K_+(\bar{\eta}, \cos\phi)$ split function are given in Appendix B.2.

Having achieved the factorisation, we now proceed with the determination of $U(\lambda)$. On substituting (3.27) into (3.26), we find

$$\frac{U(\lambda)}{K_+(\bar{\eta}, \lambda)} = \frac{1}{2\pi j} \frac{L_1(\lambda)}{L_1(-\lambda_0)} \frac{K_-(\bar{\eta}, \lambda)}{\lambda + \lambda_0} + K_-(\bar{\eta}, \lambda) L_2(\lambda)$$

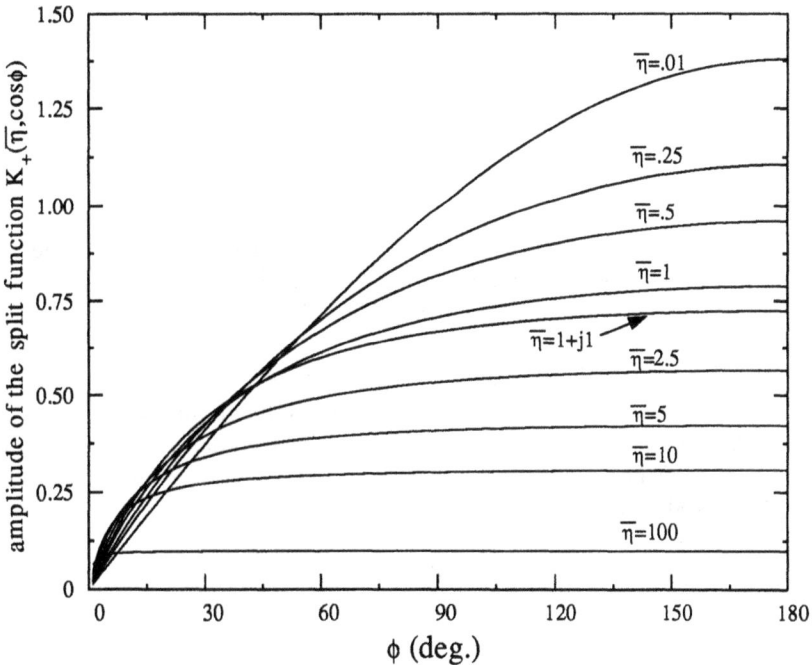

Figure 3–7: *The upper half-plane split function $K_+(\bar{\eta}, \cos\phi)$ as a function of ϕ for different $\bar{\eta}$*

which can be written alternatively as

$$\frac{U(\lambda)}{K_+(\bar{\eta}, \lambda)} - \frac{1}{2\pi j} \frac{K_+(\bar{\eta}, \lambda_0)}{\lambda + \lambda_0} = \frac{1}{2\pi j} \left[\frac{L_1(\lambda)}{L_1(-\lambda_0)} K_-(\bar{\eta}, \lambda) - K_-(\bar{\eta}, -\lambda_0) \right]$$
$$\cdot \frac{1}{\lambda + \lambda_0} + K_-(\bar{\eta}, \lambda) L_2(\lambda) \qquad (3.37)$$

We observe that the left hand side of (3.37) contains only upper functions, whereas the right hand side consists of lower functions only. From Liouville's theorem, both sides must then be equal to a polynomial $B(\lambda)$, whose order matches the asymptotic behaviour of the right and left hand sides. That is,

$$U(\lambda) = \frac{1}{2\pi j} \frac{K_+(\bar{\eta}, \lambda) K_+(\bar{\eta}, \lambda_0)}{\lambda + \lambda_0} A(\lambda) \qquad (3.38)$$

where $A(\lambda)$ is a polynomial of order one higher than $B(\lambda)$. To examine the asymptotic behaviour of the left hand side of (3.37) we note that (MALI-UZHINETS, 1958)

$$\lim_{|\text{Im. } (\alpha)| \to \infty} \psi_\pi(\alpha) = O\left(\exp\left\{ \frac{|\text{Im. } (\alpha)|}{8\pi} \right\} \right) \qquad (3.39)$$

and consequently

$$K_+(\bar{\eta}, \lambda \to \infty) = O(1) \qquad (3.40)$$

where $O(\cdot)$ is the Landau symbol. Also,

$$U(\lambda \to \infty) = O(\lambda^{-1})$$

since the edge condition dictates that the electric current be finite at the edge ($x \to 0$). This is in accordance with the Abelian theorem, which states that if $P_e(\lambda) = O(\lambda^\nu)$ with E_z as given in (3.19), then E_z grows as $O(x^{-\nu-1})$ as $x \to 0$.

From the above we deduce that the left hand side of (3.37) grows as $O(|\lambda|^{-1})$ as $|\lambda| \to \infty$, and the same is true for the right hand side, provided $L_1(\lambda)$ behaves no worse than $O(1)$. Consequently, in accordance with Liouville's theorem, the left hand side of (3.37) must be zero, implying that $A(\lambda) = 1$ so that (3.25) is also satisfied. Thus,

$$U(\lambda) = \frac{1}{2\pi j} \frac{K_+(\bar{\eta}, \lambda) K_+(\bar{\eta}, \lambda_0)}{\lambda + \lambda_0} \qquad (3.41)$$

and from (3.24) we obtain

$$P_e(\cos \alpha) = \frac{1}{2\pi j} \frac{K_+(\bar{\eta}, \cos \alpha) K_+(\bar{\eta}, \cos \phi_0)}{\cos \alpha + \cos \phi_0} \qquad (3.42)$$

The determination of E_z^s from its spectral representation can be accomplished using the steepest descent method. As part of this process, the integration path C is deformed into the steepest descent path $S(\phi)$ illustrated in Fig. 3–8. Obviously, one must be careful with the poles at

$$
\begin{aligned}
\alpha_{p1} &= \pi - \phi_0 \\
\alpha_{p2} &= \pi + \phi_0 \\
\alpha_{p3} &= -\theta
\end{aligned}
\tag{3.43}
$$

where α_{p1} is associated with the reflected field, α_{p2} is associated with the transmitted (and incident) field and α_{p3} is referred to as the surface wave pole. For capacitive resistive sheets, α_{p3} gives rise to propagating waves bound to the surface of the sheet, and corresponds to the zero of the term $[1 + \sqrt{2}\, \cos \frac{1}{2}\,(3\pi/2 - \alpha - \theta)]$ appearing in the denominator of the expression (3.30) for $K_+(\bar{\eta}, \cos \alpha)$. If, for the moment, we ignore the presence of these poles, then a simple application of the steepest descent method (see Appendix C) yields

$$
E_z^s \sim E_z^d = \sqrt{\frac{2\pi}{k_0}}\, e^{j\pi/4}\, \frac{e^{-jk_0\rho}}{\sqrt{\rho}}\, P_e(\cos \phi)
\tag{3.44}
$$

or

$$
E_z^d = \frac{e^{-jk\rho}}{\sqrt{\rho}}\, D_{Ee}^{nu}(\phi, \phi_0, \bar{\eta})
\tag{3.45}
$$

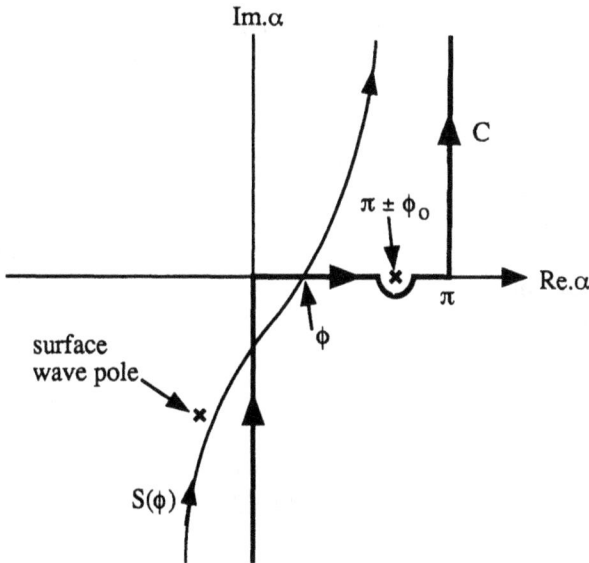

Figure 3–8: *Deformation of the contour into the steepest descent path in the α-plane, with the geometrical optics and surface wave poles shown*

for large $k_0\rho$, where

$$D_{\text{Ee}}^{\text{nu}}(\phi, \phi_0, \bar{\eta}) = \frac{e^{-j\pi/4}}{\sqrt{2\pi k_0}} \frac{K_+(\bar{\eta}, \cos \phi) \, K_+(\bar{\eta}, \cos \phi_0)}{\cos \phi + \cos \phi_0} \qquad (3.46)$$

is the (non-uniform) edge diffraction coefficient of the resistive half-plane. Correspondingly, E_z^{d} is referred to as the diffracted field, and is the scattered field without any specular contributions. As required by reciprocity, $D_{\text{E}}(\bar{\eta}, \phi, \phi_0)$ is symmetric in ϕ and ϕ_0.

Obviously, (3.44)–(3.46) are valid in the far zone only and in those regions where the reflected, transmitted and surface wave fields are not present. This is indeed true for backscatter calculations ($\phi = \phi_0$) and typical plots of the diffracted field for those situations are given in Fig. 3–9. Specifically, Fig. 3–9 shows plots of the E-polarisation echowidth given by

$$\sigma_{\text{E}} = \underset{\rho \to \infty}{2\pi\rho \lim} \frac{|E_z^{\text{s}}|^2}{|E_z^{\text{i}}|^2} = 2\pi \, |D_{\text{Ee}}^{\text{nu}}(\phi, \phi_0, \bar{\eta})|^2 \qquad (3.47)$$

or

$$\frac{\sigma_{\text{E}}}{\lambda_0} = 2\pi \, |P_{\text{e}}(\cos \phi)|^2 \qquad (3.48)$$

The plots of $\sigma_{\text{E}}/\lambda_0$ in Fig. 3–9(a) are for resistive half-planes of different resistivities. As illustrated, the main feature of these patterns (when compared with the metallic half-plane) is the lower echowidth near edge-on ($\phi \approx 180°$). For all values of $\bar{\eta} = 2R_{\text{e}}/Z_0$ (complex or real), the edge-on echowidth is lower than -8 dB, i.e. the edge-on echowidth of the metallic half-plane. Moreover, the echowidth decreases with increasing $\bar{\eta}$ and this is the reason for using resistive loading to reduce the radar cross-section of airborne vehicles. The decrease in echowidth is attributed to a reduction of the half-plane current density near the edge as illustrated in Fig. 3–10 (SENIOR, 1979). Corresponding echowidth curves for thin dielectric layers are depicted in Fig. 3–9(b). These were also generated from (3.47) with $\bar{\eta}$ defined in (2.51). It should be noted that the thickness of the simulated dielectric layer was kept less than $\lambda/10$, where λ denotes the wavelength in the material.

3.3.2   Uniform diffracted field

To evaluate the half-plane diffracted field at an arbitrary point (ρ, ϕ) with ρ finite, it is necessary to perform a uniform asymptotic evaluation of the integral (3.15). On using the identity

$$\frac{4}{\cos \alpha + \cos \phi_0} \left\{ \begin{array}{l} \sin(\alpha/2) \sin(\phi_0/2) \\ \cos(\alpha/2) \cos(\phi_0/2) \end{array} \right\} = \sec \frac{\alpha + \phi_0}{2} \mp \sec \frac{\alpha - \phi_0}{2} \qquad (3.49)$$

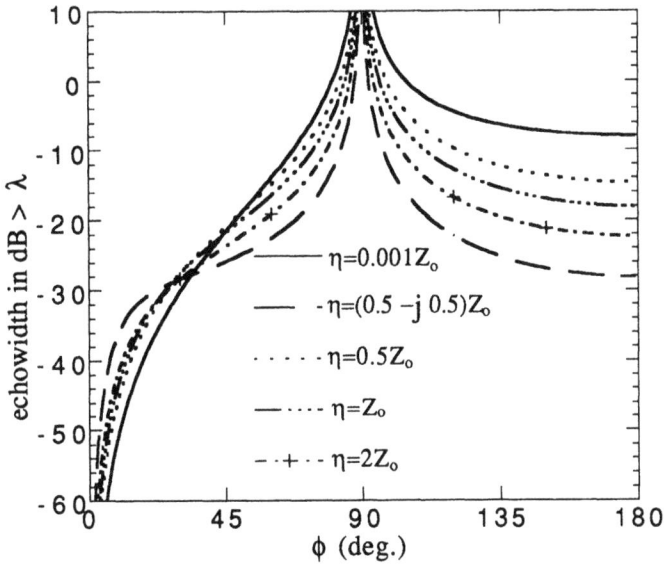

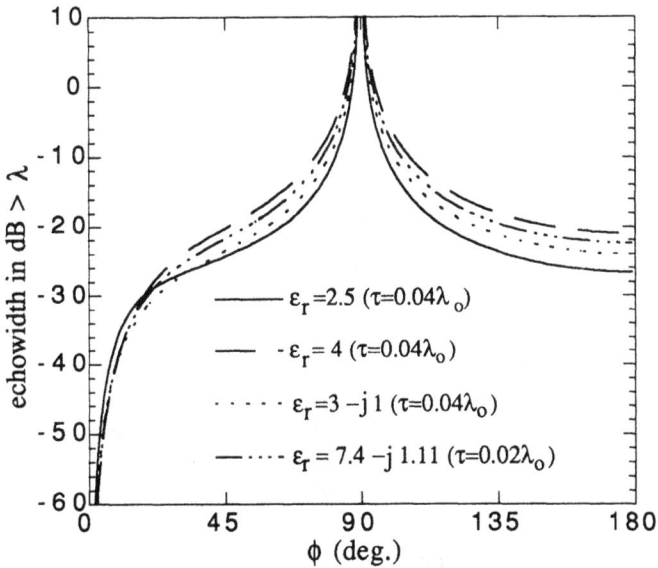

Figure 3-9: *E-polarisation backscatter echowidth plots for resistive half-planes with (a) different resistivities $R_e = \eta/2$, (b) different dielectric constants (see (2.51))*

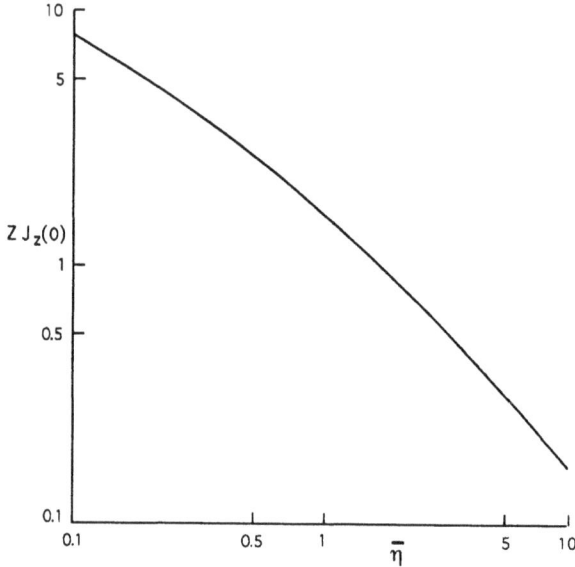

Figure 3–10: *Current at the edge of a resistive half-plane as a function of $\bar{\eta}$ for edge-on illumination*

in (3.42), (3.15) takes the form

$$
\begin{aligned}
E_z^s &= \frac{1}{2\pi j} \int_C \left[\sec \frac{\alpha + \phi_0}{2} - \sec \frac{\alpha - \phi_0}{2} \right] \\
&\quad \cdot \frac{K_+(\bar{\eta}, \cos \alpha)\, K_+(\bar{\eta}, \cos \phi_0)}{4 \sin \dfrac{\alpha}{2} \sin \dfrac{\phi_0}{2}} e^{-jk_0 \rho \cos(\phi \mp \alpha)}\, d\alpha \qquad (3.50)
\end{aligned}
$$

and several approaches exist for the uniform evaluation of this integral. One of the two most popular methods involves the mapping of the integration path onto the real axis, regularising the integrand by multiplying and dividing by a polynomial having zeros at the poles, and subsequently expanding the non-singular part of the integrand in a Maclaurin series, before performing a closed form evaluation of the first non-vanishing term of the series. This procedure (to be referred to as the multiplicative method) was described by KOUYOUMJIAN AND PATHAK (1974) as the modified Pauli-Clemmow method (PATHAK AND KOUYOUMJIAN, 1970; CLEMMOW, 1950) who used it in their development of the uniform geometrical theory of diffraction (UTD). However, the method is most suited for integrands with poles on the real axis of the α-plane unless additional terms of the Maclaurin series expansion are kept (GENNARELLI AND PALUMBO, 1984). In our case, the surface wave pole α_{p3}, defined in (3.43), is not real unless $\bar{\eta}$ is real and greater than unity (in which case α_{p3} is not captured). To perform a uniform evaluation of (3.50) valid for complex and real poles, we will instead employ a method originally used for diffraction problems by CLEMMOW (1966) and later expounded by VOLAKIS AND HERMAN (1986) (see also VAN DER WAERDEN (1951), FELSEN AND MARCUVITZ (1973), and YIP AND CHIAVETTA (1987)). In accordance with this method, the integrand is first regularised by adding and subtracting a function that has the same singularity(ies) as the original integrand, but is cast in a form that enables us to carry out the integration exactly. The details of this procedure (to be referred to as the additive method) are given later in this section, and it has been shown that the additive and multiplicative methods lead to identical expansions when all terms are considered (ROJAS, 1987).

Regardless of the method used, the goal of a uniform evaluation is to obtain an expression in which the total field is continuous across the reflection and shadow boundaries illustrated in Fig. 3–11. Before proceeding with the evaluation of (3.50), it is instructive to consider the transition functions that are required for this solution. With this in mind, we begin with the special case of (3.50) for a perfectly conducting half-plane. Only the first two poles in (3.43) are then pertinent, and in addition the integration can be carried out exactly.

Metallic half-plane

For the perfectly conducting half-plane, $\bar{\eta} = 0$, and by using (3.34), we obtain

$$E^{s}_{zp.c.} = \frac{1}{4\pi j} \int_{C} \left[\sec \frac{\alpha + \phi_0}{2} - \sec \frac{\alpha - \phi_0}{2} \right] e^{-jk_0\rho\cos(\phi-\alpha)} \, d\alpha \qquad (3.51)$$

for $\phi < \pi$, where the subscript "p.c." indicates that this expression is valid only for the perfectly conducting half-plane. Deforming the path C to the steepest descent path illustrated in Fig. 3–8 gives

$$E^{s}_{zp.c.} = \frac{1}{4\pi j} \int_{S(\phi)} \left[\sec \frac{\alpha + \phi_0}{2} - \sec \frac{\alpha - \phi_0}{2} \right] e^{-jk_0\rho\cos(\phi-\alpha)} \, d\alpha + E^{r}_{z} \qquad (3.52)$$

where

$$E^{r}_{z} = \begin{cases} -e^{jk\rho\cos(\phi+\phi_0)} & \phi_0 + \phi < \pi \\ 0 & \text{elsewhere} \end{cases} \qquad (3.53)$$

is the reflected field. E^{r} was recovered from the residue of the pole $\alpha_1 = \pi - \phi_0$ that is captured during the deformation of C to $S(\phi)$. To evaluate the integral in (3.52), we note that (CLEMMOW, 1966)

$$\int_{S(\phi)} \sec\left(\frac{\alpha - \alpha_0}{2}\right) e^{-jk_0\rho\cos(\alpha-\phi)} \, d\alpha$$

$$= \int_{S(0)} \sec\left(\frac{\alpha \pm \alpha_0 \mp \phi}{2}\right) e^{-jk_0\rho\cos\beta} \, d\beta$$

(a)

(b)

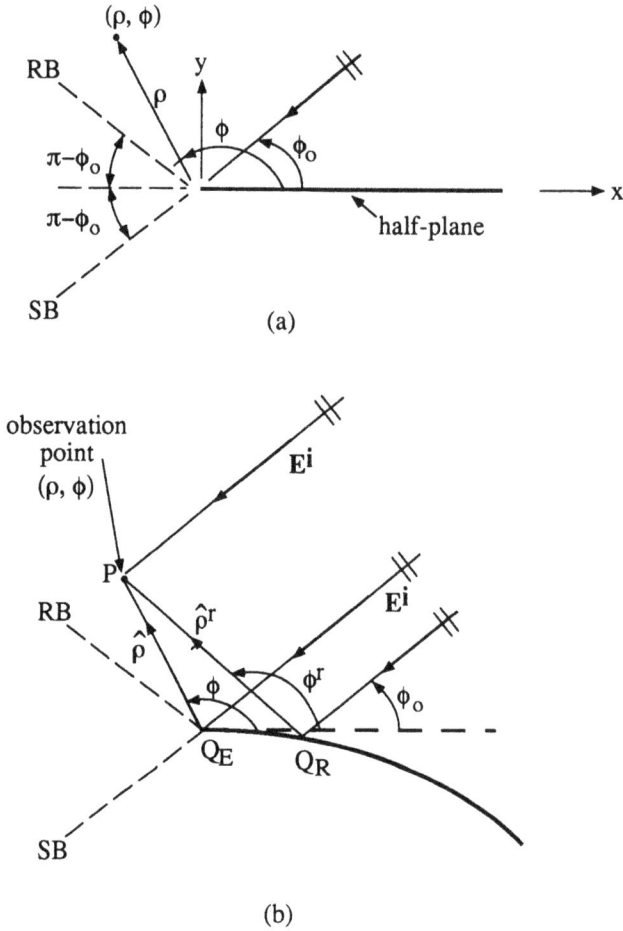

Figure 3-11: *Illustration of the reflection (RB) and shadow (SB) boundaries for plane wave diffraction by the edge of a (a) half-plane, (b) curved surface*

$$= 2\cos\left(\frac{\alpha_0 - \phi}{2}\right) \int_{S(0)} \frac{\cos\frac{\beta}{2}}{\cos\beta + \cos(\alpha_0 - \phi)} e^{-jk_0\rho\cos\beta}\, d\beta$$

$$= -2\sqrt{2}\, e^{-j\pi/4} e^{-jk_0\rho} \cos\frac{\phi - \alpha_0}{2} \int_{-\infty}^{\infty} \frac{e^{-k_0\rho\tau^2}}{\tau^2 + 2j\cos^2\left[(\phi - \alpha_0)/2\right]}\, d\tau$$

$$= \mp 4\sqrt{\pi}\, e^{-j\pi/4} e^{-jk_0\rho} F_C\left[\pm\sqrt{2k_0\rho}\cos\left(\frac{\phi - \alpha_0}{2}\right)\right] \qquad (3.54)$$

where

$$F_C(z) = e^{jz^2} \int_z^{\infty} e^{-j\tau^2}\, d\tau \qquad (3.55)$$

with the lower sign in (3.54) when $\frac{\pi}{4} < \arg z < \frac{5\pi}{4}$ (i.e. for $\phi - \alpha_0 > \pi$). This modified Fresnel integral, which will be referred to as the Clemmow transition function, satisfies the identity

$$F_C(-z) + F_C(z) = \sqrt{\pi}\, e^{-j\pi/4} e^{jz^2} \qquad (3.56)$$

which is crucial in maintaining continuity of the total field across the reflection and shadow boundaries. That is, the discontinuity of the integral (3.54) simply balances the abrupt "shadowing" of the incident or reflected fields so as to maintain total field continuity. To observe this let us consider the evaluation of (3.52) via (3.54). We find

$$E^s_{zp.c.} + E^i_z = -\frac{e^{j\pi/4}}{\sqrt{\pi}} e^{-jk_0\rho}$$

$$\cdot \left[\mp F_C \left(\pm\sqrt{2k_0\rho}\, \cos\frac{\phi+\phi_0}{2} \right) \pm F_C \left(\pm\sqrt{2k_0\rho}\, \cos\frac{\phi-\phi_0}{2} \right) \right]$$

$$+ e^{jk_0\rho\cos(\phi+\phi_0)} u(\pi - \phi - \phi_0) + e^{jk_0\rho\cos(\phi-\phi_0)} u(\pi - \phi + \phi_0) \qquad (3.57)$$

where $u(x)$ is the unit step function, and in the last term it accounts for the shadowing of the incident field for $\phi > \pi + \phi_0$. To the right of the reflection boundary, $\cos[(\phi \pm \phi_0)/2] > 0$, and it then follows that the total field can be written as

$$E_{zp.c.} = E^s_{zp.c.} + E^i_{zp.c.}$$

$$= \left[-\frac{e^{j\pi/4}}{\sqrt{\pi}} e^{-jk_0\rho} F_C \left(\sqrt{2k_0\rho}\, \cos\frac{\phi-\phi_0}{2} \right) + e^{jk_0\rho\cos(\phi-\phi_0)} \right]$$

$$+ \left[\frac{e^{j\pi/4}}{\sqrt{\pi}} e^{-jk_0\rho} F_C \left(\sqrt{2k_0\rho}\, \cos\frac{\phi+\phi_0}{2} \right) - e^{jk_0\rho\cos(\phi+\phi_0)} \right] \qquad (3.58)$$

Also, on using the identity (3.56) and (3.53), we find that (3.58) is valid even when $\phi + \phi_0 > \pi$, implying that (3.57) is continuous across the reflection and shadow boundaries. Interestingly, by using the identity (3.56), the total field can be written in a more compact form as

$$E_{zp.c.} = \frac{e^{j\pi/4}}{\sqrt{\pi}} e^{-jk_0\rho} \left[F_C \left(-\sqrt{2k_0\rho}\, \cos\frac{\phi-\phi_0}{2} \right) \right.$$

$$\left. - F_C \left(-\sqrt{2k_0\rho}\, \cos\frac{\phi+\phi_0}{2} \right) \right] \qquad (3.59)$$

and this holds for all ϕ and ϕ_0. We remark that (3.59) is an exact expression of the total field in the presence of a perfectly conducting half-plane. It is often written in terms of the UTD transition function defined as (KOUYOUMJIAN AND PATHAK, 1974)

$$F_{KP}(z^2) = \pm 2jz\, F_C(\pm z) \qquad (3.60)$$

where again the minus sign is chosen when $\frac{\pi}{4} < \arg z < \frac{5\pi}{4}$. Using (3.60), the total field for the perfectly conducting half-plane can be rewritten as

$$E_{z\text{p.c.}} = \frac{e^{-jk_0\rho}}{\sqrt{\rho}} \, D_{\text{E}}^{\text{u}}(\phi, \phi_0) \tag{3.61}$$

where

$$D_{\text{E}}^{\text{u}}(\phi, \phi_0) = -\frac{e^{-j\pi/4}}{2\sqrt{2\pi k_0}} \left[\sec\left(\frac{\phi - \phi_0}{2}\right) F_{\text{KP}}\left(2k_0\rho \cos^2 \frac{\phi - \phi_0}{2}\right) \right.$$
$$\left. - \sec\left(\frac{\phi + \phi_0}{2}\right) F_{\text{KP}}\left(2k_0\rho \cos^2 \frac{\phi + \phi_0}{2}\right) \right] \tag{3.62}$$

is the (soft) uniform diffraction coefficient and F_{KP} must be evaluated using the lower sign in (3.60). FORTRAN subroutines for the evaluation of $F_{\text{KP}}(z^2)$ and $F_C(z)$ are given in Appendix B.3. Magnitude and phase plots of $F_{\text{KP}}(z^2)$ for real $z > 0$ are shown in Fig. 3–12.

It is apparent from the above that for an edged structure the total field can be expressed in several different ways, but the one most commonly used is

$$E_z = E_z^{\text{i}} + E_z^{\text{r}} + E_z^{\text{sw}} + E_z^{\text{d}} = E_z^{\text{go}} + E_z^{\text{sw}} + E_z^{\text{d}} \tag{3.63}$$

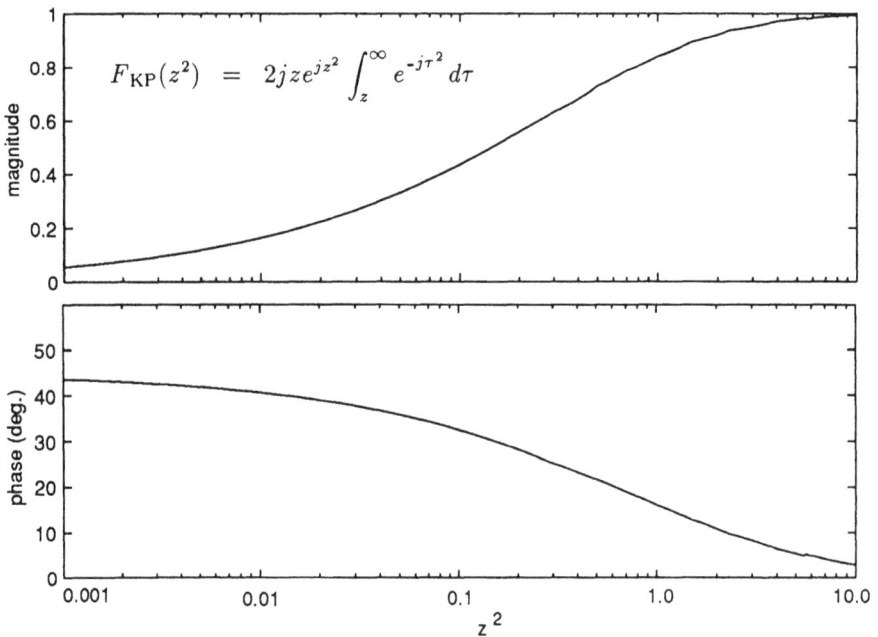

Figure 3–12: *The transition function $F_{\text{KP}}(z^2)$ for z real and > 0*

where E_z^{d} denotes the diffracted field and E_z^{go} is the geometrical optics field. E_z^{sw} includes the residue contribution of any surface wave poles captured during the deformation of C to $S(\phi)$ and is always zero only for metallic structures. This decomposition of the total field is in accordance with the UTD ansatz and is particularly convenient for generalising (3.62) to the edge of the curved surface illustrated in Fig. 3–11(b). In the case of the metallic half-plane shown in Fig. 3–11(a), E^{d} is given by (3.62) provided $F_{\mathrm{KP}}(z)$ is evaluated using the upper sign, namely:

$$E_{z\,\mathrm{p.c.}}^{\mathrm{d}} = \frac{e^{-jk_0\rho}}{\sqrt{\rho}}\, D_{\mathrm{E}}^{\mathrm{u}}(\phi, \phi_0)$$

with

$$
D_{\mathrm{E}}^{\mathrm{u}}(\phi, \phi_0) = -\frac{e^{-j\pi/4}}{2\sqrt{2\pi k_0}}\left[\sec\left(\frac{\phi - \phi_0}{2}\right) F_{\mathrm{KP}}\left(\left|\sqrt{2k_0\rho}\,\cos\left(\frac{\phi - \phi_0}{2}\right)\right|^2\right)\right.
$$
$$
\left. - \sec\left(\frac{\phi + \phi_0}{2}\right) F_{\mathrm{KP}}\left(\left|\sqrt{2k_0\rho}\,\cos\left(\frac{\phi + \phi_0}{2}\right)\right|^2\right)\right] \quad (3.64)
$$

Also, as before,

$$E_z^{\mathrm{go}} = e^{jk_0\rho\cos(\phi - \phi_0)}\,u(\pi - \phi + \phi_0) - e^{jk_0\rho\cos(\phi + \phi_0)}\,u(\pi - \phi - \phi_0)$$

which is discontinuous. Thus, to obtain a continuous total field, it is necessary that the diffracted field counteracts the discontinuities of the geometrical optics terms at the reflection and shadow boundaries.

To see how total field continuity is maintained in the context of UTD, we consider the evaluation of the diffracted field to the left and right of the reflection boundary, i.e. at $\phi = \pi - \phi_0 \mp \epsilon$, as $\epsilon \to 0$. At this angle $\cos\left[(\phi + \phi_0)/2\right] = \cos\left[(\pi \mp \epsilon)/2\right] = \pm\sin(\epsilon/2)$, and on using the approximation

$$F_{\mathrm{KP}}(z)|_{z\to 0} \approx \sqrt{\pi z}\, e^{j(\pi/4 + z)} \quad (3.65)$$

it follows that

$$E_{z\,\mathrm{p.c.}}^{\mathrm{d}}(\phi = \pi - \phi_0 \mp \epsilon) =$$
$$\frac{e^{-jk_0\rho}}{\sqrt{\rho}}\left\{-\frac{e^{-j\pi/4}}{2\sqrt{2\pi k_0}}\sec\frac{\phi - \phi_0}{2} F_{\mathrm{KP}}\left(\left|\sqrt{2k_0\rho}\,\cos\left(\frac{\phi - \phi_0}{2}\right)\right|^2\right)\right\} \mp \frac{1}{2}E_z^{\mathrm{r}}$$

where E_z^{r} is the reflected field given above. Thus, E^{d} is discontinuous at the reflection boundary by an amount equal to the reflected field and consequently $E_{z\,\mathrm{p.c.}}^{\mathrm{d}} + E_z^{\mathrm{r}}$ is continuous at $\phi = \pi - \phi_0$. The same holds for the sum $E_{z\,\mathrm{p.c.}}^{\mathrm{d}} + E_z^{\mathrm{i}}$ at $\phi = \pi + \phi_0$.

Were we to consider the problem of diffraction by the curved edge in Fig. 3–11(b), the edge diffraction coefficient would have to be modified to read

$$D_{\mathrm{E}}^{\mathrm{u}}(\phi, \phi_0) = -\frac{e^{-j\pi/4}}{2\sqrt{2\pi k_0}} \left[\sec\frac{\phi - \phi_0}{2} F_{\mathrm{KP}} \left(\left| \sqrt{2k_0 L^{\mathrm{i}}} \cos\left(\frac{\phi - \phi_0}{2}\right) \right|^2 \right) \right.$$

$$\left. - \sec\frac{\phi + \phi_0}{2} F_{\mathrm{KP}} \left(\left| \sqrt{2k_0 L^{\mathrm{r}}} \cos\left(\frac{\phi + \phi_0}{2}\right) \right|^2 \right) \right] \tag{3.66}$$

where $L^{\mathrm{i,r}}$ are heuristic distance parameters introduced by KOUYOUMJIAN AND PATHAK (1974) to ensure the continuity of the total field at the reflection and shadow boundaries. For a curved edge with plane wave illumination the parameters are

$$L^{\mathrm{i}} = \rho, \qquad L^{\mathrm{r}} = \frac{\rho R_1^{\mathrm{r}}}{R_1^{\mathrm{r}} + \rho} \tag{3.67}$$

in which $1/R_1^{\mathrm{r}} = 2/(R_1 \cos\theta_i)$ is the curvature of the reflected wave at the reflection point Q_{E}, R_1 is the surface radius at Q_{E} and θ_i is the angle between the normal at Q_{E} and the incident/reflected ray. Since the reflected field at Q_{E} has the form

$$E_z^{\mathrm{r}} = -E_z^{\mathrm{i}}(Q_{\mathrm{E}}) \sqrt{\frac{R_1^{\mathrm{r}}}{R_1^{\mathrm{r}} + \rho^{\mathrm{r}}}} \, e^{-jk_0\rho^{\mathrm{r}}} \, u(\pi - \phi - \phi_0)$$

(3.65) shows that the uniform diffraction coefficient (3.66) renders the total field continuous at the reflection boundary. We note that KOUYOUMJIAN AND PATHAK (1974) have developed more general expressions for the distance parameters $L^{\mathrm{i,r}}$ that account for cylindrical wave or other illumination, and for a non-zero curvature of the edge in the direction normal to the plane of incidence. Alternative distance parameters in the context of the uniform asymptotic theory (UAT) have been given by LEE AND DESCHAMPS (1976).

Resistive half-plane

Multiplicative approach. The above uniform evaluation of the solution for the metallic half-plane relied on the identity (3.54), which, of course, cannot be used in the case of a resistive half-plane. The scattered field (3.50) for the resistive half-plane is comprised of integrals of the form

$$I = \int_C G(\alpha) \sec\left(\frac{\alpha - \alpha_{\mathrm{pi}} \pm \pi}{2}\right) e^{-jk_0\rho\cos(\alpha - \phi)} \, d\alpha \tag{3.68}$$

where $G(\alpha)$ is slowly varying near the pole α_{pi}, provided the surface wave pole α_{p3} is not near the steepest descent path. In this case the multiplicative approach (FELSEN AND MARCUVITZ, 1973; GENNARELLI AND PALUMBO, 1984) can be used. By retaining the first term of the Maclaurin series expansion for $G(\alpha)$ (see Appendix C) we have

$$
\int_{S(\alpha_s)} g(\alpha)\, e^{k_0\rho\, f(\alpha)}\, d\alpha \sim g(\alpha_s)\sqrt{\frac{-2\pi}{k_0\rho\, f''(\alpha)}}\, e^{k_0\rho\, f(\alpha_s)}\, F_{\mathrm{KP}}(k_0\rho a) \tag{3.69}
$$

where

$$
\begin{aligned}
\alpha_s &= \text{saddle point}\\
\alpha_p &= \text{geometrical optics pole}\\
a &= -j[f(\alpha_p) - f(\alpha_s)],
\end{aligned}
$$

and it is readily seen that

$$
g(\alpha) = G(\alpha)\sec\left(\frac{\alpha - \alpha_{\mathrm{pi}} \pm \pi}{2}\right)
$$

whereas $f(\alpha) = -j\cos(\alpha - \phi)$. Note that if $g(\alpha)$ is free of pole singularities (i.e. $\alpha_{\mathrm{pi}} \to \infty$) then $F_{\mathrm{KP}}(k_0\rho a) = 1$ and the same holds as $k_0\rho \to \infty$.

Making use of (3.69) for the evaluation of (3.50), we find that

$$
E^{\mathrm{d}} \sim \frac{e^{-jk_0\rho}}{\sqrt{\rho}}\, D_{\mathrm{Ee}}^{\mathrm{u}}(\phi, \phi_0, \bar{\eta}) \tag{3.70}
$$

where

$$
\begin{aligned}
D_{\mathrm{Ee}}^{\mathrm{u}}(\phi, \phi_0, \bar{\eta}) &= -\frac{e^{-j\pi/4}}{2\sqrt{2\pi k_0}}\Bigg[\sec\left(\frac{\phi - \phi_0}{2}\right) F_{\mathrm{KP}}\left(\left|\sqrt{2k_0\rho}\cos\left(\frac{\phi - \phi_0}{2}\right)\right|^2\right)\\
&\quad - \sec\left(\frac{\phi + \phi_0}{2}\right) F_{\mathrm{KP}}\left(\left|\sqrt{2k_0\rho}\cos\left(\frac{\phi + \phi_0}{2}\right)\right|^2\right)\Bigg]\\
&\quad\cdot \frac{K_+(\bar{\eta}, \cos\phi)\, K_+(\bar{\eta}, \cos\phi_0)}{2\sin\dfrac{\phi}{2}\sin\dfrac{\phi_0}{2}}
\end{aligned} \tag{3.71}
$$

is the resistive half-plane diffraction coefficient that is uniform at the reflection and shadow boundaries, but neglects the surface wave singularities contained in the split function. To compute the total field, E^{d} in (3.70) must be substituted into (3.63) where E_z^{go} is now associated with the fields incident, reflected and transmitted through the resistive half-plane. To demonstrate the uniformity of (3.70), we make use of (3.27) and (3.65) to obtain

$$
\begin{aligned}
E_z^{\mathrm{d}}(\phi = \pi - \phi_0 \mp \epsilon) &=\\
\frac{1}{2}\frac{e^{-jk_0\rho}}{\sqrt{\rho}}&R_\perp\frac{e^{-j\pi/4}}{\sqrt{2\pi k_0}}\sec\left(\frac{\phi - \phi_0}{2}\right) F_{\mathrm{KP}}\left(\left|\sqrt{2k_0\rho}\cos\frac{\phi - \phi_0}{2}\right|^2\right)\\
&\mp \frac{1}{2}R_\perp e^{jk_0\rho\cos(\phi + \phi_0)}
\end{aligned}
$$

where $R_\perp = -(1 + \bar\eta \sin \phi_0)^{-1}$ is the reflection coefficient of the resistive sheet. Since $E^{\mathrm{r}} = R_\perp e^{jk_0\rho\cos(\phi+\phi_0)}$ it follows that $E^{\mathrm{d}} + E^{\mathrm{r}}$ is continuous across the reflection boundary. At the shadow boundary we find that

$$E_z^{\mathrm{d}}(\phi = \pi + \phi_0 \mp \epsilon) = \mp\frac{1}{2}\frac{1}{1 + \bar\eta \sin \phi_0} e^{jk_0\rho\cos(\phi-\phi_0)} + \text{continuous terms}$$

and thus

$$E_z^{\mathrm{i}} + E_z^{\mathrm{d}}|_{\phi=\pi+\phi_0-\epsilon} = E_z^{\mathrm{trans}} + E_z^{\mathrm{d}}|_{\phi=\pi+\phi_0+\epsilon}$$

where

$$E_z^{\mathrm{trans}} = \frac{\bar\eta \sin \phi_0}{1 + \bar\eta \sin \phi_0} u(-\pi + \phi - \phi_0)$$

is the field transmitted through the resistive sheet. We remark that because of (3.60) and the identity (3.56), the total field in the presence of the resistive half-plane can be obtained from (3.70) once the absolute values are removed from the arguments of $F_{\mathrm{KP}}(z^2)$ and the latter is evaluated using the lower sign of (3.60). This is equivalent to using (3.59) for the perfectly conducting half-plane.

Additive approach. The surface wave pole is captured during the deformation of C to $S(\phi)$ and this occurs when

$$0 < \mathrm{Re}.\ \theta < |\mathrm{gd}(\mathrm{Im}.\ \theta)|, \qquad \mathrm{Im}.\ \theta > 0$$

where $\mathrm{gd}(\alpha) = \cos^{-1}[1/\cosh\alpha]\,\mathrm{sgn}(\alpha)$ is the Gudermann function that defines the path $S(0)$ in the α-plane. For E-polarisation, the pole is captured whenever the resistivity is capacitive (negative reactance), and in that case, or if $-\theta$ lies near $S(\phi)$, the function $G(\alpha)$ in (3.68) can no longer be considered slowly varying.

This presents us with the possibility of an integrand having more than one pole singularity and consequently the asymptotic result (3.69) is not appropriate. Moreover, the result (3.69) is not uniform when the pole crosses the steepest descent path away from the saddle point. The multiplicative method can still be applied to obtain a uniform evaluation of the integrand (see Appendix C), but the additive approach is now simpler to use. In accordance with this, the integral (3.50) is rewritten as

$$E_z^{\mathrm{d}} = \int_C \left[Q(\alpha) - \sum_{i=1}^{3} \tilde{Q}(\alpha_{\mathrm{pi}}) \sec\left(\frac{\alpha - \alpha_{\mathrm{pi}} \pm \pi}{2}\right) \right] e^{-jk_0\rho\cos(\alpha-\phi)}\, d\alpha$$

$$+ \sum_{i=1}^{3} \tilde{Q}(\alpha_{\mathrm{pi}}) \int_C \sec\left(\frac{\alpha - \alpha_{\mathrm{pi}} \pm \pi}{2}\right) e^{-jk_0\rho\cos(\alpha-\phi)}\, d\alpha \qquad (3.72)$$

where

$$\tilde{Q}(\alpha_{\mathrm{pi}}) = \left.\frac{Q(\alpha)}{\sec\left(\dfrac{\alpha - \alpha_{\mathrm{pi}} \pm \pi}{2}\right)}\right|_{\alpha \to \alpha_{\mathrm{pi}}} \qquad (3.73)$$

and

$$Q(\alpha) = \frac{1}{2\pi j}\left[\sec\left(\frac{\alpha+\phi_0}{2}\right) - \sec\left(\frac{\alpha-\phi_0}{2}\right)\right]$$
$$\cdot\left[1 + \sqrt{2}\cos\frac{1}{2}\left(\frac{3\pi}{2} - \alpha - \theta\right)\right]^{-1}\frac{K_{u+}(\bar{\eta},\cos\alpha)\,K_+(\bar{\eta},\cos\phi_0)}{4\sin\dfrac{\alpha}{2}\cdot\sin\dfrac{\phi_0}{2}} \quad (3.74)$$

where

$$K_{u+}(\bar{\eta},\cos\alpha) = K_+(\bar{\eta},\cos\alpha)\left[1 + \sqrt{2}\cos\frac{1}{2}\left(\frac{3\pi}{2} - \alpha - \theta\right)\right] \quad (3.75)$$

We observe that the terms added and subtracted contain the same pole singularities as the integrand of (3.50). The integrand in the square brackets is now free of singularities and can be evaluated via the usual steepest descent method, whereas the last integral(s) of (3.72) can be evaluated exactly using the identity (3.54). Note that the $\pm$ signs in (3.72) are arbitrary and either one can be selected for convenience. For the poles given in (3.43), we find that

$$D_{\text{Ee}}^{\text{u}}(\phi,\phi_0,\bar{\eta}) = D_{\text{Ee}}^{\text{nu}}(\phi,\phi_0,\bar{\eta}) - \sqrt{\frac{2\pi}{k_0}}\,e^{j\pi/4}$$
$$\cdot\sum_{n=1}^{3}\tilde{Q}(\alpha_{\text{pi}})\sec\left(\frac{\phi-\alpha_{\text{pi}}+\pi}{2}\right)\left[1 - F_{\text{KP}}\left(2k_0\rho\cos^2\frac{\phi-\alpha_{\text{pi}}+\pi}{2}\right)\right] \quad (3.76)$$

or more explicitly (HERMAN AND VOLAKIS, 1987)

$$D_{\text{Ee}}^{\text{u}}(\phi,\phi_0,\bar{\eta}) = D_{\text{Ee}}^{\text{nu}}(\phi,\phi_0,\bar{\eta}) + D_{\text{Ee}}^{\text{go}}(\phi,\phi_0,\bar{\eta}) + D_{\text{Ee}}^{\text{sw}}(\phi,\phi_0,\bar{\eta}) \quad (3.77)$$

where

$$D_{\text{Ee}}^{\text{nu}}(\phi,\phi_0,\bar{\eta}) =$$
$$-\frac{e^{-j\pi/4}}{2\sqrt{2\pi k_0}}\left(\sec\frac{\phi-\phi_0}{2} - \sec\frac{\phi+\phi_0}{2}\right)\frac{K_+(\bar{\eta},\cos\phi)\,K_+(\bar{\eta},\cos\phi_0)}{2\sin\dfrac{\phi}{2}\sin\dfrac{\phi_0}{2}}$$

$$D_{\text{Ee}}^{\text{go}}(\phi,\phi_0,\bar{\eta}) =$$
$$-\frac{e^{-j\pi/4}}{2\sqrt{2\pi k_0}}R_\perp\left\{\sec\left(\frac{\phi-\phi_0}{2}\right)\left[1 - F_{\text{KP}}\left(2k_0\rho\cos^2\frac{\phi-\phi_0}{2}\right)\right]\right.$$
$$\left. - \sec\left(\frac{\phi+\phi_0}{2}\right)\left[1 - F_{\text{KP}}\left(2k_0\rho\cos^2\frac{\phi+\phi_0}{2}\right)\right]\right\} \quad (3.78)$$

$$D_{\text{Ee}}^{\text{sw}}(\phi,\phi_0,\bar{\eta}) =$$
$$-\sqrt{\frac{2\pi}{k_0}}\,e^{j\pi/4}\tilde{Q}(-\theta)\sec\left(\frac{\mp\pi\pm\phi+\theta}{2}\right)\left[1 - F_{\text{KP}}\left(2k_0\rho\cos^2\frac{\theta\pm\phi\mp\pi}{2}\right)\right]$$
$$y \gtrless 0 \quad (3.79)$$

The last two expressions can be identified as the diffracted field contributions due to the geometrical optics and surface wave poles, respectively. Note that the sign change in (3.79) occurs because $\alpha_s = 2\pi - \phi$ for $y < 0$, whereas $\alpha_s = \phi$ for $y > 0$. We again remark that when the uniform diffraction coefficient (3.77) is inserted in (3.70), the resulting expression represents the diffracted field if $F_{\rm KP}$ is evaluated with a positive argument only, but can also give the total field if $F_{\rm KP}$ is evaluated in accordance with (3.60). Then (3.56) can be used to recover the geometrical optics and surface wave fields.

Specimen bistatic patterns of the total field for capacitive ($\bar{\eta} = 0.5 - j5.0$) and inductive ($\bar{\eta} = 0.5 + j5.0$) resistive half-planes are given in Figs. 3–13 and 3–14. For these patterns the observation distance was kept at $\rho = 1.6\lambda_0$ and the incidence angle was $\phi_0 = 120°$. Consequently, the reflection boundary was at $\phi = 60°$ and the shadow boundary was at $\phi = 300°$. It is seen that the patterns are continuous at these angles, demonstrating the uniformity of (3.77)–(3.79). Another point of importance is the effect of the surface wave pole. If $\bar{\eta}$ is real, the pole is never near the steepest descent path, nor is it captured by the deformation. However, when $\bar{\eta}$ is capacitive, the surface wave pole is captured during the deformation of the path C and its contribution to the diffracted field may then be noticeable when $\phi \approx 0°$ or $360°$, depending on the value of Im. $(-\theta)$. As remarked above, by using the identities (3.56) and (3.60) in (3.79), we can recover the residue contribution, the surface wave field $E_z^{\rm sw}$ is found to be

$$-\frac{K_+(\bar{\eta}, \cos\phi_0)\, K_{u+}(\bar{\eta}, \cos\theta)}{2\sin\dfrac{\theta}{2}\sin\dfrac{\phi_0}{2}}\left[\sec\left(\frac{-\theta+\phi_0}{2}\right) - \sec\left(\frac{\theta+\phi_0}{2}\right)\right] e^{-jk_0\rho\cos(\theta\pm\phi)}$$

(3.80)

for $y \gtrless 0$. The slight rise in the total field shown in Fig. 3–13 is due to the surface wave field and its associated diffraction. Obviously, the surface wave effect is more noticeable for a conductive half-plane, which is discussed next. One should also distinguish between the contributions due to the residue of the surface wave pole and to surface wave diffraction. The first is commonly referred to as the surface wave field. However, even if the surface wave field is absent (i.e. the surface wave pole is not captured), the diffracted field given in (3.79) must still be included for a complete uniform representation of the total field.

We close this section by making a comparison of the uniform diffraction coefficient (3.71) and $D_{\rm Ee}^{\rm nu}(\phi, \phi_0, \bar{\eta}) + D_{\rm Ee}^{\rm go}(\phi, \phi_0, \bar{\eta})$ appearing in (3.77). Both of these represent a uniform evaluation of the same integral in the absence of the surface pole(s). Consequently, one would expect that they should be nearly identical. This is indeed true since

$$F_{\rm KP}(z) \approx 1 + j\frac{1}{2z} \qquad (z \gtrsim 5.5)$$

(3.81)

and if $\rho \gtrsim \lambda_0$, $D_{\rm Ee}^{\rm go} \approx 0$ a few degrees away from the optical discontinuity

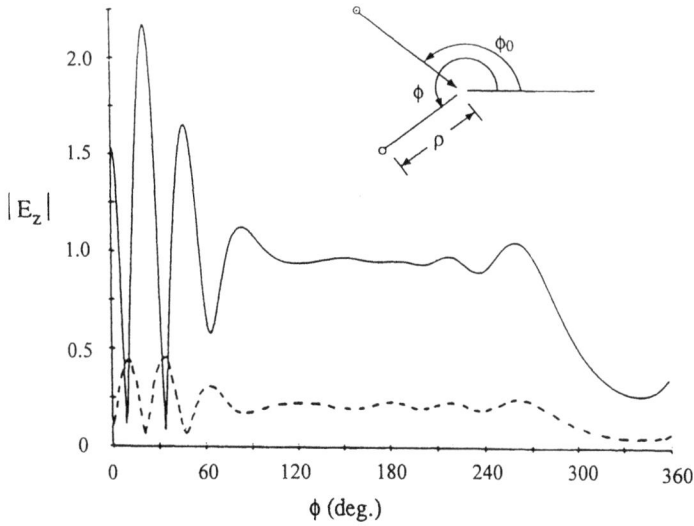

Figure 3–13: *Total field due to plane wave excitation at $\rho = 1.6\lambda_0$ from the edge of conductive (——) and resistive (- - - -) half-planes for $\bar{\eta} = 0.5 - j5$ and $\phi_0 = 120°$: E-polarisation*

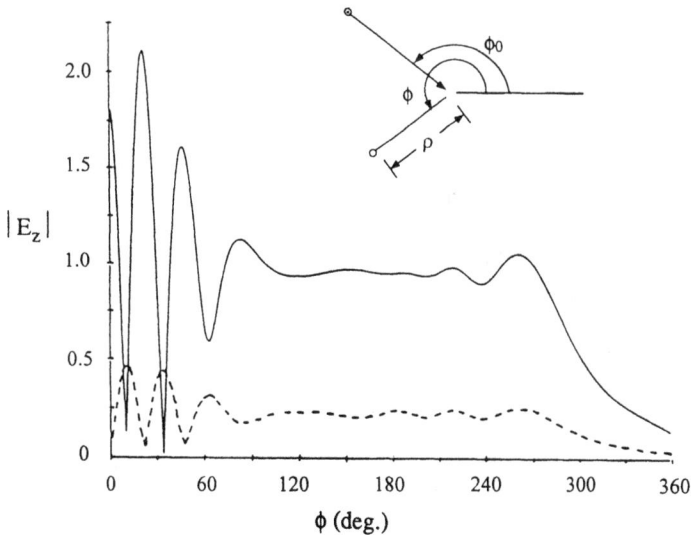

Figure 3–14: *Total field due to plane wave excitation at $\rho = 1.6\lambda_0$ from the edge of conductive (——) and resistive (- - - -) half-planes for $\bar{\eta} = 0.5 + j5$ and $\phi_0 = 120°$: E-polarisation*

boundaries. At the optical boundaries themselves, the two expressions are identical. Thus, any difference between the two expressions is in the angular sector a few degrees either side of the optical boundaries with the two being equal at those boundaries. Since the uniform diffraction coefficients provide a smooth transition between their identical values at either end of the sectors, they are practically identical unless ρ is very small. For $\rho \lesssim 0.5\lambda_0$ the validity of both diffraction coefficients is, of course, questionable. In any case, the coefficient (3.77) based on the additive method is more accurate than (3.71).

3.4   Conductive half-plane (E-polarisation)

Let us again consider the half-plane in Fig. 3–6, which is now assumed to satisfy the conductive sheet boundary condition (2.61). As in the previous section, the half-plane is illuminated by the E-polarised plane wave (3.13). The boundary condition (2.61) or (2.55) then simplifies to

$$H_x(x, y = 0^+) = H_x(x, y = 0^-) = -R_m[E_z]_-^+ \qquad (3.82)$$

for $y = 0$ and $x > 0$, where R_m denotes the conductivity of the sheet in S/square.

To solve for the field diffracted by the conductive half-plane using the dual integral equation method, we introduce the spectral representations

$$E_z^s(x, y \gtrless 0) \;=\; \pm \int_C P_m(\cos\alpha)\, e^{-jk_0\rho\cos(\phi\mp\alpha)}\, d\alpha \qquad (3.83)$$

$$H_x^s(x, y \gtrless 0) \;=\; Y_0 \int_C \sin\alpha\, P_m(\cos\alpha)\, e^{-jk_0\rho\cos(\phi\mp\alpha)}\, d\alpha \qquad (3.84)$$

where (E^s, H^s) are again the scattered fields. The spectral function $P_m(\cos\alpha)$ is now proportional to the spectrum of the magnetic current density supported by the conductive half-plane. As required, the magentic current is associated with the discontinuity in the tangential electric field across the half-plane, and this is evident in the representation (3.83).

To solve for the angular spectrum $P_m(\cos\alpha)$, we first proceed with the application of the boundary conditions:

1. Continuity of E_z across the half-plane $y = 0$, $x < 0$

2. Discontinuity of E_z across the half-plane $y = 0$, $x > 0$ as dictated by (3.82)

Application of these conditions in conjunction with (3.13), (3.83), and (3.84) leads to the dual integral equations

$$\int_{-\infty}^{\infty} \frac{P_m(\lambda)}{\sqrt{1-\lambda^2}}\, d\lambda \;=\; 0 \qquad x < 0 \qquad (3.85)$$

$$\int_{-\infty}^{\infty} \left(\frac{1}{\sqrt{1-\lambda^2}} + \frac{Y_0}{2R_{\mathrm{m}}} \right) P_{\mathrm{m}}(\lambda) \, e^{-jk_0 x \lambda} \, d\lambda \;=\; \frac{Y_0}{2R_{\mathrm{m}}} \sqrt{1-\lambda_0^2} \; e^{jk_0 x \lambda_0}$$

$$x > 0 \qquad (3.86)$$

where, as before, $\lambda = \cos\alpha$ and $\lambda_0 = \cos\phi_0$. The solution of these equations can be carried out using the procedure discussed in the previous section. Specifically, application of Cauchy's theorem leads to

$$\frac{P_{\mathrm{m}}(\lambda)}{\sqrt{1-\lambda^2}} \;=\; U(\lambda) \qquad (3.87)$$

$$\{ K_+(\bar{\eta}_{\mathrm{m}}, \lambda) \, K_-(\bar{\eta}_{\mathrm{m}}, \lambda) \}^{-1} \, P_{\mathrm{m}}(\lambda) \;=\; -\frac{1}{2\pi j} \bar{\eta}_{\mathrm{m}} \frac{\sqrt{1-\lambda_0^2}}{\lambda + \lambda_0} \frac{L(\lambda)}{L(-\lambda_0)} \qquad (3.88)$$

where

$$\bar{\eta}_{\mathrm{m}} = 1/(2R_{\mathrm{m}} Z_0) \qquad (3.89)$$

and $K_\pm(\bar{\eta}_{\mathrm{m}}, \lambda)$ are defined in (3.27) and (3.30). From (3.87) and (3.88) it follows that

$$U(\lambda) = -\frac{\bar{\eta}_{\mathrm{m}}}{2\pi j} \frac{\sqrt{1+\lambda_0}}{\sqrt{1-\lambda}} \frac{K_+(\bar{\eta}_{\mathrm{m}}, \lambda_0) \, K_+(\bar{\eta}_{\mathrm{m}}, \lambda)}{\lambda + \lambda_0}$$

Consequently,

$$P_{\mathrm{m}}(\cos\alpha) \;=\; -\frac{\bar{\eta}_{\mathrm{m}}}{2\pi j} \sqrt{1+\cos\phi_0} \, \sqrt{1+\cos\alpha}$$

$$\cdot \frac{K_+(\bar{\eta}_{\mathrm{m}}, \cos\alpha) \, K_+(\bar{\eta}_{\mathrm{m}}, \cos\phi_0)}{\cos\alpha + \cos\phi_0} \qquad (3.90)$$

from which we deduce that the far zone diffracted field is

$$E_z^{\mathrm{d}} \sim \frac{e^{-jk_0\rho}}{\sqrt{\rho}} \, D_{\mathrm{Em}}^{\mathrm{nu}}(\phi, \phi_0, \bar{\eta}_{\mathrm{m}}) \qquad (3.91)$$

where

$$D_{\mathrm{Em}}^{\mathrm{nu}}(\phi, \phi_0, \bar{\eta}_{\mathrm{m}}) \;=\; -\frac{e^{-j\pi/4}}{\sqrt{2\pi k_0}} \left(2\bar{\eta}_{\mathrm{m}} \cos\frac{\phi_0}{2} \cos\frac{\phi}{2} \right)$$

$$\cdot \frac{K_+(\bar{\eta}_{\mathrm{m}}, \cos\phi) \, K_+(\bar{\eta}_{\mathrm{m}}, \cos\phi_0)}{\cos\phi + \cos\phi_0}$$

$$=\; -\frac{e^{-j\pi/4}}{2\sqrt{2\pi k_0}} \bar{\eta}_{\mathrm{m}} \left[\sec\left(\frac{\phi + \phi_0}{2} \right) + \sec\left(\frac{\phi - \phi_0}{2} \right) \right]$$

$$\cdot K_+(\bar{\eta}_{\mathrm{m}}, \cos\phi) \, K_+(\bar{\eta}_{\mathrm{m}}, \cos\phi_0) \qquad (3.92)$$

is the (non-uniform) diffraction coefficient.

We observe that $D_{\mathrm{Em}}^{\mathrm{nu}}$ is very similar in form to the resistive sheet diffraction coefficient $D_{\mathrm{Ee}}^{\mathrm{nu}}$ given in (3.46). Indeed

$$D_{\mathrm{Em}}^{\mathrm{nu}}(\phi, \phi_0, \bar{\eta}_{\mathrm{m}}) = -2\bar{\eta}_{\mathrm{m}} \cos\frac{\phi_0}{2} \cos\frac{\phi}{2} \, D_{\mathrm{Ee}}^{\mathrm{nu}}(\phi, \phi_0, \bar{\eta}_{\mathrm{m}}) \qquad (3.93)$$

and, more importantly, $P_{\mathrm{m}}(\cos\alpha)$ has the same singularities as $P_{\mathrm{e}}(\cos\alpha)$ given in (3.43). Thus, (3.93) also holds for the uniform diffraction coefficient. However, $D_{\mathrm{Em}}^{\mathrm{u}}$ is typically written as

$$D_{\mathrm{Em}}^{\mathrm{u}}(\phi,\phi_0,\bar{\eta}_{\mathrm{m}}) = D_{\mathrm{Em}}^{\mathrm{nu}}(\phi,\phi_0,\bar{\eta}_{\mathrm{m}}) + D_{\mathrm{Em}}^{\mathrm{go}}(\phi,\phi_0,\bar{\eta}_{\mathrm{m}}) + D_{\mathrm{Em}}^{\mathrm{sw}}(\phi,\phi_0,\bar{\eta}_{\mathrm{m}}) \quad (3.94)$$

with

$$D_{\mathrm{Em}}^{\mathrm{go}}(\phi,\phi_0,\bar{\eta}_{\mathrm{m}}) = \frac{e^{-j\pi/4}}{2\sqrt{2\pi k_0}} \left(\frac{\bar{\eta}_{\mathrm{m}}}{1 + \bar{\eta}_{\mathrm{m}}\sin\phi_0} \right)$$
$$\cdot \left[\sec\frac{\phi+\phi_0}{2} \left(1 - F_{\mathrm{KP}}\left[2k_0\rho\cos^2\frac{\phi+\phi_0}{2} \right] \right) \right.$$
$$\left. + \sec\frac{\phi-\phi_0}{2} \left(1 - F_{\mathrm{KP}}\left[2k_0\rho\cos^2\frac{\phi-\phi_0}{2} \right] \right) \right] \quad (3.95)$$

$$D_{\mathrm{Em}}^{\mathrm{sw}}(\phi,\phi_0,\bar{\eta}_{\mathrm{m}}) = \pm\frac{e^{-j\pi/4}}{2\sqrt{2\pi k_0}}\bar{\eta}_{\mathrm{m}}\, K_{u+}(\bar{\eta}_{\mathrm{m}},\cos\theta)\, K_{+}(\bar{\eta}_{\mathrm{m}},\cos\phi_0)$$
$$\cdot \sec\left(\frac{\mp\pi\pm\phi+\theta}{2} \right) \left[\sec\left(\frac{-\theta+\phi_0}{2} \right) + \sec\left(\frac{\theta+\phi_0}{2} \right) \right]$$
$$\cdot \left(1 - F_{\mathrm{KP}}\left[2k_0\rho\cos^2\frac{\pm\phi\mp\pi+\theta}{2} \right] \right) \qquad y \gtrless 0 \quad (3.96)$$

where K_{u+} is again defined in (3.75).

Sample bistatic patterns for the total field for capacitive and inductive η_m are shown in Figs. 3–13 and 3–14. The surface wave effect is indeed quite noticeable for capacitive $\bar{\eta}_{\mathrm{m}}$ near $\phi \approx 360°$ (see discussion in previous section).

3.5   Resistive and conductive half-planes (H-polarisation)

The H-polarisation diffraction by a half-plane satisfying the transition conditions (2.59) or (2.61) can be readily recovered from the E-polarisation results by invoking duality. Specifically, the field diffracted by a resistive half-plane due to the plane wave illumination (see Fig. 3–6)

$$\mathbf{H}^{\mathrm{i}} = \hat{z}\, e^{jk_0(x\cos\phi_0 + y\sin\phi_0)} \quad (3.97)$$

is given by

$$H_z^{\mathrm{d}} \sim \frac{e^{-jk_0\rho}}{\sqrt{\rho}} D_{\mathrm{Em}}^{\mathrm{nu}}\left(\phi,\phi_0,\frac{1}{\bar{\eta}_{\mathrm{e}}} \right) \quad (3.98)$$

The diffraction coefficient $D_{\mathrm{Em}}^{\mathrm{nu}}$ is defined in (3.92) and $\bar{\eta}_{\mathrm{e}} = 2R_{\mathrm{e}}/Z_0$, as given by (3.28). Also, we can write the total uniform H-polarisation field as

$$H_z \sim \frac{e^{-jk_0\rho}}{\sqrt{\rho}} D_{\mathrm{Em}}^{\mathrm{u}}\left(\phi,\phi_0,\frac{1}{\bar{\eta}_{\mathrm{e}}} \right) \quad (3.99)$$

with D^u_{Em} defined in (3.92)–(3.96). Plots of the H-polarisation resistive half-plane echowidth

$$\sigma_H = 2\pi\rho \lim_{\rho\to\infty} \frac{|H^s_z|^2}{|H^i_z|^2} = 2\pi \left| D^{nu}_{Em}\left(\phi, \phi_0, \frac{1}{\bar\eta_e}\right) \right|^2$$

for different resistivities are given in Fig. 3–15. In contrast to Fig. 3–9(a), we

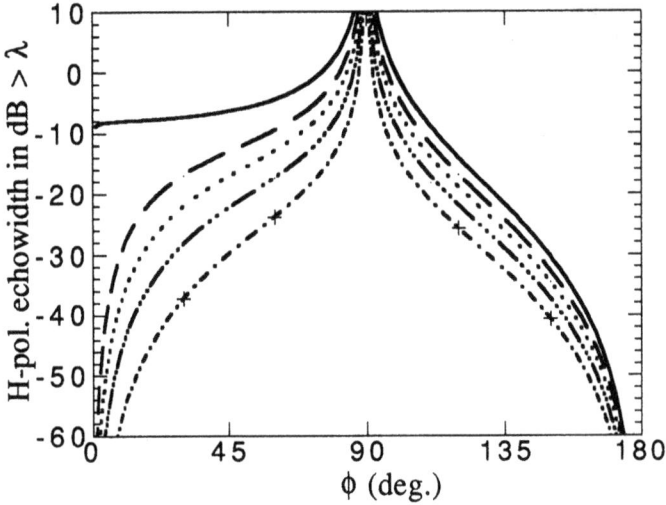

Figure 3–15: *H-polarisation backscatter echowidth plots for resistive half-planes with different resistivities $\bar\eta_e = 0.001$ (——), 0.5 (- - - -), 1 ($\cdots$), 2 (— $\cdots$ —) and $0.5 - j5$ (— $\cdot$ + $\cdot$ —)*

observe that for H polarisation the returns are higher in the region $\phi < \pi/2$ corresponding to more scattering from the trailing edge. Overall, larger values of $\bar\eta_e$ yield lower returns, as was the case for E-polarisation.

The H-polarisation diffracted field for a conductive half-plane is given by

$$H^d_z \sim \frac{e^{-jk_0\rho}}{\sqrt\rho} D^{nu}_{Ee}\left(\phi, \phi_0, \frac{1}{\bar\eta_m}\right) \qquad (3.100)$$

The diffraction coefficient D^{nu}_{Ee} is now defined in (3.46) and $\bar\eta_m = 1/(2R_m Z_0)$. Similarly, the total uniform H-polarisation field is given by

$$H_z \sim \frac{e^{-jk_0\rho}}{\sqrt\rho} D^u_{Ee}\left(\phi, \phi_0, \frac{1}{\bar\eta_m}\right) \qquad (3.101)$$

with D^u_{Ee} defined in (3.71) or (3.77)–(3.79).

3.6   Impedance half-plane

As indicated in Section 2.4, the impedance half-plane can be considered as a special case of a combination sheet, and from (2.69)–(2.73), the impedance half-plane can be replaced by a pair of resistive and conductive sheets with

$$R_{\mathrm{e}} = \frac{\eta}{2}, \qquad R_{\mathrm{m}} = \frac{1}{2\eta}$$

in which η denotes the impedance of the half-plane. Consider the half-plane shown in Fig. 3–6 illuminated by the plane wave

$$\mathbf{E} = \hat{z}\, e^{jk_0(x\cos\phi_0 + y\sin\phi_0)}$$

and satisfying the SIBC

$$E_z(x, \pm 0) = \mp \eta\, H_x(x, \pm 0) \tag{3.102}$$

The E-polarisation diffracted field is then given by

$$
\begin{aligned}
D_{\mathrm{E}}(\phi, \phi, \bar{\eta}) &= D_{\mathrm{Ee}}(\phi, \phi_0, \bar{\eta}) + D_{\mathrm{Em}}(\phi, \phi_0, \bar{\eta}) \\
&= \left(1 - 2\bar{\eta}\cos\frac{\phi}{2}\cos\frac{\phi_0}{2}\right) D_{\mathrm{Ee}}(\phi, \phi_0, \bar{\eta})
\end{aligned} \tag{3.103}
$$

where $\bar{\eta} = \eta/Z_0$ is the normalised surface impedance, D_{Ee} was defined in Sections 3.3 and 3.4, and

$$\bar{\eta}_{\mathrm{e}} = \frac{2R_{\mathrm{e}}}{Z_0} = \bar{\eta}, \qquad \bar{\eta}_{\mathrm{m}} = \frac{Z_0}{2R_{\mathrm{m}}} = \bar{\eta}$$

Thus, the E-polarisation non-uniform diffraction coefficient for the impedance half-plane is

$$D_{\mathrm{E}}^{\mathrm{nu}}(\phi, \phi_0, \bar{\eta}) =$$

$$\frac{e^{-j\pi/4}}{\sqrt{2\pi k_0}} \frac{1 - 2\bar{\eta}\cos\dfrac{\phi_0}{2}\cos\dfrac{\phi}{2}}{\cos\phi + \cos\phi_0}\, K_+(\bar{\eta}, \cos\phi)\, K_+(\bar{\eta}, \cos\phi_0) \tag{3.104}$$

In the case of the H-polarised incident plane wave

$$\mathbf{H}^{\mathrm{i}} = \hat{z}\, e^{jk_0(x\cos\phi_0 + y\sin\phi_0)}$$

the diffraction coefficient is the dual of (3.104), i.e.

$$D_{\mathrm{H}}(\phi, \phi_0, \bar{\eta}) = D_{\mathrm{E}}\left(\phi, \phi_0, \frac{1}{\bar{\eta}}\right) =$$

$$\frac{e^{-j\pi/4}}{\sqrt{2\pi k_0}} \frac{1 - \dfrac{2}{\bar{\eta}}\cos\dfrac{\phi_0}{2}\cos\dfrac{\phi}{2}}{\cos\phi + \cos\phi_0}\, K_+\left(\frac{1}{\bar{\eta}}, \cos\phi\right) K_+\left(\frac{1}{\bar{\eta}}, \cos\phi_0\right) \tag{3.105}$$

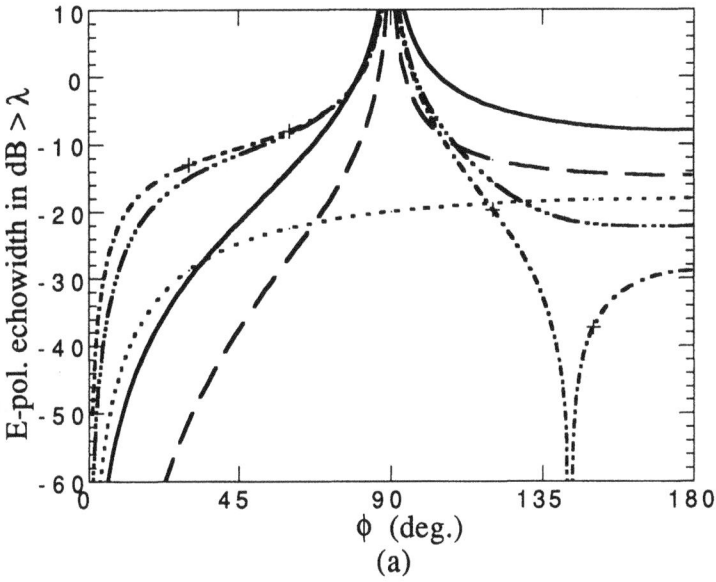

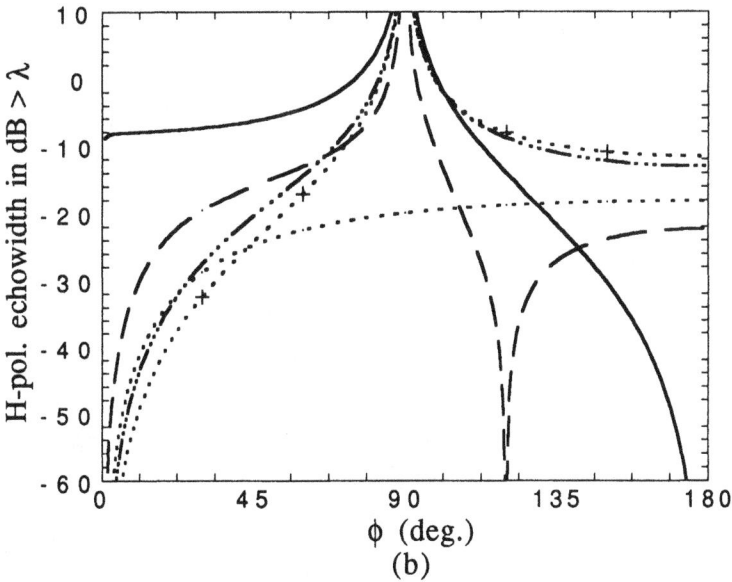

Figure 3–16: *Backscatter echowidth plots for impedance half-planes having* $\bar{\eta} =$ 0.001 *(——)*, 0.5 *(— —)*, 1 *(· · ·)*, $1 - j2$ *(— · · · —) and* 5 *(— · + · —): (a) E-polarisation, (b) H-polarisation*

Plots of the E- and H-polarisation backscatter echowidths

$$\begin{aligned} \sigma_{\mathrm{E}} &= 2\pi \left| D_{\mathrm{E}}^{\mathrm{nu}}(\phi, \phi_0, \bar{\eta}) \right|^2 \\ \sigma_{\mathrm{H}} &= 2\pi \left| D_{\mathrm{H}}^{\mathrm{nu}}(\phi, \phi_0, \bar{\eta}) \right|^2 \end{aligned} \tag{3.106}$$

are shown in Fig. 3–16. As expected, the edge-on H-polarisation echowidth vanishes when $\bar{\eta} = 0$, and σ_{E} and σ_{H} are equal when $\bar{\eta} = 1$. Also, note that for $\bar{\eta} = 1$, the far zone diffracted field is finite at the reflection boundary because the reflected field is zero. In any case, as $\bar{\eta}$ increases, the echowidth is reduced and this is the primary reason for using radar absorbing material coatings to reduce the echowidth of leading edges.

Uniform expressions for the impedance half-plane diffraction coefficients have been given by HERMAN AND VOLAKIS (1987) using the additive method and by ROJAS (1988) using the multiplicative method. These results account for the geometrical optics and surface wave poles. If the surface wave contributions are not of interest (or if the surface wave pole is never near the steepest descent path), then (3.69) can be employed to obtain a diffraction coefficient that is uniform at the optical boundaries. Doing this and making use of (3.49) yields

$$D_{\mathrm{E}}^{\mathrm{u}}(\phi, \phi_0, \bar{\eta}) =$$

$$-\frac{e^{-j\pi/4}}{2\sqrt{2\pi k_0}} \left[C_1(\phi, \phi_0, \bar{\eta}) \sec\left(\frac{\phi - \phi_0}{2}\right) F_{\mathrm{KP}}\left(2k_0\rho \cos^2\frac{\phi - \phi_0}{2}\right) \right.$$

$$\left. + C_2(\phi, \phi_0, \bar{\eta}) \sec\left(\frac{\phi + \phi_0}{2}\right) F_{\mathrm{KP}}\left(2k_0\rho \cos^2\frac{\phi + \phi_0}{2}\right) \right] \tag{3.107}$$

where

$$C_1(\phi, \phi_0, \bar{\eta}) = \frac{2\bar{\eta}\sin\dfrac{\phi}{2}\sin\dfrac{\phi_0}{2} + 1}{2\sin\dfrac{\phi}{2}\sin\dfrac{\phi_0}{2}} \, K_+(\bar{\eta}, \cos\phi)\, K_+(\bar{\eta}, \cos\phi_0)$$

and

$$C_2(\phi, \phi_0, \bar{\eta}) = \frac{2\bar{\eta}\sin\dfrac{\phi}{2}\sin\dfrac{\phi_0}{2} - 1}{2\sin\dfrac{\phi}{2}\sin\dfrac{\phi_0}{2}} \, K_+(\bar{\eta}, \cos\phi)\, K_+(\bar{\eta}, \cos\phi_0)$$

We remark that for $\bar{\eta} = 0$, corresponding to a perfectly conducting half-plane, this uniform diffraction coefficient reduces to the UTD diffraction coefficient (KOUYOUMJIAN AND PATHAK, 1974) for a metallic half-plane. More importantly, at the shadow boundary (see Fig. 3–11)

$$C_1(\phi = \pi + \phi_0, \phi_0, \bar{\eta}) \rightarrow 1$$

and at the reflection boundary

$$C_2(\phi = \pi - \phi_0, \phi_0, \bar{\eta}) \rightarrow R_\perp = \frac{\bar{\eta} \sin \phi_0 - 1}{\bar{\eta} \sin \phi_0 + 1}$$

where $R_\perp$ is the plane wave reflection coefficient given in (2.81). On the basis of these results and the behaviour of the transition function $F_{KP}(z)$, it is evident that the sums $E_z^d + E^i$ and $E_z^d + E_z^r$ are continuous across the shadow and reflection boundaries, respectively. This is illustrated in Fig. 3–17, which shows the total field in the presence of an impedance half-plane when excited by a plane wave. In these plots the field is computed with $\phi_0 = 30°$ at a distance $\rho = 1.6\lambda$ from the edge. Not surprisingly, the total field decreases with increasing $\bar{\eta}$ and this is true for both E- and H- polarisations.

3.7   Sheet and impedance junctions

So far we have considered the diffraction by a single half-plane satisfying either a sheet transition condition (STC) or a first order impedance boundary condition (SIBC). We now go on to develop the solution for the field diffracted by the junction of two thin material layers or by metal-backed coatings and half-planes on dielectric interfaces (see Fig. 3–18). These are all geometries that can be simulated by a junction of STC or SIBC half-planes as illustrated in Fig. 3–19. In the following we first describe the diffraction analysis for the junction of two resistive half-planes. This analysis is then generalised to the junction formed by two conductive half-sheets and subsequently to the junction of two impedance half-planes. The latter is the coastal diffraction problem that was originally considered by CLEMMOW (1953). However, Clemmow chose not to replace the original configuration by impedance junctions and consequently he was not able to derive closed form expressions for the pertinent split functions.

Diffraction by the junction of two impedance half-planes was first treated by MALIUZHINETS (1958) as a special case of wedge diffraction. The Maliuzhinets procedure will be discussed in the next chapter, in which we present the solution of the impedance wedge problem. However, for the case of a planar impedance junction, the simpler dual integral equation method can be employed with only minor modifications to that previously described. Moreover, the dual integral equation or Wiener-Hopf method can yield the solutions for mixed STC and SIBC junctions (see Fig. 3–19), whereas the Maliuzhinets method is not readily applicable to these configurations. Among published solutions for SIBC/STC junctions, ROJAS (1988) employed the Wiener-Hopf method to develop diffraction coefficients for impedance junctions illuminated at skew incidence. Also, SENIOR (1991) used the dual integral equation method to derive similar coefficients for resistive/conductive sheet junctions. In this section we present the diffraction coefficients for resistive, conductive and impedance junctions illuminated at normal incidence. The more general case of

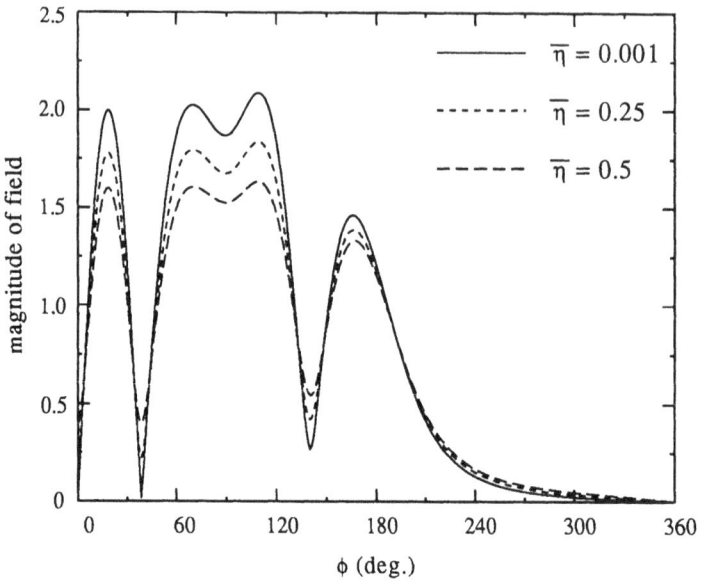

(a)

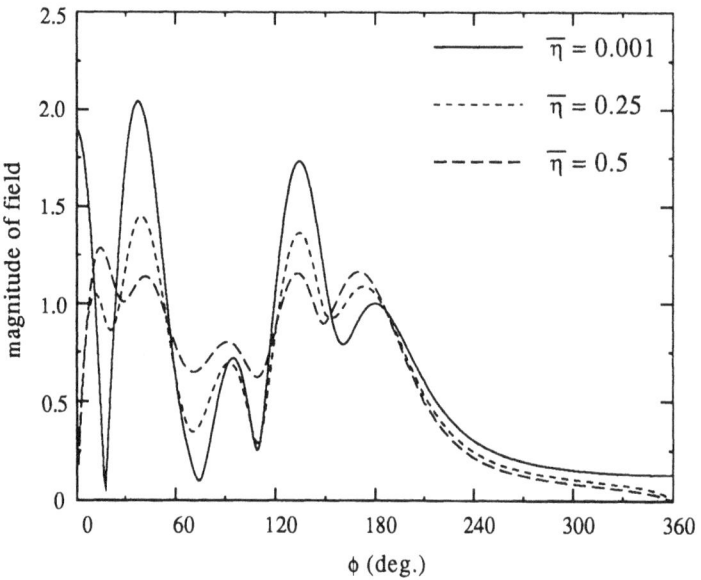

(b)

Figure 3–17: *The total field at a distance $\rho = 5\lambda_0/\pi$ from the edge of an impedance half-plane for plane wave illumination at $\phi_0 = 30°$: (a) E-polarisation, (b) H-polarisation*

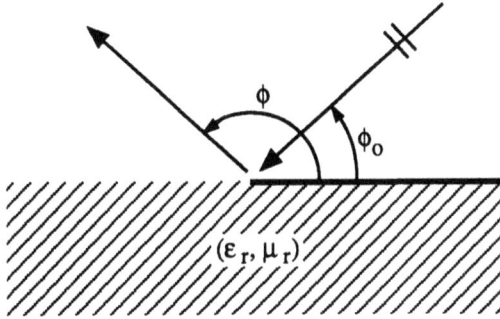

Figure 3–18: *Half-plane on a dielectric interface*

skew incidence is treated in Section 3.8.

3.7.1   Resistive sheet junction (E-polarisation)

Consider again the E-polarised plane wave (3.13) that now impinges on the resistive sheet junction shown in Figure 3–19(a). As illustrated, the left half-plane satisfies the sheet condition

$$E_z = -R_{e1}[H_x]_-^+ \qquad y = 0, \quad x < 0 \tag{3.108}$$

whereas for the right half-plane

$$E_z = -R_{e2}[H_x]_-^+ \qquad y = 0, \quad x > 0 \tag{3.109}$$

The parameters R_{e1} and R_{e2} are, of course, the resistivities of the left and right half-sheets.

To apply the dual integral equation method it is convenient to regard the structure as a single sheet of resistivity R_{e1} occupying the entire plane $y = 0$ with a second sheet superimposed on the portion $x > 0$ (see Fig. 3–20). The second sheet is added in parallel and has resistivity

$$R'_{e2} = \frac{R_{e1} R_{e2}}{R_{e1} - R_{e2}}$$

so that $R_{e1} \parallel R'_{e2} = R_{e2}$. With this set-up, the total field in the presence of the junction can be written as

$$E_z = E_z^{go} + E_z^{s}, \qquad \mathbf{H} = \mathbf{H}^{go} + \mathbf{H}^{s} \tag{3.110}$$

where

$$E_z^{go} = E_z^{i} + E_z^{r1} = \begin{cases} e^{jk_0(x\cos\phi_0 + y\sin\phi_0)} - \dfrac{e^{jk_0(x\cos\phi_0 - y\sin\phi_0)}}{1 + \bar{\eta}_{e1}\sin\phi_0} & y > 0 \\[3mm] \dfrac{\bar{\eta}_{e1}\sin\phi_0}{1 + \bar{\eta}_{e1}\sin\phi_0} e^{jk_0(x\cos\phi_0 + y\sin\phi_0)} & y < 0 \end{cases} \tag{3.111}$$

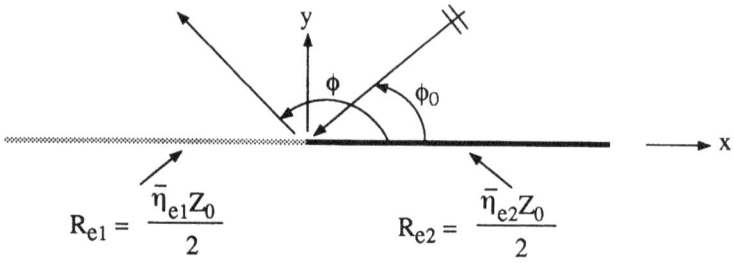

(a)

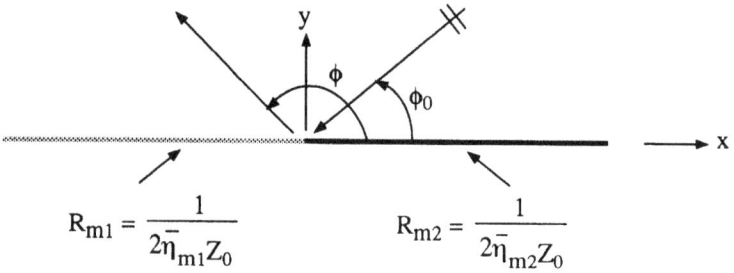

(b)

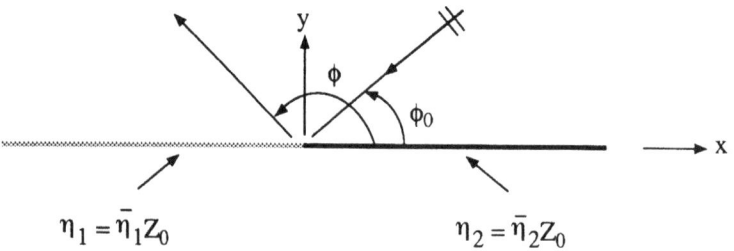

(c)

Figure 3–19: *Half-plane junction problems for (a) resistive sheets, (b) conductive sheets, (c) impedance half-planes*

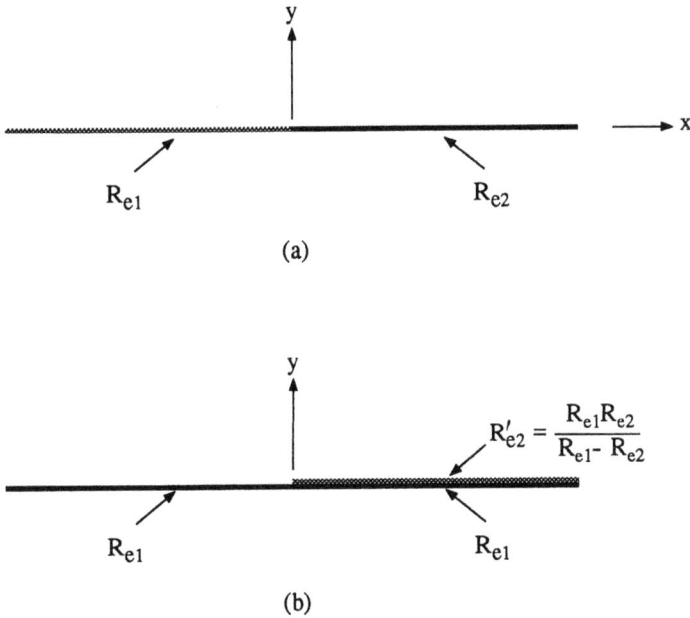

(a)

(b)

Figure 3–20: *Electrical break-up of the resistive sheet junction for analysis via the dual integral equation method: (a) original junction, (b) equivalent problem*

$$H_x^{go} = H_x^i + H_x^{r1} =$$

$$-Y_0 \sin \phi_0 \begin{cases} e^{jk_0(x \cos \phi_0 + y \sin \phi_0)} + \dfrac{e^{jk_0(x \cos \phi_0 - y \sin \phi_0)}}{1 + \bar{\eta}_{e1} \sin \phi_0} & y > 0 \\[3mm] \dfrac{\bar{\eta}_{e1} \sin \phi_0}{1 + \bar{\eta}_{e1} \sin \phi_0} e^{jk_0(x \cos \phi_0 + y \sin \phi_0)} & y < 0 \end{cases} \qquad (3.112)$$

are the fields in the absence of the half-sheet of resistivity R'_{e2} and

$$\bar{\eta}_{e1} = \frac{2R_{e1}}{Z_0} \qquad (3.113)$$

Clearly, E_z^{r1} corresponds to the field reflected by the uniform sheet of resistivity R_{e1}. It remains to determine the scattered field E_z^s attributable to the half-sheet having resistivity R'_{e2}. In accordance with the discussion in Section 3.2, this can be represented as

$$E_z^s(x, y \gtrless 0) = \int_C P_e(\cos \alpha) e^{-jk_0 \rho \cos(\phi \mp \alpha)} d\alpha \qquad (3.114)$$

$$H_x^s(x, y \gtrless 0) = \pm Y_0 \int_C \sin \alpha \, P_e(\cos \alpha) e^{-jk_0 \rho \cos(\phi \mp \alpha)} d\alpha \qquad (3.115)$$

where, as before, $P_e(\cos \alpha)$ is the unknown angular spectrum to be determined in accordance with the boundary conditions (3.108) and (3.109).

By virtue of the chosen representations (3.110)–(3.115), E_z is continuous across $y = 0$ and we can proceed with the enforcement of the conditions (3.108) and (3.109) to set up the dual integral equations. From (3.108) we have

$$E_z^s + E_z^{g\circ} = -R_{e1}[H_x(x, 0^+) - H_x(x, 0^-)]$$

for $x < 0$, and this yields

$$\int_{-\infty}^{\infty} \left(\bar{\eta}_{e1} + \frac{1}{\sqrt{1 - \lambda^2}} \right) P_e(\lambda)\, e^{-jk_0 x \lambda}\, d\lambda = 0, \qquad x < 0 \qquad (3.116)$$

Similarly from the STC (3.109) we deduce that

$$\int_{-\infty}^{\infty} \left(\bar{\eta}_{e2} + \frac{1}{\sqrt{1 - \lambda^2}} \right) P_e(\lambda)\, e^{-jk_0 x \lambda}\, d\lambda = \frac{(\bar{\eta}_{e2} - \bar{\eta}_{e1})}{\left[\bar{\eta}_{e1} + \left(1 / \sqrt{1 - \lambda_0^2} \right) \right]} e^{jk_0 x \lambda_0} \qquad (3.117)$$

for $x > 0$, where $\lambda_0 = \cos \phi_0$ and

$$\bar{\eta}_{e2} = \frac{2R_{e2}}{Z_0} \qquad (3.118)$$

The pair of equations (3.116) and (3.117) are similar to (3.21) and (3.23), and can be solved for $P_e(\lambda)$ in a like manner using Cauchy's residue theorem. From (3.116) by closing the path of integration with a semi-infinite contour in the upper half-plane it follows that the non-exponential portion of the integrand must be a function regular in the upper half-plane. Consequently,

$$P_e(\lambda) = K_-(\bar{\eta}_{e1}, \lambda)\, U(\lambda) \qquad (3.119)$$

where $U(\lambda)$ is an upper function and $K_-(\bar{\eta}_{e1}, \lambda)$ has been defined in (3.27)–(3.30). Similarly, by closing the path of integration in (3.117) with a semi-infinite contour in the lower half-plane we deduce

$$\frac{P_e(\lambda)}{K_+(\bar{\eta}_{e2}, \lambda)\, K_-(\bar{\eta}_{e2}, \lambda)} =$$
$$-\frac{1}{2\pi j} \frac{(\bar{\eta}_{e2} - \bar{\eta}_{e1})}{\lambda + \lambda_0} K_+(\bar{\eta}_{e1}, \lambda_0)\, K_-(\bar{\eta}_{e1}, \lambda_0) \frac{L_1(\lambda)}{L_1(-\lambda_0)} + L_2(\lambda)$$

where $L_{1,2}(\lambda)$ are arbitrary lower functions. In conjunction with (3.119) we then have

$$U(\lambda) = -\frac{1}{2\pi j} \frac{(\bar{\eta}_{e2} - \bar{\eta}_{e1})}{\lambda + \lambda_0} K_-(\bar{\eta}_{e1}, \lambda_0)\, K_+(\bar{\eta}_{e2}, \lambda)\, K_+(\bar{\eta}_{e2}, \lambda_0)\, A(\lambda)$$

where $A(\lambda)$ is a polynomial whose order is such that $P_e(\lambda) = O(\lambda^{-1})$ in addition to ensuring that the right hand side of (3.117) is recovered. Since $K_+(\eta, \lambda) = O(1)$, it follows that $A(\lambda) = 1$ and thus

$$P_e(\lambda) = -\frac{1}{2\pi j}(\bar{\eta}_{e2} - \bar{\eta}_{e1}) \frac{K_-(\bar{\eta}_{e1}, \lambda)\, K_-(\bar{\eta}_{e1}, \lambda_0)\, K_+(\bar{\eta}_{e2}, \lambda)\, K_+(\bar{\eta}_{e2}, \lambda_0)}{\lambda + \lambda_0} \qquad (3.120)$$

This is symmetric in λ and λ_0 as required, and $P_e(\lambda)$ vanishes when $\bar{\eta}_{e2} = \bar{\eta}_{e1}$. Furthermore, (3.120) reduces to the half-plane result (3.42) when $\bar{\eta}_{e1} \to \infty$. This is easily verified by making use of the split function behaviour given in (3.33).

On substituting (3.120) into the integral (3.114) and performing a steepest descent path evaluation for large $k_0\rho$, we find

$$E_z^{\rm d} \sim \frac{e^{-jk\rho}}{\sqrt{\rho}}\, D_{\rm Ee}^{\rm nu}(\phi, \phi_0, \bar{\eta}_{e1}, \bar{\eta}_{e2}) \qquad (3.121)$$

where

$$D_{\rm Ee}^{\rm nu}(\phi, \phi_0, \bar{\eta}_{e1}, \bar{\eta}_{e2}) = -\frac{e^{-j\pi/4}}{\sqrt{2\pi k_0}}(\bar{\eta}_{e2} - \bar{\eta}_{e1})$$

$$\cdot \frac{K_-(\bar{\eta}_{e1}, \cos\phi)\, K_-(\bar{\eta}_{e1}, \cos\phi_0)\, K_+(\bar{\eta}_{e2}, \cos\phi)\, K_+(\bar{\eta}_{e2}, \cos\phi_0)}{\cos\phi + \cos\phi_0} \qquad (3.122)$$

is the E-polarisation diffraction coefficient of the resistive sheet junction. Corresponding uniform diffraction coefficients can be derived by following the procedure described in Section 3.3.2. The pertinent surface wave poles are now $\alpha = -\theta_2$ and $\alpha = \pi + \theta_1$, where $\sin\theta_1 = 1/\bar{\eta}_{e1}$ and $\sin\theta_2 = 1/\bar{\eta}_{e2}$.

In comparing (3.122) with the resistive half-plane diffraction coefficient (3.46), it is immediately evident that we can express the junction diffraction coefficient as a product of the half-plane coefficients. Specifically, we can rewrite (3.122) as

$$\begin{aligned} D_{\rm Ee}^{\rm nu}(\phi, \phi_0, \bar{\eta}_{e1}, \bar{\eta}_{e2}) &= \sqrt{2\pi k}\, e^{j\pi/4}(\bar{\eta}_{e2} - \bar{\eta}_{e1})(\cos\phi + \cos\phi_0) \\ &\quad \cdot D_{\rm Ee}^{\rm nu}(\phi, \phi_0, \bar{\eta}_{e1}, \infty)\, D_{\rm Ee}^{\rm nu}(\phi, \phi_0, \infty, \bar{\eta}_{e2}) \\ &= \sqrt{2\pi k}\, e^{j\pi/4}(\bar{\eta}_{e2} - \bar{\eta}_{e1})(\cos\phi + \cos\phi_0) \\ &\quad \cdot D_{\rm Ee}^{\rm nu}(\pi \mp \phi_0, \pi - \phi_0, \bar{\eta}_{e1})\, D_{\rm Ee}^{\rm nu}(\phi, \phi_0, \bar{\eta}_{e2}) \end{aligned} \qquad (3.123)$$

where $D_{\rm Ee}^{\rm nu}(\phi, \phi_0, \bar{\eta})$ is given in (3.46) and is the E-polarisation diffraction coefficient for an isolated resistive half-plane. Note, however, that this last form does not hold when the uniform expressions for $D_{\rm Ee}^{\rm nu}(\phi, \phi_0, \bar{\eta})$ are inserted, and the uniform coefficient for the junction must be derived directly from (3.122). Representative echowidth plots of the resistive sheet junction computed from (3.47) and (3.122) are given in Fig. 3–21 with the resistivity of the left half-plane kept at $R_{e1} = 250\ \Omega$ ($\bar{\eta}_{e1} = 1.326$).

3.7.2 Conductive sheet junction (E-polarisation)

The diffraction coefficient for a junction formed by two conductive half-planes can be constructed by following the same procedure described in the previous subsection and in Section 3.4. Referring to Fig. 3–19(b), we have

$$E_z^{\rm d} \sim \frac{e^{-jk\rho}}{\sqrt{\rho}}\, D_{\rm Em}^{\rm nu}(\phi, \phi_0, \bar{\eta}_{m1}, \bar{\eta}_{m2}) \qquad (3.124)$$

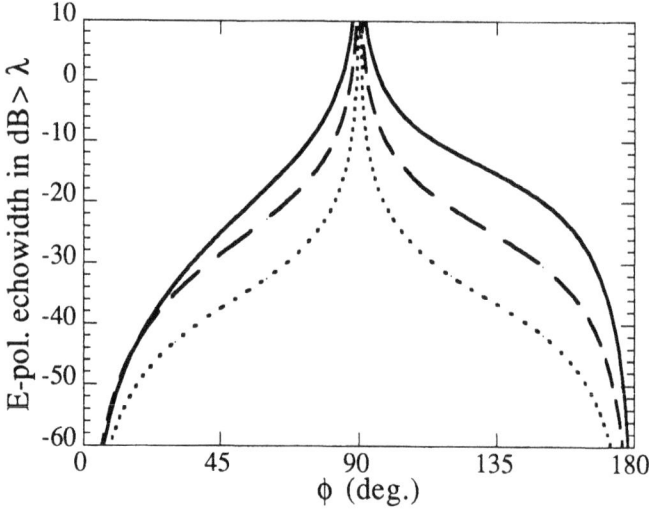

Figure 3–21: *E-polarisation backscatter echowidth plots for a resistive sheet junction with $\bar{\eta}_{e1} = 1.326$ ($R_{e1} = 250\ \Omega$) and $\bar{\eta}_{e2} = 0.001$ (——), 0.5 (– – –), 1.0 (- - - -)*

with

$$D_{Em}^{nu}(\phi, \phi_0, \bar{\eta}_{m1}, \bar{\eta}_{m2}) = \mp \frac{e^{-j\pi/4}}{\sqrt{2\pi k_0}}(\bar{\eta}_{m2} - \bar{\eta}_{m1})$$
$$\cdot \frac{K_-(\bar{\eta}_{m1}, \cos\phi)\, K_-(\bar{\eta}_{m1}, \cos\phi_0)\, K_+(\bar{\eta}_{m2}, \cos\phi)\, K_+(\bar{\eta}_{m2}, \cos\phi_0)}{\cos\phi + \cos\phi_0} \qquad (3.125)$$

in which the upper sign is for $\phi < \pi$ and the lower sign for $\phi > \pi$ (lower half space). Consistent with the definition (3.89), the normalised quantities η_{m1} and η_{m2} are given by

$$\bar{\eta}_{m1} = \frac{1}{2R_{m1}Z_0}, \qquad \bar{\eta}_{m2} = \frac{1}{2R_{m2}Z_0} \qquad (3.126)$$

where R_{m1} and R_{m2} are the conductivities of the left and right half-planes forming the junction.

As in the case of the resistive sheet junction, the diffraction coefficient (3.125) can be written as a product of the corresponding half-plane diffraction coefficients. Thus

$$\begin{aligned}D_{Ee}^{nu}(\phi, \phi_0, \bar{\eta}_{m1}, \bar{\eta}_{m2}) &= \mp\sqrt{2\pi k}\, e^{-j\pi/4}(\bar{\eta}_{m2} - \bar{\eta}_{m1})(\cos\phi + \cos\phi_0) \\ &\quad \cdot D_{Ee}^{nu}(\pi \mp \phi, \pi - \phi_0, \bar{\eta}_{m1})\, D_{Ee}^{nu}(\phi, \phi_0, \bar{\eta}_{m2}) \qquad (3.127)\end{aligned}$$

and this should be compared with the result (3.123) for the resistive sheet junction.

3.7.3   Resistive and conductive junctions (H-polarisation)

By invoking duality, the H-polarisation diffraction coefficients for the resistive and conductive junctions are readily recovered. Specifically, when the junctions in Fig. 3–19(a) and (b) are illuminated by the plane wave

$$\mathbf{H}_z^i = \hat{z}\, e^{jk_0(\cos\phi_0 + y\sin\phi_0)}$$

(see also Section 3.5) we find that

$$H_z^d \sim \frac{e^{-jk_0\rho}}{\sqrt{\rho}}\, D_{Em}^{nu}\left(\phi, \phi_0, \frac{1}{\bar{\eta}_{e1}}, \frac{1}{\bar{\eta}_{e2}}\right) \tag{3.128}$$

for the resistive sheet junction and

$$H_z^d \sim \frac{e^{-jk_0\rho}}{\sqrt{\rho}}\, D_{Ee}^{nu}\left(\phi, \phi_0, \frac{1}{\bar{\eta}_{m1}}, \frac{1}{\bar{\eta}_{m2}}\right) \tag{3.129}$$

for the conductive sheet junction, where the normalised sheet parameters are defined in (3.113), (3.118) and (3.126). The non-uniform expressions for the diffraction coefficients are given in (3.122), (3.123), (3.125) and (3.127).

3.7.4   Impedance and combination sheet junctions

Consider now the diffraction by a junction formed by two impedance half-planes as shown in Fig. 3–19(c). The left half of the junction has impedance $\eta_1 = \bar{\eta}_1 Z_0$, whereas the right half has $\eta_2 = \bar{\eta}_2 Z_0$. As discussed in Section 3.6, the diffraction coefficient of planar impedance problems can be constructed by combining the corresponding diffraction coefficients for the resistive and conductive sheets. In this case the resistivity and conductivity are related to the impedances $\eta_{1,2}$ in accordance with (2.73).

Generalising (3.103), we can write the diffraction coefficient for the impedance junction as

$$D_E(\phi, \phi_0, \bar{\eta}_1, \bar{\eta}_2) = D_{Ee}(\phi, \phi_0, \bar{\eta}_1, \bar{\eta}_2) + D_{Em}(\phi, \phi_0, \bar{\eta}_1, \bar{\eta}_2) \tag{3.130}$$

Alternatively,

$$\begin{aligned}
D_E(\phi, \phi_0, \bar{\eta}_1, \bar{\eta}_2) &= 2D_{Ee}(\phi, \phi_0, \bar{\eta}_1, \bar{\eta}_2) = -2\frac{e^{-j\pi/4}}{\sqrt{2\pi k_0}}(\bar{\eta}_2 - \bar{\eta}_1) \\
&\cdot \frac{K_-(\bar{\eta}_1, \cos\phi)\, K_-(\bar{\eta}_1, \cos\phi_0)\, K_+(\bar{\eta}_2, \cos\phi)\, K_+(\bar{\eta}_2, \cos\phi_0)}{\cos\phi + \cos\phi_0}
\end{aligned} \tag{3.131}$$

valid for $0 \leq \phi \leq \pi$. Because the impedance sheets are opaque the diffracted field is zero in the lower half space and, by duality, the diffraction coefficient for H-polarisation is

$$D_H(\phi, \phi_0, \bar{\eta}_1, \bar{\eta}_2) = D_E\left(\phi, \phi_0, \frac{1}{\bar{\eta}_1}, \frac{1}{\bar{\eta}_2}\right)$$

So far all of the diffraction coefficients have been for two abutting *like* sheets (impedance, resistive or conductive), but there may be situations where an impedance or metallic surface is butted to, for example, a resistive sheet. This could occur with an artificial structure if a metallic section were joined to a non-metallic frame, and resistive sheet extensions are often used to reduce the scattering from a metal termination. It is therefore of interest to construct the diffraction coefficients for junctions formed by combination sheets. One such junction is that formed by a resistive half sheet and an impedance half-plane, and this was considered by UZGÖREN ET AL. (1989). The diffraction coefficient for this junction is easily obtained in terms of the coefficients given previously by noting that the impedance half-plane is equivalent to a combination of coincident resistive and conductive half-planes. With this substitution, the resulting structure is comprised of a resistive sheet junction and a conductive half-plane. Because the conductive sheet supports magnetic currents whereas the resistive sheet supports electric currents, coincident resistive and conductive sheets do not interact. This was discussed in Section 2.4 and is evident in the result given in (3.103). Thus, for the given resistive-impedance junction with the impedance half-plane to the right, the E-polarisation diffraction coefficient is

$$
\begin{aligned}
D_{\mathrm{E}}(\phi, \phi_0, \bar{\eta}_{e1}, \bar{\eta}_2) &= D_{\mathrm{Ee}}(\phi, \phi_0, \bar{\eta}_{e1}, \bar{\eta}_2) + D_{\mathrm{Em}}(\phi, \phi_0, \bar{\eta}_2) \\
&= -2\bar{\eta}_2 \cos\frac{\phi}{2} \cos\frac{\phi_0}{2} \, D_{\mathrm{Ee}}(\phi, \phi_0, \bar{\eta}_2) \\
&\quad + D_{\mathrm{Ee}}(\phi, \phi_0, \bar{\eta}_{e1}, \bar{\eta}_2)
\end{aligned}
\tag{3.132}
$$

where $\bar{\eta}_{e1} = 2R_{e1}/Z_0$, $\bar{\eta}_2$ is the normalised impedance of the right half-plane and D_{Ee} is defined in (3.46). For H-polarisation the corresponding diffraction coefficient is

$$
D_{\mathrm{H}}(\phi, \phi_0, \bar{\eta}_{e1}, \bar{\eta}_2) = D_{\mathrm{Em}}\left(\phi, \phi_0, \frac{1}{\bar{\eta}_{e1}}, \frac{1}{\bar{\eta}_2}\right) + D_{\mathrm{Ee}}\left(\phi, \phi_0, \frac{1}{\bar{\eta}_2}\right)
\tag{3.133}
$$

By a similar argument, we can readily obtain the diffraction coefficient for a junction formed by abutting resistive and conductive half-planes. Again, the key to these derivations is the fact that the resistive and conductive sheet currents do not couple when in the same plane. Thus, the diffraction coefficient for a resistive-conductive junction can be written as the sum of the coefficients for each half-plane. Assuming that the resistive half-plane is to the left we have

$$
\begin{aligned}
D_{\mathrm{E}}(\phi, \phi_0, \bar{\eta}_{e1}, \bar{\eta}_{m2}) &= D_{\mathrm{Ee}}(\pi - \phi, \pi - \phi_0, \bar{\eta}_{e1}) + D_{\mathrm{Em}}(\phi, \phi_0, \bar{\eta}_{m2}) \\
&= -2\bar{\eta}_{m2} \cos\frac{\phi}{2} \cos\frac{\phi_0}{2} \, D_{\mathrm{Ee}}(\phi, \phi_0, \bar{\eta}_{m2}) \\
&\quad + D_{\mathrm{Ee}}(\pi - \phi, \pi - \phi_0, \bar{\eta}_{e1})
\end{aligned}
\tag{3.134}
$$

for E-polarisation and

$$D_{\mathrm{H}}(\phi, \phi_0, \bar{\eta}_{\mathrm{e}1}, \bar{\eta}_{\mathrm{m}2}) = D_{\mathrm{Em}}\left(\pi - \phi, \pi - \phi_0, \frac{1}{\bar{\eta}_{\mathrm{e}1}}\right) + D_{\mathrm{Ee}}\left(\phi, \phi_0, \frac{1}{\bar{\eta}_{\mathrm{m}2}}\right) \quad (3.135)$$

for H-polarisation.

3.8 Skew incidence on junctions

For three-dimensional applications it is necessary to consider the problem of diffraction by junctions illuminated at skew incidence as illustrated in Fig. 3–22. In this case the incident plane wave has the form

$$\begin{aligned}
\mathbf{E}^{\mathrm{i}} &= (\hat{x}\, e_x + \hat{y}\, e_y + \hat{z}\, e_z) e^{-jk_0 \hat{\imath} \cdot \mathbf{r}} & (3.136) \\
Z_0 \mathbf{H}^{\mathrm{i}} &= \hat{\imath} \times \mathbf{E}^{\mathrm{i}}
\end{aligned}$$

where

$$\hat{\imath} = -\hat{x} \cos\phi_0 \sin\beta - \hat{y} \sin\phi_0 \sin\beta + \hat{z} \cos\beta \quad (3.137)$$

specifies the direction of incidence,

$$\mathbf{r} = x\hat{x} + y\hat{y} + z\hat{z}$$

and β is a measure of the obliquity or skewness of the incidence. For $\beta = \pi/2$, (3.136) reduces to (3.13), which represents a plane wave impinging normally on the half-plane or junction. Consequently, for $\beta = \pi/2$ the results stated in Section 3.7 are immediately applicable.

In this section we develop the diffraction coefficients for the junctions in Fig. 3–19 under the excitation (3.136). The dual integral equation method is again employed but, since the transition conditions (2.52) or (2.55) imply the existence of two coupled currents on the sheet(s), the analysis is more involved. However, it has been found that a certain combination of these currents leads to a decoupling, and WILLIAMS (1960) noted that the decoupling is associated with the field components normal to the surface. It is therefore advantageous to work with the transition conditions (2.59) and (2.61) that involve only the normal (E_y and H_y) components. From Maxwell's equations, two like field components are sufficient for the complete specification of the field, and this enables us to consider E_y and H_y as independent components in this analysis.

In the following we solve for the field diffracted by a resistive sheet junction using the method described by SENIOR (1991). The solution is then generalised to the case of conductive and impedance junctions.

3.8.1 Resistive sheet junctions

Consider the planar resistive junction shown in Fig. 3–19(a), where the left half has a resistivity $R_{\mathrm{e}1}$ and the right half has a resistivity $R_{\mathrm{e}2}$. From (2.59),

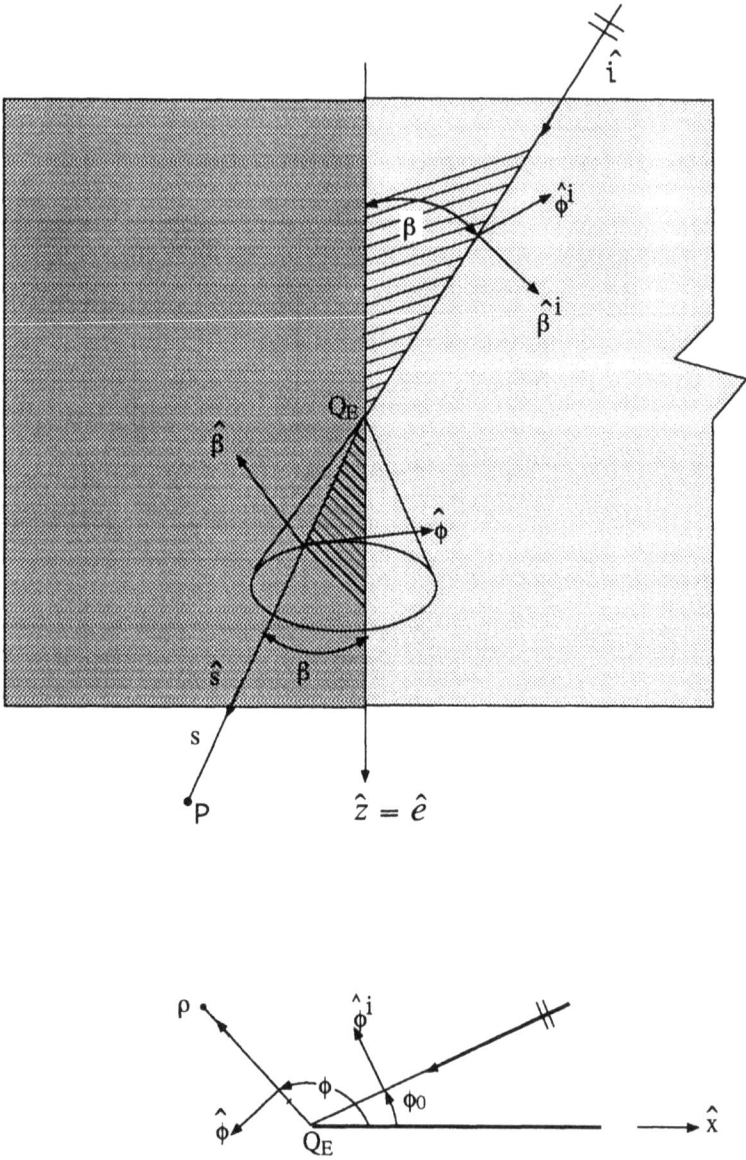

Figure 3–22: *Geometry for a junction illuminated at skew incidence*

the transition conditions satisfied by the left half sheet are

$$[H_y]_-^+ = 0 \qquad \left[\frac{\partial H_y}{\partial y}\right]_-^+ - \frac{2jk_0}{\bar{\eta}_{e1}} H_y = 0$$

$$\left[\frac{\partial E_y}{\partial y}\right]_-^+ = 0 \qquad \frac{\partial E_y}{\partial y} - \frac{jk_0\bar{\eta}_{e1}}{2}[E_y]_-^+ = 0$$

$$(3.138)$$

whereas the transition conditions satisfied by the right half sheet are

$$[H_y]_-^+ = 0 \qquad \left[\frac{\partial H_y}{\partial y}\right]_-^+ - \frac{2jk_0}{\bar{\eta}_{e2}} H_y = 0$$

$$\left[\frac{\partial E_y}{\partial y}\right]_-^+ = 0 \qquad \frac{\partial E_y}{\partial y} - \frac{jk_0\bar{\eta}_{e2}}{2}[E_y]_-^+ = 0$$

$$(3.139)$$

As usual, $\bar{\eta}_{e1} = 2R_{e1}/Z_0$ and $\bar{\eta}_{e2} = 2R_{e2}/Z_0$. The junction is illuminated by the plane wave (3.136) whose y components are

$$E_y^i = e_y e^{-jk_0 \mathbf{i} \cdot \mathbf{r}}, \qquad H_y^i = Y_0 h_y e^{-jk \mathbf{i} \cdot \mathbf{r}} \qquad (3.140)$$

Typically, for soft or parallel illumination

$$e_y = \cos\beta \sin\phi_0, \qquad h_y = \cos\phi_0 \qquad (3.141)$$

whereas for hard or perpendicular illumination

$$e_y = \cos\phi_0, \qquad h_y = -\cos\beta \sin\phi_0 \qquad (3.142)$$

We seek the scattered/diffracted fields due to this illumination.

To proceed with the analysis via the dual integral equation method, we again break up the sheet junction as illustrated in Fig. 3–20. The total field is then the sum of the geometrical optics field of the sheet occupying the entire $y = 0$ plane and the field produced by the equivalent half-plane to the right. We have

$$E_y = E_y^{go} + E_y^s, \qquad H_y = H_y^{go} + H_y^s \qquad (3.143)$$

and from (3.137) we readily find that

$$E_y^{go} = e_y \left[e^{j\kappa(x\cos\phi_0 + y\sin\phi_0)} + \frac{\sin\beta\sin\phi_0}{\bar{\eta}_{e1} + \sin\beta\sin\phi_0} e^{j\kappa(x\cos\phi_0 - y\sin\phi_0)} \right]$$

$$(3.144)$$

$$H_y^{go} = Y_0 h_y \left[e^{j\kappa(x\cos\phi_0 + y\sin\phi_0)} - \frac{1}{1 + \bar{\eta}_{e1}\sin\beta\sin\phi_0} e^{j\kappa(x\cos\phi_0 - y\sin\phi_0)} \right]$$

for $y > 0$ and

$$
E_y^{go} = e_y \frac{\bar{\eta}_{e1}}{\bar{\eta}_{e1} + \sin\beta\sin\phi_0} e^{j\kappa(x\cos\phi_0 + y\sin\phi_0)}
$$
(3.145)

$$
H_y^{go} = Y_0 h_y \frac{\bar{\eta}_{e1}\sin\beta\sin\phi_0}{1 + \bar{\eta}_{e1}\sin\beta\sin\phi_0} e^{j\kappa(x\cos\phi_0 + y\sin\phi_0)}
$$

for $y < 0$. In these expressions $\kappa = k_0\sin\beta$ and we have suppressed the z dependence of the fields, which amounts to multiplying (3.144) and (3.145) by the factor $e^{-jk_0 z\cos\beta}$. Because the geometry is independent of z, the scattered fields have the same z-dependence as the incident field. Consequently, a factor $e^{-jk_0 z\cos\beta}$ is associated with all field components and can be suppressed. This is done throughout the section.

In accordance with the dual integral equation method, the scattered fields are expressed as

$$
E_y^s = \pm\int_C P(\cos\alpha)\, e^{-j\kappa\rho\cos(\phi\mp\alpha)}\, d\alpha
$$
(3.146)

$$
H_y^s = Y_0 \int_C Q(\cos\alpha)\, e^{-j\kappa\rho\cos(\phi\mp\alpha)}\, d\alpha
$$
(3.147)

where $P(\cos\alpha)$ and $Q(\cos\alpha)$ are treated as independent spectra. To solve for $Q(\cos\alpha)$ we proceed with the enforcement of the transition conditions that involve H_y. This leads to the dual integral equations

$$
\int_{-\infty}^{\infty}\left(\bar{\eta}_{e1}\sin\beta + \frac{1}{\sqrt{1-\lambda^2}}\right) Q(\lambda)\, e^{-j\kappa\lambda x}\, d\lambda = 0 \qquad x < 0 \qquad (3.148)
$$

$$
\int_{-\infty}^{\infty}\left(\bar{\eta}_{e2}\sin\beta + \frac{1}{\sqrt{1-\lambda^2}}\right) Q(\lambda)\, e^{-j\kappa\lambda x}\, d\lambda =
$$

$$
(\bar{\eta}_{e2} - \bar{\eta}_{e1}) h_y \frac{\sin\beta\sin\phi_0}{1 + \bar{\eta}_{e1}\sin\beta\sin\phi_0} e^{j\kappa x\lambda_0} \qquad x > 0 \qquad (3.149)
$$

which are in all respects identical in form to (3.116) and (3.117). It therefore follows that $Q(\lambda)$ has the same form as $P_e(\lambda)$ in (3.116) and (3.117). The only difference is that the polynomial $A(\lambda)$ is not unity but instead $A(\lambda) = a + b\lambda$, so that $Q(\lambda) = O(1)$. This is required since the normal field components grow as $O(x^{-1})$ as $x \to 0$ and, from the Abelian theorem, if $Q(\lambda) = O(\lambda^\nu)$, then H_y grows as $O(x^{-\nu-1})$ as $x \to 0$. Hence

$$
Q(\lambda) = -\frac{1}{2\pi j}(\bar{\eta}_{e2} - \bar{\eta}_{e1})\sin\beta\, K_-(\bar{\eta}_{e1}\sin\beta, \lambda)\, K_-(\bar{\eta}_{e1}\sin\beta, \lambda_0)
$$

$$
\cdot K_+(\bar{\eta}_{e2}\sin\beta, \lambda)\, K_+(\bar{\eta}_{e2}\sin\beta, \lambda_0)\left(\frac{h_y}{\lambda + \lambda_0} + C_1\right) \qquad (3.150)
$$

where C_1 is a constant still to be determined and, from (3.27),

$$
K_+(\bar{\eta}\sin\beta, \cos\phi_0)\, K_-(\bar{\eta}\sin\beta, \cos\phi_0) = \frac{\sin\phi_0}{1 + \bar{\eta}\sin\beta\sin\phi_0}
$$

Following a similar procedure for the E_y component we find that $P(\lambda)$ satisfies the dual integral equations

$$\int_{-\infty}^{\infty}\left(\frac{\sin\beta}{\bar{\eta}_{e1}}+\frac{1}{\sqrt{1-\lambda^2}}\right)P(\lambda)\,e^{-j\kappa\lambda x}\,d\lambda = 0 \qquad x < 0 \qquad (3.151)$$

$$\int_{-\infty}^{\infty}\left(\frac{\sin\beta}{\bar{\eta}_{e2}}+\frac{1}{\sqrt{1-\lambda^2}}\right)P(\lambda)\,e^{-j\kappa\lambda x}\,d\lambda =$$

$$e_y\left(\frac{1}{\bar{\eta}_{e2}}-\frac{1}{\bar{\eta}_{e1}}\right)\frac{\sin\beta\sin\phi_0}{1+(1/\bar{\eta}_{e1})\sin\beta\sin\phi_0}e^{j\kappa x\lambda_0} \qquad x > 0 \qquad (3.152)$$

which are the electromagnetic duals of (3.148) and (3.149) if $\bar{\eta}_{e1}$ and $\bar{\eta}_{e2}$ are treated as impedances. Accordingly,

$$P(\lambda) = -\frac{1}{2\pi j}\sin\beta\left(\frac{1}{\bar{\eta}_{e2}}-\frac{1}{\bar{\eta}_{e1}}\right)K_-\left(\frac{\sin\beta}{\bar{\eta}_{e1}},\lambda\right)K_-\left(\frac{\sin\beta}{\bar{\eta}_{e1}},\lambda_0\right)$$

$$\cdot K_+\left(\frac{\sin\beta}{\bar{\eta}_{e2}},\lambda\right)K_+\left(\frac{\sin\beta}{\bar{\eta}_{e2}},\lambda_0\right)\left(\frac{e_y}{\lambda+\lambda_0}+C_2\right) \qquad (3.153)$$

where C_2 is again a constant to be determined.

The meaning of the constants C_1 and C_2 becomes apparent when we use Maxwell's equations

$$\left(k_0^2+\frac{\partial^2}{\partial y^2}\right)E_x^s = k_0^2\cos\beta\,Z_0 H_y^s+\frac{\partial^2}{\partial x\,\partial y}E_y^s \qquad (3.154)$$

$$\left(k_0^2+\frac{\partial^2}{\partial y^2}\right)E_z^s = -jk_0\left[Z_0\frac{\partial H_y^s}{\partial x}+\cos\beta\frac{\partial E_y^s}{\partial y}\right] \qquad (3.155)$$

to obtain the tangential field components from the normal field components. By expressing (3.154) and (3.155) in terms of the angular spectra of the field and then inverse transforming we get

$$E_x^s = \int_{-\infty}^{\infty}\frac{\cos\beta\,Q(\lambda)-\sin^2\beta\,\lambda\sqrt{1-\lambda^2}\,P(\lambda)}{(\cos^2\beta+\lambda^2\sin^2\beta)\sqrt{1-\lambda^2}}e^{-j\kappa(\lambda x+|y|\sqrt{1-\lambda^2})}\,d\lambda \quad (3.156)$$

$$E_y^s = -\sin\beta\int_{-\infty}^{\infty}\frac{\lambda Q(\lambda)+\cos\beta\sqrt{1-\lambda^2}\,P(\lambda)}{(\cos^2\beta+\lambda^2\sin^2\beta)\sqrt{1-\lambda^2}}e^{-j\kappa(\lambda x+|y|\sqrt{1-\lambda^2})}\,d\lambda \quad (3.157)$$

In the evaluation of these integrals the poles $\lambda = \pm j\cot\beta$ may be captured and, since these give rise to non-physical (exponentially growing) waves, their residues must be zero. Using either (3.156) or (3.157), the resulting conditions are

$$P(\pm j\cot\beta) = \mp j\,Q(\pm j\cot\beta) \qquad (3.158)$$

which are sufficient to specify $C_{1,2}$. Solving for the constants and carrying out the steepest descent evaluation of the integrals (3.146) and (3.147) yields

$$E_y^d \sim \frac{e^{-jk_0 s}}{\sqrt{s}}D_{Ee}(\phi,\phi_0,\bar{\eta}_{e1},\bar{\eta}_{e2}) \qquad (3.159)$$

$$Z_0 H_y^{\rm d} \sim \frac{e^{-jk_0 s}}{\sqrt{s}} D_{\rm He}(\phi, \phi_0, \bar{\eta}_{\rm e1}, \bar{\eta}_{\rm e2}) \tag{3.160}$$

where $s = z \cos\beta + \rho \sin\beta$ (and $\rho = s \sin\beta$) is the measurement distance from the diffraction point $Q_{\rm E}$ (see Fig. 3–22). The diffraction coefficients $D_{\rm Ee}$ and $D_{\rm He}$ are (SENIOR, 1991)

$$
\begin{aligned}
D_{\rm Ee}(\phi, \phi_0, \bar{\eta}_{\rm e1}, \bar{\eta}_{\rm e2}) &= \pm\frac{e^{-j\pi/4}}{\sqrt{2\pi k_0}}(1 - \sin^2\beta \sin^2\phi_0)^{-1} \\
&\quad \cdot \left[e_y\, U\left(\frac{1}{\bar{\eta}_{\rm e1}}, \frac{1}{\bar{\eta}_{\rm e2}}\right) - h_y\, V\left(\frac{1}{\bar{\eta}_{\rm e1}}, \frac{1}{\bar{\eta}_{\rm e2}}\right) \right] \tag{3.161}
\end{aligned}
$$

$$
\begin{aligned}
D_{\rm He}(\phi, \phi_0, \bar{\eta}_{\rm e1}, \bar{\eta}_{\rm e2}) &= \frac{e^{-j\pi/4}}{\sqrt{2\pi k_0}}(1 - \sin^2\beta \sin^2\phi_0)^{-1} \\
&\quad \cdot \left[h_y\, U(\bar{\eta}_{\rm e1}, \bar{\eta}_{\rm e2}) + e_y\, V(\bar{\eta}_{\rm e1}, \bar{\eta}_{\rm e2}) \right] \tag{3.162}
\end{aligned}
$$

in which

$$
\begin{aligned}
U(\bar{\eta}_{\rm e1}, \bar{\eta}_{\rm e2}) &= \\
(\bar{\eta}_{\rm e1} &- \bar{\eta}_{\rm e2}) \left\{ \frac{\cos^2\beta + \sin^2\beta \cos^2\phi_0}{\cos\phi + \cos\phi_0} - \sin\beta \cos\beta\,(\tan\delta + \tan\beta \cos\phi_0) \right\} \\
&\quad \cdot K_-(\bar{\eta}_{\rm e1} \sin\beta, \cos\phi)\, K_-(\bar{\eta}_{\rm e1} \sin\beta, \cos\phi_0) \\
&\quad \cdot K_+(\bar{\eta}_{\rm e2} \sin\beta, \cos\phi)\, K_+(\bar{\eta}_{\rm e2} \sin\beta, \cos\phi_0) \tag{3.163}
\end{aligned}
$$

$$
\begin{aligned}
V(\bar{\eta}_{\rm e1}, \bar{\eta}_{\rm e2}) &= -(\bar{\eta}_{\rm e1} - \bar{\eta}_{\rm e2}) \frac{\sin\beta \cos\beta}{\cos\delta\, N(\bar{\eta}_{\rm e1}, \bar{\eta}_{\rm e2})} \\
&\quad \cdot K_-(\bar{\eta}_{\rm e1} \sin\beta, \cos\phi)\, K_-(\bar{\eta}_{\rm e1} \sin\beta, \cos\phi_0) \\
&\quad \cdot K_+(\bar{\eta}_{\rm e2} \sin\beta, \cos\phi)\, K_+(\bar{\eta}_{\rm e2} \sin\beta, \cos\phi_0) \tag{3.164}
\end{aligned}
$$

$$
e^{j\delta} = e^{j\,\delta(\bar{\eta}_{\rm e1}, \bar{\eta}_{\rm e2})} = \sqrt{\bar{\eta}_{\rm e1}\bar{\eta}_{\rm e2}}\, \frac{K_-(\bar{\eta}_{\rm e1} \sin\beta, \cos\vartheta)\, K_+(\bar{\eta}_{\rm e2} \sin\beta, \cos\vartheta)}{K_-\left(\dfrac{\sin\beta}{\bar{\eta}_{\rm e1}}, \cos\vartheta\right) K_+\left(\dfrac{\sin\beta}{\bar{\eta}_{\rm e2}}, \cos\vartheta\right)} \tag{3.165}
$$

$$
N(\bar{\eta}_{\rm e1}, \bar{\eta}_{\rm e2}) = \sqrt{\bar{\eta}_{\rm e1}\bar{\eta}_{\rm e2}}\, \frac{K_-(\bar{\eta}_{\rm e1} \sin\beta, \cos\phi_0)\, K_+(\bar{\eta}_{\rm e2} \sin\beta, \cos\phi_0)}{K_-\left(\dfrac{\sin\beta}{\bar{\eta}_{\rm e1}}, \cos\phi_0\right) K_+\left(\dfrac{\sin\beta}{\bar{\eta}_{\rm e2}}, \cos\phi_0\right)} \tag{3.166}
$$

and

$$\vartheta = \frac{\pi}{2} + j \ln\left(\tan\frac{\beta}{2}\right) \tag{3.167}$$

so that $\vartheta = \pi/2$ when $\beta = \pi/2$.

By virtue of the definitions for the split functions, it follows that

$$\frac{1}{N(\bar{\eta}_{e1}, \bar{\eta}_{e2})} = N\left(\frac{1}{\bar{\eta}_{e1}}, \frac{1}{\bar{\eta}_{e2}}\right) \tag{3.168}$$

and

$$\delta\left(\frac{1}{\bar{\eta}_{e1}}, \frac{1}{\bar{\eta}_{e2}}\right) = -\delta(\bar{\eta}_{e1}, \bar{\eta}_{e2}) \tag{3.169}$$

Also, as $\bar{\eta}_{e1} \to \infty$, implying a resistive half-plane, we find that

$$U(\infty, \bar{\eta}_{e2}) = \left\{ \frac{\cos^2 \beta - \sin^2 \beta \cos \phi \cos \phi_0}{\sin \beta (\cos \phi + \cos \phi_0)} + \cos \beta \frac{\cos \beta + \sin 2\gamma}{\sin \beta + \cos 2\gamma} \right\}$$

$$\cdot K_+(\bar{\eta}_{e2} \sin \beta, \cos \phi) \, K_+(\bar{\eta}_{e2} \sin \beta, \cos \phi_0) \tag{3.170}$$

and

$$V(\infty, \bar{\eta}_{e2}) = -\sqrt{\frac{2 \sin \beta}{\bar{\eta}_{e2}}} \frac{\cos \beta \cos \dfrac{\phi_0}{2}}{\cos\left(\dfrac{\pi}{4} - \dfrac{\beta}{2} + \gamma\right)} K_+(\bar{\eta}_{e2} \sin \beta, \cos \phi) \, K_+\left(\frac{\sin \beta}{\bar{\eta}_{e2}}, \cos \phi_0\right) \tag{3.171}$$

where

$$\gamma = -\left(\frac{\pi}{4} - \frac{\beta}{2}\right) - \delta(\infty, \bar{\eta}_{e2}) \tag{3.172}$$

is the variable used by SENIOR (1975b) and VOLAKIS (1986) in their analyses of the impedance half-plane at skew incidence. In deriving (3.170) from (3.163), we used (3.33) and the simplification

$$\tan \delta(\infty, \bar{\eta}_{e2}) = \frac{\sin 2\delta}{1 + \cos 2\delta} = -\frac{\cos(\beta - 2\gamma)}{1 + \sin(\beta - 2\gamma)} = -\frac{\cos \beta + \sin 2\gamma}{\sin \beta + \cos 2\gamma}$$

As discussed in Section 3.6, the diffraction coefficients for impedance half-planes and junctions can be obtained directly from those for the resistive and conductive junctions. The conductive junction diffraction coefficients can be obtained from (3.161) and (3.162) by invoking duality. Specifically, we have

$$D_{Em}(\phi, \phi_0, \bar{\eta}_{m1}, \bar{\eta}_{m2}) = \frac{e^{-j\pi/4}}{\sqrt{2\pi k_0}} (1 - \sin^2 \beta \sin^2 \phi_0)^{-1}$$

$$\cdot \left[e_y U\left(\frac{1}{\bar{\eta}_{m1}}, \frac{1}{\bar{\eta}_{m2}}\right) - h_y V\left(\frac{1}{\bar{\eta}_{m1}}, \frac{1}{\bar{\eta}_{m2}}\right) \right] \tag{3.173}$$

$$D_{Hm}(\phi, \phi_0, \bar{\eta}_{m1}, \bar{\eta}_{m2}) = \pm \frac{e^{-j\pi/4}}{\sqrt{2\pi k_0}} (1 - \sin^2 \beta \sin^2 \phi_0)^{-1}$$

$$\cdot \left[h_y U(\bar{\eta}_{m1}, \bar{\eta}_{m2}) + e_y V(\bar{\eta}_{m1}, \bar{\eta}_{m2}) \right] \tag{3.174}$$

The corresponding diffraction coefficients for the impedance junction are then (see (3.130))

$$
\begin{aligned}
D_{\mathrm{E}}(\phi, \phi_0, \bar\eta_1, \bar\eta_2) &= D_{\mathrm{Ee}}(\phi, \phi_0, \bar\eta_1, \bar\eta_2) + D_{\mathrm{Em}}(\phi, \phi_0, \bar\eta_1, \bar\eta_2) \\
D_{\mathrm{H}}(\phi, \phi_0, \bar\eta_1, \bar\eta_2) &= D_{\mathrm{He}}(\phi, \phi_0, \bar\eta_1, \bar\eta_2) + D_{\mathrm{Hm}}(\phi, \phi_0, \bar\eta_1, \bar\eta_2)
\end{aligned}
$$

where, as before, $\bar\eta_{1,2}$ denote the normalised surface impedances of the left and right hand sides of the junction, respectively. When $\bar\eta_1 \to \infty$, the impedance junction becomes an impedance half-plane and in this case the diffraction coefficients reduce to

$$
D_{\mathrm{E}}(\phi, \phi_0, \infty, \bar\eta) =
$$
$$
\frac{e^{-j\pi/4}}{\sqrt{2\pi k_0}}(1 - \sin^2\beta \sin^2\phi_0)^{-1}\left[e_y\, U_{\mathrm{I}}\left(\frac{1}{\bar\eta}\right) - h_y\, V_{\mathrm{I}}\left(\frac{1}{\bar\eta}\right)\right] \quad (3.175)
$$
$$
D_{\mathrm{H}}(\phi, \phi_0, \infty, \bar\eta) =
$$
$$
\frac{e^{-j\pi/4}}{\sqrt{2\pi k_0}}(1 - \sin^2\beta \sin^2\phi_0)^{-1}\left[h_y\, U_{\mathrm{I}}(\bar\eta) + e_y\, V_{\mathrm{I}}(\bar\eta)\right] \quad (3.176)
$$

where

$$
U_{\mathrm{I}}(\bar\eta) =
$$
$$
\left\{
\frac{\left(\cos^2\beta - \sin^2\beta \cos\phi \cos\phi_0\right)\left(1 - 2\bar\eta \sin\beta \cos\dfrac{\phi}{2}\cos\dfrac{\phi_0}{2}\right)}{\sin\beta(\cos\phi + \cos\phi_0)}
\right.
$$
$$
\left.
+ \frac{\cos\beta\left[\cos\beta + \sin 2\gamma + \left(2\bar\eta \sin\beta \cos\dfrac{\phi}{2}\cos\dfrac{\phi_0}{2}\right)(\cos\beta - \sin 2\gamma)\right]}{\sin\beta + \cos 2\gamma}
\right\}
$$
$$
\cdot K_+(\bar\eta \sin\beta, \cos\phi)\, K_+(\bar\eta \sin\beta, \cos\phi_0) \quad (3.177)
$$

and

$$
V_{\mathrm{I}}(\bar\eta) = \cos\beta\sqrt{2\sin\beta}\left[\frac{\sqrt{\bar\eta}\cos\dfrac{\phi}{2}}{\cos\left(\dfrac{\pi}{4} - \dfrac{\beta}{2} - \gamma\right)} - \frac{\cos\dfrac{\phi_0}{2}}{\sqrt{\bar\eta}\cos\left(\dfrac{\pi}{4} - \dfrac{\beta}{2} + \gamma\right)}\right]
$$
$$
\cdot K_+(\bar\eta \sin\beta, \cos\phi)\, K_+\left(\frac{\sin\beta}{\bar\eta}\cos\phi_0\right) \quad (3.178)
$$

These results were derived by SENIOR (1975b) (see also VOLAKIS (1986) for a correction to the U_{I} function).

So far we have only given the expressions for the normal components of the diffracted field, but in general we are interested in the complete vector fields.

These can be obtained by substituting (3.161) and (3.162) into (3.154) and (3.155). Doing so, we find

$$\mathbf{E}^{\mathrm{d}} = -\frac{e^{-jk_0 s}}{\sqrt{s}} A_2(\phi, \phi_0) \left\{ \left[e_y U \left(\frac{1}{\bar{\eta}_{e1}}, \frac{1}{\bar{\eta}_{e2}} \right) - h_y V \left(\frac{1}{\bar{\eta}_{e1}}, \frac{1}{\bar{\eta}_{e2}} \right) \right] \hat{s} \times \hat{s} \times \hat{y} \right.$$

$$\left. + \left[h_y U \left(\bar{\eta}_{e1}, \bar{\eta}_{e2} \right) + e_y V \left(\bar{\eta}_{e1}, \bar{\eta}_{e2} \right) \right] \hat{s} \times \hat{y} \right\} \qquad (3.179)$$

where

$$A_2(\phi, \phi_0) = \frac{e^{-j\pi/4}}{\sqrt{2\pi k}} (1 - \sin^2 \beta \sin^2 \phi_0)^{-1} (1 - \sin^2 \beta \sin^2 \phi)^{-1} \qquad (3.180)$$

and from Fig. 3–21

$$\hat{s} = \hat{x} \sin \beta \cos \phi + \hat{y} \sin \beta \sin \phi + \hat{z} \cos \beta$$

Expression (3.179) is quite compact but is rather difficult to use. In practice, it is more convenient to express the field in the ray-fixed coordinate system introduced by KOUYOUMJIAN AND PATHAK (1974). By using ray-fixed coordinates, the incident and diffracted fields can be resolved into components parallel and perpendicular to the planes of incidence and diffraction, and only a 2 × 2 diffraction matrix is needed for the complete specification of the field.

To rewrite (3.179) in ray-fixed coordinates, we introduce the edge-fixed planes of incidence and diffraction. The first contains the unit tangent to the edge $\hat{e} = \hat{z}$ at Q_E and the incident ray, and the latter contains the diffracted ray and $\hat{e}$. These planes are shaded in Fig. 3–22. The unit vectors parallel to the planes of incidence and diffraction are

$$\begin{aligned} \hat{\beta}^{\mathrm{i}} &= \hat{\imath} \times \hat{\phi}^{\mathrm{i}} \\ \hat{\beta} &= \hat{s} \times \hat{\phi} \end{aligned} \qquad (3.181)$$

where $\hat{\phi}$ and $\hat{\phi}^{\mathrm{i}}$ are unit vectors perpendicular to their respective edge-fixed planes. From geometrical considerations they are

$$\begin{aligned} \hat{\phi}^{\mathrm{i}} &= \hat{\imath} \times \hat{e}/\sin \beta \\ \hat{\phi} &= \hat{s} \times \hat{e}/\sin \beta \end{aligned} \qquad (3.182)$$

Using the edge-fixed coordinates, the diffracted field (3.179) can be rewritten as (VOLAKIS, 1986)

$$\left\{ \begin{array}{c} E_\beta^{\mathrm{d}} \\ E_\phi^{\mathrm{d}} \end{array} \right\} = -A_1(s)$$

$$\cdot \left[\begin{array}{c:c} D_{\mathrm{CO}}(\phi, \phi_0, \bar{\eta}_{e1}, \bar{\eta}_{e2}) & D_{\mathrm{X}} \left(\phi, \phi_0, \dfrac{1}{\bar{\eta}_{e1}}, \dfrac{1}{\bar{\eta}_{e2}} \right) \\ \hdashline -D_{\mathrm{X}}(\phi, \phi_0, \bar{\eta}_{e1}, \bar{\eta}_{e2}) & D_{\mathrm{CO}} \left(\phi, \phi_0, \dfrac{1}{\bar{\eta}_{e1}}, \dfrac{1}{\bar{\eta}_{e2}} \right) \end{array} \right] \left\{ \begin{array}{c} E_{\beta^{\mathrm{i}}}^{\mathrm{i}} \\ E_{\phi^{\mathrm{i}}}^{\mathrm{i}} \end{array} \right\} \qquad (3.183)$$

where $\mathbf{E}^i = \hat{\beta}^i E^i_{\beta^i} + \hat{\phi}^i E^i_{\phi^i}$ is the incident field evaluated at Q_E and $\mathbf{E}^d = \hat{\beta} E_\beta + \hat{\phi} E_\phi$ is the diffracted field evaluated at P. The components of the diffraction matrix are given by

$$D_{CO}(\phi, \phi_0, \bar{\eta}_{e1}, \bar{\eta}_{e2}) = A_2(\phi, \phi_0) \left[E\left(\bar{\eta}_{e1}, \bar{\eta}_{e2}\right) + \cos\beta \, F\left(\frac{1}{\bar{\eta}_{e1}}, \frac{1}{\bar{\eta}_{e2}}\right) \right] \quad (3.184)$$

$$D_X(\phi, \phi_0, \bar{\eta}_{e1}, \bar{\eta}_{e2}) = A_2(\phi, \phi_0) \left[G\left(\bar{\eta}_{e1}, \bar{\eta}_{e2}\right) + \cos\beta \, H\left(\frac{1}{\bar{\eta}_{e1}}, \frac{1}{\bar{\eta}_{e2}}\right) \right] \quad (3.185)$$

with

$$E(\bar{\eta}_{e1}, \bar{\eta}_{e2}) =$$
$$- \left[\cos\phi \cos\phi_0 \, U\left(\bar{\eta}_{e1}, \bar{\eta}_{e2}\right) - \cos\beta \sin\phi \cos\phi_0 \, V\left(\frac{1}{\bar{\eta}_{e1}}, \frac{1}{\bar{\eta}_{e2}}\right) \right]$$

$$G(\bar{\eta}_{e1}, \bar{\eta}_{e2}) =$$
$$- \left[\cos\beta \sin\phi \cos\phi_0 \, U\left(\bar{\eta}_{e1}, \bar{\eta}_{e2}\right) + \cos\phi \cos\phi_0 \, V\left(\frac{1}{\bar{\eta}_{e1}}, \frac{1}{\bar{\eta}_{e2}}\right) \right]$$

$$F\left(\frac{1}{\bar{\eta}_{e1}}, \frac{1}{\bar{\eta}_{e2}}\right) =$$
$$- \left[\cos\beta \sin\phi \sin\phi_0 \, U\left(\frac{1}{\bar{\eta}_{e1}}, \frac{1}{\bar{\eta}_{e2}}\right) + \cos\phi \sin\phi_0 \, V\left(\bar{\eta}_{e1}, \bar{\eta}_{e2}\right) \right]$$

$$H\left(\frac{1}{\bar{\eta}_{e1}}, \frac{1}{\bar{\eta}_{e2}}\right) =$$
$$- \left[-\cos\phi \sin\phi_0 \, U\left(\frac{1}{\bar{\eta}_{e1}}, \frac{1}{\bar{\eta}_{e2}}\right) + \cos\beta \sin\phi \sin\phi_0 \, V\left(\bar{\eta}_{e1}, \bar{\eta}_{e2}\right) \right]$$

The spread (and phase) factor

$$A_1(s) = \sqrt{\frac{\rho_c}{s(\rho_c + s)}} \, e^{-jks} \quad (3.186)$$

was introduced so that (3.183) is also applicable to junctions and edges that have a curvature (see Fig. 3–23). In (3.186) the parameter ρ_c is the distance between the two caustics of the diffracted ray tube (one being at the edge). It is given by (KOUYOUMJIAN AND PATHAK, 1974)

$$\frac{1}{\rho_c} = \frac{1}{\rho^i_e} - \frac{\hat{n}_e \cdot (\hat{i} - \hat{s})}{a \sin^2\beta}$$

where a is the radius of curvature of the edge, $\hat{n}_e$ is the unit vector normal to the edge, in the plane of the edge and directed away from the centre of curvature, and ρ^i_e is the radius of curvature of the incident wave in the plane of incidence. For plane wave incidence, $\rho^i_e = \infty$ and if, in addition, the edge or junction is straight, then $a \to \infty$, and consequently $A_1(s)$ reduces to $e^{-jk_0 s}/s$.

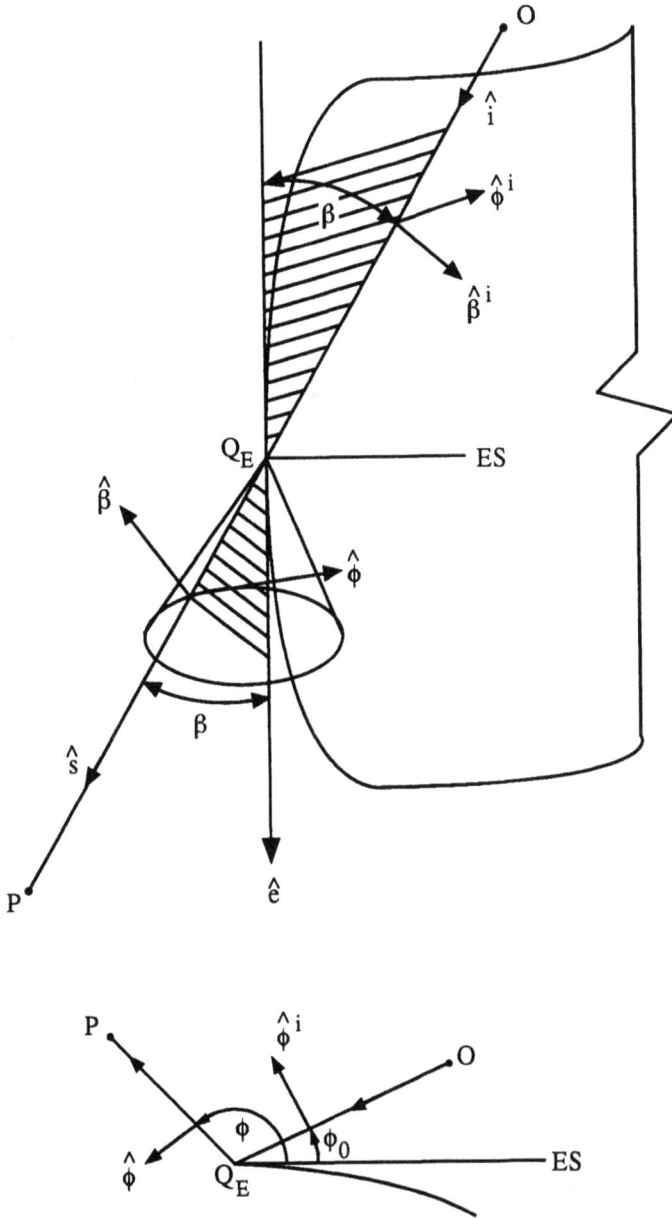

Figure 3–23: *Geometry of a curved edge illuminated at skew incidence*

Uniform expressions for the dyadic diffraction coefficient in (3.183) were derived by VOLAKIS (1986) and later by ROJAS (1988), who also included surface wave contributions. They can be obtained using the procedure described in Appendix C, but the expressions are rather lengthy, and the reader is referred to the references for the explicit forms.

3.9   Tapered impedance junctions

The solutions presented in this chapter are applicable to SIBC and STC junctions, where both sides of the junction have a constant resistivity or impedance. However, analytical solutions to planar junction problems are possible whenever the resistivity/impedance is linearly varying as illustrated in Fig. 3–24. In

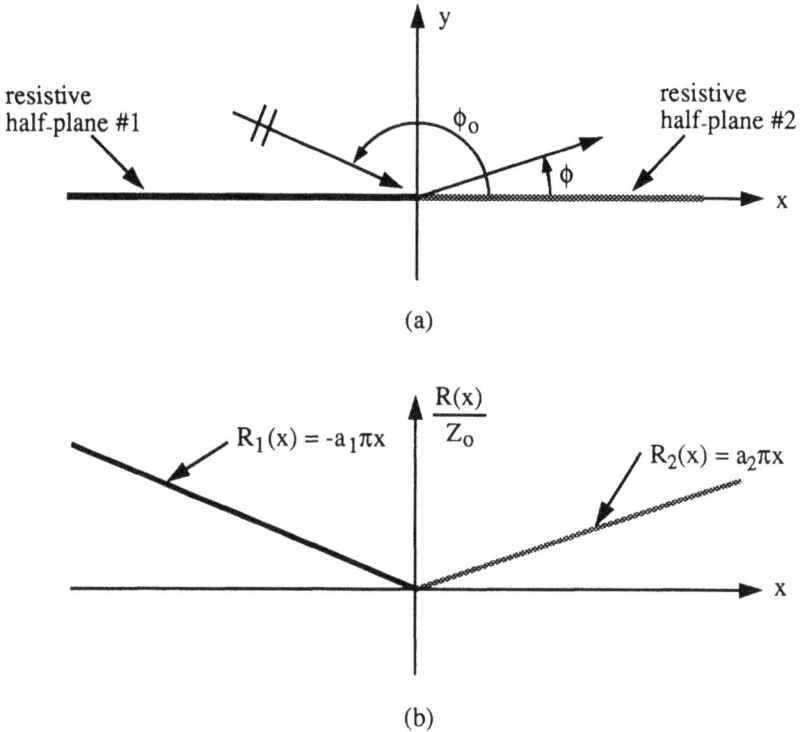

(a)

(b)

Figure 3–24: *Geometry (a) and resistivity profiles (b) for two abutting coplanar resistive half-planes*

addition, the resistivity/impedance of the half-planes must be increasing away from the junction for E-polarisation and decreasing in the same direction for H-polarisation. This is necessary for the separability of the equations obtained upon application of the transition/boundary conditions.

The first solution to problems involving varying impedance walls was given by FELSEN AND MARCUVITZ (1973), who examined the diffraction by a wedge with linearly varying impedance boundary conditions on the faces. When the wedge angle is zero, this solution can of course be specialised to the case of half-planes whose face impedances are linearly varying away from the edge. The diffraction by an isolated resistive half-plane with a linearly varying resistivity was considered by YANG ET AL. (1988) who employed the Kontorovich-Lebedev (1939) transform to solve the wave equation subject to the pertinent sheet conditions. KEMPEL ET AL. (1993) generalised this solution to obtain closed form diffraction coefficients for the resistive junction illustrated in Fig. 3–24. The results in Fig. 3–25 are due to KEMPEL ET AL. (1993) and refer to a junction

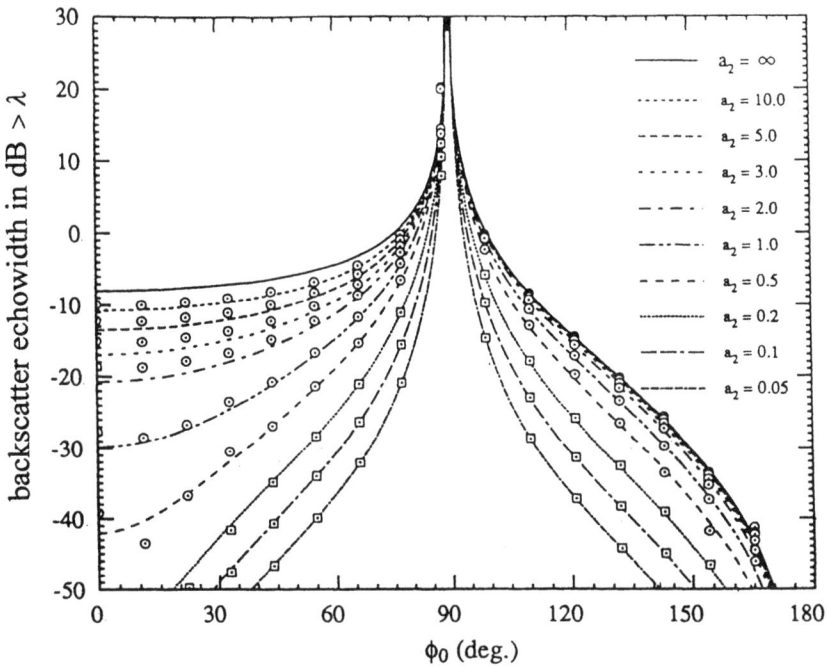

Figure 3–25: *Parametric curves for the backscatter echowidth of the tapered resistive junction shown in Fig. 3–24 with $a_1 = 0$: numerical model $\odot\odot$; PO approximation $\boxdot\boxdot$*

whose left side is a metallic half-plane whereas the right side is a resistive half-plane of resistivity $R_2 = a_2\pi x$, with x measured from the junction point. From the parametric curves in Fig. 3–25 and illustrated comparison data (shown as symbols), it is concluded that the physical optics (PO) approximation can yield accurate results when the resistivity is slowly varying away from the junction (typically for $a_2 < 0.3$). Not surprisingly, the curves in Fig. 3–25 show that the echowidth of the junction decreases as the slope of the half-plane resistivity becomes smaller, implying less contrast between the right and left half-planes.

We close this chapter by noting that solutions to diffraction problems involving resistive half-planes in the presence of another reflecting surface are also possible. In this case the dual integral equation method can be employed with minor modifications. The problem of diffraction by a resistive half-plane placed at some distance above a uniform resistive sheet has been solved in this manner (NATZKE AND VOLAKIS, 1993). Also, COLLINS AND VOLAKIS (1990) considered diffraction by a resistive half-plane on a dielectric interface (see Fig. 3–18), and the special case of a metallic half-plane residing on a dielectric interface (see CLEMMOW (1953)) was considered by COBLIN AND PEARSON (1984) using the Wiener-Hopf method.

References

Booker, H. G. and Clemmow, P. C. (1950), "On the concept of an angular spectrum of plane waves and its relation to that of polar diagram and aperture distribution", *Proc. IEE*, Pt. III, **97**, pp. 11–17.

Carlson, J. F. and Heins, A. E. (1947), "The reflection of an electromagnetic plane wave by an infinite set of plates", *Quart. Appl. Math.*, **4**, pp. 313–329.

Clemmow, P. C. (1950), "Some extensions of the method of integration by steepest descent", *Quart. J. Mech. Appl. Math.*, **3**, pp. 241–256.

Clemmow, P. C. (1951), "A method for the exact solution of a class of two-dimensional diffraction problems", *Proc. Roy. Soc. London, Ser. A*, **205**, pp. 286–308.

Clemmow, P. C. (1953), "Radio wave propagation over a flat earth across a boundary separating two different media", *Phil. Trans. Roy. Soc. London*, **246-A**, pp. 1–55.

Clemmow, P. C. (1966), *The Plane Wave Spectrum Representation of Electromagnetic Fields*, Pergamon Press, New York.

Coblin, R. D. and Pearson, L. W. (1984), "A geometrical theory of diffraction for a half plane residing on a dielectric interface between dissimilar media: transverse magnetic polarization", *Radio Sci.*, **19**, pp. 1277–1288.

Collins, J. and Volakis, J. L. (1990), "Electromagnetic scattering from a resistive half plane on a dielectric interface", *Wave Motion*, **12**, pp. 81–96.

Copson, E. T. (1946), "On an integral equation arising in the theory of diffraction", *Quart. J. Math.*, **17**, pp. 19–34.

Felsen, L. B. and Marcuvitz, N. (1973), *Radiation and Scattering of Waves*, Prentice Hall, Englewood Cliffs, NJ, pp. 674 *et seq.*

Gennarelli, C. and Palumbo, L. (1984), "A uniform asymptotic expansion of typical diffraction integrals with many coalescing simple pole singularities and a first order saddle point", *IEEE Trans. Antennas Propagat.*, **AP-32**, pp. 1122–1124.

Herman, M. I. and Volakis, J. L. (1987), "High frequency scattering by a resistive strip and extensions to conductive and impedance strips", *Radio Sci.*, **22**, pp. 335–349.

Kempel, L. C., Volakis, J. L. and Senior, T. B. A. (1993), "Transverse magnetic diffraction from tapered resistive junctions", *Radio Sci.*, **28**, pp. 129–138.

Kobayashi, K. (1990), "Wiener-Hopf method and modified residue calculus techniques", in *Analysis Methods for Electromagnetic Wave Problems*, (E. Yamashita, Ed.), Artech House, Norwood, Ch. 8.

Kontorovich, M. J. and Lebedev, N. N. (1939), "On the method of solution of some problems of the diffraction theory", *J. Phys. USSR*, **1**, pp. 229–241.

Kouyoumjian, R. G. and Pathak, P. H. (1974), "A uniform geometrical theory of diffraction for an edge in a perfectly conducting surface", *Proc. IEEE*, **62**, pp. 1448–1461.

Lee, S.-W. and Deschamps, G. A. (1976), "A uniform asymptotic theory of electromagnetic diffraction by a curved wedge", *IEEE Trans. Antennas Propagat.*, **AP-24**, pp. 25–34.

Maliuzhinets, G. D. (1958), "Excitation, reflection and emission of surface waves from a wedge with given face impedances", *Soviet Phys. Dokl.*, **3**, pp. 752–755 (transl. of *Dokl. Akad. Nauk. SSSR*, **121**, pp. 436–439).

Miles, J. W. (1949), "On the diffraction of an electromagnetic wave through a plane screen", *J. Appl. Phys.*, **20**, pp. 760–771. See also errata, **21**, p. 468 (1950).

Mittra, R. and Lee, S.-W. (1971), *Analytical Techniques in the Theory of Guided Waves*, Macmillan Co., New York.

Natzke, J. R. and Volakis, J. L. (1993), "Characterization of a resistive half plane over a resistive sheet", *IEEE Trans. Antennas Propagat.*, **41**, pp. 1063–1068.

Noble, B. (1958), *Methods Based on the Wiener-Hopf Technique*, Pergamon Press, New York.

Osipov, A. V. (1990), "Calculation of the Maliuzhinets function in a complex region", *Sov. Phys. Acoust.*, **36**, pp. 63–66 (English trans.).

Pathak, P. H. and Kouyoumjian, R. G. (1970), "The dyadic diffraction coefficient for a perfectly conducting wedge", Sci. Report 2183-4, ElectroScience Lab., Dept. of Elec. Eng., Ohio State Univ., Columbus.

Ricoy, M. A. and Volakis, J. L. (1989), "E-polarization diffraction by a thick metal-dielectric join", *J. Electromagn. Waves Applics.*, **3**, pp. 383–407.

Rojas, R. G. (1987), "Comparison between two asymptotic methods", *IEEE Trans. Antennas Propagat.*, **AP-35**, pp. 1489–1492.

Rojas, R. G. (1988), "Wiener-Hopf analysis of the EM diffraction by an impedance discontinuity in a planar surface and by an impedance half plane", *IEEE Trans. Antennas Propagat.*, **36**, pp. 71–83.

Sancer, M., personal communication.

Senior, T. B. A. (1952), "Diffraction by a semi-infinite metallic sheet", *Proc. Roy. Soc. London, Ser. A*, **213**, pp. 436–458.

Senior, T. B. A. (1959), "Diffraction by an imperfectly conducting half plane at oblique incidence", *Appl. Sci. Res.*, **B8**, pp. 35–61.

Senior, T. B. A. (1975a), "Half plane edge diffraction", *Radio Sci.*, **10**, pp. 645–650.

Senior, T. B. A. (1975b), "Diffraction tensors for imperfectly conducting edges", *Radio Sci.*, **10**, pp. 911–919.

Senior, T. B. A. (1979), "Backscattering from resistive strips", *IEEE Trans. Antennas Propagat.*, **AP-27**, pp. 808–813.

Senior, T. B. A. (1991), "Skew incidence on a material junction", *Radio Sci.*, **26**, pp. 305–311.

Uzgören, G., Büyükaksoy, A. and Serbest, A. H. (1989), "Diffraction coefficient related to a discontinuity formed by impedance and resistive half planes", *Proc. IEE*, **136**, Pt. H, pp. 19–23.

Van der Waerden, B. L. (1951), "On the method of saddle points", *Appl. Sci. Res.*, **B2**, pp. 33–45.

Volakis, J. L. (1986), "A uniform geometrical theory of diffraction for an imperfectly conducting half plane", *IEEE Trans. Antennas Propagat.*, **AP-34**, pp. 172–180.

Volakis, J. L. and Herman, M. I. (1986), "A uniform asymptotic evaluation of integrals", *Proc. IEEE*, **74**, pp. 1043–1044.

Volakis, J. L. and Senior, T. B. A. (1985), "Simple expressions for a function occurring in diffraction theory", *IEEE Trans. Antennas Propagat.*, **AP-33**, pp. 678–680.

Wiener, N. and Hopf, E. (1931), "Über eine klasse singulärer Integralgleichungen", *Sitz. Berlin, Preuss. Acad. Wiss.*, pp. 696–706.

Williams, W. E. (1960), "Diffraction of an electromagnetic plane wave by a metallic sheet", *Proc. Roy. Soc. London, Ser. A*, **257**, pp. 413–419.

Yang, S.-I., Ra, J. W. and Senior, T. B. A. (1988), "E-polarized scattering by a resistive half plane with linearly varying resistivity", *Radio Sci.*, **23**, pp. 463–469.

Yip, E. and Chiavetta, R. (1987), "Comparison of uniform asymptotic expansions of diffraction integrals", *IEEE Trans. Antennas Propagat.*, **AP-35**, pp. 1179–1180.

Chapter 4

Application to impedance wedges

4.1 Introduction

The half-plane which was the subject of the preceding chapter is the special case of a wedge whose interior angle is zero, but for a complex target such as an aircraft (see Fig. 4-1), a high frequency simulation requires the knowl-

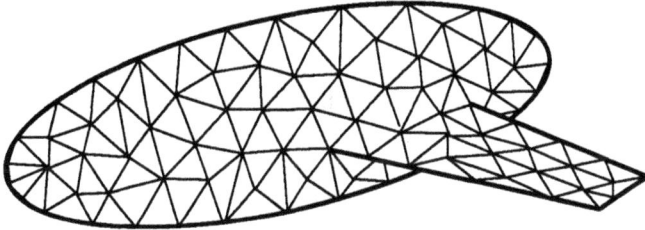

Figure 4–1: *Faceting of an airframe geometry. The junctions between facets are treated as wedges in simulating radar scattering or on-board antenna radiation*

edge of the diffraction coefficient for a wedge of arbitrary angle. Indeed, the diffraction coefficient for a perfectly conducting wedge is at the heart of such well-known techniques as the uniform geometrical theory of diffraction (UTD) (KOUYOUMJIAN AND PATHAK, 1974; HANSEN, 1981; KNOTT ET AL., 1985; Mc-NAMARA ET AL., 1990; PATHAK, 1992) and the physical theory of diffraction (PTD) (UFIMTSEV, 1971; MITZNER, 1974; LEE, 1990), and to extend these to impedance surfaces, it is essential to consider diffraction by the impedance wedge illustrated in Fig. 4-2.

For a plane wave incident in a plane perpendicular to the edge (referred to as *normal* incidence), a method valid for arbitrary (constant) face impedances was developed by MALIUZHINETS (1951, 1958b) and bears his name. The method involves the solution of a first order functional difference equation for a spectral function, and is described in the next section. A uniform

107

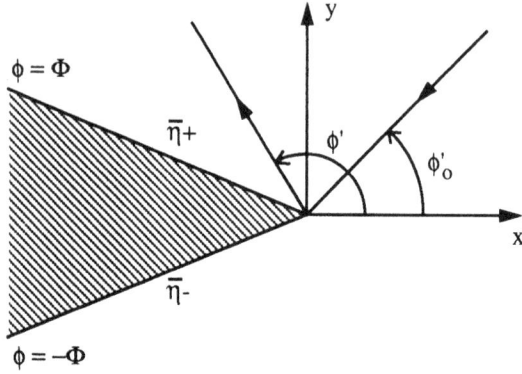

Figure 4–2:   *Wedge geometry and the coordinate system for Maliuzhinets'*
method

expression for the diffraction coefficient (TIBERIO ET AL., 1985; HERMAN AND
VOLAKIS, 1988) is then presented. Unfortunately, the more general case of skew
incidence is much more difficult. Two spectral functions now appear and these
satisfy first order difference equations which are in general coupled. Elimination of either function leads to a second order equation and, as in the case of
a differential equation with variable coefficients, there is no standard method
of solution. To date an exact solution has been obtained only for those wedge
angles and/or impedances for which the difference equations can be decoupled
(VACCARO, 1980; SENIOR, 1986), but an approximate solution is possible which
is here described. We conclude by presenting a PTD formulation for an impedance wedge (SYED AND VOLAKIS, 1992a), along with its application to some
simple bodies.

4.2   Normal incidence

Consider the wedge shown in Fig. 4–2 whose faces satisfy the impedance boundary condition (2.36). In contrast to the coordinates employed in Chapter 3,
the polar angle is now measured from the *continuation* of the plane bisecting
the structure and, to emphasise this fact, the cylindrical polar coordinates are
taken to be ρ, ϕ', z as indicated in Fig. 4–2. Thus, in the particular case of a
half-plane corresponding to a wedge of zero included angle, $\phi' = \pi - \phi$, where
ϕ is the angle in Fig. 3-6. For E-polarisation the boundary condition (2.36)
can be written as

$$\frac{1}{\rho}\frac{\partial E_z}{\partial \phi'} \pm \frac{jk_0}{\bar{\eta}_\pm}E_z = 0 \qquad (4.1)$$

at $\phi' = \pm\Phi$, where $\bar{\eta}_\pm$ are the normalised surface impedances of the two faces,
and we have used the fact that $H_\rho = -Y_0/(jk_0\rho)\,\partial E_z/\partial\phi'$. Similarly, for H-

polarisation we have

$$\frac{1}{\rho}\frac{\partial H_z}{\partial \phi'} \pm jk_0\bar{\eta}_\pm H_z = 0 \tag{4.2}$$

at $\phi' = \pm\Phi$.

4.2.1 Formulation and non-uniform diffraction coefficients

The wedge is illuminated by the plane wave

$$\mathbf{E}^i \text{ or } \mathbf{H}^i = \hat{z}\, e^{jk_0\rho\cos(\phi'-\phi_0')} \tag{4.3}$$

and we seek the scattered field. It is obvious that the problem is a scalar one for E_z or H_z and, since (4.1) and (4.2) are the electromagnetic duals of each other, the corresponding solutions also have this property. It is therefore sufficient to confine attention to E-polarisation.

The first step is to introduce an appropriate representation for the field. The plane wave representation (3.15) is no longer suitable, and MALIUZHINETS (1951, 1958b) chose instead the form used by Sommerfeld (see CARSLAW (1898)) for the analogous problem of a perfectly conducting wedge. Specifically, the total field in the presence of the wedge is written as

$$E_z(\rho,\phi') = \frac{1}{2\pi j} \int_\gamma e^{jk_0\rho\cos\alpha} s(\alpha + \phi')\, d\alpha \tag{4.4}$$

where $\gamma = \gamma_1 + \gamma_2$ is the Sommerfeld contour shown in Fig. 4–3. The two loops γ_1 and γ_2 differ in the sign of α and, if $s(\alpha)$ is analytic in the strip $|\mathrm{Re.}\ \alpha| < \Phi + \epsilon$ for $\epsilon > 0$, it can be verified that E_z satisfies the radiation condition. The key result was proved by MALIUZHINETS (1958a) and is as follows. Given

$$E_z(\rho,\phi') < \frac{e^{b\rho}}{\rho^{1-a}} \tag{4.5}$$

for $\rho > 0$ where $a\ (> 0)$ and b are constants, then within the class of functions $s(\alpha)$ which are analytic on γ and in the region enclosed (except possibly at infinity) and such that

$$s(\alpha) = O\left\{\exp\left[(1 - a)\,|\mathrm{Im.}\ \alpha|\right]\right\} \tag{4.6}$$

as $|\mathrm{Im.}\ \alpha| \to \infty$, there is a unique *odd* function of α which satisfies (4.4) regarded as an integral equation. In particular, if $E_z = 0$, $s(\alpha + \phi')$ must be an *even* function of α. We note that for small $k_0\rho$ (4.5) is simply the edge condition, and if $\eta_\pm \neq 0$ it turns out that $E_z = O(1)$ as $\rho \to 0$, implying $a = 1$. To recover the incident field (4.3), $s(\alpha)$ must have a simple pole at $\alpha = \phi_0'$ with appropriate residue. This does not affect the one-to-one correspondence

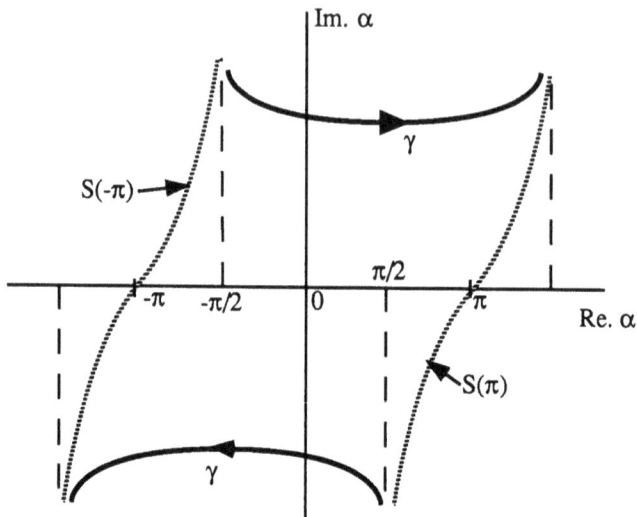

Figure 4–3: *The Sommerfeld contour γ and the steepest descent paths $S(-\pi)$ and $S(\pi)$ needed to close it*

between E_z and $s(\alpha)$ and, apart from this pole, $s(\alpha)$ must be analytic in the strip $|\mathrm{Re.}\ \alpha| < \Phi + \epsilon$.

The spectrum $s(\alpha)$ is determined by the boundary conditions on the wedge faces. On substituting (4.4) into (4.1) and using the fact that

$$
\begin{aligned}
\frac{\partial E_z}{\partial \phi'} &= \frac{1}{2\pi j} \int_\gamma e^{jk_0\rho\cos\alpha} \frac{\partial}{\partial\phi'} s(\alpha + \phi')\, d\alpha \\
&= \frac{1}{2\pi j} \int_\gamma e^{jk_0\rho\cos\alpha} \frac{\partial}{\partial\alpha} s(\alpha + \phi')\, d\alpha \\
&= \frac{1}{2\pi j} \left\{ \left[e^{jk_0\rho\cos\alpha} s(\alpha + \phi') \right]_{3\pi/2+j\infty}^{-\pi/2+j\infty} + \left[e^{jk_0\rho\cos\alpha} s(\alpha + \phi') \right]_{-3\pi/2-j\infty}^{\pi/2-j\infty} \right. \\
&\qquad \left. - \int_\gamma e^{jk_0\rho\cos\alpha}(-jk_0\rho\sin\alpha)\, s(\alpha + \phi')\, d\alpha \right\} \\
&= \frac{k_0\rho}{2\pi} \int_\gamma e^{jk_0\rho\cos\alpha} \sin\alpha\, s(\alpha + \phi')\, d\alpha
\end{aligned}
$$

we find

$$
\int_\gamma e^{jk_0\rho\cos\alpha}(\sin\alpha + \sin\theta_+)\, s(\alpha + \Phi)\, d\alpha = 0 \tag{4.7}
$$

and similarly

$$
\int_\gamma e^{jk_0\rho\cos\alpha}(\sin\alpha - \sin\theta_-)\, s(\alpha - \Phi)\, d\alpha = 0 \tag{4.8}
$$

where

$$\sin \theta_{\pm} = \begin{cases} 1/\bar{\eta}_{\pm} & \text{(E-polarisation)} \\[2ex] \bar{\eta}_{\pm} & \text{(H-polarisation)} \end{cases} \tag{4.9}$$

The necessary and sufficient conditions for these to be satisfied are that the integrands are even functions of α, that is

$$(\sin \alpha \pm \sin \theta_{\pm}) s(\alpha \pm \Phi) - (- \sin \alpha \pm \sin \theta_{\pm}) s(-\alpha \pm \Phi) = 0 \tag{4.10}$$

together with the order condition (4.6) and the analyticity requirement.

To solve the pair of equations (4.10) it is convenient to write

$$s(\alpha) = \sigma(\alpha) \frac{\Psi(\alpha)}{\Psi(\phi_0')} \tag{4.11}$$

where $\sigma(\alpha)$ is chosen to satisfy

$$\sigma(\alpha \pm \Phi) = \sigma(-\alpha \pm \Phi) \tag{4.12}$$

and incorporates the geometrical optics pole at $\alpha = \phi_0'$. The remaining function $\Psi(\alpha)$ is then required to be free of poles and zeros in the strip $|\text{Re. } \alpha| < \Phi + \epsilon$, and plays a similar role to the split function in a Wiener-Hopf analysis. From (4.10)–(4.12)

$$(\sin \alpha \pm \sin \theta_{\pm}) \Psi(\alpha \pm \Phi) - (- \sin \alpha \pm \sin \theta_{\pm}) \Psi(-\alpha \pm \Phi) = 0 \tag{4.13}$$

On replacing α by $\alpha + \Phi$ in the equation with the upper $(+)$ signs, and α by $\alpha - \Phi$ in the other, we obtain

$$\begin{aligned} \{\sin(\alpha + \Phi) + \sin \theta_{+}\} \{\sin(\alpha - \Phi) + \sin \theta_{-}\} \, \Psi(\alpha + 2\Phi) \\ = \{\sin(\alpha + \Phi) - \sin \theta_{+}\} \{\sin(\alpha - \Phi) - \sin \theta_{-}\} \, \Psi(\alpha - 2\Phi) \end{aligned} \tag{4.14}$$

which is a true first order functional difference equation. The construction of the solution $\Psi(\alpha)$ in terms of the Maliuzhinets function $\psi_{\Phi}(\alpha)$ is described in Appendix B.4, where it is shown that

$$\begin{aligned} \Psi(\alpha) &= \psi_{\Phi}\left(\alpha + \Phi + \theta_{+} - \frac{\pi}{2}\right) \psi_{\Phi}\left(\alpha + \Phi - \theta_{+} + \frac{\pi}{2}\right) \\ &\quad \cdot \psi_{\Phi}\left(\alpha - \Phi + \theta_{-} - \frac{\pi}{2}\right) \psi_{\Phi}\left(\alpha - \Phi - \theta_{-} + \frac{\pi}{2}\right) \end{aligned} \tag{4.15}$$

In general

$$\psi_{\Phi}(\alpha) = O\left\{\exp\left(\frac{1}{4n}|\text{Im. } \alpha|\right)\right\} \tag{4.16}$$

implying

$$\Psi(\alpha) = O\left\{\exp\left(\frac{1}{n}|\text{Im. } \alpha|\right)\right\} \tag{4.17}$$

as $|\text{Im. }\alpha| \to \infty$, where $n = 2\Phi/\pi$. Thus, $0 < n \leq 2$, with $n = 2$ corresponding to a half-plane.

The only remaining task is to determine $\sigma(\alpha)$. The most general solution of (4.12) is a rational function of $\sin\frac{\alpha}{n}$, and the only one which satisfies the order and pole requirements is

$$\sigma(\alpha) = c\left(\sin\frac{\alpha}{n} - \sin\frac{\phi'_0}{n}\right)^{-1}$$

where the constant c must be such that when $\sigma(\alpha)$ is substituted into (4.11) and then (4.4), the residue at the pole $\alpha = \phi'_0$ reproduces E^i_z. Hence $c = \frac{1}{n}\cos(\phi'_0/n)$ and the resulting expression for $s(\alpha)$ is

$$s(\alpha) = \frac{\dfrac{1}{n}\cos\dfrac{\phi'_0}{n}}{\sin\dfrac{\alpha}{n} - \sin\dfrac{\phi'_0}{n}} \frac{\Psi(\alpha)}{\Psi(\phi'_0)} \tag{4.18}$$

From (4.17) we observe that $s(\alpha) = O(1)$, implying $E_z = O(1)$ as noted earlier, and this is still true if $\theta_+ = 0$ or $\theta_- = 0$, but not both. However, if $\theta_+ = \frac{\pi}{2} + j\infty$ or $\theta_- = \frac{\pi}{2} + j\infty$ (but not both),

$$\Psi(\alpha) = O\left\{\exp\left(\frac{1}{2n}|\text{Im. }\alpha|\right)\right\}$$

giving

$$s(\alpha) = O\left\{\exp\left(-\frac{1}{2n}|\text{Im. }\alpha|\right)\right\}$$

and thus $E_z = O\{\rho^{1/(2n)}\}$. Finally, if $\theta_+ = \frac{\pi}{2} + j\infty$ and $\theta_- = \frac{\pi}{2} + j\infty$, then $\Psi(\alpha) = O(1)$ and $E_z = O(\rho^{1/n})$. Since $\theta_\pm = \frac{\pi}{2} + j\infty$ implies $\bar{\eta}_\pm = 0$ corresponding to a perfectly conducting wedge for E-polarisation, the edge condition is consistent with the known behaviour of E_z at the edge of a metallic wedge (BOWMAN ET AL., 1987).

When (4.18) is substituted into (4.4), the contour can be closed using the steepest descent paths $S(\pm\pi)$ shown in Fig. 4–3, and the result is

$$E_z(\rho,\phi') = \frac{1}{2\pi j}\int_{S(-\pi)-S(\pi)} e^{jk_0\rho\cos\alpha} \frac{\dfrac{1}{n}\cos\dfrac{\phi'_0}{n}}{\sin\dfrac{\alpha+\phi'}{n} - \sin\dfrac{\phi'_0}{n}} \frac{\Psi(\alpha+\phi')}{\Psi(\phi'_0)}\, d\alpha$$

$$+ 2\pi j \sum \text{Res} \tag{4.19}$$

On the two paths Re. $\alpha = \pm\pi + \text{gd}(\text{Im. }\alpha)$ where gd is the Gudermann function defined in Section 3.3.2, and the residues are those of the poles $\alpha = \alpha_p$ such that

$$-\pi + \text{gd}(\text{Im. }\alpha) < \text{Re. }\alpha_p < \pi + \text{gd}(\text{Im. }\alpha) \tag{4.20}$$

If, for the moment, we ignore these contributions, a simple application of the steepest descent method for $k_0\rho \gg 1$ yields the diffracted field

$$E_z^{\mathrm{d}} = \frac{e^{-jk_0\rho}}{\sqrt{\rho}} \, D^{\mathrm{nu}}(\phi', \phi_0', \theta_+, \theta_-) \tag{4.21}$$

(see (3.45) and (3.63)) where

$$
\begin{aligned}
&D^{\mathrm{nu}}(\phi', \phi_0', \theta_+, \theta_-) \\
&= \frac{e^{-j\pi/4}}{\sqrt{2\pi k_0}} \{ s(\phi' - \pi) - s(\phi' + \pi) \} \\
&= \frac{e^{-j\pi/4}}{n\sqrt{2\pi k_0}} \frac{\cos\dfrac{\phi_0'}{n}}{\Psi(\phi_0')} \left\{ \frac{\Psi(\phi' - \pi)}{\sin\dfrac{\phi' - \pi}{n} - \sin\dfrac{\phi_0'}{n}} - \frac{\Psi(\phi' + \pi)}{\sin\dfrac{\phi' + \pi}{n} - \sin\dfrac{\phi_0'}{n}} \right\}
\end{aligned} \tag{4.22}
$$

is the non-uniform diffraction coefficient.

Sample plots of the backscatter echowidth for E-polarisation computed using (3.47) and (4.22) are shown in Figs. 4–4(a) and 4–5(a). These correspond to the internal wedge angles of 30° ($\Phi = 165°$) and 60° ($\Phi = 150°$) respectively, and the observation angle ϕ is measured from the upper face of the wedge as indicated in Fig. 4–6. As in the case of an impedance half-plane (see Fig. 3–16), the echowidth at angles close to edge-on decreases with increasing $\bar{\eta}$, and for $\bar{\eta} = 1$ there is no infinity at the shadow boundary since $R_\perp^\pm$ is then zero. In contrast, for near grazing incidence ($\phi_0 \lesssim 20°$) the echowidth is higher than that for a perfectly conducting wedge if $\bar{\eta} \geq 1$. The analogous results for H-polarisation are shown in Figs. 4–4(b) and 4–5(b). For near grazing angles of incidence the echowidth now decreases with increasing η and is zero when $\phi = 0$ if $\bar{\eta} \neq 0$, but near edge-on the echowidth may exceed that for a perfectly conducting wedge.

4.2.2   Residue computation

As expected the residues in (4.19) correspond to the geometrical optics (incident and reflected) and surface wave contributions to the field, and from (3.63) we have

$$2\pi j \sum \mathrm{Res} = E_z^{\mathrm{go}} + E_z^{\mathrm{sw}} = E_z^{\mathrm{i}} + E_z^{\mathrm{r}} + E_z^{\mathrm{sw}} \tag{4.23}$$

The poles responsible for E_z^{sw} depend on $\bar{\eta}_\pm$ and are, in general, complex, but those producing E_z^{go} are real and associated with the difference of sines in (4.19). Of the latter, the only ones satisfying (4.20) are

$$\alpha_{\mathrm{P}_1} = -\phi' + \phi_0', \qquad \alpha_{\mathrm{P}_2} = -\phi' - \phi_0' + n\pi, \qquad \alpha_{\mathrm{P}_3} = -\phi' - \phi_0' - n\pi \tag{4.24}$$

where $n\pi = 2\Phi$ as before. None of the others are of interest since they are neither captured nor lie close to the steepest descent paths for $|\phi'|, |\phi_0'| \leq \Phi$.

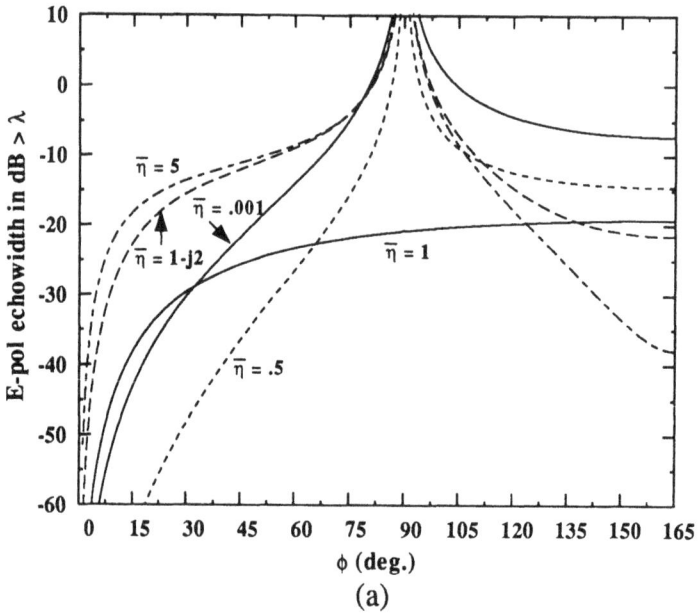

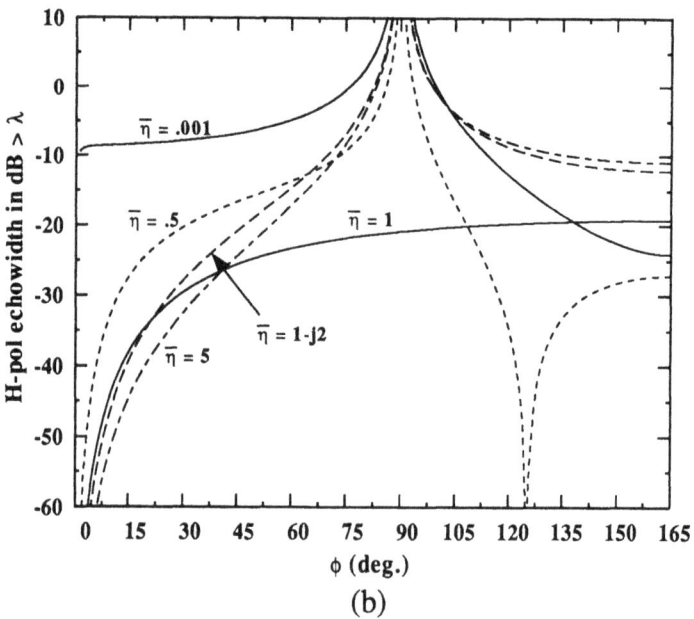

Figure 4-4: *Backscatter echowidth plots for 30° impedance wedges with $\bar{\eta}_+ = \bar{\eta}_- = \bar{\eta}$ as functions of $\phi = 165° - \phi'$: (a) E-polarisation; (b) H-polarisation*

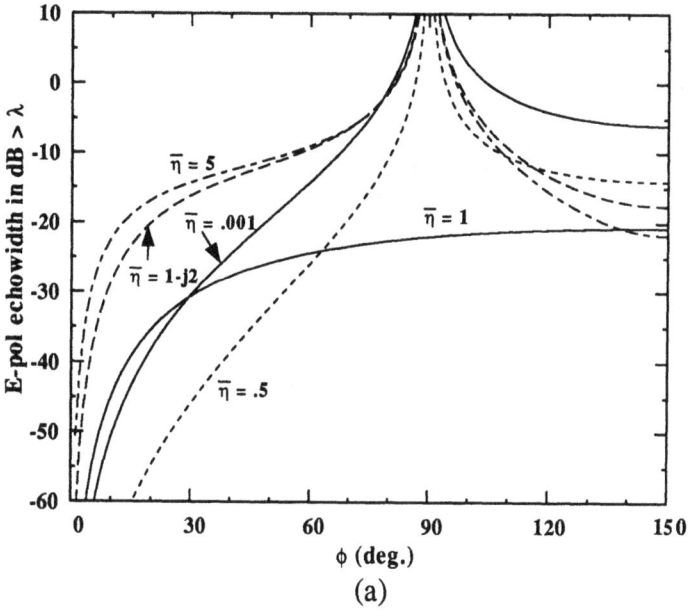

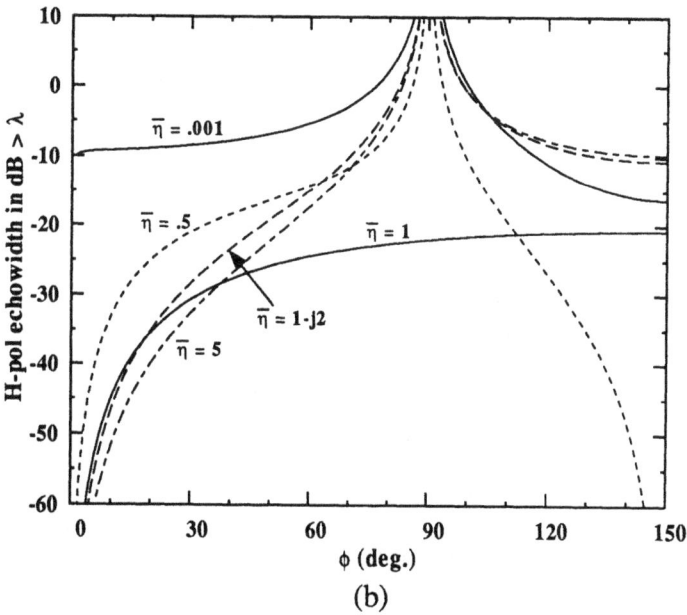

Figure 4–5: *Backscatter echowidth plots for 60° impedance wedges with $\bar{\eta}_+ = \bar{\eta}_- = \bar{\eta}$ as functions of $\phi = 150° - \phi'$: (a) E-polarisation; (b) H-polarisation*

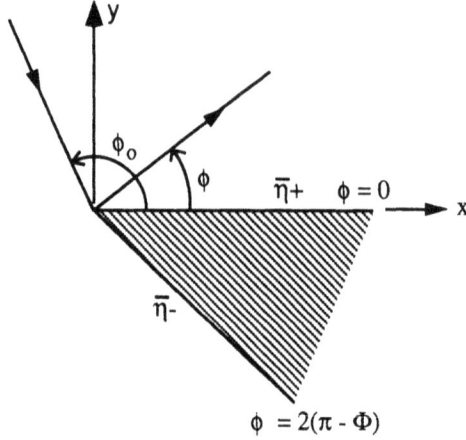

Figure 4–6: *Wedge geometry and the new coordinate system*

From α_{p_1} we recover the incident field E_z^i in the illuminated region, whereas α_{p_2} and α_{p_3} give rise to the fields reflected from the upper and lower faces of the wedge, respectively. It is easily shown that

$$E_z^r = -\frac{\Psi(n\pi - \phi_0')}{\Psi(\phi_0')}e^{jk_0\rho\cos(n\pi-\phi'-\phi_0')}u\left(\pi - |\phi' + \phi_0' - n\pi|\right)$$
$$-\frac{\Psi(-n\pi - \phi_0')}{\Psi(\phi_0')}e^{jk_0\rho\cos(n\pi+\phi'+\phi_0')}u\left(\pi - |\phi' + \phi_0' + n\pi|\right)$$

where u is the unit step function, and $\Psi(\pm n\pi - \phi_0')/\Psi(\phi_0')$ can be evaluated in closed form using the identities in Appendix B.4. We then have

$$E_z^r = R_\perp^+ e^{jk_0\rho\cos(n\pi-\phi'-\phi_0')}u\left(\pi - |\phi' + \phi_0' - n\pi|\right)$$
$$+ R_\perp^- e^{jk_0\rho\cos(n\pi+\phi'+\phi_0')}u\left(\pi - |\phi' + \phi_0' + n\pi|\right) \tag{4.25}$$

where

$$R_\perp^\pm = \frac{\sin\left(\frac{n\pi}{2} \mp \phi_0'\right) - \sin\theta_\pm}{\sin\left(\frac{n\pi}{2} \mp \phi_0'\right) + \sin\theta_\pm} \tag{4.26}$$

are simply the reflection coefficients for an impedance plane (see (2.81)) with $\sin\theta_\pm$ defined in (4.9).

To evaluate E_z^{sw} in (4.23) we must first identify the surface wave poles. From (B.17) we observe that $\psi_\Phi(\alpha)$ has poles at $\alpha = \mp(n\pi + 3\pi/2)$, and when these are examined in conjunction with (4.15) and (4.19), it is apparent that only the poles of $\Psi(\alpha + \phi')$ which occur at

$$\alpha_{p_4} = -\phi' + \left(\pi + \frac{n\pi}{2} + \theta_+\right), \qquad \alpha_{p_5} = -\phi' - \left(\pi + \frac{n\pi}{2} + \theta_-\right) \tag{4.27}$$

can satisfy (4.20), and these are produced by the second and third factors on the right hand side of (4.15). We note that $\theta_\pm$ are complex if $\bar{\eta}_\pm$ are complex or $\bar{\eta}_\pm$ real with $|\bar{\eta}_\pm| > 1$. The residues of the poles (4.27) give rise to the surface wave contribution

$$E_z^{sw} = \begin{cases} C_+^{sw} e^{-jk_0\rho\cos(n\pi/2+\theta_+ -\phi')} & 0 \le \frac{n\pi}{2} - \phi' < \text{gd}(\text{Im. }\theta_+) - \text{Re. }\theta_+ \\ C_-^{sw} e^{-jk_0\rho\cos(n\pi/2+\theta_- +\phi')} & 0 \le \frac{n\pi}{2} + \phi' < \text{gd}(\text{Im. }\theta_-) - \text{Re. }\theta_- \\ 0 & \text{otherwise} \end{cases}$$

(4.28)

with

$$C_+^{sw} = 2\sin\frac{\pi}{2n}\frac{\cos\frac{\phi_0'}{n}}{\Psi(\phi_0')}\psi_\Phi\left(n\pi - \frac{\pi}{2}\right)\psi_\Phi\left(-\phi' + \frac{\pi}{2} + n\pi + 2\theta_+\right)$$
$$\cdot \psi_\Phi\left(-\phi' + \frac{\pi}{2} + \theta_+ + \theta_-\right)\psi_\Phi\left(-\phi' + \frac{3\pi}{2} + \theta_+ - \theta_-\right)$$
$$\cdot \left\{\cos\frac{\pi+\theta_+}{n} - \sin\frac{\phi_0'}{n}\right\}^{-1}$$

(4.29)

$$C_-^{sw} = 2\sin\frac{\pi}{2n}\frac{\cos\frac{\phi_0'}{n}}{\Psi(\phi_0')}\psi_\Phi\left(n\pi - \frac{\pi}{2}\right)\psi_\Phi\left(-\phi' - \frac{3\pi}{2} + \theta_+ - \theta_-\right)$$
$$\cdot \psi_\Phi\left(-\phi' - \frac{\pi}{2} - \theta_+ - \theta_-\right)\psi_\Phi\left(-\phi' - \frac{\pi}{2} - n\pi - 2\theta_-\right)$$
$$\cdot \left\{\cos\frac{\pi+\theta_-}{n} + \sin\frac{\phi_0'}{n}\right\}^{-1}$$

(4.30)

where the subscripts $\pm$ attached to C^{sw} refer to the surface waves on the upper and lower faces of the wedge, respectively. As pointed out in Section 3.2.2 in connection with an impedance/resistive half-plane, for E-polarisation the surface wave pole is captured wherever the impedance is capacitive, whereas for H-polarisation the capture occurs when the impedance is inductive. If the impedance is real, the pole is never captured, but may lie close to the steepest descent path.

4.2.3 Special cases

A perfectly conducting wedge and an impedance half-plane (for which $n = 2$) are special cases of the general problem treated in the previous section and, since the solutions for these can be obtained by other methods, it is important to compare the results. To facilitate the comparison it is convenient to change the orientation of the wedge to that shown in Fig. 4–6, and to measure the polar angles from the top face of the wedge. Then

$$\phi = n\pi/2 - \phi', \qquad \phi_0 = n\pi/2 - \phi_0'$$

and the non-uniform diffraction coefficient (4.22) becomes

$$
D^{\mathrm{nu}}(\phi,\phi_0,\theta_+,\theta_-) = \frac{e^{-j\pi/4}}{n\sqrt{2\pi k_0}} \frac{\sin\dfrac{\phi_0}{n}}{\Psi\left(\dfrac{n\pi}{2}-\phi_0\right)}\left\{\frac{\Psi\left(\dfrac{n\pi}{2}-\phi-\pi\right)}{\cos\dfrac{\phi+\pi}{n}-\cos\dfrac{\phi_0}{n}}\right.
$$

$$
\left. -\frac{\Psi\left(\dfrac{n\pi}{2}-\phi+\pi\right)}{\cos\dfrac{\phi-\pi}{n}-\cos\dfrac{\phi_0}{n}}\right\} \tag{4.31}
$$

For a perfectly conducting wedge with H-polarisation, (4.5) gives $\theta_+ = \theta_- = 0$, and from (4.15) and (B.14)

$$
\Psi(\alpha)\big|_{\theta_+=\theta_-=0} = \frac{1}{2}\left\{\psi_\Phi\left(\frac{\pi}{2}\right)\right\}^4 \cos\frac{\alpha}{n}
$$

Hence

$$
D^{\mathrm{nu}}(\phi,\phi_0,0,0) = \frac{e^{-j\pi/4}}{n\sqrt{2\pi k_0}}\left\{\frac{\sin\dfrac{\phi+\pi}{n}}{\cos\dfrac{\phi+\pi}{n}-\cos\dfrac{\phi_0}{n}}-\frac{\sin\dfrac{\phi-\pi}{n}}{\cos\dfrac{\phi-\pi}{n}-\cos\dfrac{\phi_0}{n}}\right\}
$$

and since

$$
\cot\frac{x+y}{2}\mp\cot\frac{x-y}{2} = \frac{2}{\cos x-\cos y}\left\{\begin{array}{c}\sin y\\ -\sin x\end{array}\right\} \tag{4.32}
$$

$$
D^{\mathrm{nu}}(\phi,\phi_0,0,0) = -\frac{e^{-j\pi/4}}{2n\sqrt{2\pi k_0}}\left\{\cot\frac{\pi+(\phi-\phi_0)}{2n}+\cot\frac{\pi+(\phi+\phi_0)}{2n}\right.
$$

$$
\left. +\cot\frac{\pi-(\phi-\phi_0)}{2n}+\cot\frac{\pi-(\phi+\phi_0)}{2n}\right\} \tag{4.33}
$$

which is recognised as the H-polarisation (or "hard") diffraction coefficient for a metallic wedge (KOUYOUMJIAN AND PATHAK, 1974). In the case of E-polarisation, perfect conductivity implies $\theta_+ = \theta_- = \pi/2 + j\infty$, and from the asymptotic behaviour of the Maliuzhinets function,

$$
\frac{\Psi(\alpha)}{\Psi(\phi_0)}\bigg|_{\theta_+=\theta_-=\pi/2+j\infty} = 1
$$

Then

$$
D^{\mathrm{nu}}\left(\phi,\phi_0,\frac{\pi}{2}+j\infty,\frac{\pi}{2}+j\infty\right) =
$$

$$
\frac{e^{-j\pi/4}}{n\sqrt{2\pi k_0}}\left\{\frac{\sin\dfrac{\phi_0}{n}}{\cos\dfrac{\phi+\pi}{n}-\cos\dfrac{\phi_0}{n}}-\frac{\sin\dfrac{\phi_0}{n}}{\cos\dfrac{\phi-\pi}{n}-\cos\dfrac{\phi_0}{n}}\right\}
$$

and using (4.32) we obtain

$$
D^{\mathrm{nu}}\left(\phi, \phi_0, \frac{\pi}{2} + j\infty, \frac{\pi}{2} + j\infty\right) =
$$

$$
-\frac{e^{-j\pi/4}}{2n\sqrt{2\pi k_0}} \left\{ \cot\frac{\pi + (\phi - \phi_0)}{2n} - \cot\frac{\pi + (\phi + \phi_0)}{2n} \right.
$$

$$
\left. + \cot\frac{\pi - (\phi - \phi_0)}{2n} - \cot\frac{\pi - (\phi + \phi_0)}{2n} \right\} \tag{4.34}
$$

which is the E-polarisation (or "soft") diffraction coefficient for a metallic wedge.

Another special case of the impedance wedge is a half-plane with the same impedance on both sides. Then $n = 2$ with $\theta_+ = \theta_- = \theta$ (say), and (4.31) should reduce to the diffraction coefficients given in (3.105) and (3.106) for E- and H-polarisations, respectively. We can show this by using the identity (B.14) to simplify the various terms in (4.31). When it is used to bring the Maliuzhinets functions into the numerator we have

$$
\frac{1}{\Psi(\pi - \phi_0)} =
$$

$$
\frac{\psi_\pi\left(\frac{\pi}{2} - \phi_0 + \theta\right) \psi_\pi\left(\frac{3\pi}{2} - \phi_0 - \theta\right) \psi_\pi\left(-\frac{3\pi}{2} - \phi_0 + \theta\right) \psi_\pi\left(-\frac{\pi}{2} - \phi_0 - \theta\right)}{\left\{\psi_\pi\left(\frac{\pi}{2}\right)\right\}^8 \cos\frac{\pi - \phi_0 + \theta}{4} \cos\frac{2\pi - \phi_0 - \theta}{4} \cos\frac{\pi + \phi_0 - \theta}{4} \cos\frac{\phi_0 + \theta}{4}}
$$

and two further applications of (B.14) then give

$$
\frac{1}{\Psi(\pi - \phi_0)} = \frac{8\left\{\psi_\pi\left(\frac{\pi}{2} - \phi_0 + \theta\right) \psi_\pi\left(\frac{3\pi}{2} - \phi_0 - \theta\right)\right\}^2}{\left\{\psi_\pi\left(\frac{\pi}{2}\right)\right\}^8}
$$

$$
\cdot\left\{\left(1 + \sqrt{2}\cos\frac{\frac{\pi}{2} - \phi_0 + \theta}{2}\right)\left(1 + \sqrt{2}\cos\frac{\frac{3\pi}{2} - \phi_0 - \theta}{2}\right)\right\}^{-1}
$$

Similarly

$$
\Psi(-\phi) = \left\{\psi_\pi\left(\frac{\pi}{2} - \phi + \theta\right) \psi_\pi\left(\frac{3\pi}{2} - \phi - \theta\right)\right\}^2 \frac{\cos\frac{\pi + \phi - \theta}{4} \cos\frac{\phi + \theta}{4}}{\cos\frac{\phi - \theta}{4} \cos\frac{\pi - \phi - \theta}{4}}
$$

and

$$
\Psi(2\pi - \phi) = \left\{\psi_\pi\left(\frac{\pi}{2} - \phi + \theta\right) \psi_\pi\left(\frac{3\pi}{2} - \phi - \theta\right)\right\}^2 \frac{\cos\frac{2\pi - \phi + \theta}{4} \cos\frac{3\pi - \phi - \theta}{4}}{\cos\frac{\pi - \phi + \theta}{4} \cos\frac{2\pi - \phi - \theta}{4}}
$$

On substituting these into (4.31) and using (3.30), we obtain

$$
D^{\mathrm{nu}}(\phi, \phi_0, \theta, \theta) = \frac{e^{-j\pi/4}}{\sqrt{2\pi k_0}} \frac{K_+(\sin\theta, \cos\phi) K_+(\sin\theta, \cos\phi_0)}{\sin\frac{\phi}{2} \sin\frac{\phi_0}{2}(\cos\phi + \cos\phi_0)}
$$

$$\cdot \left\{ \cos \frac{\phi - \theta}{2} \sin \frac{\phi + \theta}{2} \left(\sin \frac{\phi}{2} - \cos \frac{\phi_0}{2} \right) \right.$$

$$\left. - \cos \frac{\phi + \theta}{2} \sin \frac{\phi - \theta}{2} \left(\sin \frac{\phi}{2} + \cos \frac{\phi_0}{2} \right) \right\} \qquad (4.35)$$

where K_+ is the Wiener-Hopf split function, and since the terms in brackets are simply

$$\sin \frac{\phi}{2} \sin \frac{\phi_0}{2} \left(\sin \theta - 2 \cos \frac{\phi}{2} \cos \frac{\phi_0}{2} \right)$$

the identification of $\sin \theta$ as shown in (4.9) reduces (4.35) to the expressions (3.105) and (3.106) for E- and H-polarisations, respectively. Because the latter were derived using the dual integral equation method, this provides an independent check of (4.31). We note in passing that (4.31) is also valid for a half-plane with different impedances on the two sides—a geometry whose solution is not easily obtained by the Wiener-Hopf technique.

4.2.4   Uniform diffraction coefficient

This can be found by carrying out a uniform asymptotic evaluation of the steepest descent integral (4.19). The geometrical optics and surface wave poles were identified in (4.24) and (4.27) respectively, and when the observation angle is close to the shadow boundary of a surface wave pole, the preferred method is the additive approach described in Appendix C. From (4.28) these shadow boundaries are

$$\phi = \text{gd}(\text{Im. } \theta_+) - \text{Re. } \theta_+$$

for the upper face of the wedge, and

$$\phi = n\pi - \{\text{gd}(\text{Im. } \theta_-) - \text{Re. } \theta_-\}$$

for the lower face, and the influence of the pole is felt whenever the shadow boundary is near to the corresponding face of the wedge. A uniform expression for the diffraction coefficient obtained using the additive approach has been given by HERMAN AND VOLAKIS (1988).

For angles such that the presence of the surface wave poles can be ignored, the multiplicative approach (see Appendix C) is simpler and more attractive because each of four cotangent terms similar to those in (4.33) and (4.34) is associated with a different pole corresponding to either a shadow or reflection boundary. In the special case of a metallic wedge, the UTD diffraction coefficient obtained in this manner has been given by KOUYOUMJIAN AND PATHAK (1974). To apply the method to (4.19) we must first express the integrand in terms of cotangents and, by using (4.32), the contribution $E_z^{\text{d}-}$ of the path $S(-\pi)$ can be written as

$$E_z^{\text{d}-} = \frac{1}{4\pi j n} \int_{S(-\pi)} e^{jk_0 \rho \cos \alpha} \left\{ \cot \frac{\alpha - (\phi - \phi_0)}{2n} - \cot \frac{\alpha - (\phi + \phi_0)}{2n} \right\}$$

$$\cdot \frac{\Psi \left(\frac{n\pi}{2} + \alpha - \phi\right)}{\Psi \left(\frac{n\pi}{2} - \phi_0\right)} \, d\alpha \tag{4.36}$$

The first and second cotangents are associated with the poles $\alpha_{p_1} = \phi - \phi_0$ and $\alpha_{p_3} = -2n\pi + \phi + \phi_0$, respectively. When, for example, α_{p_3} crosses $S(-\pi)$, E_z^{d-} must be discontinuous by an amount $R_\perp^-$ to balance the residue of the pole which must be added to E_z^d to give the total field E_z^-. We can verify that this is so by using the identity (B.16) for the Maliuzhinets function to show

$$\frac{\Psi \left(\frac{n\pi}{2} + \alpha - \phi\right)}{\Psi \left(\frac{n\pi}{2} - \phi_0\right)} = -R_\perp^- \frac{\Psi \left(\frac{n\pi}{2} + \alpha - \phi\right)}{\Psi \left(-\frac{3n\pi}{2} + \phi_0\right)} \tag{4.37}$$

where $R_\perp^-$ is the plane wave reflection coefficient (4.26) with $\phi_0' = \frac{n\pi}{2} - \phi_0$. When $\alpha = \alpha_{p_3}$ the right hand side of (4.37) is simply $-R_\perp^-$, and the continuity of E_z^- now follows.

By inserting (4.37) and (4.26) into (4.36) and proceeding in a similar manner for the contribution of the $S(\pi)$, (4.19) becomes

$$\begin{aligned}
E_z^d = \; & \frac{1}{4\pi jn} \int_{S(-\pi)} e^{jk_0 \rho \cos \alpha} \left\{ \frac{\Psi \left(\frac{n\pi}{2} + \alpha - \phi\right)}{\Psi \left(\frac{n\pi}{2} - \phi_0\right)} \cot \frac{\alpha - (\phi - \phi_0)}{2n} \right. \\
& \left. + \frac{\sin(n\pi - \phi_0) - \sin \theta_-}{\sin(n\pi - \phi_0) + \sin \theta_-} \frac{\Psi \left(\frac{n\pi}{2} + \alpha - \phi\right)}{\Psi \left(-\frac{3n\pi}{2} + \phi_0\right)} \cot \frac{\alpha - (\phi + \phi_0)}{2n} \right\} d\alpha \\
& - \frac{1}{4\pi jn} \int_{S(\pi)} e^{jk_0 \rho \cos \alpha} \left\{ \frac{\Psi \left(\frac{n\pi}{2} + \alpha - \phi\right)}{\Psi \left(\frac{n\pi}{2} - \phi_0\right)} \cot \frac{\alpha - (\phi - \phi_0)}{2n} \right. \\
& \left. + \frac{\sin \phi_0 - \sin \theta_+}{\sin \phi_0 + \sin \theta_+} \frac{\Psi \left(\frac{n\pi}{2} + \alpha - \phi\right)}{\Psi \left(\frac{n\pi}{2} + \phi_0\right)} \cot \frac{\alpha - (\phi + \phi_0)}{2n} \right\} d\alpha \tag{4.38}
\end{aligned}$$

The differences from the corresponding expression in TIBERIO ET AL. (1985) are due to misprints in the latter. The integral (4.38) can be evaluated using the method described in Appendix C (see (C.21)) and the result is

$$E_z^d = \frac{e^{-jk_0\rho}}{\sqrt{\rho}} D^u(\phi, \phi_0, \theta_+, \theta_-)$$

with

$$\begin{aligned}
D^u(\phi, \phi_0, \theta_+, \theta_-) = \\
-\frac{e^{-j\pi/4}}{2n\sqrt{2\pi k_0}} \left\{ \frac{\Psi \left(\frac{n\pi}{2} - \pi - \phi\right)}{\Psi \left(\frac{n\pi}{2} - \phi_0\right)} \cot \frac{\pi + (\phi - \phi_0)}{2n} F_{\mathrm{KP}}(k_0 \rho a_i^-) \right.
\end{aligned}$$

$$+ \frac{\Psi\left(\frac{n\pi}{2} + \pi - \phi\right)}{\Psi\left(\frac{n\pi}{2} - \phi_0\right)} \cot \frac{\pi - (\phi - \phi_0)}{2n} F_{\mathrm{KP}}(k_0 \rho a_{\mathrm{i}}^-)$$

$$+ R_\perp^- \frac{\Psi\left(\frac{n\pi}{2} - \pi - \phi\right)}{\Psi\left(-\frac{3n\pi}{2} + \phi_0\right)} \cot \frac{\pi + (\phi + \phi_0)}{2n} F_{\mathrm{KP}}(k_0 \rho a_{\mathrm{r}}^+)$$

$$+ R_\perp^+ \frac{\Psi\left(\frac{n\pi}{2} + \pi - \phi\right)}{\Psi\left(\frac{n\pi}{2} + \phi_0\right)} \cot \frac{\pi - (\phi + \phi_0)}{2n} F_{\mathrm{KP}}(k_0 \rho a_{\mathrm{i}}^+) \Bigg\} \qquad (4.39)$$

where F_{KP} is the UTD transition function (3.60) and

$$a_{\mathrm{i}}^\pm \simeq 2 \cos^2 \frac{\phi \pm \phi_0}{2} , \qquad a_{\mathrm{r}}^+ \simeq 2 \cos^2 \frac{2n\pi - (\phi + \phi_0)}{2} \qquad (4.40)$$

More accurate expressions for $a_{\mathrm{i}}^\pm$ and $a_{\mathrm{r}}^\pm$ have been given by KOUYOUMJIAN AND PATHAK (1974) (see also MCNAMARA ET AL., 1990).

Sample calculations based on (4.39) that demonstrate the uniformity of the total field E_z across the reflection ($\phi = 150°$) and shadow ($\phi = 210°$) boundaries for two different wedges are shown in Fig. 4-7. We note that for E-polarisation $\sin\theta = 4$ implies $\bar{\eta} = 0.25$, whereas $\sin\theta = \infty$ implies $\bar{\eta} = 0$, i.e. perfect conductivity. The similarity of the patterns in Fig. 4-7 to those in

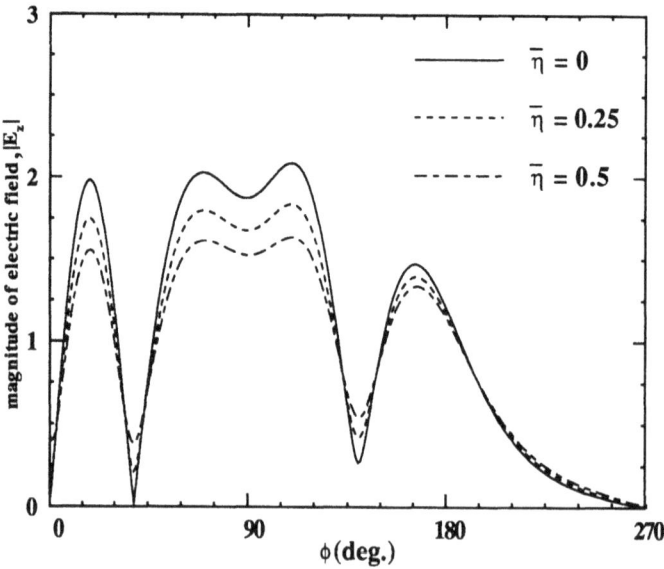

Figure 4-7: *Total electric field E_z at a distance $\rho = 5\lambda_0/\pi$ from the edge of a right-angled wedge with $\bar{\eta}_+ = \bar{\eta}_- = \bar{\eta}$ for a plane wave incident at $\phi_0 = 30°$*

Fig. 3–17(a) is because the observation angle is small and the scattered field is strongly influenced by the impedance of the nearby (upper) face.

4.3 Skew incidence

The problem of a plane wave at skew incidence on an impedance wedge is illustrated in Fig. 4–8, and special cases for which the exact solutions have been found are (i) a full plane ($\Phi = \pi/2$) impedance junction (VACCARO, 1980;

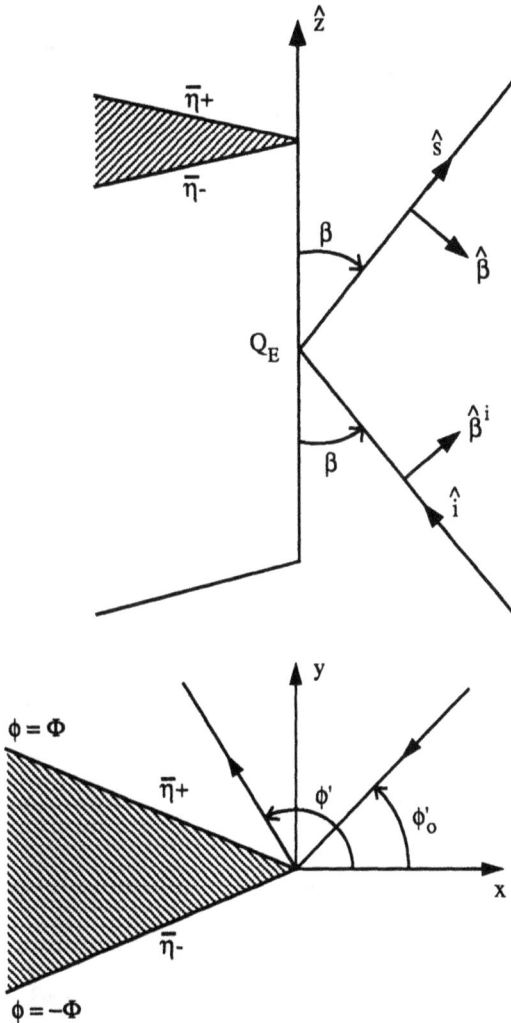

Figure 4–8: *Coordinate system for skew incidence on a wedge*

SENIOR, 1986; ROJAS, 1988a), (ii) a half plane ($\Phi = \pi$) with different face impedances (BUCCI AND FRANCESCHETTI, 1976), and (iii) right-angled interior ($\Phi = \pi/4$) and exterior ($\Phi = 3\pi/4$) wedges with one face perfectly conducting (VACCARO, 1981; SENIOR, 1986; SENIOR AND VOLAKIS, 1986; ROJAS, 1988b). The first of these is often referred to as a two-part impedance plane, and its solution was derived in Chapter 3 using dual integral equations. Obviously, the first two are special cases of a wedge and, for a half-plane with the same face impedances, the solution for skew incidence was presented in Section 3.8. The extension to different face impedances is given later in this section.

For the wedge angles $\Phi = \pi/4$, $\pi/2$, $3\pi/4$ and π, all of the skew incidence solutions mentioned above can be found using the generalised reflection method of VACCARO (1980). This is basically an extension of Maliuzhinets' method, but is not the most convenient for the development of a general-purpose computer code. The somewhat different approach which we will follow yields the exact solutions for (i) and (ii), and reduces to the normal incidence solution for any Φ. More importantly, it can be used to develop an approximate skew incidence solution for a wedge of arbitrary angle.

To simplify the resulting difference equations we start by introducing a new but equivalent form of the impedance boundary condition (2.36). By tangential differentiation of the component equations, we obtain

$$\frac{\partial E_z}{\partial z} = \pm\eta_\pm \frac{\partial H_\rho}{\partial z}, \qquad \frac{\partial}{\partial\rho}(\rho E_\rho) = \mp\eta_\pm \frac{\partial}{\partial\rho}(\rho H_z)$$

and addition then gives

$$-\frac{1}{\rho}\frac{\partial E_{\phi'}}{\partial\phi'} = \pm\eta_\pm\left(\frac{\partial H_\rho}{\partial z} - \frac{\partial H_z}{\partial\rho} - \frac{1}{\rho}H_z\right)$$

where we have used the divergence condition $\nabla.\mathbf{E} = 0$. Hence, from Maxwell's equations

$$\frac{\partial E_{\phi'}}{\partial\phi'} \pm jk_0\rho\bar{\eta}_\pm E_{\phi'} + E_\rho = 0 \qquad (\phi' = \pm\Phi) \tag{4.41}$$

and similarly

$$\frac{\partial H_{\phi'}}{\partial\phi'} \pm \frac{jk_0\rho}{\bar{\eta}_\pm}H_{\phi'} + H_\rho = 0 \qquad (\phi' = \pm\Phi) \tag{4.42}$$

where $\bar{\eta}_\pm = Y_0\eta_\pm$ as before. We note that (4.41) and (4.42) do not involve the z components.

4.3.1 Formulation and solution for special cases

Consider the impedance wedge shown in Fig. 4–8 where the external wedge angle is $2\Phi = n\pi$ and the upper ($+$) and lower ($-$) faces of the wedge are subject to the impedance boundary conditions (4.41) and (4.42). The wedge is illuminated by the plane wave (3.136) whose z components are given by

$$E_z^i = e_z e^{-jk_0 \hat{i}\cdot\mathbf{r}}, \qquad Z_0 H_z^i = h_z e^{-jk_0 \hat{i}\cdot\mathbf{r}} \qquad (4.43)$$

where

$$\hat{i} = -\hat{x}\cos\phi_0' \sin\beta - \hat{y}\sin\phi_0'\sin\beta + \hat{z}\cos\beta$$

is the direction of incidence. As in Section 3.8, the angle β is a measure of the skewness, with $\beta = \pi/2$ for incidence in a plane normal to the edge. We seek the scattered field satisfying the boundary conditions on $\phi' = \pm\Phi$ and the analyticity requirements discussed in Section 4.2.1.

In accordance with Maliuzhinets' method we write

$$E_z(\rho,\phi') = \frac{e^{-jk_0 z\cos\beta}}{2\pi j}\int_\gamma e^{jk_0\rho\sin\beta\cos\alpha}\, s_e(\alpha+\phi')\, d\alpha \qquad (4.44)$$

$$Z_0\, H_z(\rho,\phi') = \frac{e^{-jk_0 z\cos\beta}}{2\pi j}\int_\gamma e^{jk_0\rho\sin\beta\cos\alpha}\, s_h(\alpha+\phi')\, d\alpha \qquad (4.45)$$

where γ is the Sommerfeld contour shown in Fig. 4–3. These represent the total fields and are analogous to the forms (3.146) and (3.147) used for skew incidence on a plane. Since each face of the wedge supports two (orthogonal) components of the current, two spectral functions are necessary to represent the field. We note that the scattered (and therefore total) fields have the same z dependence $\exp(-jk_0 z\cos\beta)$ as the incident field and, because of this,

$$E_\rho = \frac{1}{jk_0\sin^2\beta}\left\{\cos\beta\frac{\partial E_z}{\partial\rho} + \frac{1}{\rho}\frac{\partial}{\partial\phi'}(Z_0 H_z)\right\}$$

$$E_{\phi'} = \frac{1}{jk_0\sin^2\beta}\left\{\frac{\cos\beta}{\rho}\frac{\partial E_z}{\partial\phi'} - \frac{\partial}{\partial\rho}(Z_0 H_z)\right\}$$

$$ \qquad\qquad (4.46)$$

$$Z_0 H_\rho = \frac{1}{jk_0\sin^2\beta}\left\{\cos\beta\frac{\partial}{\partial\rho}(Z_0 H_z) - \frac{1}{\rho}\frac{\partial E_z}{\partial\phi'}\right\}$$

$$Z_0 H_{\phi'} = \frac{1}{jk_0\sin^2\beta}\left\{\frac{\cos\beta}{\rho}\frac{\partial}{\partial\phi'}(Z_0 H_z) + \frac{\partial E_z}{\partial\rho}\right\}$$

When (4.44)–(4.46) are substituted into (4.41) and (4.42), we obtain

$$\int_\gamma e^{jk_0\rho\sin\beta\cos\alpha}\left(\sin\alpha \pm \frac{1}{\bar{\eta}_\pm\sin\beta}\right)\{\cos\beta\sin\alpha\, s_h(\alpha\pm\Phi)$$

$$+ \cos\alpha\, s_e(\alpha\pm\Phi)\}\, d\alpha = 0 \quad (4.47)$$

$$\int_\gamma e^{jk_0\rho\sin\beta\cos\alpha}\left(\sin\alpha \pm \frac{\bar{\eta}_\pm}{\sin\beta}\right)\{\cos\beta\sin\alpha\, s_e(\alpha\pm\Phi)$$

$$- \cos\alpha\, s_h(\alpha\pm\Phi)\}\, d\alpha = 0 \quad (4.48)$$

These are four coupled integral equations involving $s_{e,h}(\alpha)$ and, because of the form in which the boundary conditions were expressed, the impedances are confined to common factors multiplying the spectra. The necessary and sufficient conditions for (4.47) and (4.48) to be satisfied are (MALIUZHINETS, 1958a)

$$A_e^\pm(\alpha)\{\cos\beta\sin\alpha\,s_h(\alpha\pm\Phi)+\cos\alpha\,s_e(\alpha\pm\Phi)\}$$
$$=-\cos\beta\sin\alpha\,s_h(-\alpha\pm\Phi)+\cos\alpha\,s_e(-\alpha\pm\Phi) \qquad (4.49)$$

$$A_h^\pm(\alpha)\{\cos\beta\sin\alpha\,s_e(\alpha\pm\Phi)-\cos\alpha\,s_h(\alpha\pm\Phi)\}$$
$$=-\cos\beta\sin\alpha\,s_e(-\alpha\pm\Phi)-\cos\alpha\,s_h(-\alpha\pm\Phi) \qquad (4.50)$$

where

$$A_{e,h}^\pm=\frac{\sin\alpha\pm\sin\theta_{e,h}^\pm}{-\sin\alpha\pm\sin\theta_{e,h}^\pm} \qquad (4.51)$$

and

$$\sin\theta_e^\pm=\frac{1}{\bar\eta_\pm\sin\beta}\,,\qquad\sin\theta_h^\pm=\frac{\bar\eta_\pm}{\sin\beta} \qquad (4.52)$$

As expected, when $\beta=\pi/2$ corresponding to normal incidence, (4.49) and (4.50) decouple and reduce to (4.10). In special cases with $\beta\neq\pi/2$ they can also be decoupled by introducing combinations of $s_{e,h}(\alpha)$ corresponding to field components perpendicular to a wedge face (SENIOR, 1986). This is possible when $\Phi=\pi/2$, π and, provided one face is perfectly conducting, $\Phi=\pi/4$, $3\pi/4$. However, the method used by VACCARO (1981) and SENIOR (1986) does not lead to spectra having a common form, and we will therefore follow a different procedure (SYED AND VOLAKIS, 1992a, 1995) with the goal of allowing arbitrary wedge angles.

The first step is to divide (4.49) and (4.50) by $\cos\alpha$, giving

$$A_e^\pm(\alpha)\{\cos\beta\tan\alpha\,s_h(\alpha\pm\Phi)+s_e(\alpha\pm\Phi)\}$$
$$=-\cos\beta\tan\alpha\,s_h(-\alpha\pm\Phi)+s_e(-\alpha\pm\Phi) \qquad (4.53)$$

$$A_h^\pm(\alpha)\{\cos\beta\tan\alpha\,s_e(\alpha\pm\Phi)-s_h(\alpha\pm\Phi)\}$$
$$=-\cos\beta\tan\alpha\,s_e(-\alpha\pm\Phi)-s_h(-\alpha\pm\Phi) \qquad (4.54)$$

and then define the functions

$$t_1(\alpha+\Phi)=\cos\beta\tan\alpha\,s_h(\alpha+\Phi)+s_e(\alpha+\Phi)$$
$$t_2(\alpha+\Phi)=\cos\beta\tan\alpha\,s_e(\alpha+\Phi)-s_h(\alpha+\Phi) \qquad (4.55)$$

implying

$$t_1(\alpha-\Phi)=\cos\beta\tan\alpha\,s_h(\alpha-\Phi)+s_e(\alpha-\Phi)+b_1(\alpha)$$
$$t_2(\alpha-\Phi)=\cos\beta\tan\alpha\,s_e(\alpha-\Phi)-s_h(\alpha-\Phi)+b_2(\alpha) \qquad (4.56)$$

where

$$b_{1,2}(\alpha) = -\cos\beta\tan 2\Phi\,\{\tan\alpha\tan(\alpha - 2\Phi) + 1\}\,s_{h,e}(\alpha - \Phi) \qquad (4.57)$$

On substituting (4.55) and (4.56) into (4.53) and (4.54) we obtain the new difference equations

$$A_{e,h}^+(\alpha)\,t_{1,2}(\alpha + \Phi) = t_{1,2}(-\alpha + \Phi) \qquad (4.58)$$

for the upper face of the wedge, and

$$A_{e,h}^-(\alpha)\,t_{1,2}(\alpha - \Phi) = t_{1,2}(-\alpha - \Phi) + p_{1,2}(\alpha) \qquad (4.59)$$

for the lower face, with

$$p_{1,2}(\alpha) = A_{e,h}^-(\alpha)\,b_{1,2}(\alpha) - b_{1,2}(-\alpha) \qquad (4.60)$$

Since (4.58) and (4.59) are identical to (4.10) apart from the presence of $p_{1,2}(\alpha)$ in (4.59), they can be solved by the same procedure used for (4.10) if $p_{1,2}(\alpha) = 0$. From (4.57) and (4.60),

$$
\begin{aligned}
p_{1,2}(\alpha) = {}& -\frac{\cos\beta\sin 2\Phi}{\cos\alpha\cos(\alpha - 2\Phi)}\,A_{e,h}^-(\alpha)\,s_{h,e}(\alpha - \Phi) \\
& + \frac{\cos\beta\sin 2\Phi}{\cos\alpha\cos(\alpha + 2\Phi)}\,s_{h,e}(-\alpha + \Phi)
\end{aligned}
\qquad (4.61)
$$

and we observe that this is zero in the special cases noted earlier, namely

(i) $\Phi = \pi/2$ (two-part impedance plane for arbitrary β, as shown in Fig. 3–22),

(ii) $\Phi = \pi$ (half-plane for arbitrary β and face impedances),

(iii) $\beta = \pi/2$ (normal incidence for any Φ).

Throughout the rest of this section we will confine our attention to these cases for which $p_{1,2}(\alpha) = 0$ and, once $t_{1,2}(\alpha)$ have been found, the original spectra are, from (4.55),

$$s_e(\alpha) = \frac{\cos(\alpha - \Phi)}{1 - \sin^2\beta\sin^2(\alpha - \Phi)}\,\{\cos(\alpha - \Phi)\,t_1(\alpha) + \cos\beta\sin(\alpha - \Phi)\,t_2(\alpha)\}$$

$$(4.62)$$

$$s_h(\alpha) = \frac{\cos(\alpha - \Phi)}{1 - \sin^2\beta\sin^2(\alpha - \Phi)}\,\{\cos\beta\sin(\alpha - \Phi)\,t_1(\alpha) - \cos(\alpha - \Phi)\,t_2(\alpha)\}$$

Following the procedure used for normal incidence, we set

$$t_{1,2}(\alpha) = \sigma_{1,2}(\alpha)\,\frac{\Psi_{e,h}(\alpha)}{\Psi_{e,h}(\phi_0')} \qquad (4.63)$$

where

$$\Psi_{e,h}(\alpha) = \psi_\Phi\left(\alpha + \Phi + \theta_+^{e,h} - \frac{\pi}{2}\right) \psi_\Phi\left(\alpha + \Phi - \theta_+^{e,h} + \frac{\pi}{2}\right)$$
$$\cdot \psi_\Phi\left(\alpha - \Phi + \theta_-^{e,h} - \frac{\pi}{2}\right) \psi_\Phi\left(\alpha - \Phi - \theta_-^{e,h} + \frac{\pi}{2}\right) \quad (4.64)$$

The functions $\sigma_{1,2}(\alpha)$ must therefore satisfy (4.12) and have a first order pole at $\alpha = \phi_0'$ to reproduce the incident field. In addition, they must be such that $s_{e,h}(\alpha) = O(1)$ as $|\text{Im. }\alpha| \to \infty$ to give $E_z = O(1)$ as $\rho \to 0$. Both of these conditions were encountered for normal incidence, but as a consequence of the division by $\cos\alpha$ in going from (4.49) and (4.50) to (4.53) and (4.54), $\sigma_{1,2}(\alpha)$ must also include the singularities of $\sec(\alpha - \Phi)$ which lie in the strip $|\text{Re. }\alpha| < \pi$. To satisfy these requirements we write

$$\sigma_{1,2}(\alpha) = B_{1,2}\frac{\frac{1}{n}\cos\frac{\phi_0'}{n}}{\sin\frac{\alpha}{n} - \sin\frac{\phi_0'}{n}} + \frac{C_{e,h}}{\sin\frac{\alpha}{n} - \sin\frac{\alpha_1}{n}} + \frac{D_{e,h}}{\sin\frac{\alpha}{n} - \sin\frac{\alpha_2}{n}} \quad (4.65)$$

where $B_{1,2}$, $C_{e,h}$ and $D_{e,h}$ are constants to be determined, and $\alpha_{1,2}$ are the zeros of $\cos(\alpha - \Phi)$ in the strip. If $\Phi > \pi/2$ as we shall assume, $\alpha_1 = \Phi - \pi/2$ and $\alpha_2 = \Phi - 3\pi/2$.

The residues of $\sigma_{1,2}(\alpha)$ at $\alpha = \phi_0'$ must reproduce the incident field and, when (4.63) and (4.65) are inserted into (4.62), we find

$$B_1 = \cos\beta\tan(\phi_0' - \Phi)h_z + e_z$$
$$\quad (4.66)$$
$$B_2 = \cos\beta\tan(\phi_0' - \Phi)e_z - h_z$$

The remaining constants $C_{e,h}$ and $D_{e,h}$ are somewhat analogous to the constants $C_{1,2}$ in (3.150) and (3.153), and are required to eliminate the poles of $s_{e,h}(\alpha)$ at the zeros of $1 - \sin^2\beta\sin^2(\alpha - \Phi)$ in the strip. If

$$\alpha_0 = j\ln\tan\frac{\beta}{2} \quad (4.67)$$

implying $\cos\alpha_0 = \csc\beta$ and $\sin\alpha_0 = -j\cot\beta$, the four poles are at

$$\alpha = \Phi - \pi/2 \pm \alpha_0, \qquad \alpha = \Phi - 3\pi/2 \pm \alpha_0$$

From (4.62) the conditions which must be satisfied for $s_e(\alpha)$ *and* $s_h(\alpha)$ are

$$t_1\left(\Phi - \frac{\pi}{2} \pm \alpha_0\right) = \pm jt_2\left(\Phi - \frac{\pi}{2} \pm \alpha_0\right)$$
$$\quad (4.68)$$
$$t_1\left(\Phi - \frac{3\pi}{2} \pm \alpha_0\right) = \pm jt_2\left(\Phi - \frac{3\pi}{2} \pm \alpha_0\right)$$

and these are four equations from which $C_{e,h}$ and $D_{e,h}$ can be determined.

This completes the specification of $\sigma_{1,2}(\alpha)$ and, hence, $t_{1,2}(\alpha)$ and $s_{e,h}(\alpha)$. The results are, of course, exact only for the special values of Φ noted earlier, and when the expressions for $s_{e,h}(\alpha)$ are substituted into (4.44) and (4.45), the z components of the total field can be found. If the pole contributions are ignored (see Appendix C), a non-uniform evaluation of the Sommerfeld integrals produces the diffracted field components

$$E_z^d \simeq \frac{e^{-jk_0 s}}{\sqrt{s}} \frac{e^{-j\pi/4}}{\sqrt{2\pi k_0} \sin\beta} \{s_e(\phi' - \pi) - s_e(\phi' + \pi)\}$$

$$(4.69)$$

$$Z_0 H_z^d \simeq \frac{e^{-jk_0 s}}{\sqrt{s}} \frac{e^{-j\pi/4}}{\sqrt{2\pi k_0} \sin\beta} \{s_h(\phi' - \pi) - s_h(\phi' + \pi)\}$$

where $\rho = s\sin\beta$ and $s = \rho\sin\beta + z\cos\beta$ is the distance from the diffraction point Q_E (see Fig. 4–8). Since these are far field expressions,

$$E_z^d = -\sin\beta\, E_\beta^d, \qquad Z_0 H_z^d = -\sin\beta\, E_\phi^d \qquad (4.70)$$

and similarly

$$E_z^i = \sin\beta\, E_{\beta i}^i, \qquad Z_0 H_z^i = \sin\beta\, E_{\phi i}^i \qquad (4.71)$$

where $\hat{s} = \hat{\phi} \times \hat{\beta}$ and $\hat{i} = \hat{\beta}^i \times \hat{\phi}^i$, with the unit vectors as shown in Fig. 4–8. As noted in Section 3.8.1, the β and ϕ components are the ones required to represent the diffracted field in edge-fixed coordinates.

Sample calculations of the bistatic echowidth are shown in Fig. 4–9 for a half-plane ($\Phi = \pi$) with several impedance combinations.

4.3.2   Arbitrary wedge angle

To obtain the exact solution for wedge angles other than the special ones noted above, it is necessary to solve the coupled difference equations (4.58) and (4.59) without assuming that $p_{1,2}(\alpha) = 0$. Unfortunately, there is no method for doing so, but it is possible to construct an approximate solution.

From the derivation of (4.58) and (4.59) it is evident that the first is associated with the boundary condition at the upper face $\phi' = \Phi$, whereas the second, involving $p_{1,2}(\alpha)$, relates to the lower face $\phi' = -\Phi$. It follows that the solution given by (4.62)–(4.69) in which $p_{1,2}(\alpha)$ was put equal to zero gives most emphasis to the upper face and is expected to be reasonably accurate for any Φ when the incident field illuminates the upper face alone. This is illustrated in Fig. 4–10(a). If, on the other hand, the incident field illuminates only the lower face as shown in Fig. 4–10(b), we can construct a similar pair of equations which emphasise the lower face by satisfying (4.54) precisely instead

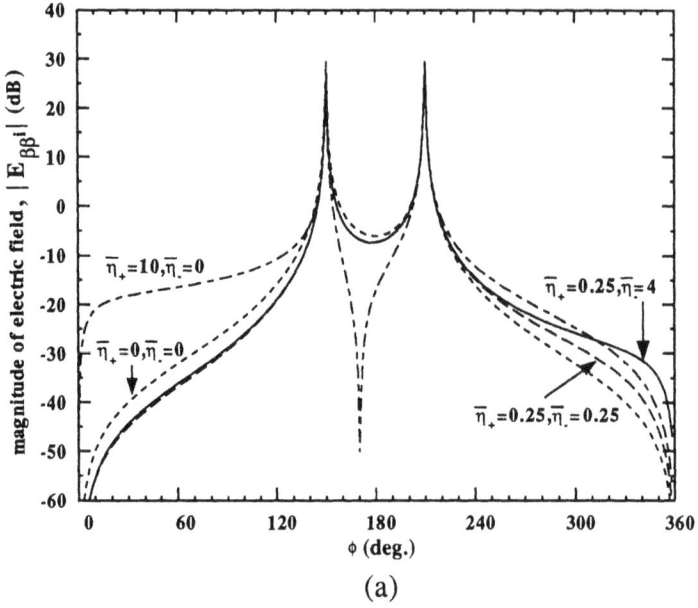

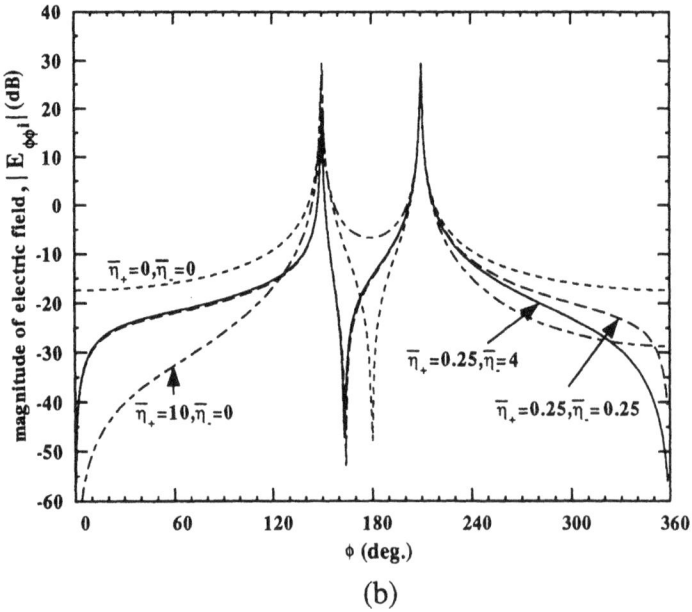

Figure 4–9: *Bistatic far field amplitude for skew incidence ($\phi_0 = 30°$, $\beta^i = 60°$) on an impedance wedge: (a) $E_{\beta\beta'}$; (b) $E_{\phi\phi'}$*

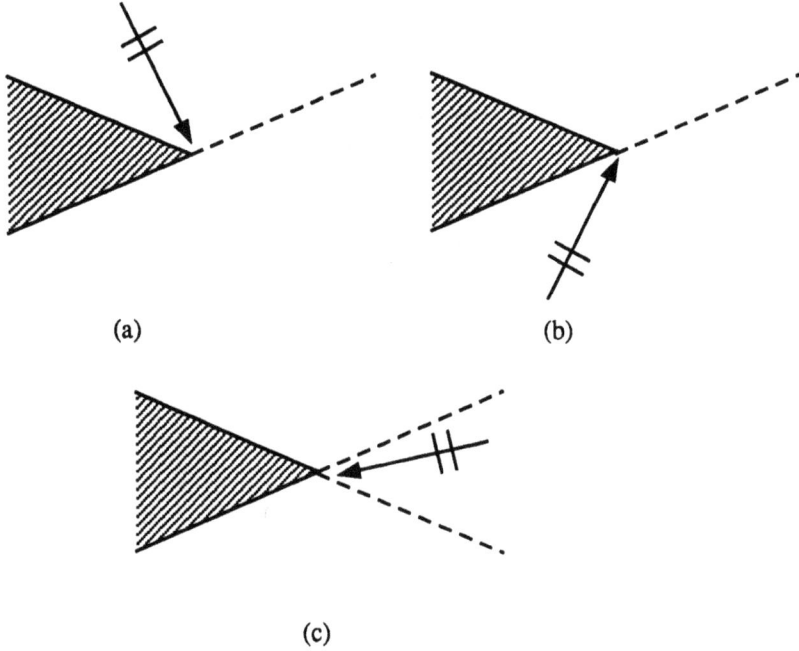

(a)   (b)

(c)

Figure 4–10: *Classification of the wedge illumination: (a) top face illuminated; (b) bottom face illuminated; (c) both faces illuminated*

of (4.53). To do so, we set

$$\tilde{t}_1(\alpha - \Phi) = \cos\beta\tan\alpha\, s_h(\alpha - \Phi) + s_e(\alpha - \Phi)$$
$$\tilde{t}_2(\alpha - \Phi) = \cos\beta\tan\alpha\, s_e(\alpha - \Phi) - s_h(\alpha - \Phi)$$

(4.72)

in place of (4.55). Then

$$\tilde{t}_1(\alpha + \Phi) = \cos\beta\tan\alpha\, s_h(\alpha + \Phi) + s_e(\alpha + \Phi) + \tilde{b}_1(\alpha)$$
$$\tilde{t}_2(\alpha + \Phi) = \cos\beta\tan\alpha\, s_e(\alpha + \Phi) - s_h(\alpha + \Phi) + \tilde{b}_2(\alpha)$$

(4.73)

(cf. (4.56)) where

$$\tilde{b}_{1,2}(\alpha) = \cos\beta\tan 2\Phi\,\{\tan\alpha\tan(\alpha + 2\Phi) + 1\}\, s_{h,e}(\alpha + \Phi) \tag{4.74}$$

and (4.53) and (4.54) become

$$A^+_{e,h}(\alpha)\,\tilde{t}_{1,2}(\alpha + \Phi) = \tilde{t}_{1,2}(-\alpha + \Phi) + \tilde{p}_{1,2}(\alpha) \tag{4.75}$$

for the upper face, and

$$A^-_{e,h}(\alpha)\,\tilde{t}_{1,2}(\alpha - \Phi) = \tilde{t}_{1,2}(-\alpha - \Phi) \tag{4.76}$$

for the lower face, where

$$\tilde{p}_{1,2}(\alpha) = \frac{\cos\beta\sin 2\Phi}{\cos\alpha\cos(\alpha + 2\Phi)} A^+_{e,h}(\alpha)\, s_{h,e}(\alpha + \Phi)$$

$$- \frac{\cos\beta\sin 2\Phi}{\cos\alpha\cos(\alpha - 2\Phi)}\, s_{h,e}(-\alpha + \Phi) \tag{4.77}$$

and $A^\pm_{e,h}(\alpha)$ are given in (4.51). The equations (4.75) and (4.76) are analogous to (4.58) and (4.59), and can be solved using Maliuzhinets' method provided $\tilde{p}_{1,2}(\alpha) = 0$, which is precisely true only if $\Phi = \pi/2$ or π, or $\beta = \pi/2$. Since $\tilde{p}_{1,2}(\alpha)$ does not appear in (4.76), an approximation based on the assumption that it is zero regardless of Φ should be reasonably accurate if the incident field illuminates only the lower face, and we can now proceed in the same manner as before. From (4.72)

$$s_e(\alpha) = \frac{\cos(\alpha + \Phi)}{1 - \sin^2\beta\sin^2(\alpha + \Phi)}\left\{\cos(\alpha + \Phi)\,\tilde{t}_1(\alpha) + \cos\beta\sin(\alpha + \Phi)\,\tilde{t}_2(\alpha)\right\}$$

$$\tag{4.78}$$

$$s_h(\alpha) = \frac{\cos(\alpha + \Phi)}{1 - \sin^2\beta\sin^2(\alpha + \Phi)}\left\{\cos\beta\sin(\alpha + \Phi)\,\tilde{t}_1(\alpha) - \cos(\alpha + \Phi)\,\tilde{t}_2(\alpha)\right\}$$

and to satisfy (4.75) and (4.76) with $\tilde{p}_{1,2}(\alpha) = 0$ we write

$$\tilde{t}_{1,2}(\alpha) = \tilde{\sigma}_{1,2}(\alpha)\,\frac{\Psi_{e,h}(\alpha)}{\Psi_{e,h}(\phi'_0)} \tag{4.79}$$

with

$$\tilde{\sigma}_{1,2}(\alpha) = \tilde{B}_{1,2}\,\frac{\dfrac{1}{n}\cos\dfrac{\phi'_0}{n}}{\sin\dfrac{\alpha}{n} - \sin\dfrac{\phi'_0}{n}} + \frac{\widetilde{C}_{e,h}}{\sin\dfrac{\alpha}{n} - \cos\dfrac{\pi}{2n}} + \frac{\widetilde{D}_{e,h}}{\sin\dfrac{\alpha}{n} - \cos\dfrac{3\pi}{2n}} \tag{4.80}$$

The constants $\tilde{B}_{1,2}$ are chosen to reproduce the incident field, requiring

$$\tilde{B}_1 = \cos\beta\tan(\phi'_0 + \Phi)\,h_z + e_z$$

$$\tilde{B}_2 = \cos\beta\tan(\phi'_0 + \Phi)\,e_z - h_z \tag{4.81}$$

and to eliminate the extraneous poles, $\widetilde{C}_{e,h}$ and $\widetilde{D}_{e,h}$ must be such that

$$\tilde{t}_1\left(\frac{\pi}{2} - \Phi \pm \alpha_0\right) = \pm j\tilde{t}_2\left(\frac{\pi}{2} - \Phi \pm \alpha_0\right)$$

$$\tilde{t}_1\left(\frac{3\pi}{2} - \Phi \pm \alpha_0\right) = \pm j\tilde{t}_2\left(\frac{3\pi}{2} - \Phi \pm \alpha_0\right) \tag{4.82}$$

where α_0 is shown in (4.67). The diffracted fields are then given by (4.69). We shall refer to this new solution as *separation method II* as opposed to *separation method I* which led to the solution (4.62)–(4.69).

The remaining task is to develop an approximate solution when both wedge faces are illuminated (see Fig. 4–10(c)). Following SYED AND VOLAKIS (1995) a separation of (4.53) and (4.54) is achieved by making the approximation

$$\sin \alpha \simeq \sin \frac{2\alpha}{n}, \qquad \cos \alpha \simeq \cos \frac{2\alpha}{n} - \cos \frac{\pi}{n} \qquad (4.83)$$

We note that these are exact when $n = 2$ corresponding to $\Phi = \pi$, and it is therefore anticipated that the results will be most accurate for wedges of small interior angle. When (4.83) are inserted into (4.53) and (4.54), we obtain

$$
\begin{aligned}
A_{e,h}^{+}(\alpha) \, t'_{1,2}(\alpha + \Phi) &= t'_{1,2}(-\alpha + \Phi) \\
A_{e,h}^{-}(\alpha) \, t'_{1,2}(\alpha - \Phi) &= t'_{1,2}(-\alpha - \Phi)
\end{aligned}
\qquad (4.84)
$$

for the upper and lower wedge faces respectively, where

$$
\begin{aligned}
t'_1(\alpha) &= \left(\cos \frac{\pi}{n} + \cos \frac{2\alpha}{n} \right) s_e(\alpha) + \cos \beta \sin \frac{2\alpha}{n} s_h(\alpha) \\
t'_2(\alpha) &= \left(\cos \frac{\pi}{n} + \cos \frac{2\alpha}{n} \right) s_h(\alpha) - \cos \beta \sin \frac{2\alpha}{n} s_e(\alpha)
\end{aligned}
\qquad (4.85)
$$

Hence

$$
S_e(\alpha) = \frac{1}{\cos^2 \beta \sin^2 \frac{2\alpha}{n} + \left(\cos \frac{\pi}{n} + \cos \frac{2\alpha}{n} \right)^2}
$$
$$
\cdot \left\{ \left(\cos \frac{\pi}{n} + \cos \frac{2\alpha}{n} \right) t'_1(\alpha) - \cos \beta \sin \frac{2\alpha}{n} t'_2(\alpha) \right\}
\qquad (4.86)
$$

$$
S_h(\alpha) = \frac{1}{\cos^2 \beta \sin^2 \frac{2\alpha}{n} + \left(\cos \frac{\pi}{n} + \cos \frac{2\alpha}{n} \right)^2}
$$
$$
\cdot \left\{ \cos \beta \sin \frac{2\alpha}{n} t'_1(\alpha) + \left(\cos \frac{\pi}{n} + \cos \frac{2\alpha}{n} \right) t'_2(\alpha) \right\}
$$

and we can now apply Maliuzhinets' method directly to (4.84). If

$$
t'_{1,2}(\alpha) = \frac{\Psi_{e,h}(\alpha)}{\Psi_{e,h}(\phi'_0)} \left\{ B'_{1,2} \frac{\frac{1}{n} \cos \frac{\phi'_0}{n}}{\sin \frac{\alpha}{n} - \sin \frac{\phi'_0}{n}} + C'_{e,h} + D'_{e,h} \sin \frac{\alpha}{n} \right\}
\qquad (4.87)
$$

the incident field requires

$$B_1' = \left(\cos \frac{\pi}{n} + \cos \frac{2\phi_0'}{n} \right) e_z + \cos \beta \sin \frac{2\phi_0'}{n} h_z$$

$$B_2' = \left(\cos \frac{\pi}{n} + \cos \frac{2\phi_0'}{n} \right) h_z - \cos \beta \sin \frac{2\phi_0'}{n} e_z$$

(4.88)

and the constants $C_{e,h}'$ and $D_{e,h}'$ are specified by the four equations

$$\left(\cos \frac{\pi}{n} + \cos \frac{2\alpha_1'}{n} \right) t_1'(\pm\alpha_1') = \pm \cos \beta \sin \frac{2\alpha_1'}{n} t_2'(\pm\alpha_1')$$

$$\left(\cos \frac{\pi}{n} + \cos \frac{2\alpha_2'}{n} \right) t_1'(\pm\alpha_2') = \pm \cos \beta \sin \frac{2\alpha_2'}{n} t_2'(\pm\alpha_2')$$

(4.89)

where $\alpha = \pm\alpha_1'$, $\pm\alpha_2'$ are the extraneous poles in (4.86). A simple calculation gives

$$\alpha_{1,2}' = \frac{n}{2} \left\{ \pm\alpha_0 + \sin^{-1} \left(\frac{\cos \frac{\pi}{n}}{\sin \beta} \right) \right\}$$

(4.90)

with $|\text{Re. } \sin^{-1}(\ )| \leq \pi/2$ and α_0 as shown in (4.67).

Knowing $s_{e,h}(\alpha)$, the diffracted fields can be obtained from (4.69) and, because of the approximation (4.83), the expressions are only valid in a restricted angular range. Specifically, the solution based on *separation method I* is appropriate when the observation angle ϕ' is in a range whose upper limit is the upper wedge face $\phi' = \Phi$, and whose lower limit is the shadow or reflection boundary, whichever is reached first. Similarly, *separation method II* should be used for observation angles starting at the lower wedge face $\phi' = -\Phi$ and ending at the shadow or reflection boundary, whichever comes first. Finally, the solution given immediately above and based on *separation method III* is applicable when ϕ' lies between the shadow boundaries. The ranges are illustrated in Fig. 4–11 for backscattering and bistatic scattering. In view of the approximations which were made, it is expected that the overall solution will be most accurate for wedges of small interior angle (Φ close to π) and incidence angles which are not too oblique (β close to $\pi/2$). A right-angled wedge for which $\Phi = 3\pi/4$ therefore constitutes a severe test, and in Figs. 4–12 and 4–13 the approximate solution is compared with the exact solution (SENIOR AND VOLAKIS, 1986) for a right-angled wedge having $\bar{\eta}_+ = 1 - j$ and $\bar{\eta}_- = 0$. The patterns shown are for the backscattered ($\phi_0' = \phi'$) field with $\beta = \pi/3$. For the co-polarised components (see (4.70) and (4.71)) the agreement is excellent and, although the cross-polarised components in the approximate solution exhibit an erroneous oscillation at angles near to the reflection boundary for the upper face, the average field levels are close to those of the exact solution. In this region the co-polarised components dominate the total field and, because of this, the inaccuracy is of less practical significance.

(a)

(b)

(c)

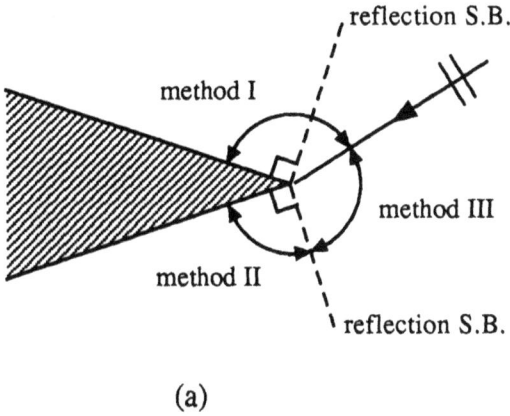

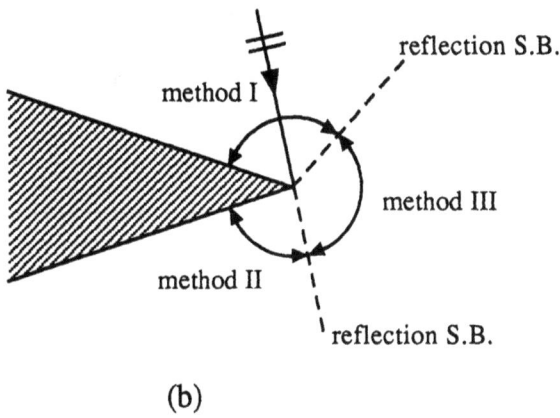

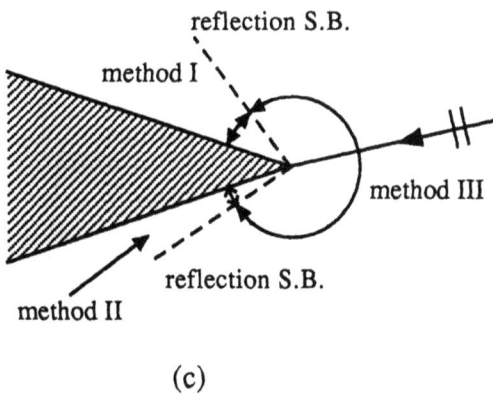

Figure 4–11: *Regions of applicability of the three solutions: (a) backscattering;
(b) and (c) bistatic scattering*

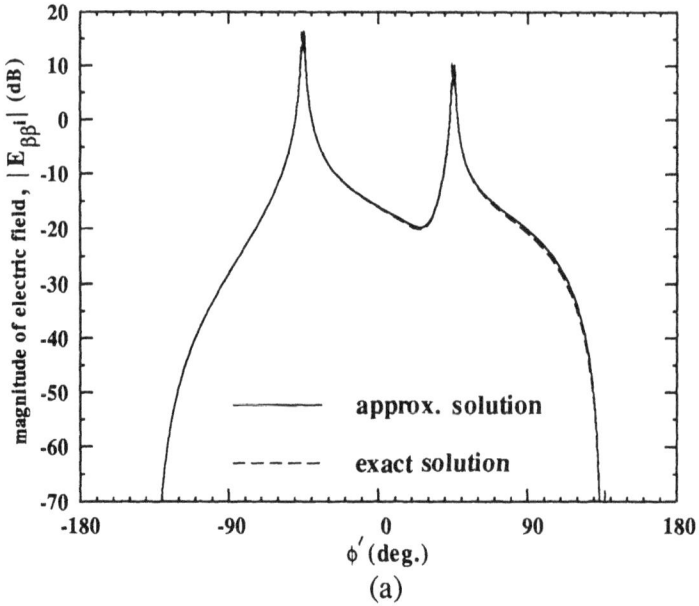

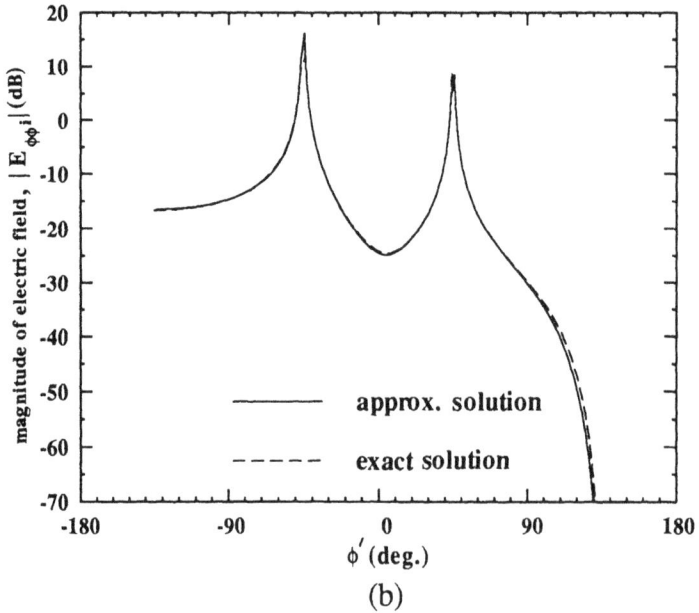

Figure 4–12: *Far zone backscatter patterns of the co-polarised components for a right-angled ($\Phi = 3\pi/4$) wedge having $\bar{\eta}_+ = 1 - j$, $\bar{\eta}_- = 0$ and $\beta = 60°$: (a) $E_{\beta\beta'}$; (b) $E_{\phi\phi'}$*

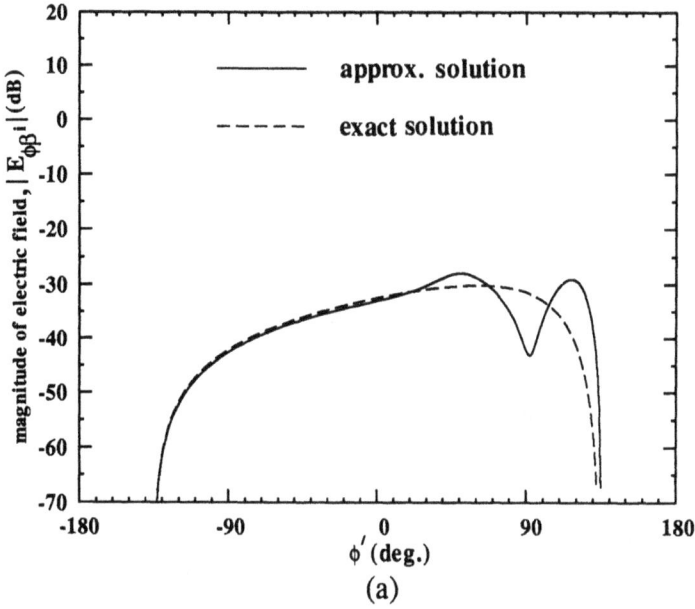

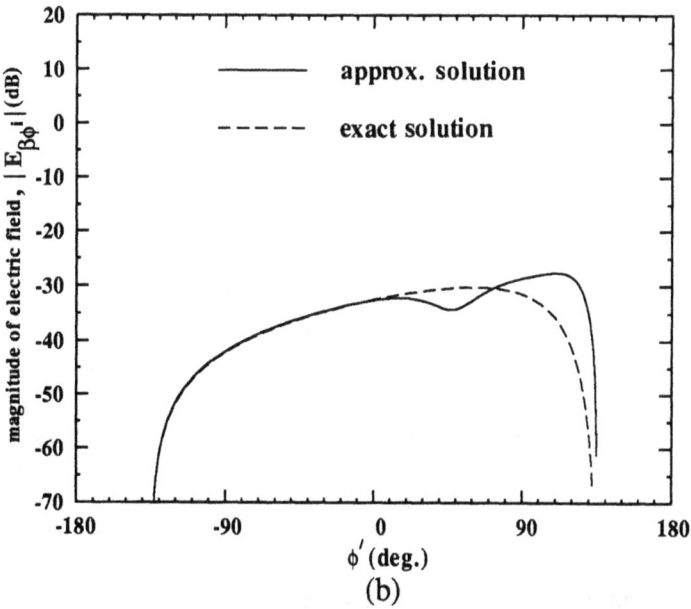

Figure 4-13: *Far zone backscatter patterns of the cross-polarised components for a right-angled* ($\Phi = 3\pi/4$) *wedge having* $\bar{\eta}_+ = 1 - j$, $\bar{\eta}_- = 0$ *and* $\beta = 60°$: *(a)* $E_{\phi\beta^i}$; *(b)* $E_{\beta\phi^i}$

Plots of the approximate expression for the backscattered field for two other wedge angles are shown in Fig. 4–14, where $\Phi = 170°$ and $165°$ with $\beta = 30°$. To test the accuracy for wedge angles like this, it is necessary to consider a finite structure such as the triangular cylinder in Fig. 4–15 whose scattering has been computed using the moment method (SYED AND VOLAKIS, 1992b). We now compare these data with a first order analytical result obtained by summing the fields diffracted by the individual edges of the cylinder, viz.

$$\mathbf{E}^s = \sum_{p=1}^{3} \mathbf{E}_p^d = -\sum_{p=1}^{3} \frac{e^{-jk_0 s_p}}{\sqrt{s}} \overline{\overline{D}}.\mathbf{E}_p^i \tag{4.91}$$

where

$$\overline{\overline{D}} = \hat{\beta}\hat{\beta}^i D_{\beta\beta^i} + \hat{\beta}\hat{\phi}^i D_{\beta\phi^i} + \hat{\phi}\hat{\beta}^i D_{\phi\beta^i} + \hat{\phi}\hat{\phi}^i D_{\phi\phi^i} \tag{4.92}$$

is the dyadic diffraction coefficient (see Section 3.8) extracted from (4.69)–(4.71). In (4.91) s is the usual far zone distance of the observation point from the diffraction point and the subscript p refers to the particular edge. A comparison of the bistatic echowidths is shown in Figs. 4–16 and 4–17 for $\alpha = 40°$ and $w = 4\lambda_0$ with $\beta = 30°$ and $\phi_0' = 70°$ (see Fig. 4–14). On the top side of the cylinder $\bar{\eta} = 0.5$, with $\bar{\eta} = 1+j$ on the other two. The patterns are in excellent agreement apart from the angles near $\phi' = 0$, π and 2π, and the small discrepancies here can be attributed to the omission of any multiply-diffracted fields.

4.4   PTD for impedance structures

The impedance wedge diffraction coefficient can be used to compute the scattering from a finite impedance structure such as a re-entry vehicle, satellite or aircraft. The application of the diffraction coefficient to a wedge of finite length is accomplished using the method of equivalent currents (MEC) or the physical theory of diffraction (PTD) in a manner similar to that employed for metallic structures (KNOTT ET AL., 1985; YOUSSEF 1989; LEE, 1990). The MEC was developed by RYAN AND PETERS (1969) (see also MILLAR (1957)), and was later improved by MICHAELI (1984, 1986) for angles off the Keller cone. The equivalent currents are fictitious electric and magnetic line currents $I^{e,m}$ placed at an edge as shown in Fig. 4–18, and are required to produce the known far zone diffracted field for a straight edge of infinite length. For a finite (and possibly curved) edge, the currents are limited to the finite edge alone, and it can be demonstrated that this approach yields a good approximation to the scattered field when the angle of incidence is not too close to grazing. In the case of an arbitrary structure, the scattering is computed by faceting the surface as illustrated in Fig. 4–1, and then integrating the equivalent currents placed at the edges of the facets.

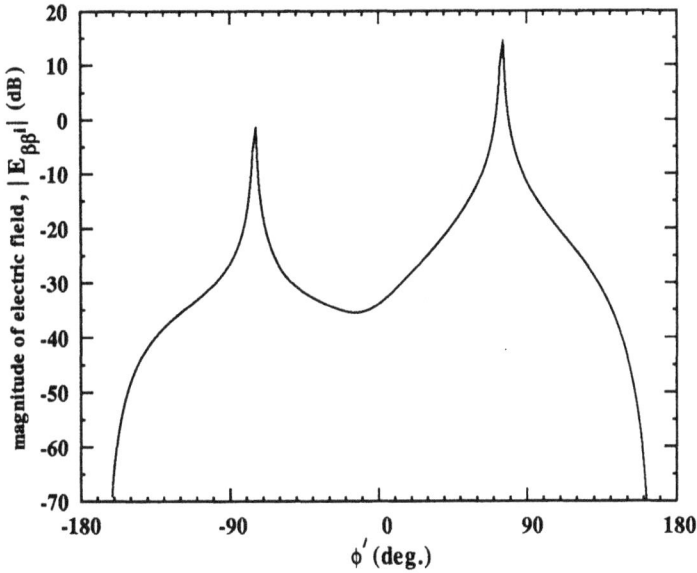

(a)

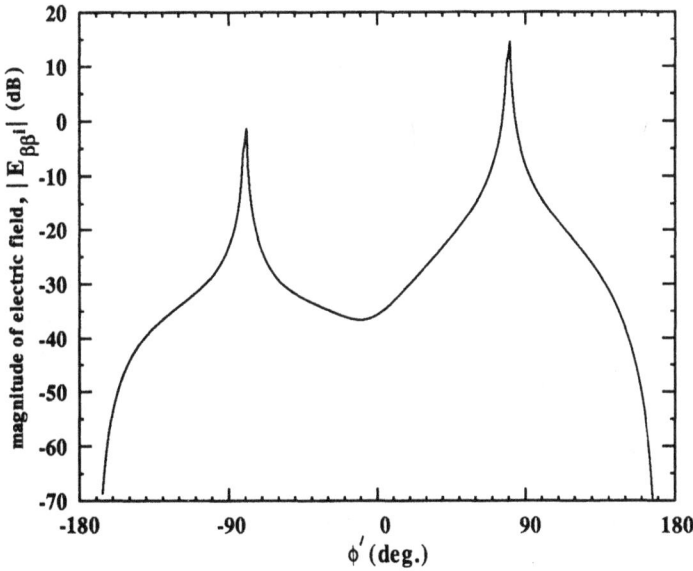

(b)

Figure 4-14: *Far zone backscatter pattern for a wedge having $\bar{\eta}_+ = 1 - j$, $\bar{\eta}_- = 0.5 - j0.1$ with $\beta = 30°$: (a) 20° ($\Phi = 170°$) wedge; (b) 30° ($\Phi = 165°$) wedge*

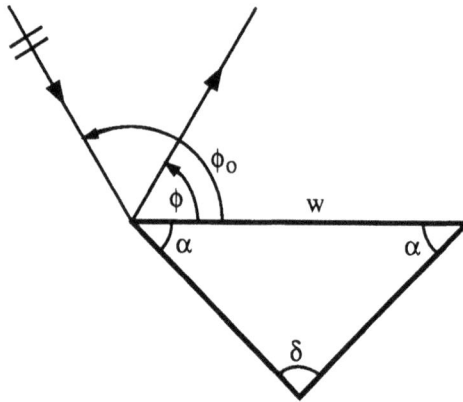

Figure 4-15: *Geometry for a triangular impedance cylinder*

In the context of the PTD (UFIMTSEV, 1971) the role of the equivalent currents is played by the so-called incremental length diffraction coefficient (ILDC) introduced by MITZNER (1974) (see also BUTONIN AND UFIMTSEV (1986)). The ILDC is applied to each element of an edge and generates the diffracted field which must be added to the physical optics contribution from the adjacent surfaces to produce the field scattered by the structure. The MEC and PTD formulations were developed independently in the 1960s, the first in the US and the second in Russia, and were initially regarded as different approaches. However, as shown by KNOTT (1985), Michaeli's equivalent currents and Mitzner's ILDC are identical when the physical optics contribution is removed from the equivalent currents. As a result, the two methods differ only in concept, but in general the PTD is more convenient. In contrast to the reflected field of geometrical optics for which there is the problem of caustics, the physical optics field is easier to compute for curved as well as flat surfaces. In practice, the physical optics field is computed first by dividing the structure into triangular facets (see Fig. 4-1) and integrating the physical optics currents over each one. The diffracted field contributions are then added by integrating the ILDCs over the specified edges of the structure.

In the following we use the simple approach employed by KNOTT AND SENIOR (1973) to derive the ILDCs for impedance wedges from the corresponding equivalent currents. Compared with the ILDCs based on Michaeli's more rigourous formulation of the currents, the expressions are less cumbersome and easier to compute and, as we note later, the accuracy is comparable in most practical applications. The results are then incorporated into a modified version of a program written by LEE (1990) for metallic structures, leading to the first implementation and validation of ILDCs for impedance structures.

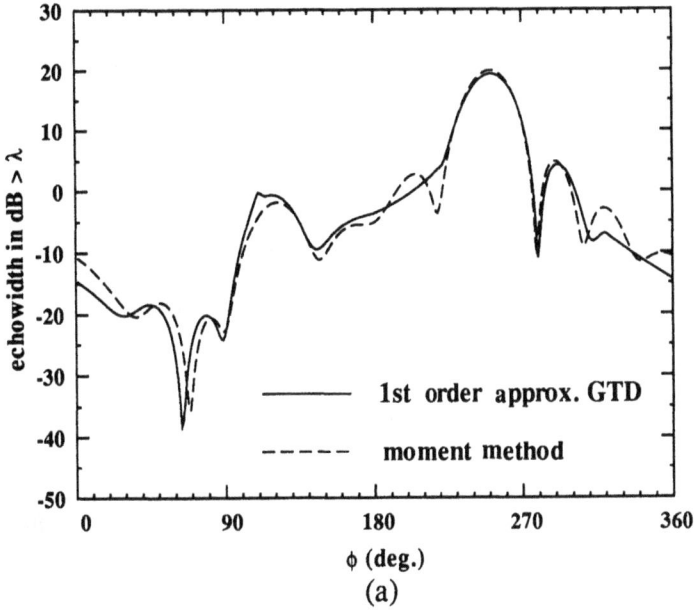

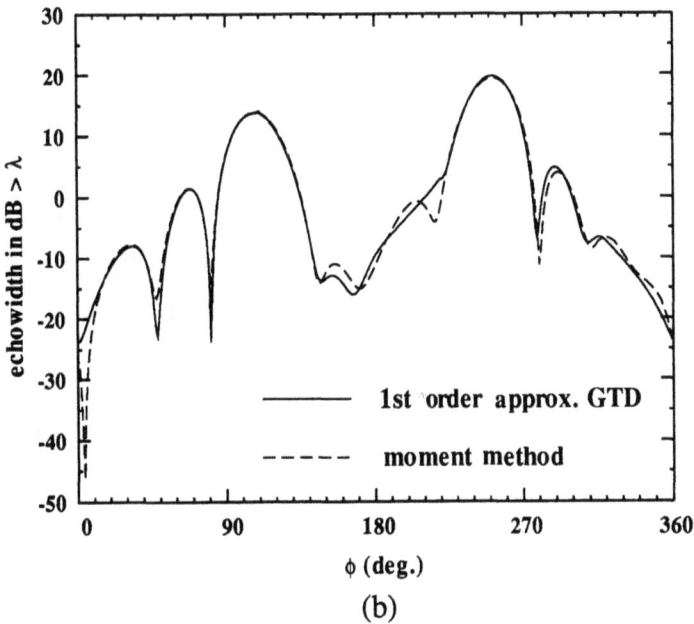

Figure 4–16: *Far zone bistatic scattering pattern for the triangular cylinder of Fig. 4-15 with w = 4λ₀, α = 40°, β = 30° and φ₀ = 70°. On the top face η̄ = 0.5 and on the other two faces η̄ = 1 + j: (a) Eββ'; (b) Eφφ'*

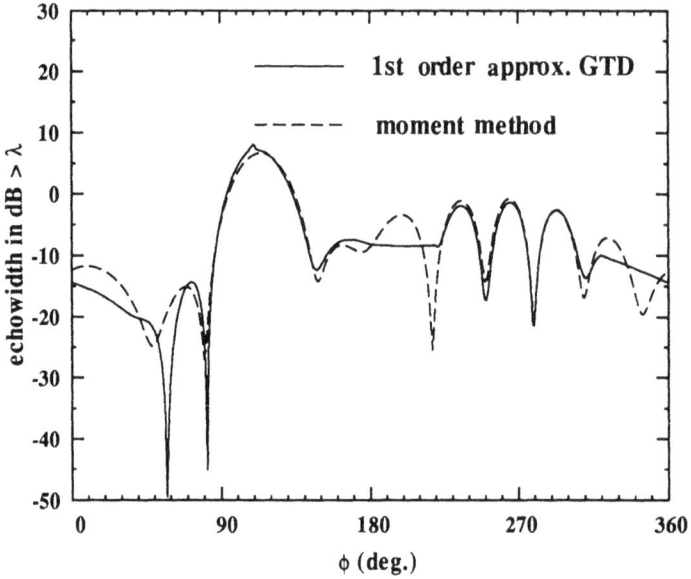

(a)

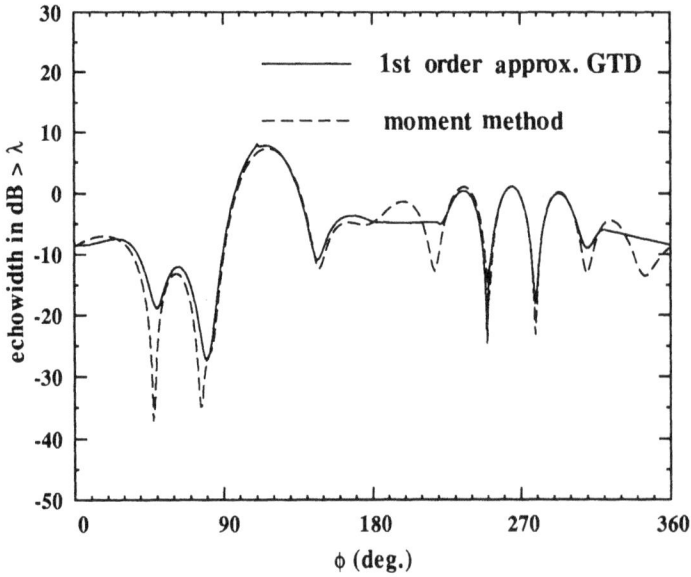

(b)

Figure 4-17: *Far zone bistatic scattering pattern for the triangular cylinder of Fig. 4-15 with $w = 4\lambda_0$, $\alpha = 40°$, $\beta = 30°$ and $\phi_0 = 70°$. On the top face $\bar{\eta} = 0.5$ and on the other two faces $\bar{\eta} = 1 + j$: (a) $E_{\phi\beta'}$; (b) $E_{\beta\phi'}$.*

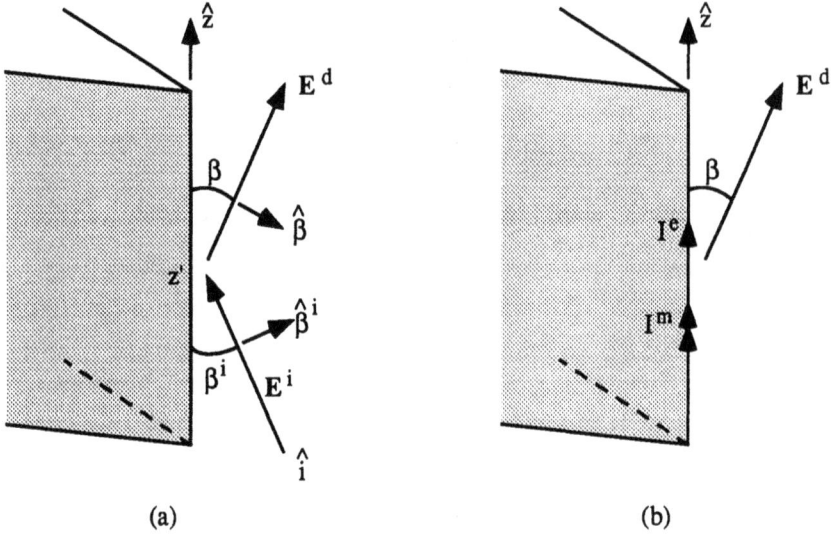

Figure 4–18: *Location and radiation properties of the equivalent currents*

4.4.1   Derivation of equivalent currents

Consider the infinite wedge shown in Fig. 4–18(a), where the edge coincides with the z axis and, as before, the normalised impedances of the upper $(+)$ and lower $(-)$ faces of the wedge are $\bar{\eta}_+$ and $\bar{\eta}_-$, respectively. The wedge is iluminated by the plane wave

$$\mathbf{E}^{i} = \hat{e}^{i}e^{-jk_{0}\hat{i}.\mathbf{r}} \tag{4.93}$$

where

$$\hat{i} = -\hat{x}\cos\phi_{0}\sin\beta^{i} - \hat{y}\sin\phi_{0}\sin\beta^{i} + \hat{z}\cos\beta^{i}$$

in the direction of incidence and ϕ_{0} is measured from the upper face of the wedge. Thus, at the location $z = z_{s}$ on the edge

$$\mathbf{E}^{i} = \hat{e}^{i}e^{-jk_{0}z_{s}\cos\beta^{i}}$$

and from (4.91) and (4.92), the β and ϕ components of the diffracted field are

$$E_{\beta}^{d} = \mathbf{E}^{d}.\hat{\beta} = -\left\{ D_{\beta\beta^{i}}\,E_{\beta^{i}}^{i}(z_{s}) + D_{\beta\phi^{i}}\,E_{\phi^{i}}^{i}(z_{s}) \right\} e^{-jk_{0}z_{s}\cos\beta^{i}}\,\frac{e^{-jk_{0}\,s(z_{s})}}{\sqrt{s(z_{s})}} \tag{4.94}$$

$$E_{\phi}^{d} = \mathbf{E}^{d}.\hat{\phi} = -\left\{ D_{\phi\beta^{i}}\,E_{\beta^{i}}^{i}(z_{s}) + D_{\phi\phi^{i}}\,E_{\phi^{i}}^{i}(z_{s}) \right\} e^{-jk_{0}z_{s}\cos\beta^{i}}\,\frac{e^{-jk_{0}\,s(z_{s})}}{\sqrt{s(z_{s})}} \tag{4.95}$$

In accordance with the law of diffraction (KELLER, 1962), z_{s} is the diffraction point such that $\beta = \beta^{i}$, and $s(z_{s})$ is the distance between the observation point and z_{s}.

To apply the MEC or PTD we must first derive a pair of equivalent currents which, when integrated over the infinite length of the edge, radiate the fields (4.94) and (4.95). For the equivalent electric current $I_{\mathrm{EC}}^{\mathrm{e}}(z)$ we write

$$I_{\mathrm{EC}}^{\mathrm{e}}(z) = \hat{z}\, I^{\mathrm{e}}(z)\, e^{-jk_0 z \cos \beta^{\mathrm{i}}} \tag{4.96}$$

where the exponential accounts for the relative phase of the incident field. The far zone field radiated by this current is

$$E_\beta = jk_0 Z_0 \sin \beta \, A_z \tag{4.97}$$

in which

$$A_z = \int_{-\infty}^{\infty} I^{\mathrm{e}}(z')\, e^{-jk_0 z' \cos \beta^{\mathrm{i}}} \frac{e^{-jk_0 s(z')}}{4\pi\, s(z')}\, dz'$$

is the magnetic vector potential, and $s(z') = s_0 - z' \cos \beta$ where s_0 is measured from the origin of coordinates on the edge. Assuming $I^{\mathrm{e}}(z')/s(z')$ is slowly varying, A_z can be evaluated using the stationary phase or SDP method (see Appendix C) to give

$$A_z \simeq \frac{e^{-j\pi/4}}{2\sqrt{2\pi k_0}\, \sin \beta^{\mathrm{i}}}\, I^{\mathrm{e}}(z_{\mathrm{s}})\, e^{-jk_0 z_{\mathrm{s}} \cos \beta^{\mathrm{i}}} \frac{e^{-jk_0 s(z_{\mathrm{s}})}}{\sqrt{s(z_{\mathrm{s}})}}$$

From (4.97) we now have

$$E_\beta = jk_0 Z_0\, I^{\mathrm{e}}(z_{\mathrm{s}}) \frac{e^{-j\pi/4}}{2\sqrt{2\pi k_0}}\, e^{-jk_0 z_{\mathrm{s}} \cos \beta^{\mathrm{i}}} \frac{e^{-jk_0 s(z_{\mathrm{s}})}}{\sqrt{s(z_{\mathrm{s}})}} \tag{4.98}$$

and since the edge of the wedge is infinite and straight, this must equal (4.94). Hence

$$I^{\mathrm{e}}(z) = -\frac{2}{Z_0} \sqrt{\frac{2\pi}{k_0}}\, e^{-j\pi/4} \left\{ D_{\beta\beta^{\mathrm{i}}}\, E_{\beta^{\mathrm{i}}}^{\mathrm{i}}(z) + D_{\beta\phi^{\mathrm{i}}}\, E_{\phi^{\mathrm{i}}}^{\mathrm{i}}(z) \right\} \tag{4.99}$$

and by following a similar procedure for the magnetic current, we find

$$I^{\mathrm{m}}(z) = 2 \sqrt{\frac{2\pi}{k_0}}\, e^{-j\pi/4} \left\{ D_{\phi\beta^{\mathrm{i}}}\, E_{\beta^{\mathrm{i}}}^{\mathrm{i}}(z) + D_{\phi\phi^{\mathrm{i}}}\, E_{\phi^{\mathrm{i}}}^{\mathrm{i}}(z) \right\} \tag{4.100}$$

We note that $I^{\mathrm{e}}(z)$ are functions of the direction of observation and therefore cannot be regarded as true currents. They are simply mathematical quantities which yield the proper diffracted field when integrated over the infinite edge of the wedge but, since they are physically located at the edge, they can be truncated or allowed to follow a curved edge. On integration these currents can then provide the field diffracted by the finite edge of a straight or curved wedge, and this is discussed next.

4.4.2 ILDC for a straight wedge

All structures of practical interest have finite dimensions, and Fig. 4–19 shows the straight edge of a wedge of finite extent ℓ, where $\hat{z}' = \hat{t}$ is a unit vector along the edge. To compute the diffracted field, it is sufficient to integrate (4.99) and (4.100) over the length ℓ, with the diffraction coefficients expressed in terms of the local (primed) coordinates of a point on the edge. The far zone diffracted field is then (KNOTT AND SENIOR, 1973)

$$\mathbf{E}^d = jk_0 \frac{e^{-jk_0 r}}{4\pi r} \int_{-\ell/2}^{\ell/2} \left\{ Z_0\, I^e(\mathbf{r}')\, \hat{s} \times (\hat{s} \times \hat{t}) + I^m(\mathbf{r}')\, \hat{s} \times \hat{t} \right\} e^{jk_0 \hat{s}\cdot\mathbf{r}'}\, dz' \quad (4.101)$$

where $\mathbf{r}'$ denotes the integration point,

$$\hat{s} = \hat{x} \sin\theta \cos\phi + \hat{y} \sin\theta \sin\phi + \hat{z} \cos\theta$$

is a unit vector in the scattering direction, and the angles θ, ϕ are those of a global coordinate system with origin 0. By inserting the expressions (4.99) and (4.100) for the filamentary currents, and noting that

$$E^i_{\beta^i}(\mathbf{r}') = \frac{\hat{t}\cdot\hat{e}^i}{\sin\beta^i} e^{-jk_0 \hat{i}\cdot\mathbf{r}'}, \qquad E^i_{\phi^i}(\mathbf{r}') = -\frac{(\hat{t}\times\hat{i})\cdot\hat{e}^i}{\sin\beta^i} e^{-jk_0 \hat{i}\cdot\mathbf{r}'}$$

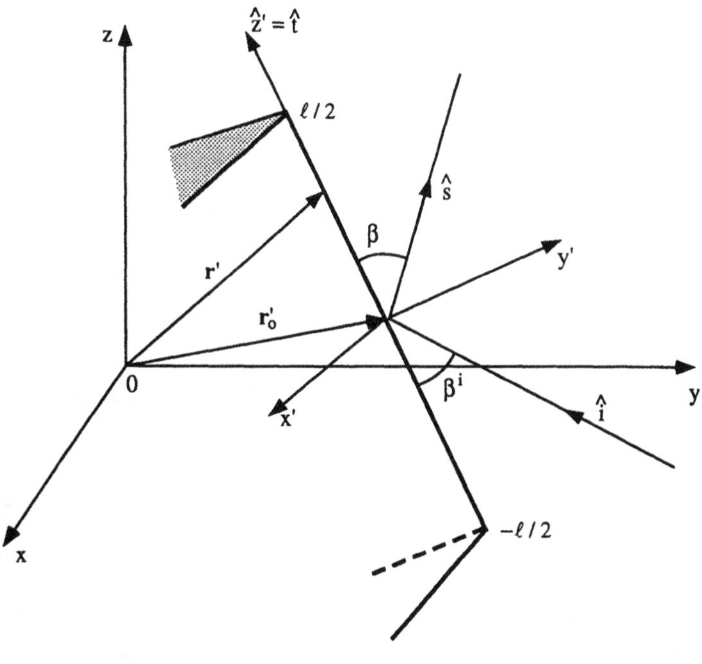

Figure 4–19: *Finite impedance wedge geometry*

we obtain

$$
\mathbf{E}^d = \frac{e^{-jk_0 r}}{r \sin \beta^i} \sqrt{\frac{k_0}{2\pi}} e^{j\pi/4} \int_{-\ell/2}^{\ell/2} \left\{ \hat{s} \times (\hat{s} \times \hat{t}) \left[(\hat{t}.\hat{e}^i) D_{\beta\beta^i} - (\hat{t} \times \hat{i}).\hat{e}^i D_{\beta\phi^i} \right] \right.
$$
$$
\left. - (\hat{s} \times \hat{t}) \left[(\hat{t}.\hat{e}^i) D_{\phi\beta^i} - (\hat{t} \times \hat{i}).\hat{e}^i D_{\phi\phi^i} \right] \right\} e^{jk_0(\hat{s}-\hat{i}).\mathbf{r}'} dz' \qquad (4.102)
$$

The incident electric field is shown in (4.93), and in terms of the global coordinates,

$$
\hat{i} = -\hat{x} \sin \theta^{\text{inc}} \cos \phi^{\text{inc}} - \hat{y} \sin \theta^{\text{inc}} \sin \phi^{\text{inc}} + \hat{z} \cos \theta^{\text{inc}}
$$

Because the edge is straight, the diffraction coefficients in (4.102) are independent of z'. Hence, if $\mathbf{r}' = \mathbf{r}_0' + z'\hat{z}$, where $\mathbf{r}_0'$ is the midpoint of the edge (see Fig. 4–19), the component of the diffracted field in the direction $\hat{u}$ is

$$
\hat{u}.\mathbf{E}^d = \frac{e^{-jk_0 r}}{r \sin \beta^i} \sqrt{\frac{k_0}{2\pi}} e^{j\pi/4} e^{jk_0(\hat{s}-\hat{i}).\mathbf{r}_0'} \left\{ (\hat{u}.\hat{t}) \left[(\hat{t}.\hat{e}^i) D_{\beta\beta^i} - (\hat{t} \times \hat{i}).\hat{e}^i D_{\beta\phi^i} \right] \right.
$$
$$
\left. + \hat{u}.(\hat{s} \times \hat{t}) \left[(\hat{t}.\hat{e}^i) D_{\phi\beta^i} - (\hat{t} \times \hat{i}).\hat{e}^i D_{\phi\phi^i} \right] \right\} \int_{-\ell/2}^{\ell/2} e^{jk_0 z'(\hat{s}-\hat{i}).\hat{i}} dz'
$$

where we have used the fact that $\hat{s}.\hat{u} = 0$ in the far field. On carrying out the integration,

$$
\hat{u}.\mathbf{E}^d = \frac{e^{-jk_0 r}}{r} (\overline{\overline{D}}.\hat{e}^i)\ell \, \text{sinc} \left\{ \tfrac{1}{2}k_0 \ell (\hat{s} - \hat{i}).\hat{i} \right\} \qquad (4.103)
$$

with

$$
\overline{\overline{D}}.\hat{e}^i = \frac{1}{\sin \beta^i} \sqrt{\frac{k_0}{2\pi}} e^{j\pi/4} e^{jk_0(\hat{s}-\hat{i}).\mathbf{r}_0'} \left\{ (\hat{u}.\hat{t}) \left[(\hat{t}.\hat{e}^i) \widetilde{D}_{\beta\beta^i} - (\hat{t} \times \hat{i}).\hat{e}^i \widetilde{D}_{\beta\phi^i} \right] \right.
$$
$$
\left. + \hat{u}.(\hat{s} \times \hat{t}) \left[(\hat{t}.\hat{e}^i) \widetilde{D}_{\phi\beta^i} - (\hat{t} \times \hat{i}).\hat{e}^i \widetilde{D}_{\phi\phi^i} \right] \right\} \qquad (4.104)
$$

In (4.104) the tilde was introduced in preparation for the definition of the ILDC. If we simply take

$$
\widetilde{D}_{uv} = D_{uv} \qquad (4.105)
$$

where $\{D_{uv}\} = \overline{\overline{D}}$ is the dyadic diffraction coefficient defined in (4.91), then (4.103) is the diffracted field in the context of the MEC, and the total field is obtained by adding the reflected field to $\mathbf{E}^d$. On the other hand, if

$$
\widetilde{D}_{uv} = D_{uv} - D_{uv}^{\text{po}} \qquad (4.106)
$$

where D_{uv}^{po} is the physical optics diffraction coefficient, (4.103) is the PTD diffracted field and $\overline{\overline{\mathcal{D}}}$ represents the ILDC. In this case (4.103) is referred to as the *fringe wave field* (UFIMTSEV, 1971), and $\overline{\overline{D}}^{\text{po}}$ serves to eliminate the contribution of the physical optics current at the edge from the GTD diffraction coefficient $\overline{\overline{D}}$. The total scattered field is then obtained by adding the physical optics and fringe wave fields, and this is the essence of the PTD. The physical optics diffraction coefficient D_{uv}^{po} for an impedance wedge is developed in Appendix D using edge-fixed coordinates.

The starting point for the above analysis was the equivalent current formulation of RYAN AND PETERS (1969), but a more rigourous derivation of the ILDC has been given by PELOSI ET AL. (1992) based on the formulation by MICHAELI (1984, 1986), MITZNER (1974) and KNOTT (1985). Unfortunately, the resulting expressions are much more cumbersome. In practice, both ILDCs yield approximations of the field diffracted by a finite edge and, in those cases where the edge is not a closed contour, it is not evident that the improvement provided by the Michaeli-Mitzner ILDC is sufficient to justify the increased computational burden (LEE AND JENG, 1991). We remark that ILDCs and equivalent currents similar to those of Michaeli and Mitzner have also been given by SHORE AND YAGHJIAN (1988) and GOKAN ET AL. (1990).

4.4.3   ILDC for a curved wedge

We now consider the curved wedge shown in Fig. 4–20 illuminated by the plane wave (4.93). To compute the diffracted field, the edge is subdivided into N small segments. The illumination of a typical *incremental length* segment is written as

$$\mathbf{E}^{\text{i}} = e^{-jk_0 \hat{\mathbf{i}} \cdot \mathbf{r_m}} \left(\hat{\theta}^{\text{inc}} a_{\theta\text{inc}} + \hat{\phi}^{\text{inc}} a_{\phi\text{inc}} \right) \tag{4.107}$$

where $\mathbf{r_m}$ is the midpoint of the segment from $\mathbf{r_s}$ to $\mathbf{r_e}$ (see Fig. 4–20) and

$$a_{\theta\text{inc}} = \hat{e}^{\text{i}} . \hat{\theta}^{\text{inc}}, \qquad a_{\phi\text{inc}} = \hat{e}^{\text{i}} . \hat{\phi}^{\text{inc}}$$

where

$$
\begin{aligned}
\hat{\theta}^{\text{inc}} &= \hat{x} \cos \theta^{\text{inc}} \cos \phi^{\text{inc}} + \hat{y} \cos \theta^{\text{inc}} \sin \phi^{\text{inc}} - \hat{z} \sin \theta^{\text{inc}} \\
\hat{\phi}^{\text{inc}} &= -\hat{x} \sin \phi^{\text{inc}} + \hat{y} \cos \phi^{\text{inc}}
\end{aligned}
$$

are spherical unit vectors in a global coordinate system. The far zone diffracted field contributed by a segment of length $\Delta \ell$ is then

$$\Delta \mathbf{E}^{\text{d}} = \frac{e^{-jk_0 r}}{r} \left(\hat{\theta} b_\theta + \hat{\phi} b_\phi \right) \tag{4.108}$$

where $\hat{\theta}, \hat{\phi}$ are spherical unit vectors in the global system. They are, of course, perpendicular to the scattering direction $\hat{s}$ and, as before, r is measured from the origin of the global coordinates. The coefficients b_θ and b_ϕ can be deduced from (4.103) and (4.104). By comparing (4.108) with (4.103) we find

$$\begin{pmatrix} b_\theta \\ b_\phi \end{pmatrix} = \begin{pmatrix} f_{\theta\theta\text{inc}} & f_{\theta\phi\text{inc}} \\ f_{\phi\theta\text{inc}} & f_{\phi\phi\text{inc}} \end{pmatrix} \begin{pmatrix} a_{\theta\text{inc}} \\ a_{\phi\text{inc}} \end{pmatrix} \tag{4.109}$$

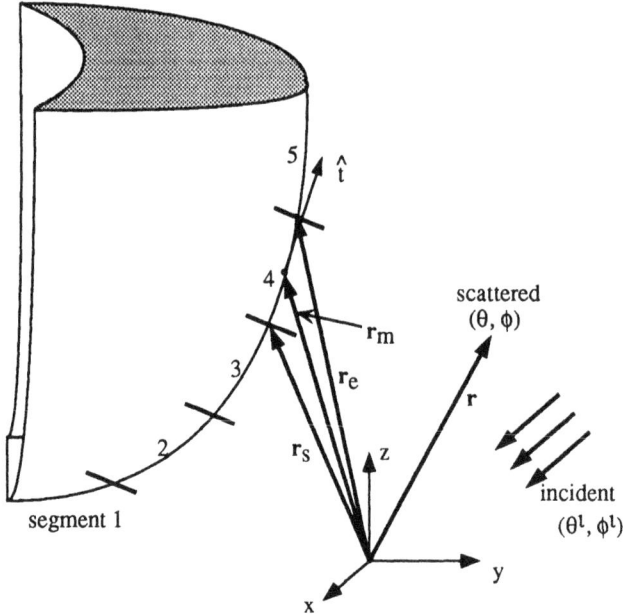

Figure 4–20: *Curved edge illuminated by a plane wave*

with

$$
f_{uv^i} = \frac{\Delta\ell}{\sin\beta^i}\sqrt{\frac{k_0}{2\pi}}\, e^{j\pi/4} e^{jk_0(\hat{s}-\hat{i})\cdot \mathbf{r}_m} \left\{ (\hat{u}.\hat{t}) \left[(\hat{t}.\hat{v}^i)\widetilde{D}_{\beta\beta^i} - (\hat{t}\times\hat{i}).\hat{v}^i\widetilde{D}_{\beta\phi^i} \right] \right.
$$
$$
\left. + \hat{u}.(\hat{s}\times\hat{t}) \left[(\hat{t}.\hat{v}^i)\widetilde{D}_{\phi\beta^i} - (\hat{t}\times\hat{i}).\hat{v}^i\widetilde{D}_{\phi\phi^i} \right] \right\} \tag{4.110}
$$

and since $\Delta\ell$ is small, we have set the sinc function in (4.103) to unity. The elements of the diffraction matrix are evaluated using local edge-fixed coordinates as if the segment were infinite in length. In particular, for the ϕ angles in the wedge diffraction coefficients of Section 4.3, we have the coordinate-free expressions

$$
\phi_0 = \begin{cases} \cos^{-1}\left[\dfrac{\hat{i}.(\hat{t}\times\hat{n})}{\sin\beta} \right] & \text{if } \hat{i}.\hat{n} \le 0 \\[4mm] 2\pi - \cos^{-1}\left[\dfrac{\hat{i}.(\hat{t}\times\hat{n})}{\sin\beta} \right] & \text{otherwise} \end{cases}
$$

$$
\phi = \begin{cases} \cos^{-1}\left[\dfrac{\hat{s}.(\hat{n}\times\hat{t})}{\sin\beta} \right] & \text{if } \hat{s}.\hat{n} \ge 0 \\[4mm] 2\pi - \cos^{-1}\left[\dfrac{\hat{s}.(\hat{n}\times\hat{t})}{\sin\beta} \right] & \text{otherwise} \end{cases}
$$

If $\hat{n}$ is the unit vector normal to the upper face of the wedge and the direction of $\hat{t}$ is such that $\hat{n} \times \hat{t}$ is a vector lying on that face, ϕ_0 and ϕ are then the angles shown in Fig. 4–6 with $\beta = \cos^{-1}(\hat{s}.\hat{t})$.

For the entire edge, the total diffracted field obtained by summing the contributions (4.108) from the individual segments is

$$\mathbf{E^d} = \sum_{n=1}^{N} \Delta \mathbf{E^d} \qquad (4.111)$$

and because (4.108)–(4.110) are expressed in global coordinates, the evaluation of (4.111) is a straightforward task if the diffraction coefficients are known. In the context of the PTD the scattered field is

$$\mathbf{E^s} = \mathbf{E^d} + \mathbf{E^{po}}$$

where $\mathbf{E^{po}}$ is the physical optics field discussed in Appendix D.

To illustrate the utility and accuracy of the PTD formulation, computations were made by modifying the PTD code of LEE (1990). Fig. 4–21 shows the radar cross-section (RCS)

$$\sigma = \lim_{r \to \infty} 4\pi r^2 \left| \frac{\mathbf{E^s}}{\mathbf{E^i}} \right|^2$$

of a solid circular metallic cylinder, $4\lambda_0$ in length, before and after the application of a thin coating simulated using the boundary condition (2.27) with (2.92). We observe that the effect of the coating is to reduce the RCS by 5–10 dB, and comparisons with the method of moments (MoM) solutions confirm the accuracy of the PTD results. Similar comparisons for the more complex geometry of a finite right circular cone are presented in Fig. 4–22. For both geometries, the ILDC was employed only at the circular edges, and the PO currents were integrated over the smooth surfaces.

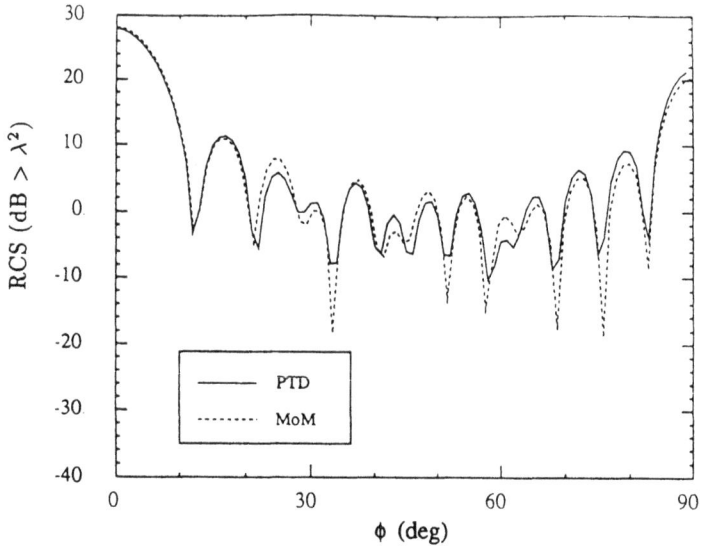

(a)

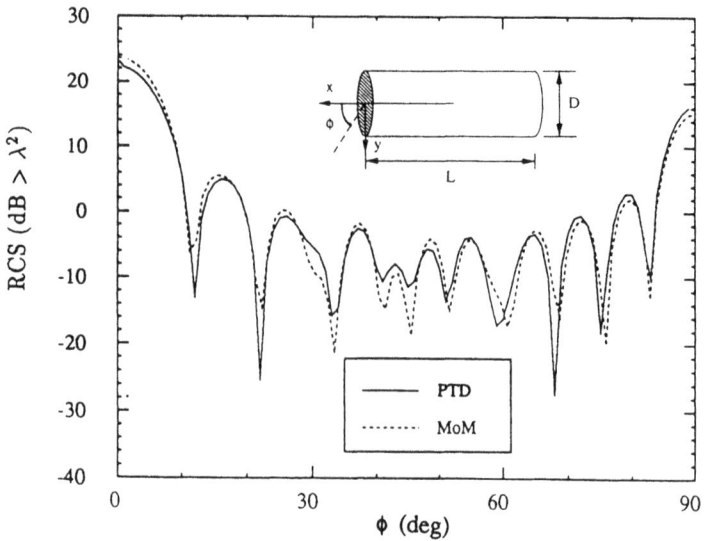

(b)

Figure 4–21: *Principal plane backscatter RCS based on E_ϕ^i, for a finite cylinder with $D = 3\lambda_0$ and $L = 4\lambda_0$: (a) perfectly conducting cylinder, (b) coated cylinder with coating thickness $0.04\lambda_0$, $\epsilon_r = 4 - j1.5$, $\mu_r = 2 - j$*

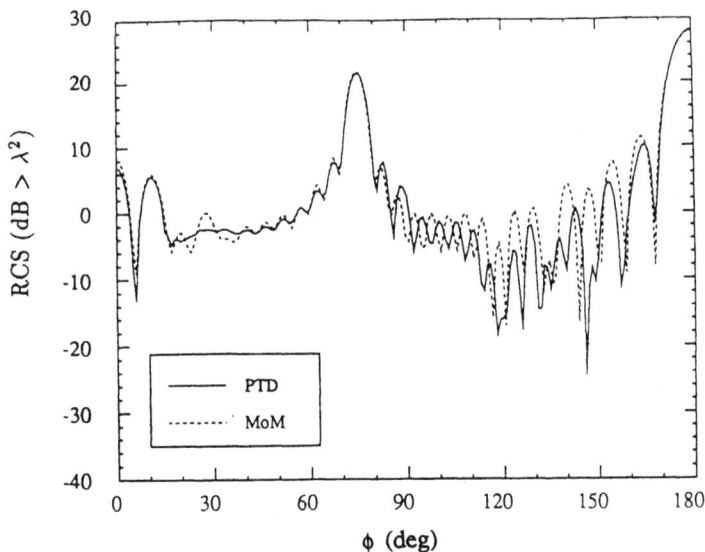

(a)

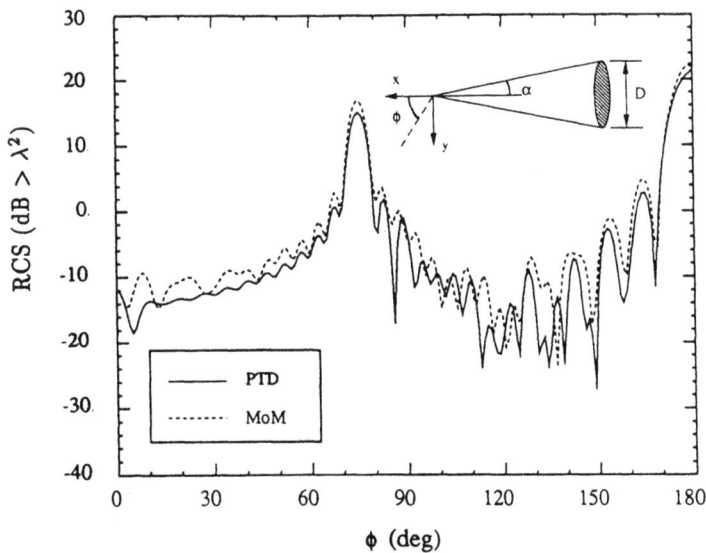

(b)

Figure 4–22: *Principal plane backscatter RCS based on $E_{\phi i}^i$ for a flat base cone with $D = 3\lambda_0$ and $\alpha = 15°$: (a) perfectly conducting cone, (b) coated cone with coating thickness $0.04\lambda_0$, $\epsilon_r = 7 - j1.5$, $\mu_r = 2 - j$*

References

Bowman, J. J., Senior, T. B. A. and Uslenghi, P. L. E., *Electromagnetic and Acoustic Scattering by Simple Shapes*, Hemisphere Pub. Co., New York, 1987.

Bucci, O. M. and Franceschetti, G. (1976), "Electromagnetic scattering by a half plane with two face impedances", *Radio Sci.*, **11**, pp. 49–59.

Butonin, D. I. and Ufimtsev, P. Ya. (1985), "Explicit expression for an acoustic edge wave scattered by an infinitesimal edge element", *Sov. Phys. Acoust.*, **32**, pp. 283–287.

Carslaw, H. S. (1898), "Multiform solutions of certain partial differential equations", *Proc. London Math. Soc.*, **30**, pp. 121–161.

Gokan, T., Ando, M. and Kinoshita, T. (1990), "A new definition of equivalent edge currents in a diffraction analysis", *Proc. Japan-China Joint Meeting on OFSET '90*, Fukuoka, Japan, pp. 31–40.

Hansen, R. C. (Ed.) (1981), *Geometrical Theory of Diffraction*, IEEE Press, New York NY.

Herman, M. I. and Volakis, J. L. (1988), "High frequency scattering by a double impedance wedge", *IEEE Trans. Antennas Propagat.*, **AP-36**, pp. 664–678.

Keller, J. B. (1962), "Geometrical theory of diffraction", *J. Opt. Soc. Amer.*, **52**, pp. 116–130.

Knott, E. F. (1985), "The relationship between Mitzner's ILDC and Michaeli's equivalent currents", *IEEE Trans. Antennas Propagat.*, **AP-33**, pp. 112–114.

Knott, E. F. and Senior, T. B. A. (1973), "Equivalent currents for a ring discontinuity", *IEEE Trans. Antennas Propagat.*, **AP-21**, pp. 693–695.

Knott, E. F., Shaeffer, J. F. and Tuley, M. T. (1985), *Radar Cross Section*, Artech House Inc., Dedham MA.

Kouyoumjian, R. G. and Pathak, P. H. (1974), "A uniform geometrical theory of diffraction for an edge in a perfectly conducting surface", *Proc. IEEE*, **62**, pp. 1448–1461.

Lee, S.-W. (1990), "McPTD-1.4: a high-frequency RCS computation code based on physical theory of diffraction", DEMACO.

Lee, S.-W. and Jeng, S. K. (1991), "Numerical computation of wedge diffraction", University of Illinois Report.

Maliuzhinets, G. D. (1951), "Some generalizations of the method of reflections in the theory of sinusoidal wave diffraction", Doctoral Dissertation, Fiz. Inst. Lebedev, Acad. Nauk. SSR (in Russian).

Maliuzhinets, G. D. (1958a), "Inversion formula for the Sommerfeld integral", *Sov. Phys. Doklady*, **3**, pp. 52–56.

Maliuzhinets, G. D. (1958b), "Excitation, reflection and emission of surface waves from a wedge with given face impedances", *Sov. Phys. Doklady*, **3**, pp. 752–755.

McNamara, D. A., Pistorius, C. W. I. and Malherbe, J. A. G. (1990), *Introduction to the Geometrical Theory of Diffraction*, Artech House, Norwood MA.

Michaeli, A. (1984), "Equivalent edge currents for arbitrary aspects of observation", *IEEE Trans. Antennas Propagat.*, **AP-32**, pp. 252–258.

Michaeli, A. (1986), "Elimination of infinities in equivalent edge currents, part I: fringe current components", *IEEE Trans. Antennas Propagat.*, **AP-34**, pp.

912–918.

Millar, R. F. (1957), The diffraction of an electromagnetic wave by a large aperture", *Proc. IEE*, **104C**, pp. 240–250.

Mitzner, K. M. (1974), "Incremental length diffraction coefficients", Aircraft Division Northrop Corporation Technical Report No. AFAL-TR-73-296; see also Knott, E. F. (1985), "The relationship between Mitzner's ILDC and Michaeli's equivalent currents", *IEEE Trans. Antennas Propagat.*, **AP-33**, pp. 112–114.

Pathak, P. H. (1992), "High frequency techniques for antenna analysis", *Proc. IEEE*, **80**, pp. 44–65.

Pelosi, G., Maci, S., Tiberio, R. and Michaeli, A. (1992), "Incremental length diffraction coefficients for an impedance wedge", *IEEE Trans. Antennas Propagat.*, **40**, pp. 1201–1210.

Rojas, R. G. (1988a), "Wiener-Hopf analysis of the EM diffraction by an impedance discontinuity in a planar surface and by an impedance half-plane", *IEEE Trans. Antennas Propagat.*, **AP-36**, pp. 71-83.

Rojas, R. G. (1988b), "Electromagnetic diffraction of an obliquely incident plane wave field by a wedge with impedance faces", *IEEE Trans. Antennas Propagat.*, **AP-36**, pp. 956–970.

Ryan, C. E. and Peters, L., Jr. (1969), "Evaluation of edge-diffracted fields including equivalent currents for the caustic regions", *IEEE Trans. Antennas Propagat.*, **AP-17**, pp. 292–299.

Senior, T. B. A. (1986), "Solution of a class of imperfect wedge problems for skew incidence", *Radio Sci.*, **21**, pp. 185–191.

Senior, T. B. A. and Volakis, J. L. (1986), "Scattering by an imperfect right-angled wedge", *IEEE Trans. Antennas Propagat.*, **AP-34**, pp. 681–689.

Shore, R. A. and Yaghjian, A. D. (1988), "Incremental length diffraction coefficients for planar surfaces", *IEEE Trans. Antennas Propagat.*, **AP-36**, pp. 55–70.

Syed, H. and Volakis, J. L. (1992a), "Electromagnetic scattering by coated convex surfaces and wedges simulated by approximate boundary conditions", University of Michigan Radiation Laboratory Report No. 025921-30-T.

Syed, H. H. and Volakis, J. L. (1992b), "An approximate skew incidence diffraction coefficient for an impedance wedge", *Electromagnetics*, **12**, pp. 33–55.

Syed, H. H. and Volakis, J. L. (1995), "An approximate solution for scattering by an impedance wedge at skew incidence", *Radio Sci.*, **30** (to appear).

Tiberio, R., Pelosi, G. and Manara, G. (1985), "A uniform GTD formulation for the diffraction by a wedge with impedance faces", *IEEE Trans. Antennas Propagat.*, **AP-33**, pp. 867–873.

Ufimtsev, P. Ya. (1971), "Method of edge waves in the physical theory of diffraction", U.S. Air Force Systems Command Foreign Technology Office Document ID No. FTD-HC-23-259-71 (translation of Russian original).

Vaccaro, V. G. (1980), "The generalized reflection method in electromagnetism", *AEÜ*, **34**, pp. 493–500.

Vaccaro, V. G. (1981), "Electromagnetic diffraction from a right-angled wedge with soft conditions on one face", *Optica Acta*, **28**, pp. 293–311.

Youssef, N. N. (1989), "Radar cross section of complex targets", *Proc. IEEE*, **77**, pp. 722–734.

Chapter 5

Second order conditions

5.1   Background

A first order boundary condition contains a single scalar or tensor impedance η which can be used to simulate the scattering properties of the surface. It represents a refinement of the conditions for ideal (pec or pmc) surfaces, and includes these as special cases, but the accuracy produced is not always sufficient. One way to improve the accuracy is to include additional derivatives of the field to create a higher order boundary condition with more degrees of freedom. The order is specified by the highest derivative present in one manifestation of the condition.

The earliest example is the second order condition developed by RYTOV (1940) to model the planar surface of a highly conducting material, but there is no evidence that the condition was ever used other than to help establish the validity of the first order one. A quarter of a century later KARP AND KARAL (1965) proposed a class of generalised impedance boundary conditions (now abbreviated to GIBCs) which they used to model a metal-backed dielectric layer whose thickness was sufficient to support more than one surface wave. They showed how to determine the coefficients and how the accuracy is affected by the layer thickness and the order of the condition but, prior to this, higher order conditions had come into use in mechanics.

For a hydroacoustic wave incident on a thin elastic plate, the assumption of an infinitesimally thin plate leads to a boundary (or transition) condition which is actually a fifth order one for the acoustic potential. It was recognised that if there is a discontinuity in the surface the condition no longer assures a unique solution of the boundary value problem, and at the discontinuity there are additional constraints (or contact conditions) which are necessary for the continuity of the surface displacement, flexure, bending moment etc. KOUZOV (1963a) considered the problem of a plane wave incident on the junction of two co-planar plates and developed the general solution containing four arbitrary constants. In a subsequent paper (KOUZOV, 1963b), the analysis was specialised to the case of a crack in an otherwise uniform plate, and it was shown how the contact conditions specify the constants. The more general problem of a right-angled wedge consisting of two plates subject to boundary conditions of arbitrary order was treated by BELINSKII ET AL. (1973), and it is only recently that the existence of such work has been appreciated by the electromagnetics community.

For a thin dielectric layer, second order transition conditions were developed by WEINSTEIN (1969) and used (LEPPINGTON, 1983) to determine the field diffracted by an abrupt change in layer thickness. Since then there have been numerous applications of second (and higher) order boundary conditions in electromagnetics, but some of the solutions are either incomplete or in error through a failure to address the uniqueness. A second order condition is the simplest one that has this property and, because of the improved accuracy that can be achieved on going to a second order condition, it is convenient to start with this.

5.2   Alternative forms

Consider first a planar surface $y = 0$. Following KARP AND KARAL (1965), a logical extension of the first order conditions (2.23) and (2.25) to the second order is

$$\prod_{m=1}^{2} \left(\Gamma_m - \frac{1}{jk_0} \frac{\partial}{\partial y} \right) E_y = 0, \qquad \prod_{m=1}^{2} \left(\Gamma'_m - \frac{1}{jk_0} \frac{\partial}{\partial y} \right) H_y = 0 \qquad (5.1)$$

for some Γ_m and Γ'_m, or equivalently

$$\sum_{m=0}^{2} a_{2-m} \frac{1}{(-jk_0)^m} \frac{\partial^m}{\partial y^m} E_y = 0, \qquad \sum_{m=0}^{2} a'_{2-m} \frac{1}{(-jk_0)^m} \frac{\partial^m}{\partial y^m} H_y = 0 \qquad (5.2)$$

and although we can obviously choose $a_0 = a'_0 = 1$, it is more convenient not to do so. For the incident plane wave

$$E_y \text{ or } H_y = e^{jk_0(x \cos \phi + y \sin \phi)} \qquad (5.3)$$

the corresponding reflection coefficients are

$$R_\parallel = - \prod_{m=1}^{2} \frac{\Gamma_m - \sin \phi}{\Gamma_m + \sin \phi} = - \frac{\sum_{m=0}^{2} a_{2-m} (- \sin \phi)^m}{\sum_{m=0}^{2} a_{2-m} (\sin \phi)^m} \qquad (5.4)$$

and

$$R_\perp = - \prod_{m=1}^{2} \frac{\Gamma'_m - \sin \phi}{\Gamma'_m + \sin \phi} = - \frac{\sum_{m=0}^{2} a'_{2-m} (- \sin \phi)^m}{\sum_{m=0}^{2} a'_{2-m} (\sin \phi)^m} \qquad (5.5)$$

The boundary condition for E_y can be written as

$$\left(a_0 \frac{\partial^2}{\partial y^2} - jk_0 a_1 \frac{\partial}{\partial y} - k_0^2 a_2 \right) E_y = 0$$

where

$$a_1/a_0 = \Gamma_1 + \Gamma_2, \qquad a_2/a_0 = \Gamma_1\Gamma_2 \tag{5.6}$$

and using the wave equation we obtain

$$\frac{\partial E_y}{\partial y} = jk_0 \left\{ \frac{a_0 + a_2}{a_1} + \frac{a_0}{k_0^2 a_1} \left(\frac{\partial^2}{\partial x^2} + \frac{\partial^2}{\partial z^2} \right) \right\} E_y \tag{5.7}$$

For the component H_y the analogous result is

$$\frac{\partial H_y}{\partial y} = jk_0 \left\{ \frac{a_2' + a_2'}{a_1'} + \frac{a_0'}{k_0^2 a_1'} \left(\frac{\partial^2}{\partial x^2} + \frac{\partial^2}{\partial z^2} \right) \right\} H_y \tag{5.8}$$

where

$$a_1'/a_0' = \Gamma_1' + \Gamma_2', \qquad a_2'/a_0' = \Gamma_1'\Gamma_2' \tag{5.9}$$

and now the only second derivatives are tangential ones. We can also express the conditions in terms of tangential field components. From Maxwell's equations and the fact that $\nabla.\mathbf{E} = 0$, (5.7) can be written as

$$\frac{\partial}{\partial x} \left(E_x - \frac{a_0 + a_2}{a_1} Z_0 H_z - \frac{a_0}{jk_0 a_1} \frac{\partial E_y}{\partial x} + \frac{\partial f}{\partial z} \right) =$$
$$-\frac{\partial}{\partial z} \left(E_z + \frac{a_0 + a_2}{a_1} Z_0 H_x - \frac{a_0}{jk_0 a_1} \frac{\partial E_y}{\partial z} - \frac{\partial f}{\partial x} \right) \tag{5.10}$$

for any function $f = f(x, z)$. Similarly, from (5.8),

$$\frac{\partial}{\partial x} \left(H_x + \frac{a_0' + a_2'}{a_1'} Y_0 E_z - \frac{a_0'}{jk_0 a_1'} \frac{\partial H_y}{\partial x} + \frac{\partial g}{\partial z} \right) =$$
$$-\frac{\partial}{\partial z} \left(H_z - \frac{a_0' + a_2'}{a_1'} Y_0 E_x - \frac{a_0'}{jk_0 a_1'} \frac{\partial H_y}{\partial z} - \frac{\partial g}{\partial x} \right)$$

for any function $g = g(x, z)$, and therefore

$$\frac{\partial}{\partial z} \left(E_x - \frac{a_1'}{a_0' + a_2'} Z_0 H_z + \frac{a_1'}{a_0' + a_2'} Z_0 \frac{\partial g}{\partial x} + \frac{a_0'}{jk_0(a_0' + a_2')} Z_0 \frac{\partial H_y}{\partial z} \right) =$$
$$\frac{\partial}{\partial x} \left(E_z + \frac{a_1'}{a_0' + a_2'} Z_0 H_x + \frac{a_1'}{a_0' + a_2'} Z_0 \frac{\partial g}{\partial z} - \frac{a_0'}{jk_0(a_0' + a_2')} Z_0 \frac{\partial H_y}{\partial x} \right) \tag{5.11}$$

Choose

$$f = \frac{a_0'}{jk_0(a_0' + a_2')} Z_0 H_y, \qquad g = -\frac{a_0' + a_2'}{a_1'} \frac{a_0}{jk_0 a_1} Y_0 E_y$$

Then if

$$\frac{a_1'}{a_0' + a_2'} = \frac{a_0 + a_2}{a_1} \tag{5.12}$$

so that

$$\frac{\Gamma_1' + \Gamma_2'}{1 + \Gamma_1'\Gamma_2'} = \frac{1 + \Gamma_1\Gamma_2}{\Gamma_1 + \Gamma_2} \tag{5.13}$$

(5.10) and (5.11) imply

$$\left(\frac{\partial^2}{\partial x^2} + \frac{\partial^2}{\partial z^2}\right)\left\{E_x - \frac{a_0 + a_2}{a_1}Z_0 H_z - \frac{a_0}{jk_0 a_1}\frac{\partial E_y}{\partial x} + \frac{a_0'}{jk_0(a_0' + a_2')}Z_0\frac{\partial H_y}{\partial z}\right\} = 0$$

$$\left(\frac{\partial^2}{\partial x^2} + \frac{\partial^2}{\partial z^2}\right)\left\{E_z + \frac{a_0 + a_2}{a_1}Z_0 H_x - \frac{a_0}{jk_0 a_1}\frac{\partial E_y}{\partial z} - \frac{a_0'}{jk_0(a_0' + a_2')}Z_0\frac{\partial H_y}{\partial x}\right\} = 0$$

and by the same argument as that used in Section 2.3, we obtain

$$E_x - \frac{a_0}{jk_0 a_1}\frac{\partial E_y}{\partial x} = \frac{a_0 + a_2}{a_1}Z_0\left(H_z - \frac{a_0'}{jk_0 a_1'}\frac{\partial H_y}{\partial z}\right)$$

$$E_z - \frac{a_0}{jk_0 a_1}\frac{\partial E_y}{\partial z} = -\frac{a_0 + a_2}{a_1}Z_0\left(H_x - \frac{a_0'}{jk_0 a_1'}\frac{\partial H_y}{\partial x}\right) \tag{5.14}$$

on $y = 0+$. Hence (SENIOR AND VOLAKIS, 1989)

$$\hat{y}\times\left\{\mathbf{E} - \frac{a_0}{jk_0 a_1}\nabla(\hat{y}.\mathbf{E})\right\} = \frac{a_0 + a_2}{a_1}Z_0\,\hat{y}\times\left(\hat{y}\times\left\{\mathbf{H} - \frac{a_0'}{jk_0 a_1'}\nabla(\hat{y}.\mathbf{H})\right\}\right) \tag{5.15}$$

which can also be written as

$$\hat{y}\times\left\{\mathbf{H} - \frac{a_0'}{jk_0 a_1'}\nabla(\hat{y}.\mathbf{H})\right\} = -\frac{a_0' + a_2'}{a_1'}Y_0\,\hat{y}\times\left(\hat{y}\times\left\{\mathbf{E} - \frac{a_0}{jk_0 a_1}\nabla(\hat{y}.\mathbf{E})\right\}\right) \tag{5.16}$$

Provided the coefficients satisfy (5.12), the condition (5.15) is equivalent to (5.1) and, as the above analysis shows, can be obtained from (5.1) by a process of tangential integration. Conversely, (5.1) follows from (5.15) by tangential differentiation. Accordingly, (5.15) is less singular than (5.1) and differs from the SIBC (2.27) in the presence of the field derivatives.

The corresponding transition conditions are easily derived. For a second order resistive sheet in the plane $y = 0$, supporting only an electric current,

$$\hat{y}\times\left\{\mathbf{E}^+ + \mathbf{E}^- - \frac{a_0}{jk_0 a_1}\nabla[\hat{y}.\mathbf{E}]_-^+\right\} = \frac{a_0 + a_2}{a_1}Z_0\,\hat{y}$$

$$\times\left\{[\hat{y}\times\mathbf{H}]_-^+ - \frac{a_0'}{jk_0 a_1'}\nabla\left(\hat{y}.\left(\mathbf{H}^+ + \mathbf{H}^-\right)\right)\right\} \tag{5.17}$$

with

$$[\hat{y}\times\mathbf{E}]_-^+ = 0 \tag{5.18}$$

and for a second order conductive sheet supporting a magnetic current,

$$\hat{y} \times \left\{ \mathbf{H}^+ + \mathbf{H}^- - \frac{a_0'}{jk_0 a_1'} \nabla[\hat{y}.\mathbf{H}]_-^+ \right\} = -\frac{a_0' + a_2'}{a_1'} Y_0 \, \hat{y}$$

$$\times \left\{ [\hat{y} \times \mathbf{E}]_-^+ - \frac{a_0}{jk_0 a_1} \nabla \left(\hat{y}. \left(\mathbf{E}^+ + \mathbf{E}^- \right) \right) \right\} \tag{5.19}$$

with

$$[\hat{y} \times \mathbf{H}]_-^+ = 0 \tag{5.20}$$

The equivalent forms based on (5.1) are

$$\prod_{m=1}^{2} \left(\Gamma_m - \frac{1}{jk_0} \frac{\partial}{\partial y} \right) E_y^+ - \prod_{m=1}^{2} \left(\Gamma_m + \frac{1}{jk_0} \frac{\partial}{\partial y} \right) E_y^- = 0$$

$$\prod_{m=1}^{2} \left(\Gamma_m' - \frac{1}{jk_0} \frac{\partial}{\partial y} \right) H_y^+ + \prod_{m=1}^{2} \left(\Gamma_m' + \frac{1}{jk_0} \frac{\partial}{\partial y} \right) H_y^- = 0$$

$$\tag{5.21}$$

with

$$\left[\frac{\partial E_y}{\partial y} \right]_-^+ = 0, \qquad [H_y]_-^+ = 0 \tag{5.22}$$

for the resistive sheet and, in the case of the incident plane wave (5.3), the reflection and transmission coefficients are

$$R_\parallel = \frac{1}{2} \left\{ 1 - \prod_{m=1}^{2} \frac{\Gamma_m - \sin\phi}{\Gamma_m + \sin\phi} \right\}$$

$$T_\parallel = \frac{1}{2} \left\{ 1 + \prod_{m=1}^{2} \frac{\Gamma_m - \sin\phi}{\Gamma_m + \sin\phi} \right\}$$

$$\tag{5.23}$$

and

$$R_\perp = -\frac{1}{2} \left\{ 1 + \prod_{m=1}^{2} \frac{\Gamma_m' - \sin\phi}{\Gamma_m' + \sin\phi} \right\}$$

$$T_\perp = \frac{1}{2} \left\{ 1 - \prod_{m=1}^{2} \frac{\Gamma_m' - \sin\phi}{\Gamma_m' + \sin\phi} \right\}$$

$$\tag{5.24}$$

based on E_y and H_y respectively. For the conductive sheet

$$\prod_{m=1}^{2} \left(\Gamma_m - \frac{1}{jk_0} \frac{\partial}{\partial y} \right) E_y^+ + \prod_{m=1}^{2} \left(\Gamma_m + \frac{1}{jk_0} \frac{\partial}{\partial y} \right) E_y^- = 0$$

$$\prod_{m=1}^{2} \left(\Gamma_m' - \frac{1}{jk_0} \frac{\partial}{\partial y} \right) H_y^+ - \prod_{m=1}^{2} \left(\Gamma_m' + \frac{1}{jk_0} \frac{\partial}{\partial y} \right) H_y^- = 0$$

$$\tag{5.25}$$

with

$$[E_y]_-^+ = 0, \qquad \left[\frac{\partial H_y}{\partial y}\right]_-^+ = 0 \qquad (5.26)$$

and the reflection and transmission coefficients are the obvious counterparts of (5.23) and (5.24).

Since (5.15) represents a refinement of the SIBC (2.27), it is expected that in any practical application $|a_1|/|a_0|$, $|a_1'|/|a_0'| \gg 1$. If this is so

$$\hat{y}.\mathbf{H} = -\frac{Y_0}{jk_0}\,\hat{y}.\nabla \times \mathbf{E} \simeq -\frac{1}{jk_0}\frac{a_0 + a_2}{a_1}\nabla_{\!s}.\mathbf{H}$$

to the same accuracy, where $\nabla_{\!s}.$ is the surface divergence, and (5.15) can then be written as

$$\hat{y} \times \mathbf{E} = \frac{a_0 + a_2}{a_1}Z_0\,\hat{y} \times \left(\hat{y} \times \left\{\mathbf{H} - \frac{a_0'}{k_0^2(a_0' + a_2')}\nabla(\nabla_{\!s}.\mathbf{H})\right\}\right.$$
$$\left. - \frac{a_0}{k_0^2(a_0 + a_2)}\nabla(\hat{y}.\nabla \times \mathbf{H})\right) \qquad (5.27)$$

The fact that this is a second order GIBC involving second order tangential derivatives of the field components is now evident, and its components are

$$E_x = \frac{a_0 + a_2}{a_1}Z_0\left\{\left[1 + \frac{1}{k_0^2}\left(\frac{a_0}{a_0 + a_2}\frac{\partial^2}{\partial x^2} - \frac{a_0'}{a_0' + a_2'}\frac{\partial^2}{\partial z^2}\right)\right]H_z\right.$$
$$\left. - \frac{1}{k_0^2}\left(\frac{a_0}{a_0 + a_2} + \frac{a_0'}{a_0' + a_2'}\right)\frac{\partial^2 H_x}{\partial x\,\partial z}\right\}$$

$$(5.28)$$

$$E_z = -\frac{a_0 + a_2}{a_1}Z_0\left\{\left[1 + \frac{1}{k_0^2}\left(\frac{a_0}{a_0 + a_2}\frac{\partial^2}{\partial z^2} - \frac{a_0'}{a_0' + a_2'}\frac{\partial^2}{\partial x^2}\right)\right]H_x\right.$$
$$\left. - \frac{1}{k_0^2}\left(\frac{a_0}{a_0 + a_2} + \frac{a_0'}{a_0' + a_2'}\right)\frac{\partial^2 H_z}{\partial x\,\partial z}\right\}$$

For a resistive sheet the analogous expressions are

$$E_x^+ + E_x^- = \frac{a_0 + a_2}{a_1}Z_0\left\{\left[1 + \frac{1}{k_0^2}\left(\frac{a_0}{a_0 + a_2}\frac{\partial^2}{\partial x^2} - \frac{a_0'}{a_0' + a_2'}\frac{\partial^2}{\partial z^2}\right)\right][H_z]_-^+\right.$$
$$\left. - \frac{1}{k_0^2}\left(\frac{a_0}{a_0 + a_2} + \frac{a_0'}{a_0' + a_2'}\right)\frac{\partial^2}{\partial x\,\partial z}[H_x]_-^+\right\}$$

$$(5.29)$$

$$E_z^+ + E_z^- = -\frac{a_0 + a_2}{a_1}Z_0\left\{\left[1 + \frac{1}{k_0^2}\left(\frac{a_0}{a_0 + a_2}\frac{\partial^2}{\partial z^2} - \frac{a_0'}{a_0' + a_2'}\frac{\partial^2}{\partial x^2}\right)\right][H_x]_-^+\right.$$
$$\left. - \frac{1}{k_0^2}\left(\frac{a_0}{a_0 + a_2} + \frac{a_0'}{a_0' + a_2'}\right)\frac{\partial^2}{\partial x\,\partial z}[H_z]_-^+\right\}$$

with

$$[E_x]_-^+ = [E_z]_-^+ = 0 \tag{5.30}$$

and for a conductive sheet

$$
\begin{aligned}
H_x^+ + H_x^- &= -\frac{a_0' + a_2'}{a_1'} Y_0 \left\{ \left[1 + \frac{1}{k_0^2} \left(\frac{a_0'}{a_0' + a_2'} \frac{\partial^2}{\partial x^2} - \frac{a_0}{a_0 + a_2} \frac{\partial^2}{\partial z^2} \right) \right] [E_z]_-^+ \right. \\
&\left. - \frac{1}{k_0^2} \left(\frac{a_0'}{a_0' + a_2'} + \frac{a_0}{a_0 + a_2} \right) \frac{\partial^2}{\partial x \, \partial z} [E_x]_-^+ \right\}
\end{aligned}
\tag{5.31}
$$

$$
\begin{aligned}
H_z^+ + H_z^- &= \frac{a_0' + a_2'}{a_1'} Y_0 \left\{ \left[1 + \frac{1}{k_0^2} \left(\frac{a_0'}{a_0' + a_2'} \frac{\partial^2}{\partial z^2} - \frac{a_0}{a_0 + a_2} \frac{\partial^2}{\partial x^2} \right) \right] [E_x]_-^+ \right. \\
&\left. - \frac{1}{k_0^2} \left(\frac{a_0'}{a_0' + a_2'} + \frac{a_0}{a_0 + a_2} \right) \frac{\partial^2}{\partial x \, \partial z} [E_z]_-^+ \right\}
\end{aligned}
\tag{5.32... }
$$

with

$$[H_x]_-^+ = [H_z]_-^+ = 0 \tag{5.32}$$

The natural extension of (5.15) to a curved anisotropic material surface S is

$$\hat{n} \times \left\{ \mathbf{E} + \frac{1}{jk_0} \nabla(A\,\hat{n}.\mathbf{E}) \right\} = \hat{n} \times \left(\hat{n} \times \left\{ \overline{\overline{\eta}}.\mathbf{H} - \frac{1}{jk_0} \nabla(B\,\hat{n}.\mathbf{H}) \right\} \right) \tag{5.33}$$

where $\overline{\overline{\eta}}$, A and B are functions of position on S and $\hat{n}$ is the outward unit vector normal. This is a rather general form of a second order GIBC but, if the parameters vary over the surface and/or η is a tensor, the condition no longer reduces to (5.7) for a planar surface.

5.3 Examples

The purpose in going from a first order to a second order boundary condition is to improve the accuracy. We now consider a few simple but important cases to show how the coefficients can be determined and examine the accuracy that can be achieved.

5.3.1 Thin dielectric layer

We take first a thin layer consisting of a homogeneous dielectric whose permittivity and permeability are $\epsilon = \epsilon_r \epsilon_0$ and $\mu = \mu_r \mu_0$, respectively, occupying the region $-\frac{\tau}{2} < y < \frac{\tau}{2}$, $-\infty < x, z < \infty$, with free space surrounding the layer. It is assumed that the dielectric provides a low contrast, so that $|\epsilon_r|$ and $|\mu_r|$ are not large compared with unity. As illustrated in Fig. 5–1, we seek to represent the effect of the layer using a current sheet in the plane $y = \delta$ where $-\frac{\tau}{2} \le \delta \le \frac{\tau}{2}$.

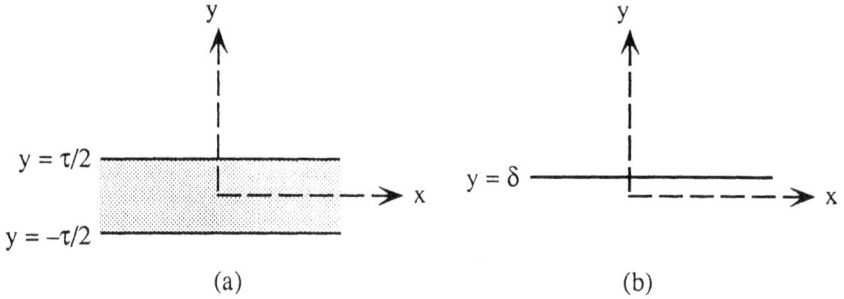

Figure 5–1: *Thin dielectric layer and its simulating sheet*

Following WEINSTEIN (1969) the tangential electric field components within the layer are expanded in Taylor series in y. Thus

$$E'_x(\delta+) = E'_x\left(\frac{\tau}{2}-\right) - \left(\frac{\tau}{2}-\delta\right)\frac{\partial}{\partial y}E'_x\left(\frac{\tau}{2}-\right) + O(\tau^2)$$

$$= E'_x\left(\frac{\tau}{2}-\right) - \left(\frac{\tau}{2}-\delta\right)$$
$$\cdot\left\{\frac{\partial}{\partial x}E'_y\left(\frac{\tau}{2}-\right) + jk_0 Z_0\mu_r\, H'_z\left(\frac{\tau}{2}-\right)\right\} + O(\tau^2)$$

where the prime denotes the internal field and we show only the dependence on the y coordinate perpendicular to the layer. By using the exact boundary conditions at $y = \frac{\tau}{2}$, we can write this in terms of the external fields as

$$E'_x(\delta+) = E_x\left(\frac{\tau}{2}+\right) - \left(\frac{\tau}{2}-\delta\right)\left\{\frac{1}{\epsilon_r}\frac{\partial}{\partial x}E_y\left(\frac{\tau}{2}+\right)\right.$$
$$\left. + jk_0 Z_0\mu_r\, H_z\left(\frac{\tau}{2}+\right)\right\} + O(\tau^2)$$

and a further expansion of the external fields enables us to express the right hand side in terms of the field components at $y = \delta+$ in free space, viz.

$$E'_x(\delta+) = E_x(\delta+) + \left(\frac{\tau}{2}-\delta\right)\left\{\frac{1}{\epsilon_r}(\epsilon_r-1)\frac{\partial}{\partial x}E_y(\delta+)\right.$$
$$\left. - jk_0 Z_0(\mu_r-1)\, H_z(\delta+)\right\} + O(\tau^2)$$

Similarly

$$E'_x(\delta-) = E_x(\delta-) - \left(\frac{\tau}{2}+\delta\right)\left\{\frac{1}{\epsilon_r}(\epsilon_r-1)\frac{\partial}{\partial x}E_y(\delta-)\right.$$
$$\left. - jk_0 Z_0(\mu_r-1)\, H_z(\delta-)\right\} + O(\tau^2)$$

and since $E'_x(\delta+) = E'_x(\delta-)$, it follows that to the first order in τ,

$$E_x(\delta+) - E_x(\delta-) =$$
$$-\frac{\tau}{2}\frac{1}{\epsilon_r}(\epsilon_r - 1)\left\{\frac{\partial}{\partial x}E_y(\delta+) + \frac{\partial}{\partial x}E_y(\delta-)\right\} + jk_0Z_0\frac{\tau}{2}(\mu_r - 1)$$
$$\cdot\{H_z(\delta+) + H_z(\delta-)\} + \delta\frac{1}{\epsilon_r}(\epsilon_r - 1)\left\{\frac{\partial}{\partial x}E_y(\delta+) - \frac{\partial}{\partial x}E_y(\delta-)\right\}$$
$$- jk_0Z_0\delta(\mu_r - 1)\{H_z(\delta+) - H_z(\delta-)\} \qquad (5.34)$$

The same procedure applied to the component E_z gives

$$E_z(\delta+) - E_z(\delta-) =$$
$$-\frac{\tau}{2}\frac{1}{\epsilon_r}(\epsilon_r - 1)\left\{\frac{\partial}{\partial z}E_y(\delta+) + \frac{\partial}{\partial z}E_y(\delta-)\right\} - jk_0Z_0\frac{\tau}{2}(\mu_r - 1)$$
$$\cdot\{H_x(\delta+) + H_x(\delta-)\} + \delta\frac{1}{\epsilon_r}(\epsilon_r - 1)\left\{\frac{\partial}{\partial z}E_y(\delta+) - \frac{\partial}{\partial z}E_y(\delta-)\right\}$$
$$+ jk_0Z_0\delta(\mu_r - 1)\{H_x(\delta+) - H_x(\delta-)\} \qquad (5.35)$$

Obviously, these are only valid outside the region occupied by the original layer.

If $\delta = 0$, i.e. the simulating sheet is at the centre of the layer, we can express (5.34) and (5.35) in terms of the components E_y and H_y alone. By differentiating the first with respect to x, the second with respect to z, and then adding, we obtain

$$\left\{\frac{\partial^2}{\partial y^2} + k_0^2\left[1 + \frac{\epsilon_r(\mu_r - 1)}{\epsilon_r - 1}\right]\right\}(E_y^+ + E_y^-)$$
$$+ \frac{2\epsilon_r}{\tau(\epsilon_r - 1)}\left\{\frac{\partial}{\partial y}E_y^+ - \frac{\partial}{\partial y}E_y^-\right\} = 0 \qquad (5.36)$$

where the affices denote the upper ($y = 0+$) and lower ($y = 0-$) sides of the sheet. Similarly, by differentiating (5.34) and (5.35) with respect to z and x respectively, and subtracting

$$\frac{\partial}{\partial y}H_y^+ + \frac{\partial}{\partial y}H_y^- - \frac{2}{\tau(\mu_r - 1)}\left(H_y^+ - H_y^-\right) = 0 \qquad (5.37)$$

The electromagnetic duals are

$$\left\{\frac{\partial^2}{\partial y^2} + k_0^2\left[1 + \frac{\mu_r(\epsilon_r - 1)}{\mu_r - 1}\right]\right\}(H_y^+ + H_y^-)$$
$$+ \frac{2\mu_r}{\tau(\mu_r - 1)}\left\{\frac{\partial}{\partial y}H_y^+ - \frac{\partial}{\partial y}H_y^-\right\} = 0 \qquad (5.38)$$

and

$$\frac{\partial}{\partial y} E_y^+ + \frac{\partial}{\partial y} E_y^- - \frac{2}{\tau(\epsilon_r - 1)} \left(E_y^+ - E_y^- \right) = 0 \qquad (5.39)$$

and these are identical to the results obtained by SENIOR AND VOLAKIS (1987) using a different approach based on the polarisation currents in the layer.

To clarify the meaning of these expressions, suppose first that the dielectric is non-magnetic so that $\mu_r = 1$. The conditions then reduce to

$$\left(1 + \frac{1}{k_0^2} \frac{\partial}{\partial y^2} \right) \left(E_y^+ + E_y^- \right) - \frac{2\tilde{R}_e}{jk_0 Y_0} \left\{ \frac{\partial}{\partial y} E_y^+ - \frac{\partial}{\partial y} E_y^- \right\} = 0 \qquad (5.40)$$

$$H_y^+ - H_y^- = 0 \qquad (5.41)$$

$$H_y^+ + H_y^- - \frac{2R_e}{jk_0 Z_0} \left\{ \frac{\partial}{\partial y} H_y^+ - \frac{\partial}{\partial y} H_y^- \right\} = 0 \qquad (5.42)$$

$$\frac{\partial}{\partial y} E_y^+ + \frac{\partial}{\partial y} E_y^- - \frac{2jk_0 R_e}{Z_0} \left(E_y^+ - E_y^- \right) = 0 \qquad (5.43)$$

where

$$R_e = -\frac{jZ_0}{k_0 \tau(\epsilon_r - 1)}, \qquad \tilde{R}_e = -\frac{jY_0 \epsilon_r}{k_0 \tau(\epsilon_r - 1)} \qquad (5.44)$$

Comparison with (2.59) and (2.60) shows that if $\tilde{R}_e$ were infinite, the four conditions would define a (first order) resistive sheet with resistivity R_e and lying in the plane $y = 0$. As noted in Section 2.4, such a sheet does not take into account the normal component of the polarisation current in the layer, and this is the role played by (5.40). For $\tilde{R}_e$ finite, it simulates a distribution of electric dipoles oriented perpendicular to the plane $y = 0$ using a magnetic current $\mathbf{J}_{m(s)}$ in the plane. Apart from the second derivative, it is identical to the transition condition for a conductive sheet of conductivity $\tilde{R}_e$ supporting a magnetic current

$$\mathbf{J}_{m(s)} = -[\hat{y} \times \mathbf{E}]_-^+$$

and the conditions (5.40) and (5.41) have been referred to (SENIOR AND VO-LAKIS, 1987) as those for a modified conductive sheet. It is an example of a second order transition condition, and the four conditions (5.40)–(5.43) describe a combination sheet consisting of a first order resistive sheet and a second order conductive one, with the latter simulating the effect of E_y'. In the case of H-polarisation such that $\mathbf{H} = \hat{z} H_z$ (say), only (5.40) and (5.43) are required, and these are the conditions used by LEPPINGTON (1983).

For a dielectric with $\mu_r \neq 1$ but $\epsilon_r = 1$, the transition conditions are the duals of (5.40)–(5.43) defining a combination sheet made up of a first order conductive sheet with conductivity

$$R_m = -\frac{jY_0}{k_0 \tau(\mu_r - 1)} \qquad (5.45)$$

and a second order resistive sheet with

$$\tilde{R}_m = -\frac{jZ_0\mu_r}{k_0\tau(\mu_r - 1)} \tag{5.46}$$

Finally, if $\mu_r \neq 1$ and $\epsilon_r \neq 1$ the transition conditions are as shown in (5.36)–(5.39). They can be written as

$$\left\{\frac{1}{2R_m} + \frac{1}{2\tilde{R}_e}\left(1 + \frac{1}{k_0^2}\frac{\partial^2}{\partial y^2}\right)\right\}\left(E_y^+ + E_y^-\right)$$
$$+ \frac{jZ_0}{k_0}\left\{\frac{\partial}{\partial y}E_y^+ - \frac{\partial}{\partial y}E_y^-\right\} = 0 \tag{5.47}$$

$$\frac{\partial}{\partial y}H_y^+ + \frac{\partial}{\partial y}H_y^- - \frac{2jk_0R_m}{Y_0}\left(H_y^+ - H_y^-\right) = 0 \tag{5.48}$$

$$\left\{\frac{1}{2R_e} + \frac{1}{2\tilde{R}_m}\left(1 + \frac{1}{k_0^2}\frac{\partial^2}{\partial y^2}\right)\right\}\left(H_y^+ + H_y^-\right)$$
$$+ \frac{jY_0}{k_0}\left\{\frac{\partial}{\partial y}H_y^+ - \frac{\partial}{\partial y}H_y^-\right\} = 0 \tag{5.49}$$

$$\frac{\partial}{\partial y}E_y^+ + \frac{\partial}{\partial y}E_y^- - \frac{2jk_0R_e}{Z_0}\left(E_y^+ - E_y^-\right) = 0 \tag{5.50}$$

and define a combination sheet consisting of second order conductive and resistive sheets. We observe that ϵ_r and μ_r contribute to each. Comparison of (5.47) and (5.48) with (5.25) shows that for the conductive sheet

$$\Gamma_1 + \Gamma_2 = -2Z_0\tilde{R}_e, \qquad \Gamma_1\Gamma_2 = -\left(\frac{\tilde{R}_e}{R_m} + 1\right)$$
$$\Gamma_2' = \infty, \qquad \Gamma_1' = 2Z_0R_m \tag{5.51}$$

and for the resistive sheet defined by (5.49) and (5.50)

$$\Gamma_1' + \Gamma_2' = -2Y_0\tilde{R}_m, \qquad \Gamma_1'\Gamma_2' = -\left(\frac{\tilde{R}_m}{R_e} + 1\right)$$
$$\Gamma_2 = \infty, \qquad \Gamma_1 = 2Y_0R_e \tag{5.52}$$

In each case the coefficients satisfy (5.13). The two sheets are uncoupled in the sense that each scatters independently of the other but, if the sheets are *not* located in the plane $y = 0$ at the centre of the layer, the added terms proportional to δ in (5.34) and (5.35) produce a coupling.

To see how accurately the transition conditions model the layer, it is sufficient to consider a non-magnetic material for which the conditions are (5.40)–(5.43). The corresponding reflection coefficients for the incident plane wave

(5.3) are

$$R_{\parallel} = \left(1 + \frac{2R_e}{Z_0 \sin \phi}\right)^{-1} - \left(1 + \frac{2\tilde{R}_e \sin \phi}{Y_0 \cos^2 \phi}\right)^{-1} \tag{5.53}$$

$$R_{\perp} = -\left(1 + \frac{2R_e}{Z_0} \sin \phi\right)^{-1} \tag{5.54}$$

and because the two sheets scatter independently, (5.53) is the sum of the reflection coefficients for a first order resistive sheet and a second order conductive sheet in isolation. For the dielectric layer shown in Fig. 5–1(a), the exact reflection coefficients can be derived from (2.93) and (2.90) using image theory and duality, and they are

$$R_{\parallel} = e^{jk_0\tau \sin \phi}\left\{1 + \frac{2}{(\epsilon_r - 1)(\epsilon_r \sin^2 \phi - \cos^2 \phi)}\left[\epsilon_r - \cos^2 \phi\right.\right.$$
$$\left.\left. - j\sqrt{\epsilon_r - \cos^2 \phi}\,\epsilon_r \sin \phi \cot\left(k_0\tau\sqrt{\epsilon_r - \cos^2 \phi}\right)\right]\right\}^{-1} \tag{5.55}$$

$$R_{\perp} = -e^{jk_0\tau \sin \phi}\left\{1 + \frac{2\sin \phi}{\epsilon_r - 1}\left[\sin \phi\right.\right.$$
$$\left.\left. - j\sqrt{\epsilon_r - \cos^2 \phi}\cot\left(k_0\tau\sqrt{\epsilon_r - \cos^2 \phi}\right)\right]\right\}^{-1} \tag{5.56}$$

where the phase factor has been introduced to account for the location of the upper surface at $y = \tau/2$. Assuming $|k_0\tau(\epsilon_r - \cos^2 \phi)^{1/2}|$ is small so that

$$\cot\left(k_0\tau\sqrt{\epsilon_r - \cos^2 \phi}\right) \simeq \left(k_0\tau\sqrt{\epsilon_r - \cos^2 \phi}\right)^{-1}$$

the exact reflection coefficients can be approximated as

$$R_{\parallel} = e^{jk_0\tau \sin \phi}\left\{1 + \frac{2}{(\epsilon_r - 1)(\epsilon_r \sin^2 \phi - \cos^2 \phi)}\right.$$
$$\left. \cdot \left(\epsilon_r - \cos^2 \phi - \frac{j\epsilon_r}{k_0\tau}\sin \phi\right)\right\}^{-1} \tag{5.57}$$

$$R_{\perp} = e^{jk_0\tau \sin \phi}\left\{1 + \frac{2\sin \phi}{\epsilon_r - 1}\left(\sin \phi - \frac{j}{k_0\tau}\right)\right\}^{-1} \tag{5.58}$$

If $k_0\tau \sin \phi$ is so small that it can be neglected in both the phase and the amplitude, (5.58) becomes identical to (5.54), but (5.57) cannot be simplified to the same extent because of the Brewster angle at $\phi = \cot^{-1} \epsilon_r^{1/2}$. However, (5.53) accurately predicts the Brewster angle, and provides a close approximation to (5.57).

In Figs. 5–2 and 5–3 the amplitudes and phases of the exact reflection

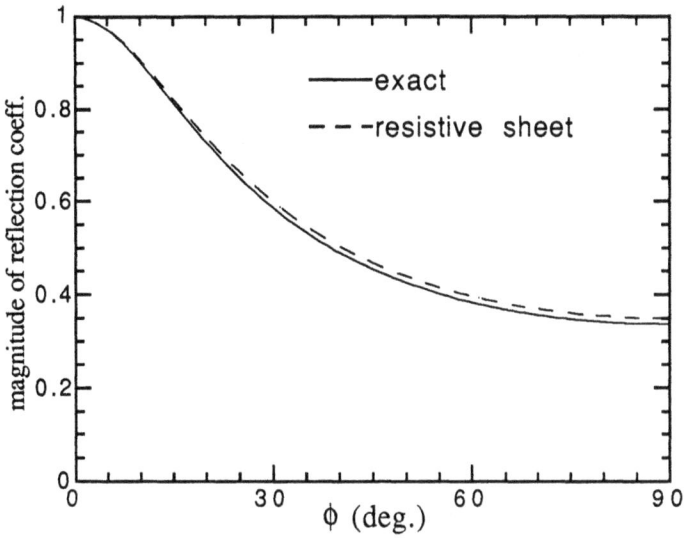

(a)

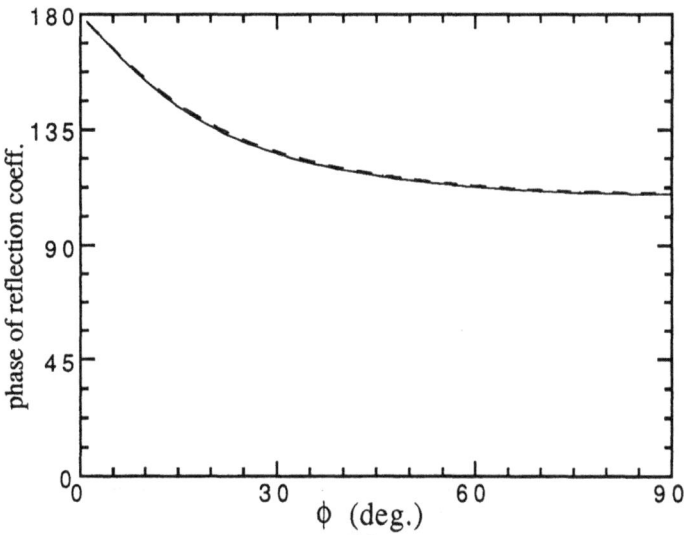

(b)

Figure 5–2: *Reflection coefficient of a planar layer with $k_0\tau = 0.25$ and $\epsilon_r = 4$ for perpendicular polarisation: (a) magnitude; (b) phase*

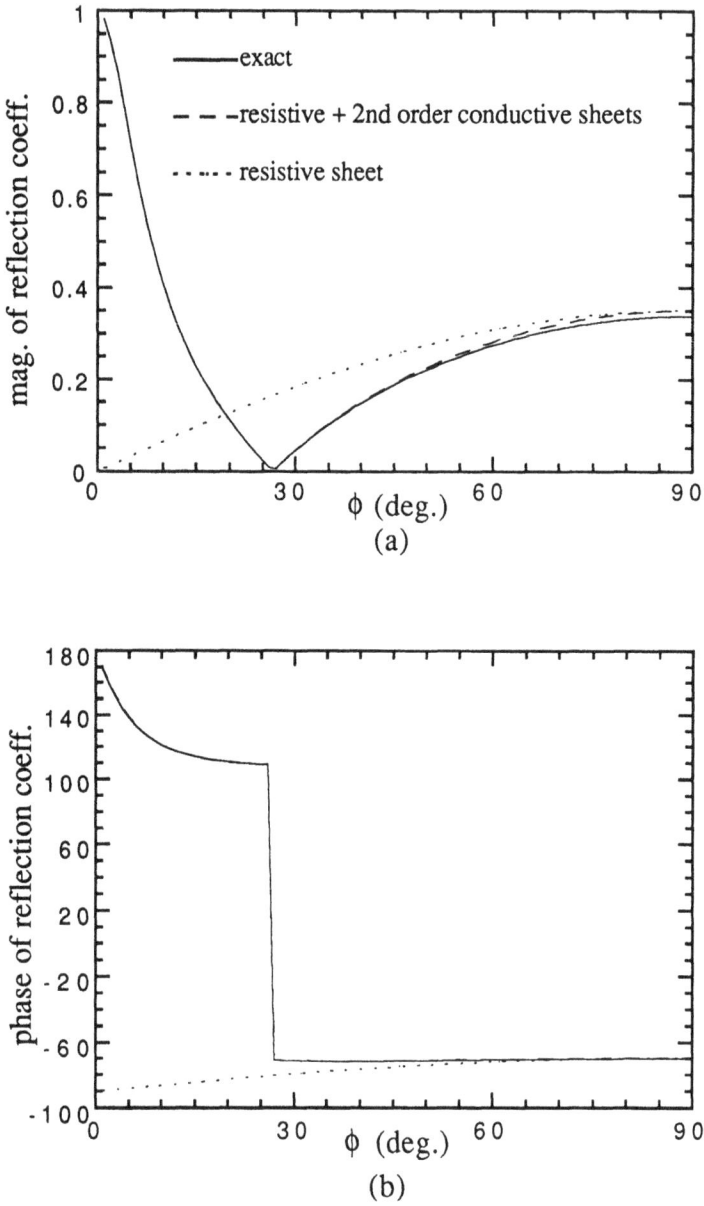

Figure 5–3: *Reflection coefficient of a planar layer with $k_0\tau = 0.25$ and $\epsilon_r = 4$ for parallel polarisation: (a) magnitude; (b) phase*

coefficients (5.55) and (5.56) are compared with those produced by the sheet simulations. Overall, the agreement is very good, and the inclusion of the conductive sheet leads to a marked improvement for parallel polarisation at all angles of incidence. A more powerful test is to take a finite slab and compare the scattered fields predicted by the equivalent sheet model with those obtained by solving the volume integral equations. For parallel polarisation the two-dimensional integral equation formulation by RICHMOND (1965, 1966) makes possible the separation of the resistive and conductive sheet contributions and, in the backscattering direction, the magnitudes are shown in Figs. 5–4 for a slab $0.1\lambda_0$ thick and $2\lambda_0$ wide. The agreement with the sheet model is excellent.

5.3.2 High contrast material

For a high contrast material such that $|N| \gg 1$ where N is the complex refractive index, a second order boundary condition applicable at the curved surface of a body is derived in Appendix A. In the particular case of the planar surface $y = 0$ of a homogeneous half space, the boundary condition reduces to (5.15) with

$$a_1 = 2N\epsilon_r, \qquad a_0 + a_2 = 2N^2$$

$$a_1' = 2N\mu_r, \qquad a_0' + a_2' = 2N^2$$

(5.59)

We can obtain the same results from the reflection coefficients (2.79) and (2.80) by making the approximation

$$\sqrt{1 - N^{-2}\cos^2\phi} \simeq 1 - \frac{1}{2N^2}(1 - \sin^2\phi)$$

and we observe that the coefficients (5.59) satisfy (5.12). For $|N| \gtrsim 1.3$ the approximate reflection coefficients are accurate to within a few percent in amplitude and a few degrees in phase for all ϕ, and this is illustrated by the comparisons in Fig. 5–5.

5.3.3 Metal-backed layer

For the metal-backed dielectric layer shown in Fig. 2–6(a), the reflection coefficients are given in (2.90) and (2.93) and, if the tangent is replaced by its argument, we recover the low contrast conditions presented in Section 5.3.1. The corresponding parameters are

$$a_0 = k_0 t, \qquad a_1 = -j\epsilon_r, \qquad a_2 = k_0 t(N^2 - 1)$$

$$a_0' = 0, \qquad a_1' = j\mu_r k_0 t, \qquad a_2' = 1$$

(5.60)

and we observe that (5.12) is satisfied. However, if $|N| \gg 1$ the boundary condition is only valid for a *very* thin layer, and SENIOR AND VOLAKIS (1989) have discussed alternative approximations leading to fourth order conditions.

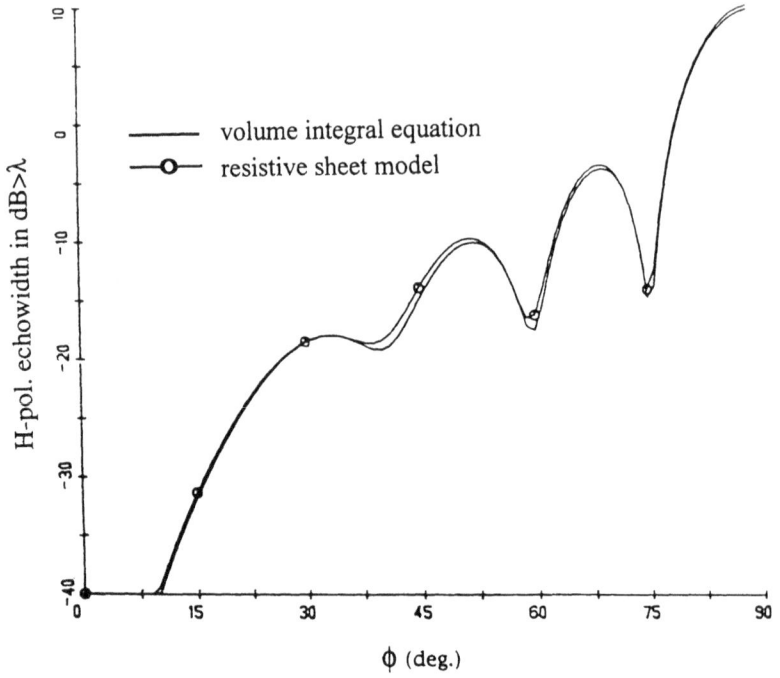

ϕ (deg.)

(a)

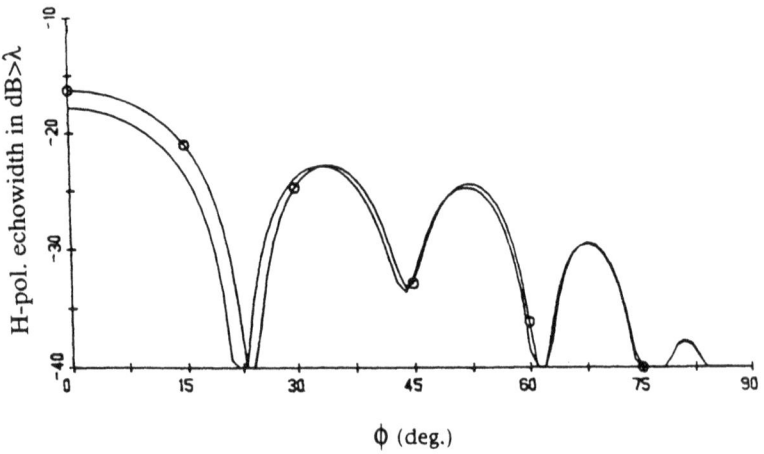

ϕ (deg.)

(b)

Figure 5–4: *Comparison of the backscattered field for parallel polarisation contributed by a single current component with that of the resistive sheet model for
a slab $2\lambda_0$ wide having $k_0\tau = 0.628$ and $\epsilon_r = 4$: (a) tangential component J_x;
(b) normal component J_y*

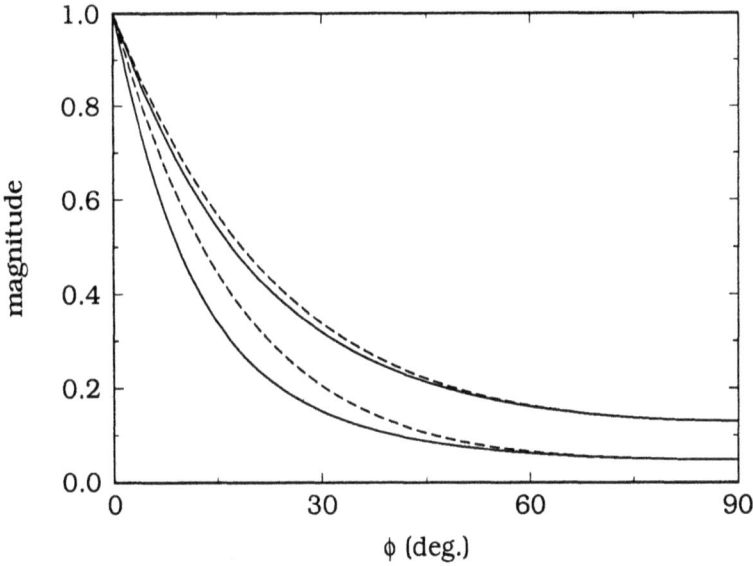

Figure 5–5: $|R_{\parallel}|$ *for a homogeneous half space (—) compared with its high contrast approximation (- - -) for* $N = 1.3$ *(upper curves) and* $N = 1.1$ *(lower curves) with* $\mu_r = 1$

To the first order in $|N|^{-1}$ the reflection coefficients reduce to (2.91) and (2.94), and correspond to an SIBC with

$$\Gamma_1 = 1/\Gamma_1' = jZ \tan(Nk_0t) \tag{5.61}$$

where Z is the intrinsic impedance of the material. If, instead, the boundary condition is applied at the location $y = -t$ of the backing, the reflection coefficients differ by a phase factor $\exp(2jk_0t\sin\phi)$. To the first order in k_0t we can account for the factor by using a second order GIBC (see Section 2.4.1) with Γ_1 and Γ_1' as indicated in (5.61) and $\Gamma_2 = \Gamma_2' = j(k_0t)^{-1}$. This is desirable for a plate which is coated on both sides, and it can be verified that to the first order in k_0t (5.13) is still satisfied.

To the next order in $|N|^{-1}$

$$\tan\left(Nk_0t\sqrt{1 - N^{-2}\cos^2\phi}\right) \simeq \tan\left\{k_0t\left(N - \frac{1}{2N}\right) + \frac{k_0t}{2N}\sin^2\phi\right\}$$

$$\simeq T\left\{1 + \frac{k_0t}{2N}\left(T + \frac{1}{T}\right)\sin^2\phi\right\}$$

where

$$T = \tan\left\{k_0t\left(N - \frac{1}{2N}\right)\right\}$$

and when this is substituted into (2.93) and (2.90) we find

$$R_{\parallel} = -\frac{\left(N - \frac{1}{2N}\right)T + \frac{1}{2N}\left\{1 + k_0 t\left(N - \frac{1}{2N}\right)\left(T + \frac{1}{T}\right)\right\}T\sin^2\phi + j\epsilon_r\sin\phi}{\left(N - \frac{1}{2N}\right)T + \frac{1}{2N}\left\{1 + k_0 t\left(N - \frac{1}{2N}\right)\left(T + \frac{1}{T}\right)\right\}T\sin^2\phi - j\epsilon_r\sin\phi}$$

$$(5.62)$$

$$R_{\perp} = -\frac{\left(N - \frac{1}{2N}\right) + \frac{1}{2N}\left\{1 - k_0 t\left(N - \frac{1}{2N}\right)\left(T + \frac{1}{T}\right)\right\}\sin^2\phi - j\mu_r T\sin\phi}{\left(N - \frac{1}{2N}\right) + \frac{1}{2N}\left\{1 - k_0 t\left(N - \frac{1}{2N}\right)\left(T + \frac{1}{T}\right)\right\}\sin^2\phi + j\mu_r T\sin\phi}$$

where we have assumed that $k_0 t/N^3$ is negligible compared with unity. The resulting second order boundary conditions have parameters

$$a_0 = \frac{1}{2N}\left\{1 + k_0 t\left(N - \frac{1}{2N}\right)\left(T + \frac{1}{T}\right)\right\}T$$

$$a_1 = -j\epsilon_r, \qquad a_2 = \left(N - \frac{1}{2N}\right)T$$

$$(5.63)$$

$$a_0' = \frac{1}{2N}\left\{1 - k_0 t\left(N - \frac{1}{2N}\right)\left(T + \frac{1}{T}\right)\right\}$$

$$a_1' = j\mu_r T, \qquad a_2' = N - \frac{1}{2N}$$

and, mathematically at least, the conditions have almost the same accuracy as the fourth order conditions developed by SENIOR AND VOLAKIS (1989). They are a more accurate version of the second order conditions employed by SYED AND VOLAKIS (1991) and reduce to these if $1/T \gg T$. The advantage of (5.63) is that if $k_0 t$ is such that the tangent can be approximated by its argument with terms $O\{(k_0 t)^2\}$ neglected, the high contrast parameters reduce to the low contrast ones given in (5.60). Thus, (5.63) includes (5.60) as a special case, and a rough criterion for when the low contrast parameters are adequate is $\left(T + \frac{1}{T}\right)\left|N - \frac{1}{2N}\right|k_0 t \lesssim 1.25$, which can be approximated as $|N|k_0 t \lesssim 0.5$. This is consistent with the results in Fig. 5-6 where the magnitude of the exact reflection coefficient for parallel polarisation is compared with the high and low contrast approximations for two different layer thicknesses. For perpendicular polarisation the comparison is similar. Versions of the high and low contrast conditions have been used by VOLAKIS AND SYED (1990) to simulate multilayer coatings applied to circular and ogival cylinders.

5.3.4   Wire grid

A wire grid or mesh, either standing alone or embedded in a dielectric layer, is a common method for providing shielding, and meshes are also used as reflectors or ground planes for antennas. The quantities of practical interest are the plane wave reflection and transmission coefficients, and a survey of available results has been given by WAIT (1978).

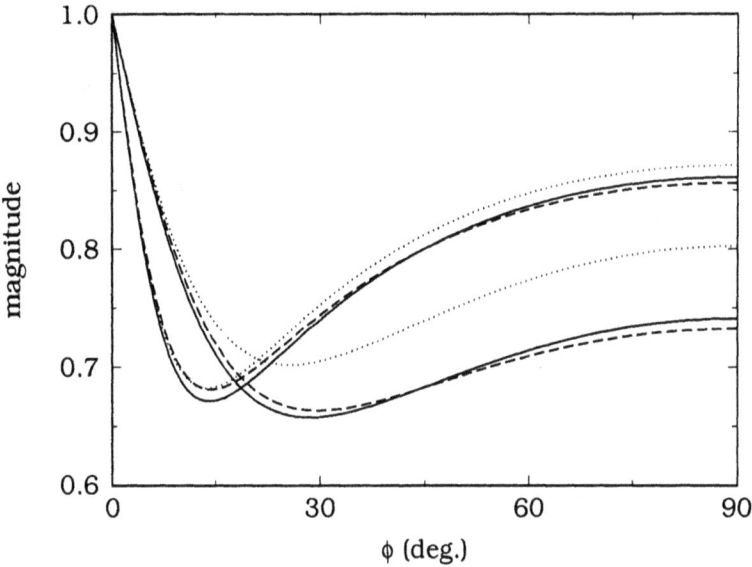

Figure 5-6: $|R_\parallel|$ (—) compared with its low (···) and high (- - -) contrast approximations for $N = 2(1 - 0.25j)$ with $\epsilon_r = \mu_r$. The upper curves are for $k_0 t = 0.15$ and the lower for $k_0 t = 0.3$

The simplest case is a square grid of perfectly conducting wires lying in the plane $y = 0$ (see Fig. 5-7). If

$$c \ll a \ll \lambda_0$$

where $b = a$ is the grid size and c is the wire radius, KONTOROVICH (1963) has shown that, for ideal contact of the wires at their intersections, the space-averaged field satisfies the transition conditions

$$E_x^+ + E_x^- = jZ_0 g \left\{ \left(1 + \frac{1}{2k_0^2}\frac{\partial^2}{\partial x^2}\right)[H_z]_-^+ - \frac{1}{2k_0^2}\frac{\partial^2}{\partial x \partial z}[H_x]_-^+ \right\}$$

$$E_z^+ + E_z^- = -jZ_0 g \left\{ \left(1 + \frac{1}{2k_0^2}\frac{\partial^2}{\partial x^2}\right)[H_x]_-^+ - \frac{1}{2k_0^2}\frac{\partial^2}{\partial x \partial z}[H_z]_-^+ \right\}$$

(5.64)

at $y = 0$ with

$$[E_x]_-^+ = [E_z]_-^+ = 0 \tag{5.65}$$

and

$$g = \frac{2a}{\lambda_0}\ln\frac{a}{2\pi c} \tag{5.66}$$

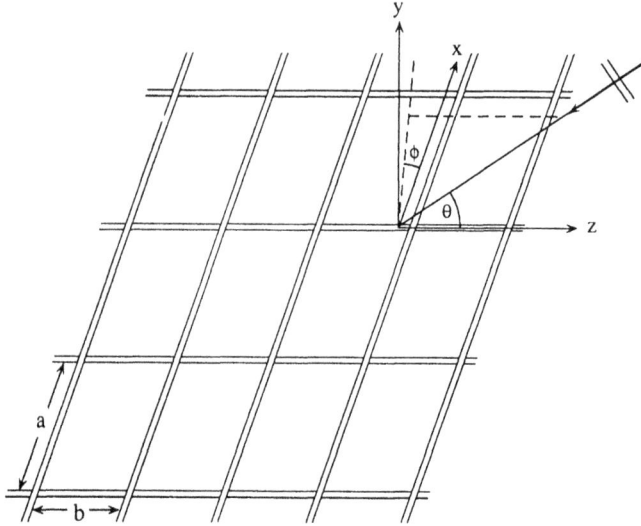

Figure 5-7: *Geometry for a rectangular wire grid*

Comparison with (5.29) and (5.30) shows that these represent a second order resistive sheet having

$$\frac{a_0 + a_2}{a_1} = jg, \qquad \frac{a_0}{a_0 + a_2} = \frac{1}{2}, \qquad \frac{a_0'}{a_0' + a_2'} = 0$$

implying

$$a_0 = a_2 = \tfrac{1}{2} jga_1, \qquad a_1' = jga_2', \qquad a_0' = 0 \qquad (5.67)$$

For an incident plane wave with

$$E_y^i\,(H_y^i) = \epsilon_y\,(h_y)\,e^{jk_0(x\cos\phi\sin\theta + y\sin\phi\sin\theta + z\cos\theta)}$$

the corresponding reflection coefficients are $R_\parallel$ ($R_\perp$). They can be obtained from (5.23) and (5.24) by replacing $\sin\phi$ with $\sin\phi\sin\theta$, and are plotted in Fig. 5-8 as functions of ϕ for five different values of g and $\theta = \pi/2$. We note that g increases as a/λ_0 increases and/or c/a decreases.

The analogous conditions for lossy wires and for a grid lying in the interface of two dielectric media have been given by CASEY (1988), and WAIT (1978) has noted that wire loss has most effect at angles close to grazing. The results are more strongly influenced by the electrical contact of the wires. If the contact is not ideal (KONTOROVICH, 1963) and/or for a rectangular grid having $b \neq a$ (ASTRAKHAN, 1964), the space-averaged field is still attributable to a second order resistive sheet, but of a more general kind than we have considered here. Experimental data for a variety of grids including an array of parallel wires have been reported by KONTOROVICH ET AL. (1962), and these support the

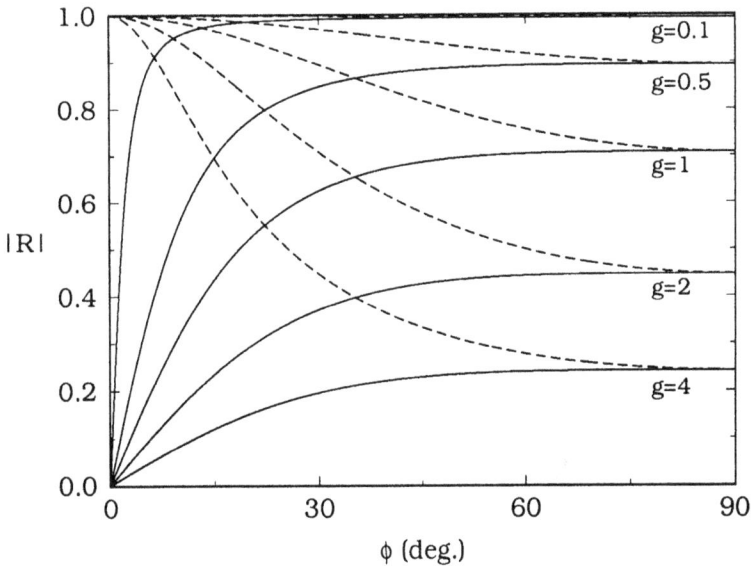

Figure 5–8: $|R_\parallel|$ *(—) and* $|R_\perp|$ *(- - -) for five different values of* g *with* $\theta = \pi/2$

theory. From a comparison with the results of a rigourous analysis, WAIT (1978) has concluded that the resistive sheet simulation accurately predicts the reflection coefficients to within 10 percent in amplitude and 10° in phase if $a \gtrsim \lambda_0/4$.

5.4 Uniqueness considerations

Although a second order GIBC is a relatively simple extension of an SIBC, the mathematical differences are significant, and it is more difficult to establish the conditions under which the boundary value problem has a unique solution. For this reason, it is convenient to treat first a scalar boundary condition.

5.4.1 Scalar GIBC

A special case of (5.7) and (5.8) is that in which E_y and H_y are independent of z corresponding to a two-dimensional field, and we start by considering the boundary condition

$$\frac{\partial U}{\partial y} = \alpha U - \beta \frac{\partial^2 U}{\partial x^2} \tag{5.68}$$

where α and β are constants. The condition is applied at the planar surface $y = 0+$ and we seek the scalar field U in $y \geq 0$ where U satisfies the two-dimensional scalar wave equation.

Following the procedure used in Section 2.7, we postulate two solutions generated by the same sources in $y > 0$. If W is the difference solution and k_0 has a small negative imaginary part, (2.139) becomes

$$\text{Im.} \int_{-\infty}^{\infty} W^* \frac{\partial W}{\partial y} \, dx = \text{Im.} \, k_0^2 \int_{-\infty}^{\infty} \int_0^{\infty} |W|^2 \, dx \, dy$$

and, since W also satisfies (5.68), insertion into the left hand side and a subsequent integration by parts gives

$$\text{Im.} \int_{-\infty}^{\infty} \left\{ \alpha |W|^2 + \beta \left| \frac{\partial W}{\partial x} \right|^2 \right\} \, dx = \text{Im.} \, k_0^2 \int_{-\infty}^{\infty} \int_0^{\infty} |W|^2 \, dx \, dy \qquad (5.69)$$

Because the right hand side is never positive and is zero only if $W = 0$ in $y \geq 0$, a *sufficient* condition for a unique solution in the case of a *passive* surface is

$$\text{Im.} \, \alpha \geq 0, \qquad \text{Im.} \, \beta \geq 0 \qquad (5.70)$$

A generalisation is to allow α and/or β to be discontinuous at (say) $x = 0$. When there is a discontinuity in the boundary condition on the surface, a knowledge of the boundary condition alone is not sufficient, and additional information is required concerning the allowed behaviour of the field at the discontinuity. This is hardly surprising since the boundary condition is not defined there, and a change from (say) $U = 0$ to $\partial U/\partial y = 0$ corresponds to an abrupt change from $\alpha = 0$ to $\alpha = \infty$ in a first order impedance boundary condition. The additional information is the edge condition (see Section 2.1) which ensures that the edge does not appear to be a true source by demanding that the energy density is integrable in the vicinity of the edge. This restricts the maximum singularity of any field component at the edge and, in the case of a discontinuity in a plane, implies that there is no field singularity greater than $\rho^{-1/2}$. Some of the consequences are

- a (total) field component parallel to the edge is finite and continuous there,

- a current component perpendicular (or a surface field component parallel) to the edge is zero there,

- a current component parallel (or a surface field component perpendicular) to the edge may be infinite there.

In most cases, the physically meaningful solution is the one with maximum allowed singularity.

For an SIBC, including $U = 0$ and $\partial U/\partial y = 0$ as special cases, the addition of the edge condition is sufficient to ensure uniqueness, and allows us to dispense with the requirement that α have no discontinuity. For the junction of two first order impedance half-planes it is found that $U(x,0)$ is continuous and finite at $x = 0$, but is non-zero unless one of the impedances is infinite, i.e. $\alpha = \infty$ on left or right. However, for a GIBC of order two or higher, additional constraints are necessary, and these are the contact conditions referred to earlier. To see why a constraint is needed, suppose β is discontinuous at $x = 0$. The integration by parts required to give (5.69) now produces an extra term

$$\text{Im.} \left[W^* \beta \frac{\partial W}{\partial x} \right]_-^+$$

where $[\ ]_-^+$ denotes the discontinuity between $x = 0+$ and $x = 0-$. Since U is continuous at the origin, so is W and, hence, W^*, and if

$$\left[\beta \frac{\partial U}{\partial x} \right]_-^+ = \Delta U(x,0) \tag{5.71}$$

for some specified Δ, (5.69) becomes

$$\text{Im.} \int_{-\infty}^{\infty} \left\{ \alpha |W|^2 + \beta \left| \frac{\partial W}{\partial x} \right|^2 \right\} dx + \text{Im.} \; \Delta |W|_0^2 = \text{Im.} \; k_0^2 \int_{-\infty}^{\infty} \int_0^{\infty} |W|^2 \, dx \, dy.$$

Uniqueness is then preserved if

$$\text{Im.} \; \Delta \geq 0 \tag{5.72}$$

and (5.71) with (5.72) is the contact condition (SENIOR, 1993a).

To specify Δ it is necessary to consider the nature of the contact between the two surfaces. This depends on the physical structure simulated by the boundary condition but, if there is no gap or insert between the surfaces, it is easy to determine Δ. From the boundary condition (5.68)

$$\int_{-\delta}^{\delta} \left(\frac{\partial U}{\partial y} - \alpha U \right) dx = - \left[\beta \frac{\partial U}{\partial x} \right]_{-\delta}^{\delta}$$

and since $\partial^2 U / \partial x^2$ has a logarithmic singularity at $x = 0$, so does $\partial U / \partial y$. Hence

$$\int_{-\delta}^{\delta} \frac{\partial U}{\partial y} \, dx \propto \delta(\ln \delta - 1) \to 0 \text{ as } \delta \to 0$$

and since U is finite and continuous at $x = 0$

$$\int_{-\delta}^{\delta} U \, dx \propto \delta \to 0 \text{ as } \delta \to 0$$

Thus

$$\left[\beta \frac{\partial U}{\partial x} \right]_-^+ = 0$$

showing that $\Delta = 0$. The contact conditions are therefore

$$U = 0 \qquad \text{at } x = 0 \tag{5.73}$$

or

$$\left[\beta \frac{\partial U}{\partial x} \right]_{-}^{+} = 0 \tag{5.74}$$

Which is appropriate depends on the physical meaning of the scalar wave function U, e.g. in the case of an electromagnetic problem, the field component that it represents, but either (5.73) or (5.74) is sufficient to ensure a unique solution of the boundary value problem. As shown in Section 5.5, the resulting solution satisfies the reciprocity condition concerning the interchange of transmitter and receiver.

Let us now return to the case of a uniform surface with constant α and β. Since (5.70) is only a sufficient condition for a passive surface, we would like to broaden it, and this is possible for a planar surface at least. If, for simplicity, we assume that k_0 is real, the most general representation of the scattered field in $y \geq 0$ is

$$U^s(x, y) = \int_{-\infty}^{\infty} P(\lambda) e^{-jk_0(x\lambda + y\sqrt{1-\lambda^2})} \, d\lambda \tag{5.75}$$

where the path of integration lies below the branch cut emanating from $\lambda = -1$ (and above below any surface wave pole at $\lambda = \lambda_0 < -1$) and above the branch cut from $\lambda = 1$ (and above any surface wave pole at $\lambda = \lambda_0 > 1$). In accordance with the radiation condition, the branch of the square root is such that $\sqrt{1 - \lambda^2}$ is real and positive if $|\lambda| < 1$, and negative pure imaginary if $|\lambda| > 1$. In the vicinity of the surface the incident field has the form

$$U^i(x, y) = \int_{-\infty}^{\infty} P_0(\lambda) e^{-jk_0(x\lambda - y\sqrt{1-\lambda^2})} \, d\lambda$$

for known $P_0(\lambda)$ and, when the boundary condition is imposed, we obtain

$$P(\lambda) = R(\lambda) P_0(\lambda)$$

where

$$R(\lambda) = \frac{\sqrt{1 - \lambda^2} + j(\alpha + \beta k_0^2 \lambda^2)/k_0}{\sqrt{1 - \lambda^2} - j(\alpha + \beta k_0^2 \lambda^2)/k_0} \tag{5.76}$$

For a unique solution it is necessary that $|R|$ is bounded for all real λ, $|\lambda| \leq 1$, and the solution then corresponds to a passive surface if $|R| \leq 1$. Clearly, $|R| \leq 1$ if Im. $(\alpha + \beta k_0^2 \lambda^2) \geq 0$, and the necessary and sufficient conditions for this to occur are

$$\text{Im. } \alpha \geq 0, \qquad \text{Im. } \beta \geq -k_0^{-2}\text{Im. } \alpha \tag{5.77}$$

Compared with (5.70), this represents an expansion of the allowed β. This is important since many simple surfaces such as the interface with a lossy

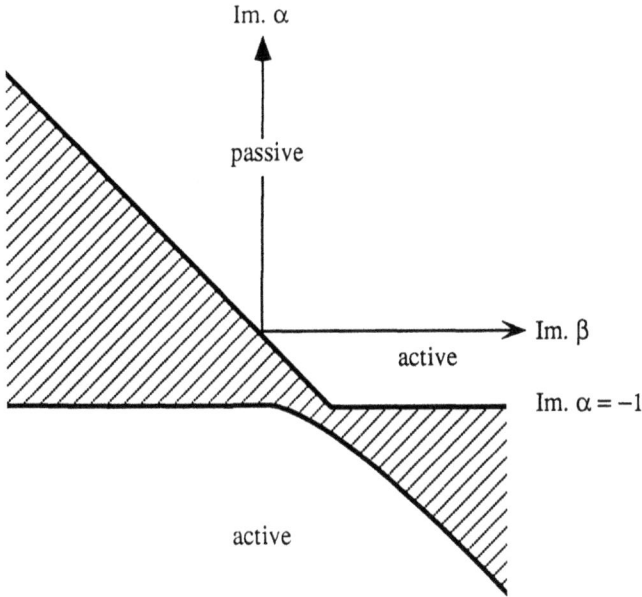

Figure 5-9: *Regions where the solution is unique for passive and active surfaces*

dielectric medium have values of α and β which satisfy (5.77) but violate (5.70). The Im. α and Im. β corresponding to the poles of $R(\lambda)$ for $|\lambda| \leq 1$ are indicated by the shaded region in Fig. 5–9, and for all other values, $|R|$ is bounded. It follows that for Im. α and Im. β lying in the *unshaded* regions the solution of the boundary value problem is unique and relates to either a passive or an active surface as shown in the figure. As in the case of an SIBC, only passive surfaces are of practical interest.

For a curved surface S, the most general form of a second order scalar boundary condition is

$$\frac{\partial U}{\partial n} = \alpha U - \frac{\partial}{\partial s}\left(\beta_{11}\frac{\partial U}{\partial s} + \beta_{12}\frac{\partial U}{\partial t}\right) - \frac{\partial}{\partial t}\left(\beta_{21}\frac{\partial U}{\partial s} + \beta_{22}\frac{\partial U}{\partial t}\right) \qquad (5.78)$$

where $\hat{s}$ and $\hat{t}$ are local tangent vectors such that $\hat{s}, \hat{n}, \hat{t}$ form a right handed system of orthogonal unit vectors. The coefficients are, in general, piecewise continuous functions of s and t but, if they are continuous on S, (2.139) gives

$$\text{Im.} \iint_S \left\{ \alpha |W|^2 + \beta_{11}\left|\frac{\partial W}{\partial s}\right|^2 + \beta_{22}\left|\frac{\partial W}{\partial t}\right|^2 \right.$$
$$\left. + \beta_{12}\frac{\partial W^*}{\partial s}\frac{\partial W}{\partial t} + \beta_{21}\frac{\partial W^*}{\partial t}\frac{\partial W}{\partial s} \right\} dS = \text{Im.}\, k_0^2 \iiint_V |W|^2\, dV \qquad (5.79)$$

Sufficient conditions for a unique solution in the case of a passive surface are then

$$\text{Im. } \alpha \geq 0, \qquad \text{Im. } \beta_{11} \geq 0, \qquad \text{Im. } \beta_{22} \geq 0 \qquad (5.80)$$

and

$$\beta_{21} = \beta_{12}^* \qquad (5.81)$$

at all points of the surface. By the same argument as that used for a flat surface, it would appear that (5.80) can be replaced by

$$\text{Im. } \alpha \geq 0, \qquad \text{Im. } \beta_{11} \geq -k_0^{-2}\text{Im. } \alpha, \qquad \text{Im. } \beta_{22} \geq -k_0^{-2}\text{Im. } \alpha \qquad (5.82)$$

At an edge where some or all of the β_{ij} are discontinuous, a contact condition is also required. If, for example, the edge is at $s = 0$, the condition is

$$U = 0 \qquad \text{at } s = 0 \qquad (5.83)$$

or

$$\left[\beta_{11} \frac{\partial U}{\partial s} + \beta_{12} \frac{\partial U}{\partial t} \right]_-^+ = 0 \qquad (5.84)$$

by analogy with (5.74). Finally, we note that a more compact presentation of (5.78) is

$$\frac{\partial U}{\partial n} = \alpha U - \nabla_s.(\overline{\overline{\beta}}.\nabla_s U) \qquad (5.85)$$

where the suffix "s" denotes the surface operator.

5.4.2   Vector GIBC

As indicated in Section 5.2, a rather general second order boundary condition for electromagnetic fields is

$$\hat{n} \times \left\{ \mathbf{E} + \frac{1}{jk_0}\nabla(A\,\hat{n}.\mathbf{E}) \right\} = \hat{n} \times \left(\hat{n} \times \left\{ \overline{\overline{\eta}}.\mathbf{H} - \frac{1}{jk_0}\nabla(B\,\hat{n}.\mathbf{H}) \right\} \right) \qquad (5.86)$$

where A, B and the elements of the tensor $\overline{\overline{\eta}}$ are functions of position on S. In the special case of the two-dimensional problem for a cylindrical surface whose generators are parallel to the z axis, (5.86) is equivalent to a scalar boundary condition of the form (5.85) for E_z or H_z, but for most surfaces the two conditions are quite distinct.

To apply the standard uniqueness theorem for passive surfaces it is necessary to express $\nabla(B.\hat{n}.\mathbf{H})$ in terms of tangential field components. We can do so on the assumption that it provides only a small correction to $\mathbf{H}$. Then

$$\hat{n}.\mathbf{H} = -\frac{Y_0}{jk_0}\hat{n}.\nabla \times \mathbf{E} \simeq -\frac{Y_0}{jk_0}\nabla_s.(\overline{\overline{\eta}}.\mathbf{H}) \qquad (5.87)$$

where $\nabla_s.$ is the surface divergence, and this enables us to write (5.86) as

$$\hat{n} \times \mathbf{E} = \hat{n} \times (\hat{n} \times \overline{\overline{\eta}}.\mathbf{H}) + \frac{j}{k_0} \hat{n} \times \nabla(A \hat{n}.\mathbf{E})$$

$$- \frac{Y_0}{k_0^2} \hat{n} \times (\hat{n} \times \nabla \{B\nabla_s.(\overline{\overline{\eta}}.\mathbf{H})\}) \qquad (5.88)$$

Henceforth we will treat $\overline{\overline{\eta}}$ as a constant scalar, independent of position on S.

If $(\mathbf{E}^{(1)}, \mathbf{H}^{(1)})$ and $(\mathbf{E}^{(2)}, \mathbf{H}^{(2)})$ are two fields generated by the same sources in the exterior region V and $\delta\mathbf{E}$, $\delta\mathbf{H}$ is the difference field, the quantity of interest is (see Section 2.7)

$$\iint_S \delta\mathbf{H}^*.(\hat{n} \times \delta\mathbf{E}) \, dS$$

and, since the difference field also satisfies (5.88), we have

$$\delta\mathbf{H}^*.(\hat{n} \times \delta\mathbf{E}) = \eta \, \delta\mathbf{H}^*.\hat{n} \times (\hat{n} \times \delta\mathbf{H}) + \frac{j}{k_0} \delta\mathbf{H}^*.\hat{n} \times \nabla(A \hat{n}.\delta\mathbf{E})$$

$$- \frac{Y_0}{k_0^2} \eta \, \delta\mathbf{H}^*.\hat{n} \times (\hat{n} \times \nabla \{B\nabla_s.\delta\mathbf{H}\})$$

But

$$\delta\mathbf{H}^*.\hat{n} \times (\hat{n} \times \delta\mathbf{H}) = -|\hat{n} \times \delta\mathbf{H}|^2$$

$$\delta\mathbf{H}^*.\hat{n} \times \nabla(A \hat{n}.\delta\mathbf{E}) = \hat{n}.\{\nabla(A \hat{n} \, \delta\mathbf{E}) \times \delta\mathbf{H}^*\}$$

$$= jk_0 Y_0 A|\hat{n}.\delta\mathbf{E}|^2 + \hat{n}.\nabla \times (\delta\mathbf{H}^* A \hat{n}.\delta\mathbf{E})$$

$$\delta\mathbf{H}^*.\hat{n} \times (\hat{n} \times \nabla \{B\nabla_s.\delta\mathbf{H}\}) = -\delta\mathbf{H}^*.\nabla_s \{B\nabla_s.\delta\mathbf{H}\}$$

$$= B|\nabla_s.\delta\mathbf{H}|^2 - \nabla_s.(\delta\mathbf{H}^* B\nabla_s.\delta\mathbf{H})$$

and therefore

$$\iint_S \delta\mathbf{H}^*.(\hat{n} \times \delta\mathbf{E}) \, dS = -\iint_S \left\{ \eta|\hat{n} \times \delta\mathbf{H}|^2 + Y_0 A|\hat{n}.\delta\mathbf{E}|^2 \right.$$

$$\left. + \frac{1}{k_0^2} Y_0 \eta B|\nabla_s.\delta\mathbf{H}|^2 \right\} dS + \frac{1}{k_0^2} I \qquad (5.89)$$

where

$$I = \iint_S \{Y_0 \eta \nabla_s.(\delta\mathbf{H}^* \, B\nabla_s.\delta\mathbf{H}) + jk_0 \hat{n}.\nabla \times (\delta\mathbf{H}^* \, A \hat{n}.\delta\mathbf{E})\} \, dS \qquad (5.90)$$

If A and B are continuous on S, the divergence theorem shows that $I = 0$. Hence

$$-\mathrm{Re}. \iint_S \left\{ \eta|\hat{n} \times \delta\mathbf{H}|^2 + Y_0 A|\hat{n}.\delta\mathbf{E}|^2 + \frac{1}{k_0^2} Y_0 \eta B|\nabla_s.\delta\mathbf{H}|^2 \right\} dS$$

$$= \omega \iiint_V \left(\mu''|\delta\mathbf{H}|^2 + \epsilon''|\delta\mathbf{E}|^2 \right) dV \qquad (5.91)$$

(see Section 2.7), and for a passive surface sufficient conditions for uniqueness are

$$\text{Re. } \eta \geq 0, \qquad \text{Re. } A \geq 0, \qquad \text{Re. } \eta B \geq 0 \tag{5.92}$$

at all points of S. If A and/or B are discontinuous across a line C, the surface is equivalent to the union of two open surfaces S_1 and S_2. For each of them

$$\iint_{S_i} \hat{n}.\nabla \times \mathbf{F} \, dS = \oint_C \mathbf{F}.\hat{\ell} \, dC \tag{5.93}$$

from Stokes' theorem, and for a vector $\mathbf{G}$ tangential to S_i

$$\iint_{S_i} \nabla_{\mathrm{s}}.\mathbf{G} \, dS = \oint_C (\hat{n} \times \mathbf{G}).\hat{\ell} \, dC \tag{5.94}$$

(VAN BLADEL, 1985) where $\hat{\ell}$ is a unit vector tangential to C and oriented in the positive direction with respect to the normal $\hat{n}$. Then

$$I = \oint_C \left[\delta\mathbf{H}^*. \left\{ Y_0\eta B(\hat{\ell} \times \hat{n})\nabla_{\mathrm{s}}.\delta\mathbf{H} + jk_0 A \, \hat{\ell} \, \hat{n}.\delta\mathbf{E} \right\} \right]_-^+ dC \tag{5.95}$$

where $[\ ]_-^+$ denotes the discontinuity across C. The second term in the integrand vanishes by virtue of the edge condition, and the first term also vanishes if the following additional constraint is imposed:

$$(\hat{\ell} \times \hat{n}).\mathbf{H} = 0 \qquad \text{on } C \tag{5.96}$$

or

$$[B\nabla_{\mathrm{s}}.\mathbf{H}]_-^+ = 0 \tag{5.97}$$

These constitute the contact conditions and we observe the similarity to (5.73) and (5.74).

It should be emphasised that (5.92) are only sufficient conditions for uniqueness and, although it seems probable that the criteria can be expanded in the same way as for the scalar boundary condition, no proof has been developed. We have also been unable to complete the proof if η is a function of position on S and/or if it is tensor.

5.5   Diffraction by half-plane junctions

A simple problem involving second order conditions is a plane wave incident on a conductive sheet lying in the plane $y = 0$ and subject to different second order transition conditions in $x < 0$ and $x > 0$. If the plane wave is incident in the xy plane, the problem is a two-dimensional scalar one which can be solved using the dual integral equation technique of CLEMMOW (1951). The method is based on the representation of the fields in terms of angular spectra. The solution is developed in the next section and we then show how the solutions of other related problems can be deduced from it.

5.5.1   Conductive sheet junction

For a planar surface it is natural to employ the transition conditions (5.25) and (5.26) involving the normal field components E_y and H_y, but there are mathematical difficulties associated with these, and the difficulties occur even in the case of first order conditions. At an edge or other line discontinuity in the surface, $[\partial E_y/\partial y]^+_-$ and $[\partial H_y/\partial y]^+_-$ may be as singular as $x^{-3/2}$ where x is the distance from the edge, and their Fourier transforms do not exist in the classical sense. One way to avoid this is to use the x-integrals of E_y and H_y, i.e.

$$\mathcal{E}_y(x,y) = \int^x E_y(x',y)\,dx', \qquad \mathcal{H}_y(x,y) = \int^x H_y(x',y)\,dx'$$

and to solve the problem for them. It can be shown (SENIOR, 1987) that the arbitrary constants that are introduced into the transition conditions for $\mathcal{E}$ and $\mathcal{H}$ play no role in the final solution, and can be omitted. In the two-dimensional problem for H- or E-polarised fields

$$\mathcal{E}_y = j\frac{Z_0}{k_0}H_z, \qquad \mathcal{H}_y = -j\frac{Y_0}{k_0}E_z$$

and the process of integration is equivalent to replacing E_y and H_y by H_z and E_z, respectively. This is what we shall do, and we note that the same effect is achieved by using the transition conditions in the form (5.17) and (5.19).

The H-polarised plane wave

$$\mathbf{H}^i = \hat{z}e^{jk_0(x\cos\phi_0+y\sin\phi_0)} \tag{5.98}$$

is incident on the surface $y = 0$ at which the transition conditions for a second order conductive sheet are imposed. In view of the above remarks, the transition conditions are written as

$$\prod_{m=1}^{2}\left(\Gamma'_m - \frac{1}{jk_0}\frac{\partial}{\partial y}\right)H^+_z + \prod_{m=1}^{2}\left(\Gamma'_m + \frac{1}{jk_0}\frac{\partial}{\partial y}\right)H^-_z = 0 \tag{5.99}$$

for $x < 0$,

$$\prod_{m=1}^{2}\left(\Gamma_m - \frac{1}{jk_0}\frac{\partial}{\partial y}\right)H^+_z + \prod_{m=1}^{2}\left(\Gamma_m + \frac{1}{jk_0}\frac{\partial}{\partial y}\right)H^-_z = 0 \tag{5.100}$$

for $x > 0$, with

$$[H_z]^+_- = 0 \tag{5.101}$$

for all x. It is assumed that the Γ'_m and Γ_m are specified. The sheet supports only a magnetic current

$$\mathbf{J}_m = \hat{z}[E_x]^+_- = -j\frac{Z_0}{k_0}\left[\frac{\partial H_z}{\partial y}\right]^+_-\hat{z} \tag{5.102}$$

and this in turn generates a magnetic field which is symmetrical about the plane $y = 0$.

The method that we shall use is equivalent to, but somewhat different from, that employed in Section 3.7. If the plane wave (5.98) were incident on a sheet having the transition condition (5.99) imposed for all x, the total field would be

$$H_z(x,y) = \begin{cases} e^{jk_0(x\cos\phi_0 + y\sin\phi_0)} + Re^{jk_0(x\cos\phi_0 - y\sin\phi_0)} & (y \geq 0) \\ \\ Te^{jk_0(x\cos\phi_0 + y\sin\phi_0)} & (y \leq 0) \end{cases} \quad (5.103)$$

where the reflection and transmission coefficients are

$$R = -\frac{1}{2}\left\{1 + \prod_{m=1}^{2} \frac{\Gamma'_m - \sin\phi_0}{\Gamma'_m + \sin\phi_0}\right\}$$

$$T = \frac{1}{2}\left\{1 - \prod_{m=1}^{2} \frac{\Gamma'_m - \sin\phi_0}{\Gamma'_m + \sin\phi_0}\right\}$$

$$(5.104)$$

respectively. Denoting this field by the superscript "o", we now write

$$H_z(x,y) = H_z^o(x,y) + H_z^1(x,y) \quad (5.105)$$

and represent H_z^1 as

$$H_z^1(x,y) = \int_{-\infty}^{\infty} P(\lambda)\, e^{-jk_0(x\lambda + |y|\sqrt{1-\lambda^2})} \frac{d\lambda}{\sqrt{1-\lambda^2}} \quad (5.106)$$

This satisfies (5.101) automatically and, in accordance with the edge condition, it is necessary that $|P(\lambda)| \to 0$ as $|\lambda| \to \infty$.

The transition conditions for H_z^1 are

$$\prod_{m=1}^{2}\left(\Gamma'_m - \frac{1}{jk_0}\frac{\partial}{\partial y}\right) H_z^{1+} + \prod_{m=1}^{2}\left(\Gamma'_m + \frac{1}{jk_0}\frac{\partial}{\partial y}\right) H_z^{1-} = 0 \quad (5.107)$$

for $x < 0$,

$$\prod_{m=1}^{2}\left(\Gamma_m - \frac{1}{jk_0}\frac{\partial}{\partial y}\right) H_z^{1+} + \prod_{m=1}^{2}\left(\Gamma_m + \frac{1}{jk_0}\frac{\partial}{\partial y}\right) H_z^{1-} = Me^{jk_0 x\lambda_0} \quad (5.108)$$

for $x > 0$ where $\lambda_0 = \cos\phi_0$ and

$$M = \prod_{m=1}^{2}(\Gamma_m + \sin\phi_0)\frac{\Gamma'_m - \sin\phi_0}{\Gamma'_m + \sin\phi_0} - \prod_{m=1}^{2}(\Gamma_m - \sin\phi_0) \quad (5.109)$$

When (5.106) is inserted into (5.107), we obtain

$$2\Gamma_1'\Gamma_2' \int_{-\infty}^{\infty} \sqrt{1-\lambda^2} \left(\frac{1}{\Gamma_1'} + \frac{1}{\sqrt{1-\lambda^2}}\right)$$

$$\cdot \left(\frac{1}{\Gamma_2'} + \frac{1}{\sqrt{1-\lambda^2}}\right) P(\lambda) e^{-jk_0 x\lambda} \, d\lambda = 0 \qquad (5.110)$$

for $x < 0$. In the terms of the split functions introduced in Section 3.2

$$\left(\frac{1}{\Gamma_1'} + \frac{1}{\sqrt{1-\lambda^2}}\right)^{-1} = K_+\left(\frac{1}{\Gamma_1'},\lambda\right) K_+\left(\frac{1}{\Gamma_1'},-\lambda\right) \qquad (5.111)$$

where $K_+ (1/\Gamma_1',\lambda)$ is analytic and free of zeros in an upper half λ-plane. From the expression given by SENIOR (1952) or, alternatively, from LEPPINGTON (1983),

$$K_+\left(\frac{1}{\Gamma_1'},\pm\lambda\right) = \sqrt{\Gamma_1'}\left\{1 \pm \frac{\Gamma_1'}{\pi\lambda} \ln 2\lambda + O\left(|\lambda|^{-1}\right)\right\} \qquad (5.112)$$

for large $|\lambda|$ provided $\Gamma_1' \neq \infty$, but if $\Gamma_1' = \infty$

$$K_+(0,\pm\lambda) = \sqrt{1\mp\lambda} \qquad (5.113)$$

If

$$K'(\lambda) = K_+\left(\frac{1}{\Gamma_1'},\lambda\right) K_+\left(\frac{1}{\Gamma_2'},\lambda\right) \qquad (5.114)$$

(5.110) can be written as

$$\int_{-\infty}^{\infty} \sqrt{1-\lambda^2} \frac{P(\lambda)}{K'(\lambda) K'(-\lambda)} e^{-jk_0 x\lambda} \, d\lambda = 0 \qquad (5.115)$$

for $x < 0$ and therefore

$$P(\lambda) = (1+\lambda)^{-1/2} K'(-\lambda) U(\lambda) \qquad (5.116)$$

where $U(\lambda)$ is a function analytic in an upper half-plane. Similarly, for $x > 0$

$$2\Gamma_1\Gamma_2 \int_{-\infty}^{\infty} \sqrt{1-\lambda^2} \frac{P(\lambda)}{K(\lambda) K(-\lambda)} e^{-jk_0 x\lambda} \, d\lambda = M e^{jk_0 x\lambda_0} \qquad (5.117)$$

where $K(\lambda)$ differs from $K'(\lambda)$ in having Γ_1 and Γ_2 in place of Γ_1' and Γ_2'. Hence

$$P(\lambda) = (1-\lambda)^{-1/2} \frac{K(\lambda)}{\lambda+\lambda_0} L(\lambda) \qquad (5.118)$$

where $L(\lambda)$ is a lower half-plane function, and the combination of (5.116) and (5.118) gives

$$P(\lambda) = (1-\lambda^2)^{-1/2} \frac{K'(-\lambda) K(\lambda)}{\lambda+\lambda_0} A(\lambda)$$

where $A(\lambda)$ is an analytic function. Since $|P(\lambda)| \to 0$ as $|\lambda| \to \infty$ it follows that $A(\lambda)$ is at most a first order polynomial in λ, and therefore

$$P(\lambda) = \left\{(1-\lambda^2)(1-\lambda_0^2)\right\}^{-1/2} K'(-\lambda)\,K'(-\lambda_0)\,K(\lambda)\,K(\lambda_0)\,\frac{c_0 + c_1\lambda}{\lambda + \lambda_0}$$
$$(5.119)$$

where c_0 and c_1 are constants. The additional factors involving λ_0 have been introduced for convenience.

When (5.119) is inserted into (5.117), we obtain

$$\frac{2}{\sin\phi_0}\Gamma_1\Gamma_2\,K'(-\lambda_0)\,K(\lambda_0)\int_{-\infty}^{\infty}\frac{K'(-\lambda)}{K(-\lambda)}\frac{c_0+c_1\lambda}{\lambda+\lambda_0}e^{-jk_0x\lambda}\,d\lambda = Me^{jk_0x\lambda_0}$$

and by closing the contour in the lower half-plane a residue evaluation gives

$$c_0 - c_1\lambda_0 = j\frac{M}{4\pi}\frac{(\Gamma_1' + \sin\phi_0)(\Gamma_2' + \sin\phi_0)}{\Gamma_1'\Gamma_2'\Gamma_1\Gamma_2\sin\phi_0}$$

i.e.

$$c_0 - c_1\lambda_0 = -\frac{j}{2\pi}\frac{a_1' - a_1}{a_2'a_2}(s^2 - \lambda_0^2)$$
$$(5.120)$$

where

$$s^2 = 1 + \frac{a_1'a_2 - a_2'a_1}{a_1' - a_1}$$
$$(5.121)$$

with

$$\begin{aligned}
a_1' &= \Gamma_1' + \Gamma_2', & a_2' &= \Gamma_1'\Gamma_2' \\
a_1 &= \Gamma_1 + \Gamma_2, & a_2 &= \Gamma_1\Gamma_2
\end{aligned}$$
$$(5.122)$$

From (5.120)

$$\frac{c_0 + c_1\lambda}{\lambda + \lambda_0} = -\frac{j}{2\pi}\frac{a_1' - a_1}{a_2'a_2}\left\{\frac{s^2 + \lambda\lambda_0}{\lambda + \lambda_0} + c_2\right\}$$

where

$$c_2 = 2\pi j\frac{a_2'a_2}{a_1' - a_1}c_1 - \lambda_0$$

is an arbitrary constant, and thus

$$\begin{aligned}
P(\lambda) &= -\frac{j}{2\pi}\frac{a_1' - a_1}{a_2'a_2}\left\{(1-\lambda^2)(1-\lambda_0^2)\right\}^{-1/2} \\
&\quad \cdot K'(-\lambda)\,K'(-\lambda_0)\,K(\lambda)\,K(\lambda_0)\left\{\frac{s^2 + \lambda\lambda_0}{\lambda + \lambda_0} + c_2\right\}
\end{aligned}$$
$$(5.123)$$

For large $|\lambda|$, $P(\lambda) = O(|\lambda|^{-1})$ if $c_2 \neq -\lambda_0$ and $P(\lambda) = O(|\lambda|^{-2})$ if $c_2 = -\lambda_0$, and the final expression for H_z^1 is

$$\begin{aligned}
H_z^1(x,y) &= -\frac{j}{2\pi}\frac{a_1' - a_1}{a_2'a_2}\frac{K'(-\lambda_0)\,K(\lambda_0)}{\sin\phi_0}\frac{y}{|y|}\int_{-\infty}^{\infty}\frac{K'(-\lambda)\,K(\lambda)}{1 - \lambda^2} \\
&\quad \cdot \left\{\frac{s^2 + \lambda\lambda_0}{\lambda + \lambda_0} + c_2\right\}e^{-jk_0(x\lambda + |y|\sqrt{1-\lambda^2})}\,d\lambda
\end{aligned}$$
$$(5.124)$$

We observe that H_z^1 is zero if $a_1' = a_1$, and the presence of the arbitrary constant c_2 shows the need for an additional constraint to ensure a unique solution. This constraint is the contact condition at the junction of the sheets.

To impose the contact condition it is necessary to determine the behaviour of H_z and $\partial H_z/\partial x$ for small $|x|$. From (5.124)

$$H_z^1(x, 0+) = B(a_1' - a_1) \int_{-\infty}^{\infty} \frac{K'(-\lambda)\, K(\lambda)}{1 - \lambda^2} \left\{ \frac{s^2 + \lambda\lambda_0}{\lambda + \lambda_0} + c_2 \right\} e^{-jk_0 x\lambda}\, d\lambda$$

$$(5.125)$$

where

$$B = -\frac{j}{2\pi} \frac{K'(-\lambda_0)\, K(\lambda_0)}{a_2' a_2 \sin\phi_0}$$

$$(5.126)$$

and, appearances to the contrary, B is finite at $\phi_0 = 0, \pi$. The first step is to additively decompose the first factor in the integrand of (5.125), and a simple analysis shows

$$K'(-\lambda)\, K(\lambda) = \frac{1}{a_1' - a_1} \frac{1 - \lambda^2}{\lambda^2 - s^2} \left\{ a_2' a_1 \frac{K(\lambda)}{K'(\lambda)} - a_1' a_2 \frac{K'(-\lambda)}{K(-\lambda)} \right\}$$

Hence

$$\frac{K'(-\lambda)\, K(\lambda)}{1 - \lambda^2} = \frac{1}{a_1' - a_1} \frac{1}{\lambda^2 - s^2} \left\{ a_2' a_1 \frac{K(\lambda)}{K'(\lambda)} + \alpha_1\lambda + \alpha_2 \right.$$

$$\left. - \left[a_1' a_2 \frac{K'(-\lambda)}{K(-\lambda)} + \alpha_1\lambda + \alpha_2 \right] \right\}$$

$$(5.127)$$

and since this is true for any α_1 and α_2, we can choose them to eliminate the poles at $\lambda = s(-s)$ from the first (second) group of terms in (5.127). Then

$$\alpha_1 = -\frac{1}{2s} \left\{ a_1' a_2 \frac{K'(s)}{K(s)} - a_2' a_1 \frac{K(s)}{K'(s)} \right\}$$

$$(5.128)$$

$$\alpha_2 = -\frac{1}{2} \left\{ a_1' a_2 \frac{K'(s)}{K(s)} + a_2' a_1 \frac{K(s)}{K'(s)} \right\}$$

implying

$$a_2^2 - \alpha_1^2 s^2 = a_1' a_2' a_1 a_2$$

and

$$H_z^1(x, 0+) = B \int_{-\infty}^{\infty} \frac{1}{\lambda^2 - s^2} \{S_+(\lambda) - S_-(\lambda)\} \left\{ \frac{s^2 + \lambda\lambda_0}{\lambda + \lambda_0} + c_2 \right\} e^{-jk_0 x\lambda}\, d\lambda$$

where

$$S_+(\lambda) = a_2' a_1 \frac{K(\lambda)}{K'(\lambda)} + \alpha_1\lambda + \alpha_2$$

$$S_-(\lambda) = a_1' a_2 \frac{K'(-\lambda)}{K(-\lambda)} + \alpha_1\lambda + \alpha_2$$

Finally, on eliminating the pole at $\lambda = -\lambda_0$ from the lower half-plane function, we have

$$H_z^1(x, 0+) = \int_{-\infty}^{\infty} T_+(\lambda)\, e^{-jk_0 x\lambda}\, d\lambda + \int_{-\infty}^{\infty} T_-(\lambda)\, e^{-jk_0 x\lambda}\, d\lambda \qquad (5.129)$$

where

$$T_+(\lambda) = B\left[\frac{S_+(\lambda)}{\lambda^2 - s^2} \left\{ \frac{s^2 + \lambda\lambda_0}{\lambda + \lambda_0} + c_2 \right\} + \frac{S_-(-\lambda_0)}{\lambda + \lambda_0} \right] \qquad (5.130)$$

is analytic in the upper half-plane, and

$$T_-(\lambda) = -B\left[\frac{S_-(\lambda)}{\lambda^2 - s^2} \left\{ \frac{s^2 + \lambda\lambda_0}{\lambda + \lambda_0} + c_2 \right\} + \frac{S_-(-\lambda_0)}{\lambda + \lambda_0} \right] \qquad (5.131)$$

is analytic in the lower half-plane.

The second term on the right hand side of (5.129) represents a function which is zero for $x > 0$ and hence, for $x > 0$,

$$H_z^1(x, 0+) = \int_{-\infty}^{\infty} T_+(\lambda)\, e^{-jk_0 x\lambda}\, d\lambda \qquad (5.132)$$

From (5.112) and (5.114)

$$\frac{S_+(\lambda)}{\lambda^2 - s^2} = \frac{1}{\lambda}\left\{ \alpha_1 + \frac{1}{\lambda}\left(\alpha_2 + a_1\sqrt{a_2' a_2} \right) \right.$$
$$\left. - a_1\sqrt{a_2' a_2}\, (a_1' - a_1)\frac{1}{\pi\lambda^2}\ln 2\lambda + O(|\lambda|^{-2}) \right\}$$

for large $|\lambda|$. Also

$$\frac{s^2 + \lambda\lambda_0}{\lambda + \lambda_0} + c_2 = c_2 + \lambda_0 + \frac{1}{\lambda}(s^2 - \lambda_0^2) + O(|\lambda|^{-2})$$

so that

$$T_+(\lambda) = \frac{B}{\lambda}\left\{ \alpha_1 c_2 + \alpha_2 + \frac{1}{\lambda}\left[\alpha_2 c_2 + \alpha_1 s^2 + (c_2 + \lambda_0)a_1\sqrt{a_2' a_2} \right] \right.$$
$$\left. - (c_2 + \lambda_0)a_1\sqrt{a_2' a_2}\,(a_1' - a_1)\frac{1}{\pi\lambda^2}\ln 2\lambda + O(|\lambda|^{-2}) \right\}$$
$$+ \frac{B}{\lambda}a_1' a_2 \frac{K'(-\lambda_0)}{K(-\lambda_0)}\left\{ 1 - \frac{\lambda_0}{\lambda} + O(|\lambda|^{-2}) \right\}$$

giving

$$H_z^1(x, 0+) = -2\pi j B\left\{ \alpha_1 c_2 + \alpha_2 - jk_0 x\left[\alpha_2 c_2 + \alpha_1 s^2 + (c_2 + \lambda_0)a_1\sqrt{a_2' a_2} \right] \right.$$
$$\left. - (c_2 + \lambda_0)a_1\sqrt{a_2' a_2}\,(a_1' - a_1)\frac{(k_0 x)^2}{2\pi}\ln k_0 x + O(x^2) \right\}$$
$$- \frac{a_1' \sin\phi_0}{(\Gamma_1' + \sin\phi_0)(\Gamma_2' + \sin\phi_0)}\left\{ 1 + jk_0 x + O(x^2) \right\}$$

for small x. We recognise the last term as the expansion of $H_z^\circ(x,0+)$ and therefore

$$
\begin{aligned}
H_z(x,0+) &= -2\pi j B\Big\{\alpha_1 c_2 + \alpha_2 - jk_0 x\Big[\alpha_2 c_2 + \alpha_1 s^2 + (c_2+\lambda_0)a_1\sqrt{a_2'a_2}\Big] \\
&\quad - (c_2+\lambda_0)a_1\sqrt{a_2'a_2}\,(a_1'-a_1)\frac{(k_0 x)^2}{2\pi}\ln k_0 x + O(x^2)\Big\}
\end{aligned} \tag{5.133}
$$

as $x \to 0+$.

Similarly, for $x < 0$,

$$
H_z^1(x,0+) = \int_{-\infty}^{\infty} T_-(\lambda)\, e^{-jk_0 x\lambda}\, d\lambda \tag{5.134}
$$

and for large $|\lambda|$

$$
\begin{aligned}
T_-(\lambda) &= -\frac{B}{\lambda}\Big\{\alpha_1 c_2 + \alpha_2 + \frac{1}{\lambda}\Big[\alpha_2 c_2 + \alpha_1 s^2 + (c_2+\lambda_0)a_1'\sqrt{a_2'a_2}\Big] \\
&\quad - (c_2+\lambda_0)a_1'\sqrt{a_2'a_2}\,(a_1'-a_1)\frac{1}{\pi\lambda^2}\ln 2\lambda + O(|\lambda|^{-2})\Big\} \\
&\quad - \frac{B}{\lambda}a_1'a_2\frac{K'(-\lambda_0)}{K(-\lambda_0)}\Big\{1 - \frac{\lambda_0}{\lambda} + O(|\lambda|^{-2})\Big\}
\end{aligned}
$$

implying

$$
\begin{aligned}
H_z(x,0+) &= -2\pi j B\Big\{\alpha_1 c_2 + \alpha_2 - jk_0 x\Big[\alpha_2 c_2 + \alpha_1 s^2 + (c_2+\lambda_0)a_1'\sqrt{a_2'a_2}\Big] \\
&\quad - (c_2+\lambda_0)a_1'\sqrt{a_2'a_2}\,(a_1'-a_1)\frac{(k_0 x)^2}{2\pi}\ln k_0|x| + O(x^2)\Big\}
\end{aligned} \tag{5.135}
$$

as $x \to 0-$. Comparison of (5.133) and (5.135) shows that H_z is finite and continuous and $\partial H_z/\partial x$ has a finite jump discontinuity at $x=0$ for all (finite) c_2. Moreover

$$
H_z(0,0+) = 0 \tag{5.136}
$$

if $c_2 = -\alpha_2/\alpha_1$ and

$$
\lim_{x\to 0-}\frac{1}{a_1'}\frac{\partial}{\partial x}H_z(x,0+) = \lim_{x\to 0+}\frac{1}{a_1}\frac{\partial}{\partial x}H_z(x,0+) \tag{5.137}
$$

if $c_2 = -\alpha_1 s^2/\alpha_2$. Equations (5.136) and (5.137) constitute the allowed contact conditions and, since the magnetic current $\mathbf{J_m}$ is proportional to H_z and is not required to be zero at the edge, (5.136) is not appropriate. Hence

$$
c_2 = -\alpha_1 s^2/\alpha_2 \tag{5.138}
$$

and the solution is now complete. From the symmetry of $P(\lambda)$ in λ and λ_0 it is evident that the reciprocity condition is satisfied.

The computation of the field H_z, including the calculation of the edge diffraction coefficient and the development of a uniform asymptotic expression valid at large distances from the edge, can be carried out using the procedures described in Section 3.3.2.

5.5.2   Other related problems

The solution just developed for the junction of two second order conductive
sheets for H-polarisation is a generic one in that the solutions for a variety
of other problems can be deduced from it. Before discussing these, however,
there are some special cases to be noted.

If the conductive sheets are used to simulate *only* a distribution of normal
electric dipoles, corresponding to the y component of the electric polarisation
current in the dielectric layer (SENIOR AND VOLAKIS, 1987), (5.51) shows that
$a'_2 = a_2 = -1$. From (5.121) we then have $s^2 = 0$, and the poles at $\lambda - \pm s$
coalesce into a double pole. Because of this, the expressions (5.128) for α_1 and
α_2 are no longer correct, but it is easy to prove that $\alpha_1 = 0$ and $\alpha_2 = (a'_1 a_1)^{1/2}$.
As a result, there is no value of c_2 for which (5.136) is satisfied, and the constant
specified by the contact condition (5.137) is

$$c_2 = 0 \qquad (5.139)$$

For the junction between two first order conductive sheets, the solution can be
found by putting $\Gamma'_2 = \Gamma_2 = \infty$. Then

$$K'(-\lambda) = K_+ \left(\frac{1}{\Gamma'_1}, -\lambda \right) \; K_+ (0, -\lambda) = \sqrt{1 + \lambda} \; K_+ \left(\frac{1}{\Gamma'_1}, -\lambda \right)$$

and similarly

$$K(\lambda) = \sqrt{1 - \lambda} \; K_+ \left(\frac{1}{\Gamma_1}, \lambda \right)$$

When these are inserted into (5.119) we obtain

$$\begin{aligned}
P(\lambda) \;=\; & K_+ \left(\frac{1}{\Gamma'_1}, -\lambda \right) K_+ \left(\frac{1}{\Gamma'_1}, -\lambda_0 \right) \\
& \cdot K_+ \left(\frac{1}{\Gamma_1}, \lambda \right) K_+ \left(\frac{1}{\Gamma_1}, \lambda_0 \right) \frac{c_0 + c_1 \lambda}{\lambda + \lambda_0}
\end{aligned} \qquad (5.140)$$

and the requirement that $P(\lambda) \to 0$ as $|\lambda| \to \infty$ forces $c_1 = 0$. The edge
condition alone is now sufficient to produce a unique solution, and the residue
evaluation gives

$$c_0 = -\frac{j}{2\pi} \left(\frac{1}{\Gamma'_1} - \frac{1}{\Gamma_1} \right) \qquad (5.141)$$

showing that $P(\lambda) = 0$ if $\Gamma'_1 = \Gamma_1$, as expected.

To obtain the solution for a second order conductive half-plane in $x > 0$,
Γ'_1 and Γ'_2 must be chosen such that

$$\left[\frac{\partial H_z}{\partial y} \right]_-^+ = 0$$

in $x < 0$, i.e. Γ_2' (or Γ_1') $= \infty$ and Γ_1' (or Γ_2') $= 0$. For $\Gamma_2' = \infty$ and small Γ_1'

$$K'(-\lambda)\,K'(-\lambda_0) = \Gamma_1'\sqrt{(1+\lambda)(1+\lambda_0)}$$

and the edge conditon still allows c_1 (and therefore c_2) to be non-zero. With this choice of Γ_2' and Γ_1', the original analysis is valid as it stands, and

$$P(\lambda) = -\frac{j}{2\pi a_2}\{(1-\lambda)(1-\lambda_0)\}^{-1/2}\,K(\lambda)\,K(\lambda_0)\left\{\frac{s^2+\lambda\lambda_0}{\lambda+\lambda_0}+c_2\right\} \quad (5.142)$$

with $s^2 = 1 + a_2$.

We now consider some other problems whose solutions can be deduced from the above. The first is the junction of two second order conductive sheets for E-polarisation. From (5.25) and (5.26) the transition conditions can be taken as

$$\prod_{m=1}^{2}\left(\bar{\Gamma}_m' - \frac{1}{jk_0}\frac{\partial}{\partial y}\right)E_z^+ - \prod_{m=1}^{2}\left(\bar{\Gamma}_m' + \frac{1}{jk_0}\frac{\partial}{\partial y}\right)E_z^- = 0 \quad (5.143)$$

for $x < 0$

$$\prod_{m=1}^{2}\left(\bar{\Gamma}_m - \frac{1}{jk_0}\frac{\partial}{\partial y}\right)E_z^+ - \prod_{m=1}^{2}\left(\bar{\Gamma}_m + \frac{1}{jk_0}\frac{\partial}{\partial y}\right)E_z^- = 0 \quad (5.144)$$

for $x > 0$, with

$$\left[\frac{\partial E_z}{\partial y}\right]_-^+ = 0 \quad (5.145)$$

for all x. The sheet supports a magnetic current

$$\mathbf{J_m} = -\hat{x}[E_z]_-^+ \quad (5.146)$$

and the resulting electric field is antisymmetric about the plane $y = 0$. By analogy with (5.98) we choose

$$\mathbf{E^i} = \hat{z}\,e^{jk_0(x\cos\phi_0 + y\sin\phi_0)} \quad (5.147)$$

and write

$$E_z = E_z^\circ + E_z^1 \quad (5.148)$$

where E_z° is the total field produced when the plane wave (5.147) is incident on a sheet having the transition condition (5.143) imposed for all x. The expression for E_z° is identical to that in (5.103) with

$$R = \frac{1}{2}\left\{1 - \prod_{m=1}^{2}\frac{\bar{\Gamma}_m' - \sin\phi_0}{\bar{\Gamma}_m' + \sin\phi_0}\right\}$$

$$ \quad (5.149)$$

$$T = \frac{1}{2}\left\{1 + \prod_{m=1}^{2}\frac{\bar{\Gamma}_m' - \sin\phi_0}{\bar{\Gamma}_m' + \sin\phi_0}\right\}$$

and the antisymmetry of E_z leads us to define

$$E_z^1 = \frac{y}{|y|} \int_{-\infty}^{\infty} Q(\lambda) \, e^{-jk_0(x\lambda + |y|\sqrt{1-\lambda^2})} \, \frac{d\lambda}{\sqrt{1-\lambda^2}} \qquad (5.150)$$

When the analysis is carried out, it is found that the expression for $Q(\lambda)$ is *identical* to that for $P(\lambda)$ given in (5.123) apart from having $\bar{\Gamma}'_m$ and $\bar{\Gamma}_m$ in place of Γ'_m and Γ_m, respectively. A contact condition is still required, and the rest of the analysis in Section 5.5.1 still applies.

The specialisation to two first order sheets is the same as for H-polarisation, but there is a difference in the case of a second order conductive half-plane in $x > 0$. We must now choose $\bar{\Gamma}'_1$ and $\bar{\Gamma}'_2$ such that

$$[E_z]_-^+ = 0 \qquad (5.151)$$

and this demands $\bar{\Gamma}'_1 = \bar{\Gamma}'_2 = \infty$. Hence

$$K'(-\lambda) = 1 + \lambda$$

When this is inserted into (5.119) we obtain

$$Q(\lambda) = \left\{ \left(\frac{1+\lambda}{1-\lambda} \right) \left(\frac{1+\lambda_0}{1-\lambda_0} \right) \right\}^{1/2} K(\lambda) \, K(\lambda_0) \, \frac{c_0 + c_1\lambda}{\lambda + \lambda_0} \qquad (5.152)$$

and the requirement that $P(\lambda) \to 0$ as $|\lambda| \to \infty$ forces $c_1 = 0$. In other words, the edge condition alone produces a unique solution. The residue evaluation gives $c_0 = j/(2\pi)$ so that

$$Q(\lambda) = \frac{j}{2\pi} \left\{ \left(\frac{1+\lambda}{1-\lambda} \right) \left(\frac{1+\lambda_0}{1-\lambda_0} \right) \right\}^{1/2} \frac{K(\lambda) \, K(\lambda_0)}{\lambda + \lambda_0} \qquad (5.153)$$

and no contact condition is either needed or possible.

For resistive sheets the solutions are the electromagnetic duals of those given above. The most general situation is the junction of two combined second order resistive and conductive sheets supporting electric and magnetic currents, respectively and, since the currents do not interact, the solution is simply the superposition of the solutions for the separate sheets. By choosing $\bar{\Gamma}'_m = \Gamma_m$ and $\bar{\Gamma}_m = \Gamma_m$ the structure becomes opaque, and the problem is that of a second order GIBC imposed at $y = 0+$. Since $Q(\lambda) = P(\lambda)$, the solution is twice that for the conductive sheet alone, and there is, of course, only one arbitrary constant to specify. For a GIBC half-plane in $x > 0$ there is likewise only one constant, since the solution for the conductive (resistive) half-plane for E-polarisation (H-polarisation) is uniquely determined by the edge condition. Thus, for the plane wave (5.98) incident on a half-plane in $x > 0$ subject to the second order GIBCs

$$\prod_{m=1}^{2} \left(\Gamma_m \mp \frac{1}{jk_0} \frac{\partial}{\partial y} \right) H_z = 0 \qquad (5.154)$$

on $y = \pm 0$, the total field is

$$H_z(x,y) = H_z^i(x,y) + H_z^s(x,y) \qquad (5.155)$$

with

$$H_z^s(x,y) = \int_{-\infty}^{\infty} \left\{ P(\lambda) + \frac{y}{|y|} Q(\lambda) \right\} e^{-jk_0(x\lambda + |y|\sqrt{1-\lambda^2})} \frac{d\lambda}{\sqrt{1-\lambda^2}} \qquad (5.156)$$

where $P(\lambda)$ and $Q(\lambda)$ are given in (5.142) and (5.153), respectively.

Instead of treating a half-plane as a special case of a surface occupying the entire plane $y = 0$, we can tackle it directly. If we do so, (5.74) is no longer the appropriate form for the contact condition. Since the surface of integration now extends over both sides of the half-plane, (5.74) must be replaced by

$$\lim_{x \to 0+} \beta_- \frac{\partial}{\partial x} U(x, 0-) = - \lim_{x \to 0+} \beta_+ \frac{\partial}{\partial x} U(x, 0+) \qquad (5.157)$$

where β_- and β_+ are the values of β on the lower and upper sides, respectively. For the GIBC (5.154), $\beta_- = \beta_+ = -j/k_0 a_1)$ with $U = H_z$, and (5.157) then becomes

$$\lim_{x \to 0+} \left\{ \frac{\partial}{\partial x} H_z(x, 0-) + \frac{\partial}{\partial x} H_z(x, 0+) \right\} = 0 \qquad (5.158)$$

(SENIOR, 1993b), from which it is evident that only $P(\lambda)$ is constrained.

The cases of most practical interest are a combination of a first order resistive sheet and a second order conductive sheet for H-polarisation, where the conductive sheet is needed to simulate the effect of the normal component of the electric polarisation current as well as any magnetic material properties; and a second order GIBC simulating a metal-backed lossy dielectric layer, where the second order is necessary to place the simulating surface at the location of the backing. These (and other) cases have all been addressed in the literature, sometimes with erroneous results.

LEPPINGTON (1983) used a combination of a first order resistive sheet and a second order conductive sheet to model a thin non-magnetic dielectric layer, and considered the diffraction of an H-polarised plane wave by an abrupt change in layer thickness. The unknown constant was determined by matching quasi-static expressions for the *interior* fields at the junction, and we recall that this is a situation for which $c_2 = 0$. The special case of a dielectric half-plane was treated by CHAKRABATI (1986), but he failed to address the uniqueness question and his solution violates even the reciprocity condition. The correct solution was provided by VOLAKIS AND SENIOR (1987) who used the continuity of the y component of the electric polarisation current to show that $c_2 = 0$. The extension to the case of a plane wave at skew angles of incidence was given by SENIOR (1989a), and BÜYÜKAKSOY ET AL. (1989) did the same for Leppington's problem.

When the dielectric has $\mu_r \neq 1$ the solution differs in the value of c_2. The necessary extension of Leppington's solution was carried out by ROJAS ET AL. (1991) who derived the contact condition (5.137) by matching the interior fields at the junction. Results for a variety of special cases were also presented, along with graphs showing the angular behaviour of the scattered field. For a single half-plane VOLAKIS (1988) had previously given the solution but with the constant c_2 omitted, and the same error was made by ROJAS AND CHOU (1990). A feeling for the importance of the constant can be obtained from Fig. 5–10, which shows the echo width for a homogeneous dielectric half-plane having $\epsilon_r = 4$ and $\mu_r = 2$ with $k_0\tau = 0.063$ and 1.26. For the thinner layer the constant has almost no effect except at angles close to edge-on and grazing but, as the thickness increases, the constant plays a more significant role. This is hardly surprising since, for a very thin layer, the first order conditions suffice.

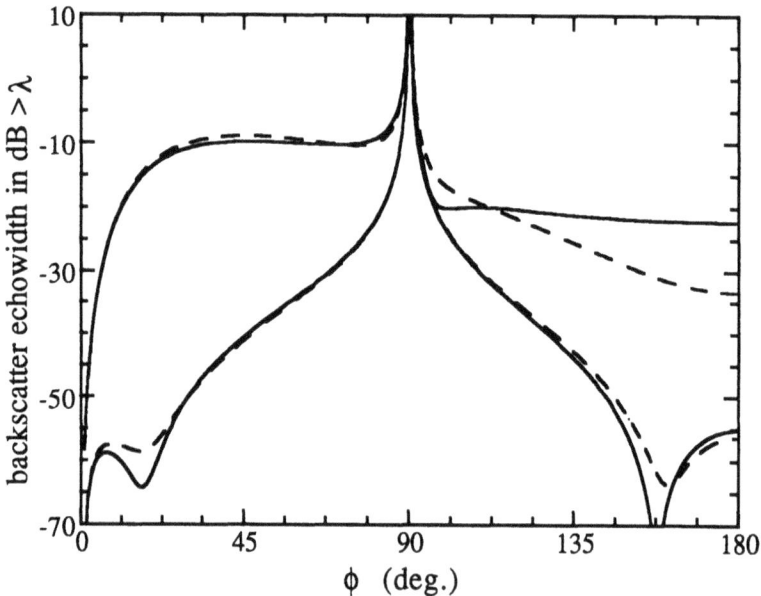

Figure 5–10: *H polarisation backscatter echowidth for a homogeneous dielectric half-plane with $\epsilon_r = 4$, $\mu_r = 2$ and $k_0\tau = 0.063$ (lower curves), 1.26 (upper curves). The dashed curves show the effect of putting $c_2 = 0$*

5.6 Wedge diffraction

For wedges of non-zero included angle a solution technique analogous to the Wiener-Hopf method was developed by MALIUZHINETS (1958c). The solution is carried out in cylindrical polar coordinates and, for a wedge subject to first order (possibly different) impedance boundary conditions on the two faces, the method is simple and elegant (see Chapter 4). Unfortunately, this is not true for second and higher order conditions, and the need to construct a particular solution of an inhomogeneous difference equation is a significant complication. Because of this, the problem has received little attention in the literature, and no correct solutions have yet appeared. Although BERNARD (1987) tackled the most general problem of a wedge of arbitrary angle subject to GIBCs of arbitrary but finite order, he was unaware of the need for contact conditions, and sought to avoid the explicit evaluation of the particular solution of the difference equation. As a result his expression for the field is both incomplete and incorrect. The special case of a right-angled wedge with identical second order conditions on the two faces was treated by SENIOR (1989b) and, although the particular solution was obtained, his final result is still wrong. The correct solution is presented in Section 5.6.2.

To illustrate the method, it is convenient to start by considering a second order impedance half-plane, and this enables us to point out the similarities to the Wiener-Hopf approach.

5.6.1   Impedance half-plane

The H-polarised plane wave (5.98) is incident on a half-plane whose faces are $\phi = \pm\pi$ where ρ, ϕ, z are cylindrical polar coordinates with the z axis at the edge. In terms of these coordinates

$$\mathbf{H}^i = \hat{z}\, e^{jk_0\rho\cos(\phi-\phi_0)} \tag{5.159}$$

and in contrast to the problem treated in the previous section, the half-plane now occupies the region $y = 0$, $x < 0$. This is the reason that, in Section 4.2.1, we used ϕ' and ϕ_0' as the polar angles in place of ϕ and ϕ_0. The boundary conditions imposed at $\phi = \pm\pi$ are

$$\prod_{m=1}^{2}\left(\Gamma_m \pm \frac{1}{jk_0\rho}\frac{\partial}{\partial\phi}\right)H_z = 0 \tag{5.160}$$

where $\Gamma_m = \sin\theta_m$ are constants specified by the material properties. Consistent with the known behaviour of the field at a lossy edge, it is required that $H_z(\rho,\phi) = O\{(k_0\rho)^\epsilon\}$ for small $k_0\rho$ with $\epsilon \geq 0$. In addition, to ensure uniqueness, a contact condition must be imposed, and this is (see (5.158))

$$\lim_{\rho\to 0}\left\{\frac{\partial}{\partial\rho}H_z(\rho,-\pi) + \frac{\partial}{\partial\rho}H_z(\rho,\pi)\right\} = 0 \tag{5.161}$$

Following MALIUZHINETS (1958c) the total field is written as

$$H_z(\rho,\phi) = \frac{1}{2\pi j}\int_\gamma e^{jk_0\rho\cos\alpha}s(\alpha+\phi)\,d\alpha \tag{5.162}$$

where γ is the double loop Sommerfeld path shown in Figure 4–3. On apply-ing the boundary conditions and using integration by parts to eliminate the derivatives with respect to ϕ, we obtain

$$\int_\gamma e^{jk_0\rho\cos\alpha}(\sin\alpha \pm \sin\theta_1)(\sin\alpha \pm \sin\theta_2)\, s(\alpha \pm \pi)\, d\alpha = 0 \qquad (5.163)$$

The edge condition demands that $s(\alpha) = O\{\exp(-\epsilon|\text{Im. }\alpha|)\}$ for large $|\text{Im. }\alpha|$, and for brevity we write this as $O(\alpha^{-\epsilon})$. Equation (5.163) then implies (MAL-IUZHINETS, 1958a)

$$(\sin\alpha \pm \theta_1)(\sin\alpha \pm \theta_2)\, s(\alpha \pm \pi) - (\sin\alpha \mp \theta_1)(\sin\alpha \mp \theta_2)\, s(-\alpha \pm \pi)$$
$$= \sin\alpha\left(A_0^\pm + A_1^\pm \cos\alpha\right) \qquad (5.164)$$

where $A_0^\pm$ and $A_1^\pm$ are four constants, as yet arbitrary.

To determine $s(\alpha)$, let

$$s(\alpha) = g(\alpha)\, t(\alpha) \qquad (5.165)$$

where

$$g(\alpha) = \frac{\Psi(\alpha,\theta_1)\,\Psi(\alpha,\theta_2)}{\Psi(\phi_0,\theta_1)\,\Psi(\phi_0,\theta_2)} \qquad (5.166)$$

and

$$\Psi(\alpha,\theta_m) = \psi_\pi\left(\alpha + \frac{3\pi}{2} - \theta_m\right)\psi_\pi\left(\alpha - \frac{3\pi}{2} + \theta_m\right)$$
$$\cdot\, \psi_\pi\left(\alpha + \frac{\pi}{2} + \theta_m\right)\psi_\pi\left(\alpha - \frac{\pi}{2} - \theta_m\right) \qquad (5.167)$$

(MALIUZHINETS, 1958c). If (as assumed) Re. $\sin\theta_m > 0$, $g(\alpha)$ is free of poles and zeros in the strip $|\text{Re. }\alpha| \leq \pi$, but if Re. $\sin\theta_m < 0$, $\Psi(\alpha,\theta_m)$ must be replaced by its reciprocal with the sign of θ_m reversed (BERNARD, 1987). This is evident from the fact that

$$\Psi(\alpha,-\theta_m) = c(\cos\alpha + \cos\theta_m)\left\{\Psi(\alpha,\theta_m)\right\}^{-1}$$

where c is a constant. For large $|\text{Im. }\alpha|$, $g(\alpha) = O(\alpha)$. Since $g(\alpha)$ is an even function and

$$g(\alpha + \pi) = \xi\, g(\alpha - \pi) \qquad (5.168)$$

where

$$\xi = \frac{(\sin\alpha - \sin\theta_1)(\sin\alpha - \sin\theta_2)}{(\sin\alpha + \sin\theta_1)(\sin\alpha + \sin\theta_2)} \qquad (5.169)$$

substitution of (5.165) into (5.164) gives

$$t(\alpha \pm \pi) - t(-\alpha \pm \pi) = \frac{\sin\alpha}{g(\alpha - \pi)}\frac{A_0^\pm + A_1^\pm\cos\alpha}{(\sin\alpha - \sin\theta_1)(\sin\alpha - \sin\theta_2)} \qquad (5.170)$$

Hence
$$t(\alpha + 2\pi) - t(\alpha - 2\pi) = -h(\alpha) \tag{5.171}$$

where

$$
\begin{aligned}
h(\alpha) &= \frac{\sin \alpha}{g(\alpha - 2\pi)} \{(\sin \alpha - \sin \theta_1)(\sin \alpha - \sin \theta_2)(A_0^+ - A_1^+ \cos \alpha) \\
&\quad - (\sin \alpha + \sin \theta_1)(\sin \alpha + \sin \theta_2)(A_0^- - A_1^- \cos \alpha)\} \\
&\quad \cdot \{(\sin \alpha + \sin \theta_1)(\sin \alpha + \sin \theta_2)\}^{-2}
\end{aligned}
\tag{5.172}
$$

A particular solution of (5.171) is

$$t_0(\alpha) = \sum_{n=0}^{\infty} h(\alpha + 2\pi + 4n\pi) = \frac{h(\alpha + 2\pi)}{1 - \xi^2} \tag{5.173}$$

and therefore

$$
\begin{aligned}
t_0(\alpha) &= \frac{1}{4\,g(\alpha)\,p(\alpha)} \{(\sin \alpha + \sin \theta_1)(\sin \alpha + \sin \theta_2)(A_0^- - A_1^- \cos \alpha) \\
&\quad - (\sin \alpha - \sin \theta_1)(\sin \alpha - \sin \theta_2)(A_0^+ - A_1^+ \cos \alpha)\} \\
&\quad \cdot (\sin \theta_1 + \sin \theta_2)^{-1}
\end{aligned}
\tag{5.174}
$$

(SENIOR, 1991) with
$$p(\alpha) = \cos^2 \alpha - \cos^2 \alpha_{\mathrm{p}} \tag{5.175}$$

and
$$\cos^2 \alpha_{\mathrm{p}} = 1 + \sin \theta_1 \sin \theta_2 \tag{5.176}$$

showing that $t_0(\alpha)$ has poles at $\alpha = \pm \alpha_{\mathrm{p}}$ and $\alpha = \pm(\pi - \alpha_{\mathrm{p}})$ within the strip $|\mathrm{Re.}\ \alpha| \le \pi$. As required, $t_0(\alpha)$ satisfies (5.170), and to meet the order requirement on $s(\alpha)$ it is necessary that $A_1^- = A_1^+ = A_1$ (say). If we also choose $A_0^- = A_0^+ (= A_0)$, $t_0(\alpha)$ becomes an odd function of α. The residues at $\alpha = \pm \alpha_{\mathrm{p}}$ are then the same, as are those at $\alpha = \pm(\pi - \alpha_{\mathrm{p}})$, and (5.174) reduces to
$$t_0(\alpha) = \frac{\sin \alpha}{2\,g(\alpha)\,p(\alpha)}(A_0 - A_1 \cos \alpha) \tag{5.177}$$

The most general expression for $t(\alpha)$ is
$$t(\alpha) = \sigma(\alpha) + t_0(\alpha) \tag{5.178}$$

where $\sigma(\alpha)$ satisfies
$$\sigma(\alpha \pm \pi) = \sigma(-\alpha \pm \pi)$$

It is therefore a function of $\sin \frac{\alpha}{2}$ and to reproduce the incident field (5.159) we choose
$$\sigma(\alpha) = \sigma_0(\alpha) + \sigma_1(\alpha) \tag{5.179}$$

with

$$\sigma_0(\alpha) = \tfrac{1}{2} \cos \frac{\phi_0}{2} \left(\sin \frac{\alpha}{2} - \sin \frac{\phi_0}{2} \right)^{-1} \tag{5.180}$$

(MALIUZHINETS, 1958c). Since $g(\alpha) = O(\alpha)$, $g(\alpha)\sigma_0(\alpha) = O(\alpha^{1/2})$, which violates the edge condition imposed. To cancel the term of excess order, an additional function $\sigma_1(\alpha)$ is introduced as shown in (5.179), and this must also serve to cancel the poles of $t_0(\alpha)$ in $|\text{Re. }\alpha| \leq \pi$. Recognising that $p(\alpha)$ is itself a function of $\sin \frac{\alpha}{2}$, we write

$$\sigma_1(\alpha) = \frac{\cos \dfrac{\phi_0}{2} \sin \dfrac{\alpha}{2}}{p(\alpha)}(c_0 \cos \alpha - c_2) \tag{5.181}$$

where c_0 and c_2 are constants, and then

$$s(\alpha) = g(\alpha)\{\sigma_0(\alpha) + \sigma_1(\alpha) + t_0(\alpha)\} \tag{5.182}$$

The elimination of the term $O(\alpha^{1/2})$ from $g(\alpha)\sigma_0(\alpha)$ requires that

$$c_0 = 1 \tag{5.183}$$

and to eliminate the poles of $t_0(\alpha)$ at $\alpha = \pm\alpha_p$ and $\pm(\pi - \alpha_p)$:

$$A_0 = \frac{1}{2}\cos\frac{\phi_0}{2}\left\{\frac{g(\pi - \alpha_p)}{\sin(\alpha_p/2)}(\cos\alpha_p + c_2) - \frac{g(\alpha_p)}{\cos(\alpha_p/2)}(\cos\alpha_p - c_2)\right\} \tag{5.184}$$

$$A_1 = \frac{1}{2}\frac{\cos(\phi_0/2)}{\cos\alpha_p}\left\{\frac{g(\pi - \alpha_p)}{\sin(\alpha_p/2)}(\cos\alpha_p + c_2) + \frac{g(\alpha_p)}{\cos(\alpha_p/2)}(\cos\alpha_p - c_2)\right\} \tag{5.185}$$

With c_0, A_0 and A_1 specified in this manner, $s(\alpha)$ is a function of order unity for large $|\text{Im. }\alpha|$, and is free of singularities in the strip $|\text{Re. }\alpha| \leq \pi$ apart from a pole at $\alpha = \phi_0$ whose residue reproduces the incident field (5.159). Accordingly, (5.182) satisfies all of the requirements except for the contact condition (5.161), and does so for arbitrary values of the constant c_2.

The remaining task is to determine c_2, and the procedure for doing so is simpler yet more subtle than with the Wiener-Hopf method. From (5.162)

$$H_z(\rho, -\pi) + H_z(\rho, \pi) = \frac{1}{2\pi j}\int_\gamma e^{jk_0\rho\cos\alpha}\{s(\alpha - \pi) + s(\alpha + \pi)\}\,d\alpha \tag{5.186}$$

where $s(\alpha)$ is given in (5.182). Using the relationships satisfied by the half-plane function $\psi_\pi(\alpha)$, we have

$$g(\alpha \pm \pi) = (\sin\alpha \mp \sin\theta_1)(\sin\alpha \mp \sin\theta_2)\,G(\alpha, \phi_0)$$

where

$$G(\alpha, \phi_0) = \frac{1}{64} \left\{ \psi_\pi \left(\frac{\pi}{2} \right) \right\}^{16} \left\{ \Psi(\alpha, \theta_1) \, \Psi(\alpha, \theta_2) \, \Psi(\phi_0, \theta_1) \, \Psi(\phi_0, \theta_2) \right\}^{-1}$$

We observe that $G(\alpha, \phi_0)$ is symmetrical in α and θ_0, and is $O(\alpha^{-1})$ for large $|\text{Im. } \alpha|$. Then

$$s(\alpha - \pi) + s(\alpha + \pi) =$$

$$\cos \frac{\phi_0}{2} G(\alpha, \phi_0) \left[(\sin \alpha - \sin \theta_1)(\sin \alpha - \sin \theta_2) \right.$$

$$\cdot \left\{ \frac{1}{2} \left(\cos \frac{\alpha}{2} - \cos \frac{\phi_0}{2} \right)^{-1} - \frac{\cos \frac{\alpha}{2}}{p(\alpha)} (\cos \alpha + c_2) \right\}$$

$$- (\sin \alpha + \sin \theta_1)(\sin \alpha + \sin \theta_2) \left\{ \frac{1}{2} \left(\cos \frac{\alpha}{2} + \cos \frac{\phi_0}{2} \right)^{-1} \right.$$

$$\left. \left. - \frac{\cos \frac{\alpha}{2}}{p(\alpha)} (\cos \alpha + c_2) \right\} \right] - \frac{\sin \alpha}{p(\alpha)} (A_0 + A_1 \cos \alpha)$$

which reduces to

$$s(\alpha - \pi) + s(\alpha + \pi) = f_1(\alpha) + f_2(\alpha) + f_3(\alpha)$$

where

$$f_1(\alpha) = 2 \cos \frac{\phi_0}{2} \sin \alpha \cos \frac{\alpha}{2} (\sin \theta_1 + \sin \theta_2) \frac{G(\alpha, \phi_0)}{p(\alpha)}$$

$$\cdot \left\{ \frac{\cos \alpha \cos \phi_0 + \cos^2 \alpha_p}{\cos \alpha + \cos \phi_0} + c_2 \right\}$$

$$f_2(\alpha) = -\sin \phi_0 \frac{G(\alpha, \phi_0) \, p(\alpha)}{\cos \alpha + \cos \phi_0}$$

$$f_3(\alpha) = -\frac{\sin \alpha}{p(\alpha)} (A_0 + A_1 \cos \alpha)$$

The function $f_1(\alpha)$ is $O(\alpha^{-3/2})$ for large $|\text{Im. } \alpha|$, and when inserted into (5.186) provides a contribution $O\left\{ (k_0\rho)^{3/2} \right\}$ for small $k_0\rho$. Since $f_2(\alpha)$ is an even function of α, its contribution is zero for all ρ (MALIUZHINETS, 1958a), and finally

$$f_3(\alpha) = -\frac{\sin \alpha}{\cos \alpha} \left(A_1 + \frac{A_0}{\cos \alpha} \right) + O(\alpha^{-2})$$

Hence

$$H_z(\rho, -\pi) + H_z(\rho, \pi) = -2(A_1 + jk_0\rho A_0) + O\left\{ (k_0\rho)^{3/2} \right\} \qquad (5.187)$$

(MALIUZHINETS, 1958b). If

$$\lim_{\rho \to 0} \left\{ H_z(\rho, -\pi) + H_z(\rho, \pi) \right\} = 0 \qquad (5.188)$$

then $A_1 = 0$ and, from (5.185),

$$c_2 = -\cos\alpha_{\mathrm{p}} \frac{g(\pi - \alpha_{\mathrm{p}})\cos\frac{\alpha_{\mathrm{p}}}{2} + g(\alpha_{\mathrm{p}})\sin\frac{\alpha_{\mathrm{p}}}{2}}{g(\pi - \alpha_{\mathrm{p}})\cos\frac{\alpha_{\mathrm{p}}}{2} - g(\alpha_{\mathrm{p}})\sin\frac{\alpha_{\mathrm{p}}}{2}} \qquad (5.189)$$

Alternatively, if (5.161) is satisfied, $A_0 = 0$ giving

$$c_2 = -\cos\alpha_{\mathrm{p}} \frac{g(\pi - \alpha_{\mathrm{p}})\cos\frac{\alpha_{\mathrm{p}}}{2} - g(\alpha_{\mathrm{p}})\sin\frac{\alpha_{\mathrm{p}}}{2}}{g(\pi - \alpha_{\mathrm{p}})\cos\frac{\alpha_{\mathrm{p}}}{2} + g(\alpha_{\mathrm{p}})\sin\frac{\alpha_{\mathrm{p}}}{2}} \qquad (5.190)$$

and since (5.161) is the required contact condition, the constant c_2 is as shown in (5.190). This completes the solution, and the straightforward determination of c_2 is in stark contrast to the arguments employed by SENIOR (1991).

The procedure for calculating the scattered field is similar to that described in Section 4.2.1. Apart from the optics terms and any surface wave contributions that may be present,

$$H_z^s(\rho, \phi) = \frac{1}{2\pi j} \int_{S(\phi)} e^{-jk_0\rho\cos(\alpha-\phi)} \left\{ s(\alpha + \pi) - s(\alpha - \pi) \right\} d\alpha \qquad (5.191)$$

where $S(\phi)$ is the steepest descent path through $\alpha = \phi$ and

$$s(\alpha + \pi) - s(\alpha - \pi) = 4\cos\frac{\phi_0}{2}\cos\frac{\alpha}{2} G(\alpha, \phi_0)$$

$$\cdot \left\{ \frac{\cos^2\alpha_{\mathrm{p}} + \cos\alpha\cos\phi_0 - 2\sin\frac{\phi_0}{2}\sin\frac{\alpha}{2}(\sin\theta_1 + \sin\theta_2)}{\cos\alpha + \cos\phi_0} + c_2 \right\} \qquad (5.192)$$

The similarity to the Wiener-Hopf solution is obvious. With the substitution $\cos\alpha = \lambda$, $\Psi(\alpha, \theta_m)$ is proportional to $K(\lambda)$ and, from (5.176), $\cos\alpha_{\mathrm{p}}$ is the same as the parameter s. We observe that A_0 and A_1 do not contribute to the solution but are needed to specify c_2. This constant is identical to the one in the previous section and, since it is independent of ϕ_0, the final expression for H_z is symmetrical in ϕ and ϕ_0 in accordance with the reciprocity condition.

5.6.2   Right-angled impedance wedge

With the experience gained from the simple problem of a half-plane we now consider the diffraction of the plane wave (5.159) by a right-angled wedge whose faces are $\phi = \pm 3\pi/4$ (see Fig. 5–11). The boundary conditions imposed are the second order GIBCs (5.160), and the contact condition is

$$\lim_{\rho \to 0} \left\{ \frac{\partial}{\partial\rho} H_z\left(\rho, -\frac{3\pi}{4}\right) + \frac{\partial}{\partial\rho} H_z\left(\rho, \frac{3\pi}{4}\right) \right\} = 0 \qquad (5.193)$$

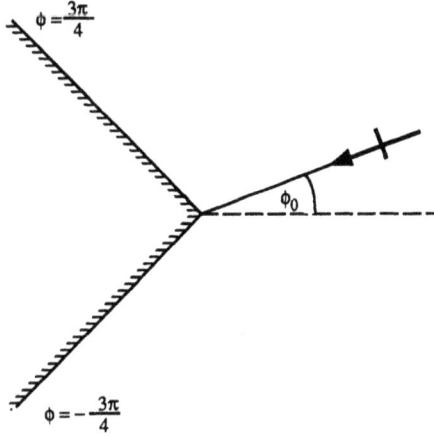

Figure 5–11: *Geometry for the wedge problem*

Initially at least the analysis is similar to that for the half-plane. When the boundary conditions are applied to the Maliuzhinets representation (5.162) for the total field, we obtain

$$\int_\gamma e^{jk_0\rho\cos\alpha}(\sin\alpha \pm \sin\theta_1)(\sin\alpha \pm \sin\theta_2)\, s\left(\alpha \pm \frac{3\pi}{4}\right)\, d\alpha = 0$$

and therefore

$$(\sin\alpha \pm \sin\theta_1)(\sin\alpha \pm \sin\theta_2)\, s\left(\alpha \pm \frac{3\pi}{4}\right)$$
$$- (\sin\alpha \mp \sin\theta_1)(\sin\alpha \mp \sin\theta_2)\, s\left(-\alpha \pm \frac{3\pi}{4}\right)$$
$$= \sin\alpha(A_0^\pm + A_1^\pm \cos\alpha) \tag{5.194}$$

where $A_0^\pm$ and $A_1^\pm$ are four arbitrary constants. It is sufficient to take $A_0^- = A_0^+ = A_0$ (say) and $A_1^- = A_1^+ = A_1$. To solve (5.194) we write

$$s(\alpha) = g(\alpha)\, t(\alpha) \tag{5.195}$$

where $g(\alpha)$ is shown in (5.166) and

$$\Psi(\alpha, \theta_m) = \psi_{3\pi/4}\left(\alpha + \frac{3\pi}{4} - \theta_m\right) \psi_{3\pi/4}\left(\alpha - \frac{3\pi}{4} + \theta_m\right)$$
$$\cdot \psi_{3\pi/4}\left(\alpha + \frac{\pi}{4} + \theta_m\right) \psi_{3\pi/4}\left(\alpha - \frac{\pi}{4} - \theta_m\right) \tag{5.196}$$

From the relationships given by MALIUZHINETS (1958c) it can be verified that $g(\alpha)$ is an even function and

$$g\left(\alpha + \frac{3\pi}{4}\right) = \frac{(\sin\alpha - \sin\theta_1)(\sin\alpha - \sin\theta_2)}{(\sin\alpha + \sin\theta_1)(\sin\alpha + \sin\theta_2)}\, g\left(\alpha - \frac{3\pi}{4}\right) \tag{5.197}$$

It follows that

$$g(\alpha + 6\pi) = g(\alpha) \qquad (5.198)$$

and, in addition, for large $|\text{Im. }\alpha|$, $g(\alpha) = O(\alpha^{4/3})$. Substitution of (5.195) into (5.194) gives

$$t\left(\alpha \pm \frac{3\pi}{4}\right) - t\left(-\alpha \pm \frac{3\pi}{4}\right) = \frac{\sin \alpha}{g\left(\alpha - \frac{3\pi}{4}\right)(\sin \alpha - \sin \theta_1)(\sin \alpha - \sin \theta_2)}$$

$$(5.199)$$

and hence

$$t\left(\alpha + \frac{3\pi}{2}\right) - t\left(\alpha - \frac{3\pi}{2}\right) = -h(\alpha) \qquad (5.200)$$

where

$$h(\alpha) = \{(s - \sin \theta_1)(s - \sin \theta_2)c(A_0 - A_1 s)$$

$$- (c + \sin \theta_1)(c + \sin \theta_2)s(A_0 + A_1 c)\}$$

$$\cdot \left\{g\left(\alpha - \frac{3\pi}{2}\right)(s - \sin \theta_1)(s - \sin \theta_2)(c - \sin \theta_1)(c - \sin \theta_2)\right\}^{-1}$$

For brevity, we have written $s = \sin(\alpha + 3\pi/4)$ and $c = \cos(\alpha + 3\pi/4)$.

Unfortunately, the method used to construct a particular solution for a half-plane is no longer applicable. As evident from (5.198), alternate terms in the expansion analogous to (5.173) are the same and the series diverges, but in cases like this it is usually possible to obtain a solution by inspection. Thus, as shown by SENIOR (1989b), a particular solution of (5.200) is

$$t_0(\alpha) = q(\alpha)\{g(\alpha)(s - \sin \theta_1)(s - \sin \theta_2)(c + \sin \theta_1)(c + \sin \theta_2)\}^{-1} \qquad (5.201)$$

where

$$q(\alpha) = \tfrac{1}{2}(s + c)\Big\{(sc + \sin \theta_1 \sin \theta_2)A_0$$

$$+ sc\left(\tfrac{1}{2}c - \tfrac{1}{2}s + \sin \theta_1 + \sin \theta_2\right)A_1\Big\}$$

$$- cs\left\{(\sin \theta_1 + \sin \theta_2)A_0 + \left(\tfrac{1}{2} + \sin \theta_1 \sin \theta_2\right)A_1\right\}\frac{2\alpha}{3\pi} \qquad (5.202)$$

This is an odd function which also satisfies (5.199), as required. For large $|\text{Im. }\alpha|$, $g(\alpha)t_0(\alpha) = O(1)$ in accordance with the edge condition, but $t_0(\alpha)$ has poles at $\alpha = \pm(3\pi/4 - \theta_m)$ and $\alpha = \pm(\pi/4 - \theta_m)$ lying in the strip $|\text{Re. }\alpha| \le 3\pi/4$.

The most general expression for $t(\alpha)$ is

$$t(\alpha) = \sigma(\alpha) + t_0(\alpha) \qquad (5.203)$$

where $\sigma(\alpha)$ satisfies

$$\sigma\left(\alpha \pm \frac{3\pi}{4}\right) = \sigma\left(-\alpha \pm \frac{3\pi}{4}\right)$$

It is therefore a function of $\sin \frac{2\alpha}{3}$ and, to reproduce the incident field (5.159), we choose

$$\sigma(\alpha) = \sigma_0(\alpha) + \sigma_1(\alpha) \tag{5.204}$$

with

$$\sigma_0(\alpha) = \frac{2}{3} \cos \frac{2\phi_0}{3} \left(\sin \frac{2\alpha}{3} - \sin \frac{2\phi_0}{3} \right)^{-1} \tag{5.205}$$

Clearly $g(\alpha)\,\sigma_0(\alpha) = O(\alpha^{2/3})$ which violates the requirement imposed by the edge condition. To cancel the term of excess order as well as the extraneous poles of $t_0(\alpha)$, an additional function $\sigma_1(\alpha)$ is introduced as shown in (5.204). This is also a function of $\sin \frac{2\alpha}{3}$, and since

$$(s - \sin\theta_m)(c + \sin\theta_m) =$$
$$\left\{ \gamma_m + \cos\left(\frac{4\alpha}{3} - \frac{2\pi}{3} \right) \right\} \left\{ \gamma_m + \cos\left(\frac{4\alpha}{3} + \frac{2\pi}{3} \right) \right\} \{81\,\Psi(\alpha,\theta_m)\}^{-1} \tag{5.206}$$

(SENIOR, 1989b) where

$$\gamma_m = \cos\left(\frac{4\theta_m}{3} - \frac{2\theta}{3} \right) \tag{5.207}$$

we write

$$\sigma_1(\alpha) = p(\alpha) \left[\left\{ \gamma_1 + \cos\left(\frac{4\alpha}{3} - \frac{2\pi}{3} \right) \right\} \left\{ \gamma_1 + \cos\left(\frac{4\alpha}{3} + \frac{2\pi}{3} \right) \right\} \right.$$
$$\left. \cdot \left\{ \gamma_2 + \cos\left(\frac{4\alpha}{3} - \frac{2\pi}{3} \right) \right\} \left\{ \gamma_2 + \cos\left(\frac{4\alpha}{3} + \frac{2\pi}{3} \right) \right\} \right]^{-1} \tag{5.208}$$

where $p(\alpha)$ is a polynomial in $\sin \frac{2\alpha}{3}$. From order considerations the polynomial must be of degree seven and, since $q(\alpha)$ is an odd function, it is sufficient if $p(\alpha)$ is also odd. We therefore choose

$$p(\alpha) = \cos \frac{2\phi_0}{3} \left(a_1 \sin \frac{2\alpha}{3} + a_2 \sin \frac{6\alpha}{3} + a_3 \sin \frac{10\alpha}{3} + a_4 \sin \frac{14\alpha}{3} \right) \tag{5.209}$$

where $a_1, \ldots, a_4$ are constants, and then $g(\alpha)\,\{\sigma_0(\alpha) + \sigma_1(\alpha)\} = O(1)$ for large $|\mathrm{Im.}\ \alpha|$ if

$$a_4 = \tfrac{1}{6} \tag{5.210}$$

We now have the three constants a_1, a_2 and a_3, plus A_0 and A_1, to eliminate the four poles $\alpha = \frac{3\pi}{4} - \theta_m$, $\frac{\pi}{4} - \theta_m$ $(m = 1,2)$. The one constant left over is specified by the contact condition.

To facilitate the elimination of the poles, (5.206) is used to write $t_0(\alpha)$ as

$$t_0(\alpha) = K \cos \frac{2\phi_0}{3}\, q(\alpha)$$
$$\cdot \left[\left\{ \gamma_1 + \cos\left(\frac{4\alpha}{3} - \frac{2\pi}{3} \right) \right\} \left\{ \gamma_1 + \cos\left(\frac{4\alpha}{3} + \frac{2\pi}{3} \right) \right\} \right.$$
$$\left. \cdot \left\{ \gamma_2 + \cos\left(\frac{4\alpha}{3} - \frac{2\pi}{3} \right) \right\} \left\{ \gamma_2 + \cos\left(\frac{4\alpha}{3} + \frac{2\pi}{3} \right) \right\} \right]^{-1} \tag{5.211}$$

where

$$K = 9^4 \, \Psi(\phi_0, \theta_1) \, \Psi(\phi_0, \theta_2) \, \sec \frac{2\phi_0}{3} \qquad (5.212)$$

The requirement is

$$a_1 \sin \frac{2\alpha}{3} + a_2 \sin \frac{6\alpha}{3} + a_3 \sin \frac{10\alpha}{3} + \frac{1}{6} \sin \frac{14\alpha}{3} =$$

$$-\tfrac{1}{2}(s+c) \left\{ (sc + \sin \theta_1 \sin \theta_2) K A_0 + sc \left(\tfrac{1}{2}c - \tfrac{1}{2}s + \sin \theta_1 + \sin \theta_2 \right) K A_1 \right\}$$

$$+ csC \frac{2\alpha}{3\pi} \qquad (5.213)$$

for $\alpha = \frac{3\pi}{4} - \theta_m$, $\frac{\pi}{4} - \theta_m$ $(m = 1, 2)$ where

$$C = (\sin \theta_1 + \sin \theta_2) K A_0 + \left(\tfrac{1}{2} + \sin \theta_1 \sin \theta_2 \right) K A_1 \qquad (5.214)$$

Since the analysis is rather tedious, the details are presented in Appendix E, where it is shown that

$$\begin{aligned} a_3 &= -\tfrac{1}{3} \left(\gamma_1 + \gamma_2 - \tfrac{1}{2} \right) + \tfrac{2}{3}\Gamma \\ a_1 &= -\tfrac{2}{3} \left(\gamma_1 \gamma_2 + \tfrac{1}{4} \right) - \tfrac{4}{3} \left(2\gamma_1 \gamma_2 + \gamma_1 + \gamma_2 + 1 \right) \Gamma \\ C &= \tfrac{\sqrt{3}}{2} \left(2\gamma_1 \gamma_2 - \gamma_1 - \gamma_2 + \tfrac{1}{2} \right) - 2\sqrt{3} \left(\gamma_1 + \gamma_2 + \tfrac{1}{2} \right) \Gamma \end{aligned} \qquad (5.215)$$

with Γ specified in (E.10). As we shall see, the constant a_2 is not needed.

The next task is to impose the contact condition (5.193). From the Maliuzhinets representation (5.162) we have

$$H_z \left(\rho, -\frac{3\pi}{4} \right) + H_z \left(\rho, \frac{3\pi}{4} \right) = \frac{1}{2\pi j} \int_\gamma e^{jk_0 \rho \cos \alpha} \left\{ s \left(\alpha - \frac{3\pi}{4} \right) + s \left(\alpha + \frac{3\pi}{4} \right) \right\} d\alpha$$

where $s(\alpha)$ is given in (5.195) with (5.203) and (5.204). From (5.205) by using (5.197)

$$g \left(\alpha - \frac{3\pi}{4} \right) \sigma_0 \left(\alpha - \frac{3\pi}{4} \right) + g \left(\alpha + \frac{3\pi}{4} \right) \sigma_0 \left(\alpha + \frac{3\pi}{4} \right) =$$

$$\frac{\frac{8}{3} \cos \frac{2\phi_0}{3} \, g \left(\alpha - \frac{3\pi}{4} \right)}{(\sin \alpha + \sin \theta_1)(\sin \alpha + \sin \theta_2)}$$

$$\cdot \frac{\sin \frac{2\phi_0}{3} (\sin^2 \alpha + \sin \theta_1 \sin \theta_2) - \sin \alpha \cos \frac{2\alpha}{3} (\sin \theta_1 + \sin \theta_2)}{\cos \frac{4\alpha}{3} + \cos \frac{4\phi_0}{3}} \qquad (5.216)$$

and (5.208) likewise gives

$$g \left(\alpha - \frac{3\pi}{4} \right) \sigma_1 \left(\alpha - \frac{3\pi}{4} \right) + g \left(\alpha + \frac{3\pi}{4} \right) \sigma_1 \left(\alpha + \frac{3\pi}{4} \right) =$$

$$-\frac{2\cos\frac{2\phi_0}{3}\,g\left(\alpha-\frac{3\pi}{4}\right)}{(\sin\alpha+\sin\theta_1)(\sin\alpha+\sin\theta_2)}(\sin\theta_1+\sin\theta_2)\sin\alpha$$

$$\cdot\left(a_1\cos\frac{2\alpha}{3}-a_2\cos\frac{6\alpha}{3}+a_3\cos\frac{10\alpha}{3}-a_4\cos\frac{14\alpha}{3}\right)$$

$$\cdot\left[\left\{\gamma_1-\cos\left(\frac{4\alpha}{3}-\frac{2\pi}{3}\right)\right\}\left\{\gamma_1-\cos\left(\frac{4\alpha}{3}+\frac{2\pi}{3}\right)\right\}\right.$$

$$\left.\cdot\left\{\gamma_2-\cos\left(\frac{4\alpha}{3}-\frac{2\pi}{3}\right)\right\}\left\{\gamma_2-\cos\left(\frac{4\alpha}{3}+\frac{2\pi}{3}\right)\right\}\right]^{-1} \qquad (5.217)$$

The sum of these is

$$\frac{2\cos\frac{2\phi_0}{3}\,g\left(\alpha-\frac{3\pi}{4}\right)}{(\sin\alpha+\sin\theta_1)(\sin\alpha+\sin\theta_2)}N \qquad (5.218)$$

where N is evident from (5.216) and (5.217). For large $|\text{Im. }\alpha|$, N contains terms $O(\alpha^{2/3})$, $O(\alpha^{1/3})$, $O(\alpha^{-2/3})$ and so on down. The choice (5.210) for a_4 ensures the cancellation of the terms $O(\alpha^{1/3})$ so that

$$N=\frac{4}{3}\sin\frac{2\phi_0}{3}\frac{\sin^2\alpha}{\cos(4\alpha/3)}\left\{1+O(\alpha^{-4/3})\right\}$$

which, to the order shown, is an even function of α. From (5.197) the factor multiplying N in (5.218) is also even, and (5.218) therefore contributes terms $O\left\{(k_0\rho)^{4/3}\right\}$ for small $k_0\rho$. The remaining part of $s(\alpha)$ is $g(\alpha)\,t_0(\alpha)$ which is given in (5.202), and it is trivial to show

$$g\left(\alpha-\frac{3\pi}{4}\right)t_0\left(\alpha-\frac{3\pi}{4}\right)+g\left(\alpha+\frac{3\pi}{4}\right)t_0\left(\alpha+\frac{3\pi}{4}\right)=$$

$$\left\{\sin\alpha(\sin^2\alpha+\sin\theta_1\sin\theta_2)(\cos\alpha+\sin\theta_1)(\cos\alpha+\sin\theta_2)\right.$$

$$\cdot(A_0+A_1\cos\alpha)+\sin\alpha\cos\alpha(\sin\theta_1+\sin\theta_2)$$

$$\cdot\left[(\sin^2\alpha+\sin\theta_1\sin\theta_2)A_0+\cos^2\alpha(\sin\theta_1+\sin\theta_2)A_1\right]$$

$$-\sin^2\alpha\cos\alpha(\sin\theta_1+\sin\theta_2)$$

$$\left.\cdot\left[(\sin\theta_1+\sin\theta_2)A_0+\left(\tfrac{1}{2}+\sin\theta_1\sin\theta_2\right)A_1\right]\frac{4\alpha}{3\pi}\right\}$$

$$\cdot\left\{(\cos\alpha+\sin\theta_1)(\cos\alpha+\sin\theta_2)(\sin^2\alpha-\sin^2\theta_1)(\sin^2\alpha-\sin^2\theta_2)\right\}^{-1}$$

Since the first set of terms dominate for large $|\text{Im. }\alpha|$, it follows that

$$s\left(\alpha-\frac{3\pi}{4}\right)+s\left(\alpha+\frac{3\pi}{4}\right)=-\frac{\sin\alpha}{\cos\alpha}\left(A_1+\frac{A_0}{\cos\alpha}\right)+O(\alpha^{-4/3})$$

implying

$$H_z\left(\rho, -\frac{3\pi}{4}\right) + H_z\left(\rho, \frac{3\pi}{4}\right) = -2(A_1 + jk_0\rho A_0) + O\left\{(k_0\rho)^{4/3}\right\} \qquad (5.219)$$

Hence, from the contact condition (5.193),

$$A_0 = 0 \qquad (5.220)$$

as in the case of the half-plane.

Although the solution is now complete, it is of interest to examine the scattered field with particular reference to the role played by the constant Γ in (5.215). If we ignore the optics terms and any surface wave contributions that may be present,

$$H_z^s(\rho, \phi) = \frac{1}{2\pi j}\int_{S(\phi)} e^{-jk_0\rho\cos(\alpha-\phi)}\left\{s(\alpha + \pi) - s(\alpha - \pi)\right\}d\alpha \qquad (5.221)$$

where $S(\phi)$ is the steepest descent path through $\alpha = \phi$. From the relationships given by MALIUZHINETS (1958c)

$$\psi_{3\pi/4}(\alpha \pm \pi) = \frac{\left\{\psi_{3\pi/4}\left(\frac{\pi}{2}\right)\right\}^2}{\psi_{3\pi/4}(\alpha)}\cos\tfrac{1}{3}\left(\alpha \pm \frac{\pi}{2}\right)$$

and therefore

$$g(\alpha \pm \pi) = G(\alpha, \phi_0)\left\{\gamma_1 + \cos\left(\frac{4\alpha}{3} \pm \frac{2\pi}{3}\right)\right\}\left\{\gamma_2 + \cos\left(\frac{4\alpha}{3} \pm \frac{2\pi}{3}\right)\right\} \qquad (5.222)$$

where

$$G(\alpha, \phi_0) = \left\{\psi_{3\pi/4}\left(\frac{\pi}{2}\right)\right\}^{16}\left\{64\,\Psi(\alpha, \theta_1)\,\Psi(\alpha, \theta_2)\,\Psi(\phi_0, \theta_1)\,\Psi(\phi_0, \theta_2)\right\}^{-1} \qquad (5.223)$$

We observe that $G(\alpha, \phi_0)$ is symmetrical in α and ϕ_0. Also, from (5.205)

$$\sigma_0(\alpha \pm \pi) = -\frac{1}{3}\cos\frac{2\pi_0}{3}\,\frac{\sin\dfrac{2\alpha}{3} + 2\sin\dfrac{2\phi_0}{3} \pm \sqrt{3}\,\cos\dfrac{2\alpha}{3}}{\sin^2\dfrac{2\alpha}{3} + \sin\dfrac{2\alpha}{3}\sin\dfrac{2\phi_0}{3} + \sin^2\dfrac{2\phi_0}{3} - \dfrac{3}{4}}$$

so that

$$g(\alpha + \pi)\sigma_0(\alpha + \pi) - g(\alpha - \pi)\sigma_0(\alpha - \pi) =$$

$$-\frac{8}{\sqrt{3}}G(\alpha, \phi_0)\cos\frac{2\alpha}{3}\cos\frac{2\phi_0}{3}$$

$$\cdot\left[\left\{\sin^2\frac{2\alpha}{3}\sin^2\frac{2\phi_0}{3} - \tfrac{1}{2}(\gamma_1 + \gamma_2 - 1)\sin\frac{2\alpha}{3}\sin\frac{2\phi_0}{3}\right.\right.$$

$$\left.+ \tfrac{1}{4}\left(\gamma_1 - \tfrac{1}{2}\right)\left(\gamma_2 - \tfrac{1}{2}\right)\right\}$$

$$\cdot\left(\sin^2\frac{2\alpha}{3} + \sin\frac{2\alpha}{3}\sin\frac{2\phi_0}{3} + \sin^2\frac{2\phi_0}{3} - \frac{3}{4}\right)^{-1} - \left.\sin^2\frac{2\alpha}{3}\right] \qquad (5.224)$$

From (5.208) and (5.222)

$$g(\alpha \pm \pi)\,\sigma_1(\alpha \pm \pi) = \frac{G(\alpha,\phi_0)\,p(\alpha \pm \pi)}{\left(\gamma_1 + \cos\frac{4\alpha}{3}\right)\left(\gamma_2 + \cos\frac{4\alpha}{3}\right)}$$

and therefore

$$g(\alpha+\pi)\,\sigma_1(\alpha+\pi) - g(\alpha-\pi)\,\sigma_1(\alpha-\pi) =$$
$$-\frac{8}{\sqrt{3}}\,G(\alpha,\phi_0)\,\cos\frac{2\alpha}{3}\,\cos\frac{2\phi_0}{3}$$
$$\cdot\left\{\sin^6\frac{2\alpha}{3} + \frac{1}{4}(6a_3-5)\sin^4\frac{2\alpha}{3}\right.$$
$$\left.-\frac{3}{8}(3a_3-1)\sin^2\frac{2\alpha}{3} + \frac{1}{64}(6a_3 - ba_1 - 1)\right\}$$
$$\cdot\left\{\sin^4\frac{2\alpha}{3} - \frac{1}{2}(\gamma_1+\gamma_2+2)\sin^2\frac{2\alpha}{3} + \frac{1}{4}(\gamma_1+1)(\gamma_2+1)\right\}^{-1} \quad (5.225)$$

Finally, from (5.211) and (5.202) with $A_0 = 0$,

$$t_0(\alpha \pm \pi) = \frac{1}{4}sc\left\{(s+c)[c - s - 2(\sin\theta_1 + \sin\theta_2)]\right.$$
$$\left.-(1+2\sin\theta_1\sin\theta_2)\left(\frac{4\alpha}{3\pi}\pm\frac{4}{3}\right)\right\}\cos\frac{2\phi_0}{3}KA_1$$
$$\cdot\left[\left\{\gamma_1 + \cos\left(\frac{4\alpha}{3}\pm\frac{2\pi}{3}\right)\right\}\left\{\gamma_2 + \cos\left(\frac{4\alpha}{3}\pm\frac{2\pi}{3}\right)\right\}\right.$$
$$\left.\cdot\left\{\gamma_1 + \cos\frac{4\alpha}{3}\right\}\left\{\gamma_2 + \cos\frac{4\alpha}{3}\right\}\right]^{-1}$$

and the combination with (5.222) gives

$$g(\alpha+\pi)\,t_0(\alpha+\pi) - g(\alpha-\pi)\,t_0(\alpha-\pi) = -\frac{1}{3}\,G(\alpha,\phi_0)\,\cos\frac{2\alpha}{3}\,\cos\frac{2\phi_0}{3}$$
$$\cdot\frac{\left(\sin^2\frac{2\alpha}{3} - \frac{1}{4}\right)(1+2\sin\theta_1\sin\theta_2)KA_1}{\sin^4\frac{2\alpha}{3} - \frac{1}{2}(\gamma_1+\gamma_2+2)\sin^2\frac{2\alpha}{3} + \frac{1}{4}(\gamma_1+1)(\gamma_2+1)} \quad (5.226)$$

The last step is the addition of (5.224), (5.225) and (5.226), and by using the relationships in (5.215) we obtain

$$s(\alpha+\pi) - s(\alpha-\pi) = -\frac{8}{\sqrt{3}}\,G(\alpha,\phi_0)\,\cos\frac{2\alpha}{3}\,\cos\frac{2\phi_0}{3}$$
$$\cdot\left[\left\{\sin^2\frac{2\alpha}{3}\sin^2\frac{2\phi_0}{3} - \frac{1}{2}(\gamma_1+\gamma_2-1)\sin\frac{2\alpha}{3}\sin\frac{2\phi_0}{3}\right.\right.$$
$$\left.+\frac{1}{4}\left(\gamma_1 - \frac{1}{2}\right)\left(\gamma_2 - \frac{1}{2}\right)\right\}$$
$$\left.\cdot\left(\sin^2\frac{2\alpha}{3} + \sin\frac{2\alpha}{3}\sin\frac{2\phi_0}{3} + \sin^2\frac{2\phi_0}{3} - \frac{3}{4}\right)^{-1} + \Gamma\right] \quad (5.227)$$

where Γ is indicated in (E.10).

The result differs from that of SENIOR (1989b) in the presence of Γ and, in contrast to the supposition made by BERNARD (1987), the particular solution of the difference equation (5.200) *does* affect the field. As we have seen, the determination of a solution which is free of singularities in the strip $|\text{Re. }\alpha| \leq \frac{3\pi}{4}$ is far from trivial, even in the "simple" case treated here. Under all circumstances $\Gamma \neq 0$ and, since it is independent of α and ϕ_0, (5.227) is symmetrical in α and ϕ_0 in accordance with reciprocity.

References

Astrakhan, M. I. (1964), "Averaged boundary conditions on the surface of a lattice with rectangular cells", *Radio Eng. and Electron. Phys.*, **8**, pp. 1239–1241.

Belinskii, B. P., Kouzov, D. P. and Chel'tsova, V. D. (1973), "On acoustic wave diffraction by plates connected at a right angle", *J. Appl. Math. Mech.*, **37**, pp. 273–281.

Bernard, J.-M. L. (1987), "Diffraction by a metallic wedge covered with a dielectric material", *Wave Motion*, **9**, pp. 543–561.

Büyükaksoy, A., Uzgören, G. and Serbest, A. H. (1989), "Diffraction of an obliquely incident plane wave by the discontinuity of a two part thin dielectric plane", *Int. J. Engng Sci.*, **27**, pp. 701–710.

Casey, K. F. (1988), "Electromagnetic shielding behavior of wire-mesh screens", *IEEE Trans. Electromagn. Compat.*, **EMC-30**, pp. 298–306.

Chakrabati, A. (1986), "Diffraction by a dielectric half-plane", *IEEE Trans. Antennas Propagat.*, **AP-34**, pp. 830–833.

Clemmow, P.C. (1951), "A method for the exact solution of a class of two-dimensional diffraction problems", *Proc. Roy. Soc. London, Ser. A*, **205**, pp. 286–308.

Karp, S. N. and Karal, F. C., Jr. (1965), "Generalized impedance boundary conditions with applications to surface wave structures", in *Electromagnetic Wave Theory*, Pt. 1 (J. Brown, Ed.), Pergamon Press, New York, pp. 479–483.

Kontorovich, M. I. (1963), "Averaged boundary conditions at the surface of a grating with a square mesh", *Radio Eng. and Electron. Phys.*, **8**, pp. 1446–1454.

Kontorovich, M. I., Petrun'kin, V. Yu., Yesepkina, N. A. and Astrakhan, M. I. (1962), "The coefficient of reflection of a plane electromagnetic wave from a plane wire mesh", *Radio Eng. and Electron. Phys.*, **7**, pp. 222–231.

Kouzov, D. P. (1963a), "Diffraction of a plane hydroacoustic wave on the boundary of two elastic plates", *J. Appl. Math. Mech.*, **27**, pp. 806–815.

Kouzov, D. P. (1963b), "Diffraction of a plane hydroacoustic wave at a crack in an elastic plate", *J. Appl. Math. Mech.*, **27**, pp. 1593–1601.

Leppington, F. G. (1983), "Travelling waves in a dielectric slab with an abrupt change in thickness", *Proc. Roy. Soc. London, Ser. A*, **386**, pp. 443–460.

Maliuzhinets, G. D. (1958a), "Inversion formula for the Sommerfeld integral", *Sov. Phys. Doklady*, **3**, pp. 52–56.

Maliuzhinets, G. D. (1958b), "Relation between the inversion formulas for the Sommerfeld integral and the formulas of Kontorovich-Lebedev", *Sov. Phys. Doklady*,

3, pp. 266–268.

Maliuzhinets, G. D. (1958c), "Excitation, reflection and emission of surface waves from a wedge with given face impedances", *Sov. Phys. Doklady*, **3**, pp. 752–755.

Richmond, J. H. (1965), "Scattering by a dielectric cylinder of arbitrary cross-section shape", *IEEE Trans. Antennas Propagat.*, **AP-13**, pp. 334–341.

Richmond, J. H. (1966), "TE-wave scattering by a dielectric cylinder of arbitrary cross-section shape", *IEEE Trans. Antennas Propagat.*, **AP-14**, pp. 460–464.

Rojas, R. G. and Chou, L. M. (1990), "Diffraction by a partially coated perfect electric conducting half plane", *Radio Sci.*, **25**, pp. 175–188.

Rojas, R. G., Ly, H. C. and Pathak, P. H. (1991), "Electromagnetic plane wave diffraction by a planar junction of two thin dielectric/ferrite half planes", *Radio Sci.*, **26**, pp. 641–660.

Rytov, S. M. (1940), "Calcul du skin-effet par la méthode des perturbations", *J. Phys. USSR*, **2**, pp. 233–242.

Senior, T. B. A. (1952), "Diffraction by a semi-infinite metallic sheet", *Proc. Roy. Soc. London, Ser. A*, **213**, pp. 436–458.

Senior, T. B. A. (1987), "A critique of certain half plane diffraction analyses", *Electromagnetics*, **7**, pp. 81–90.

Senior, T. B. A. (1989a), "Skew incidence on a dielectric half-plane", *Electromagnetics*, **9**, pp. 187–200.

Senior, T. B. A. (1989b), "Diffraction by a right-angled second order impedance wedge", *Electromagnetics*, **9**, pp. 313–330.

Senior, T. B. A. (1991), "Diffraction by a generalized impedance half plane", *Radio Sci.*, **26**, pp. 163–167.

Senior, T. B. A. (1993a), "Generalized boundary and transition conditions and the uniqueness of solution", University of Michigan Radiation Laboratory Report No. RL 891.

Senior, T. B. A. (1993b), "Diffraction by a half plane with third-order transition conditions", *Radio Sci.*, **28**, pp. 273–279.

Senior, T. B. A. and Volakis, J. L. (1987), "Sheet simulation of a thin dielectric layer", *Radio Sci.*, **22**, pp. 1261–1272.

Senior, T. B. A. and Volakis, J. L. (1989), "Derivation and application of a class of generalized boundary conditions", *IEEE Trans. Antennas Propagat.*, **AP-37**, pp. 1566–1572.

Syed, H. H. and Volakis, J. L. (1991), "High-frequency scattering by a smooth coated cylinder simulated with generalized impedance boundary conditions", *Radio Sci.*, **26**, pp. 1305–1314.

Van Bladel, J. (1985), *Electromagnetic Fields*, Hemisphere Pub. Co., New York, p. 503.

Volakis, J. L. (1988), "High-frequency scattering by a thin material half plane and strip", *Radio Sci.*, **23**, pp. 450–462.

Volakis, J. L. and Senior, T. B. A. (1987), "Diffraction by a thin dielectric half-plane", *IEEE Trans. Antennas Propagat.*, **AP-35**, pp. 1483–1487.

Volakis, J. L. and Syed, H. H. (1990), "Application of higher order boundary conditions to scattering by multilayer coated cylinders", *J. Electromagn. Waves Applics.*, **4**, pp. 1157–1180.

Wait, J. R. (1978), "Theories of scattering from wire grid and mesh structures", in *Electromagnetic Scattering* (Ed. P. L. E. Uslenghi), Academic Press, New York, pp. 253–287.

Weinstein, L. A. (1969), *The Theory of Diffraction and the Factorization Method* (translation from the Russian by Petr Beckmann), The Golem Press, Boulder, CO, pp. 295–302.

Chapter 6

Higher order conditions

By increasing the order of the boundary condition it is possible to improve the accuracy with which the surface properties are simulated, but the penalty is an increase in the complication of an analytical or numerical solution of the problem. Nevertheless, there are some instances where this is worthwhile and, since the task can be simplified somewhat by choosing appropriately the form of these higher order conditions, we address this matter first.

6.1 General forms

It is convenient to start by considering the boundary conditions on a scalar field $U(x, y, z)$ at a planar surface $y = 0$, where U represents, for example, an acoustic velocity potential or the normal component E_y (or H_y) of an electromagnetic field with H_y (or E_y) zero everywhere.

6.1.1 Scalar conditions

A general linear Mth order boundary condition is

$$\sum_{m=0}^{M} \sum_{\ell=0}^{m} \sum_{k=0}^{m-\ell} c_{k\ell m} \frac{\partial^m}{\partial x^{m-\ell-k} \, \partial y^k \, \partial z^\ell} U = 0 \tag{6.1}$$

and if the wave equation is used to eliminate all even derivatives with respect to y, we obtain

$$\left(P \frac{\partial}{\partial y} - Q \right) U = 0 \tag{6.2}$$

with

$$P = \sum_{m=0}^{M-1} \sum_{\ell=0}^{m} c_{\ell m}^{(1)} \frac{\partial^m}{\partial x^{m-\ell} \, \partial z^\ell}$$

$$\tag{6.3}$$

$$Q = \sum_{m=0}^{M} \sum_{\ell=0}^{m} c_{\ell m}^{(2)} \frac{\partial^m}{\partial x^{m-\ell} \, \partial z^\ell}$$

where the coefficients $c_{\ell m}^{(i)}$ can be expressed in terms of the $c_{k\ell m}$. Let us now examine the forms taken by the differential operators in special cases. If the surface is invariant under a 180-degree rotation about the y axis, the transformation $x, z \rightarrow -x, -z$ shows

$$c_{\ell m}^{(1)}, \; c_{\ell m}^{(2)} = 0 \qquad \text{for all odd } m$$

Invariance under a 90-degree rotation produces the additional restrictions

$$c_{12}^{(1)} = 0, \qquad c_{04}^{(1)} = c_{44}^{(1)}, \qquad c_{14}^{(1)} = -c_{34}^{(1)}$$

etc., with similar conditions on the $c_{\ell m}^{(2)}$; and finally, for invariance under an arbitrary rotation, the operators reduce to

$$P = \sum_{m=0}^{M_1} c_m^{(3)} \left(\frac{\partial^2}{\partial x^2} + \frac{\partial^2}{\partial z^2} \right)^m$$

$$Q = \sum_{m=0}^{M_2} c_m^{(4)} \left(\frac{\partial^2}{\partial x^2} + \frac{\partial^2}{\partial z^2} \right)^m$$

(6.4)

where $M_1 = [(M-1)/2]$, $M_1 = [M/2]$ and [ ] denotes the largest integer contained. This is the simplest case, corresponding to an isotropic surface. By using the wave equation once again, the boundary condition (6.2) with (6.4) can be written as

$$\sum_{m=0}^{M} a_{M-m} \frac{1}{(-jk_0)^m} \frac{\partial^m U}{\partial y^m} = 0$$

(6.5)

or

$$\prod_{m=1}^{M} \left(\Gamma_m - \frac{1}{jk_0} \frac{\partial}{\partial y} \right) U = 0$$

(6.6)

The equivalence confirms that (6.5) and (6.6) are valid only for an isotropic surface and, for a second order ($M = 2$) condition, (5.6) shows the connection between the coefficients a_m and Γ_m.

In most practical circumstances, the $c_m^{(3)}$ and $c_m^{(4)}$ in (6.4) are of increasing (with m) order in some small quantity δ, and this allows us to carry out a further simplification. By a process that is formally equivalent to a Taylor expansion of $1/P$, (6.2) becomes

$$\frac{\partial U}{\partial y} = LU$$

(6.7)

where

$$L = \sum_{m=0}^{M_1} b_m \left(\frac{\partial^2}{\partial x^2} + \frac{\partial^2}{\partial z^2} \right)^m$$

(6.8)

and this is accurate to the same order in δ. Thus, for $M > 1$, it is sufficient to restrict attention to *even* order boundary conditions for which $M_1 = \frac{1}{2}M$, and any odd order condition can be recast as an even order condition of one lower order. Moreover, in (6.5), $a_m = 0$ for all *odd* $m < M - 1$ with consequent implications regarding the Γ_m in (6.6), and this is a key feature of the fourth order boundary conditions employed by SENIOR (1993b).

To see how the simplification comes about, consider the problem of the plane wave

$$U^i = e^{jk_0(x\cos\phi + y\sin\phi)}$$

incident on the surface $y = 0$ of a homogeneous dielectric medium whose complex refractive index is N. If U is identified with the electromagnetic field component E_y, the reflection coefficient is (see (2.84))

$$R = -\frac{\left(1 + \dfrac{\sin^2\phi}{N^2 - 1}\right)^{1/2} - \epsilon_r(N^2 - 1)^{-1/2}\sin\phi}{\left(1 + \dfrac{\sin^2\phi}{N^2 - 1}\right)^{1/2} + \epsilon_r(N^2 - 1)^{-1/2}\sin\phi} \qquad (6.9)$$

For $|N| \gg 1$

$$\left(1 + \frac{\sin^2\phi}{N^2 - 1}\right)^{1/2} = 1 + \frac{1}{2}\frac{\sin^2\phi}{N^2 - 1} - \frac{1}{8}\frac{\sin^4\phi}{(N^2 - 1)^2} + O(|N|^{-6})$$

and because (6.5) implies that

$$R = -\frac{\sum_{m=0}^{M}(-1)^m a_{M-m}\sin^m\phi}{\sum_{m=0}^{M} a_{M-m}\sin^m\phi}$$

a boundary condition accurate to the fifth order in $|N|^{-1}$ is obtained by taking

$$a_0 = -\tfrac{1}{8}(N^2 - 1)^{-2}, \qquad a_2 = \tfrac{1}{2}(N^2 - 1)^{-1}, \qquad a_4 = 1$$

with

$$a_1 = 0, \qquad a_3 = \epsilon_r(N^2 - 1)^{-1/2}$$

The corresponding b_m in (6.8) are

$$b_0 = \frac{jk_0}{\epsilon_r}(N^2 - 1)^{1/2}, \qquad b_1 = -\frac{j}{2k_0\epsilon_r}(N^2 - 1)^{-1/2}$$

$$b_2 = -\frac{j}{8k_0^3\epsilon_r}(N^2 - 1)^{-3/2}$$

Boundary conditions accurate to higher orders in $|N|^{-1}$ can be developed in the same way, and it is obvious that there are other forms of (6.7) that are equivalent to each other. Since (6.9) can be written as

$$R = -\frac{\left(1 + \dfrac{\sin^2\phi}{N^2 - 1}\right)^c - \epsilon_r(N^2 - 1)^{-1/2}\left(1 + \dfrac{\sin^2\phi}{N^2 - 1}\right)^{c-1/2}\sin\phi}{\left(1 + \dfrac{\sin^2\phi}{N^2 - 1}\right)^c + \epsilon_r(N^2 - 1)^{-1/2}\left(1 + \dfrac{\sin^2\phi}{N^2 - 1}\right)^{c-1/2}\sin\phi}$$

the constant c can be chosen to transfer all or part of the differential operator L to the left hand side of (6.7) without affecting the accuracy of the boundary condition. The process has some similarity to the use of Padé approximants (see Section 8.2) but, for present purposes, (6.7) or the corresponding versions of (6.5) and (6.6) are more convenient.

Based on these considerations we now postulate the following expression for the operator L in the boundary condition (6.7) at a planar surface $y = 0$:

$$L = \alpha + \sum_{m=1}^{M/2}(-1)^m \left\{ \frac{\partial^m}{\partial x^m} \left(\beta_{11}^{(m)} \frac{\partial^m}{\partial x^m} + \beta_{12}^{(m)} \frac{\partial^m}{\partial z^m} \right) \right.$$
$$\left. + \frac{\partial^m}{\partial z^m} \left(\beta_{21}^{(m)} \frac{\partial^m}{\partial x^m} + \beta_{22}^{(m)} \frac{\partial^m}{\partial z^m} \right) \right\} \tag{6.10}$$

where the coefficients α and $\beta_{ij}^{(m)}$ are functions of x and z. If $\beta_{11}^{(m)} = \beta_{22}^{(m)}$, $\beta_{12}^{(m)} = -\beta_{21}^{(m)}$ and α, $\beta_{ij}^{(m)}$ are independent of position, (6.10) reduces to (6.8), and only then can we recast the boundary condition in the form (6.5) or (6.6). Since M is restricted to even integers, the condition is of even order M if $M > 1$. For a curved surface S the corresponding boundary condition is

$$\frac{\partial U}{\partial n} = \alpha U + \sum_{m=1}^{M/2}(-1)^m \left\{ \frac{\partial^m}{\partial s^m} \left(\beta_{11}^{(m)} \frac{\partial^m U}{\partial s^m} + \beta_{12}^{(m)} \frac{\partial^m U}{\partial t^m} \right) \right.$$
$$\left. + \frac{\partial^m}{\partial t^m} \left(\beta_{21}^{(m)} \frac{\partial^m}{\partial s^m} + \beta_{22}^{(m)} \frac{\partial^m}{\partial t^m} \right) \right\} \tag{6.11}$$

where $\hat{s}$ and $\hat{t}$ are local tangent vectors with $\hat{n}$ in the direction of the outward normal. The triplet $\hat{s}$, $\hat{n}$, $\hat{t}$ form a right-handed system of orthogonal unit vectors, and the coefficients α, $\beta_{ij}^{(m)}$ are functions of s and t. A special case of (6.11) was employed by LJALINOV (1992). The condition is also the logical extension of (5.78) to an arbitrary (even) order M, and is the most general form for which a uniqueness proof has been established.

6.1.2   Vector conditions

To develop higher order boundary conditions for an electromagnetic field that can be extended to a curved surface, it would seem natural to proceed as we did for the second order condition in Section 5.2. Starting with the scalar conditions

$$\prod_{m=1}^{M} \left(\Gamma_m - \frac{1}{jk_0}\frac{\partial}{\partial y} \right) E_y = 0, \qquad \prod_{m=1}^{M} \left(\Gamma_m' - \frac{1}{jk_0}\frac{\partial}{\partial y} \right) H_y = 0 \tag{6.12}$$

at the planar surface $y = 0$, or, equivalently (see (6.5)),

$$\sum_{m=0}^{M} a_{M-m}\frac{1}{(-jk_0)^m}\frac{\partial^m E_y}{\partial y^m} = 0, \qquad \sum_{m=0}^{M} a_{M-m}'\frac{1}{(-jk_0)^m}\frac{\partial^m H_y}{\partial y^m} = 0 \tag{6.13}$$

Maxwell's equations can be used to show

$$\hat{y} \times \left\{ \mathbf{E} - \frac{1}{jk_0} \sum_{m=0}^{M-2} \nabla \left[A_m \frac{\partial^m E_y}{\partial y_m} \right] \right\}$$

$$= \eta \hat{y} \times \left(\hat{y} \times \left\{ \mathbf{H} - \frac{1}{jk_0} \sum_{m=0}^{M-2} \nabla \left[B_m \frac{\partial^m H_y}{\partial y_m} \right] \right\} \right) \tag{6.14}$$

where

$$A_m = A_m(a_i, a_i'), \qquad B_m = A_m(a_i', a_i)$$

with

$$\eta = \eta(a_i, a_i') = \{ \eta(a_i', a_i) \}^{-1}$$

in accordance with duality. The form is evident from the work of SENIOR AND VOLAKIS (1989) where the result for $M = 3$ is given (note the present use of a_{M-m} and a_{M-m}' in place of a_m and a_m', respectively), and we observe that (6.14) involves derivatives through order $M - 1$ applied to the normal field components E_y and H_y.

In practice (6.14) is not convenient, and to use the boundary condition it is almost essential to have the derivatives applied to the tangential components of *either* the electric *or* the magnetic field. This is easy to do if we restrict attention to *even* order conditions with a_m, $a_m' = 0$ for all odd $m < M-1$, and assume that the successive derivatives have coefficients of increasing order in some small quantity δ. The procedure used to approximate (5.15) with (5.27) can then be applied to the higher order conditions. For $M = 4$, (6.14) becomes

$$\hat{y} \times \mathbf{E} = \frac{a_0 + a_2 + a_4}{a_3} Z_0 \hat{y} \times \left(\hat{y} \times \left\{ \mathbf{H} - \frac{2a_0' + a_2' + a_0'(D^2/k_0^2)}{k_0^2(a_0' + a_2' + a_4')} \nabla(\nabla_s.\mathbf{H}) \right\} \right.$$

$$\left. - \frac{2a_0 + a_2 + a_0(D^2/k_0^2)}{k_0^2(a_0 + a_2 + a_4)} \nabla(\hat{y}.\nabla \times \mathbf{H}) \right) \tag{6.15}$$

where

$$D^2 = \frac{\partial^2}{\partial x^2} + \frac{\partial^2}{\partial z^2}$$

and, in general,

$$\hat{y} \times \mathbf{E} = \eta \hat{y} \times \left(\hat{y} \times \left\{ \mathbf{H} - \sum_{m=0}^{\frac{1}{2}M-1} b_m' D^{2m} \nabla(\nabla_s.\mathbf{H}) \right\} \right.$$

$$\left. - \sum_{m=0}^{\frac{1}{2}M-1} b_m D^{2m} \nabla(\hat{y}.\nabla \times \mathbf{H}) \right) \tag{6.16}$$

involving derivatives through the Mth order.

Recognising that $D^2 = \nabla_s.\nabla_s$, where the suffix "s" denotes the surface operator, and, consistent with the needs of a uniqueness proof, we now postulate the following expression for an Mth order boundary condition at a planar surface $y = 0$:

$$\hat{y} \times \mathbf{E} = \hat{y} \times (\hat{y} \times \overline{\overline{L}}.\mathbf{H}) \qquad (6.17)$$

where

$$\overline{\overline{L}}.\mathbf{H} = \overline{\overline{\eta}}.\mathbf{H} - \nabla g(\nabla_s.\mathbf{H}) - \hat{y} \times \nabla f(\hat{y}.\nabla \times \mathbf{H}) \qquad (6.18)$$

with

$$
\begin{aligned}
f(\lambda) &= A_0\lambda - \nabla_s.(A_1\nabla_s\lambda) + \nabla_s.\nabla_s\{A_2\nabla_s.(\nabla_s\lambda)\} - \cdots \\
g(\lambda) &= B_0\lambda - \nabla_s.(B_1\nabla_s\lambda) + \nabla_s.\nabla_s\{B_2\nabla_s.(\nabla_s\lambda)\} - \cdots
\end{aligned}
\qquad (6.19)
$$

The tensor $\overline{\overline{\eta}}$ and the coefficients A_m and B_m are functions of position on the surface, and only in the special case when η is a scalar with η, A_m and B_m independent of x and z is it possible to recover (6.16). The terms displayed in (6.19) are those corresponding to $M = 6$ and, as M increases, the increasing complexity is evident.

The analogous result for a curved surface is

$$\hat{n} \times \mathbf{E} = \hat{n} \times (\hat{n} \times \overline{\overline{L}}.\mathbf{H}) \qquad (6.20)$$

where

$$\overline{\overline{L}}.\mathbf{H} = \overline{\overline{\eta}}.\mathbf{H} - \nabla g(\nabla_s.\mathbf{H}) - \hat{n} \times \nabla f(\hat{n}.\nabla \times \mathbf{H}) \qquad (6.21)$$

and f, g are as given in (6.19). For the two-dimensional problem of a field $\mathbf{H} = \hat{t}\, H_t$ incident on a cylindrical surface whose generators are parallel to the t axis, the problem is a scalar one for the field $U = H_t$, and (6.21) then reduces to a particular case of (6.11). For the corresponding resistive sheet the boundary conditions are

$$\hat{n} \times (\mathbf{E}^+ + \mathbf{E}^-) = 2\hat{n} \times \left[\hat{n} \times \overline{\overline{L}}.\mathbf{H}\right]_-^+ \qquad (6.22)$$

and

$$[\hat{n} \times \mathbf{E}]_-^+ = 0 \qquad (6.23)$$

where the affices $+$ and $-$ denote the upper and lower sides, respectively, and $[\ \ ]_-^+$ is the discontinuity across the sheet. Similarly, for a conductive sheet,

$$2\left[\hat{n} \times \mathbf{H}\right]_-^+ = \hat{n} \times \left\{\hat{n} \times \overline{\overline{L}}.(\mathbf{H}^+ + \mathbf{H}^-)\right\} \qquad (6.24)$$

and

$$[\hat{n} \times \mathbf{H}]_-^+ = 0 \qquad (6.25)$$

Although (6.20)–(6.25) are the most general boundary conditions considered here, they are not, in fact, the most general possible ones, and the tensor properties could be extended to the second and third terms on the right hand side of (6.18). Many wire grids (see Section 5.3.4), for example, require such an extension even for a second order condition and, in the case of a planar surface $y = 0$, HOPPE AND RAHMAT-SAMII (1994) have considered the second order condition

$$\hat{y} \times \mathbf{E} = \hat{y} \times \left(\hat{y} \times \left\{ \overline{\overline{\eta}} + \overline{\overline{\eta}}^{(1)} \frac{\partial^2}{\partial x^2} + \overline{\overline{\eta}}^{(2)} \frac{\partial^2}{\partial x \, \partial z} + \overline{\overline{\eta}}^{(3)} \frac{\partial^2}{\partial z^2} \right\} . \mathbf{H} \right) \qquad (6.26)$$

to simulate a thin metal-backed layer of rather general material.

6.2   Uniqueness

The next task is to establish conditions under which the Mth order scalar and vector boundary conditions (6.11) and (6.20) ensure unique solutions of the boundary value problems and, because of the forms chosen, the analysis is a simple extension of that in Section 5.4.

We address first the scalar boundary condition (6.11) and postulate two solutions generated by the same sources in the volume V exterior to S. Let k_0 have a small negative imaginary part corresponding to a slight loss in V. This means (2.139) is valid and, since the difference solution W also satisfies (6.11), we can insert the expression for $\partial W / \partial n$ into the left hand side. Assuming the coefficients $\beta_{ij}^{(m)}$ have the necessary continuity on S, integration by parts gives

$$\text{Im.} \iint_S \left\{ \alpha |W|^2 + \sum_{m=1}^{M/2} \left(\beta_{11}^{(m)} \left| \frac{\partial^m W}{\partial s^m} \right|^2 + \beta_{22}^{(m)} \left| \frac{\partial^m W}{\partial t^m} \right|^2 \right. \right.$$

$$\left. \left. + \beta_{12}^{(m)} \frac{\partial^m W^*}{\partial s^m} \frac{\partial^m W}{\partial t^m} + \beta_{21}^{(m)} \frac{\partial^m W}{\partial s^m} \frac{\partial^m W^*}{\partial t^m} \right) \right\} dS$$

$$= \text{Im.} \, k_0^2 \iiint_V |W|^2 \, dV \qquad (6.27)$$

Because the right hand side is never positive and is zero only if $W = 0$ throughout V, sufficient conditions for a unique solution in the case of a passive surface are

$$\text{Im.} \, \alpha \geq 0 \qquad (6.28)$$

$$\text{Im.} \, \beta_{11}^{(m)} \geq 0, \qquad \text{Im.} \, \beta_{22}^{(m)} \geq 0 \qquad (6.29)$$

and

$$\beta_{21}^{(m)} = \beta_{12}^{(m)*} \qquad (6.30)$$

at all points of the surface. It must be emphasised that these are only *sufficient* conditions and, based on the argument used for a planar surface in Section 5.4, it seems probable that (6.29) can be replaced by the less stringent requirement

$$\text{Im.} \left(\alpha + \sum_{m=1}^{M/2} k_0^{2m} \beta_{11}^{(m)} \right) \geq 0, \qquad \text{Im.} \left(\alpha + \sum_{m=1}^{M/2} k_0^{2m} \beta_{22}^{(m)} \right) \geq 0 \qquad (6.31)$$

At an edge or other surface singularity where some or all of the $\beta_{ij}^{(m)}$ are discontinuous, contact conditions are also needed and, for an Mth order boundary condition with $M > 1$ and even, there are $\frac{1}{2}M$ such constraints. Their form is evident from the end point contributions in the integrations by parts that led to (6.27). If, for example, the edge is at $s = 0$, the contact conditions for a fourth order boundary condition are

$$U = 0 \qquad \text{at } s = 0 \tag{6.32}$$

or

$$\left[\beta_{11}^{(1)} \frac{\partial U}{\partial s} + \beta_{12}^{(1)} \frac{\partial U}{\partial s} - \frac{\partial}{\partial s} \left\{ \beta_{11}^{(2)} \frac{\partial U^2}{\partial^2 s} + \beta_{12}^{(2)} \frac{\partial U^2}{\partial t^2} \right\} \right]_{-}^{+} = 0 \tag{6.33}$$

and

$$\frac{\partial U}{\partial s} = 0 \qquad \text{at } s = 0 \tag{6.34}$$

or

$$\left[\beta_{11}^{(2)} \frac{\partial U^2}{\partial^2 s} + \beta_{12}^{(2)} \frac{\partial U^2}{\partial t^2} \right]_{-}^{+} = 0 \tag{6.35}$$

where $[\ ]_{-}^{+}$ denotes the discontinuity between $s = 0+$ and $s = 0-$. The choice of which condition to employ from each pair depends on the physical meaning of the scalar wave function U.

We now consider the electromagnetic boundary condition (6.20). If $(\mathbf{E}^{(1)}, \mathbf{H}^{(1)})$ and $(\mathbf{E}^{(2)}, \mathbf{H}^{(2)})$ are two fields generated by the same sources in V, and $(\delta\mathbf{E}, \delta\mathbf{H})$ is the difference field, the quantity of interest is

$$\iint_S \partial\mathbf{H}^*.(\hat{n} \times \delta\mathbf{E})\, dS$$

and since the difference field also satisfies (6.20), it follows that

$$\delta\mathbf{H}^*.(\hat{n} \times \delta\mathbf{E}) = \delta\mathbf{H}^*.\hat{n} \times (\hat{n} \times \{\overline{\overline{\eta}}.\delta\mathbf{H} - \nabla g(\nabla_{\!s}.\delta\mathbf{H}) - \hat{n} \times \nabla f(\hat{n}.\nabla \times \delta\mathbf{H})\})$$

where f and g are given in (6.19). From the expression for $\overline{\overline{\eta}}$ in (2.134),

$$\delta\mathbf{H}^*.\hat{n} \times (\hat{n} \times \overline{\overline{\eta}}.\delta\mathbf{H}) = -\eta_{11}|\delta H_s|^2 - \eta_{22}|\delta H_t|^2 - \eta_{12}\,\delta H_t\,\delta H_s^* - \eta_{21}\,\delta H_s\,\delta H_t^*$$

Also

$$\begin{aligned}
&-\delta\mathbf{H}^*.\hat{n} \times (\hat{n} \times \{\hat{n} \times \nabla f\}) \\
&= \delta\mathbf{H}^*.\hat{n} \times \nabla f(\hat{n}.\nabla \times \delta\mathbf{H}) \\
&= -A_0|\hat{n}.\nabla \times \delta\mathbf{H}|^2 - A_1|\nabla_{\!s}(\hat{n}.\nabla \times \delta\mathbf{H})|^2 - \cdots \\
&\quad + \hat{n}.\nabla \times \{\delta\mathbf{H}^*\,(A_0\,\hat{n}.\nabla \times \delta\mathbf{H} - \nabla_{\!s}.[A_1\nabla_{\!s}\,(\hat{n}.\nabla \times \delta\mathbf{H})] + \cdots)\} \\
&\quad + \nabla_{\!s}.\{\hat{n}.\nabla \times \delta\mathbf{H}^*\,A_1\nabla_{\!s}\,(\hat{n}.\nabla \times \delta\mathbf{H}) + \cdots\}
\end{aligned}$$

and

$$-\delta \mathbf{H}^* . \hat{n} \times (\hat{n} \times \nabla g)$$
$$= \delta \mathbf{H}^* . \nabla_{\!s} g (\nabla_{\!s} . \delta \mathbf{H})$$
$$= -B_0 |\nabla_{\!s} . \delta \mathbf{H}|^2 - B_1 |\nabla_{\!s} (\nabla_{\!s} . \delta \mathbf{H})|^2 - \cdots$$
$$+ \nabla_{\!s} . \{ \delta \mathbf{H}^* (B_0 \nabla_{\!s} . \delta \mathbf{H} - \nabla_{\!s} . [B_1 \nabla_{\!s} (\nabla_{\!s} . \delta \mathbf{H})] + \cdots)$$
$$+ \nabla_{\!s} . \delta \mathbf{H}^* B_1 \nabla_{\!s} (\nabla_{\!s} . \delta \mathbf{H}) + \cdots \}$$

Hence

$$\iint_S \delta \mathbf{H}^* . (\hat{n} \times \delta \mathbf{E}) \, dS = - \iint_S P \, dS + I \qquad (6.36)$$

where

$$P = \eta_{11} |\delta H_s|^2 + \eta_{22} |\delta H_t|^2 + \eta_{12} \, \delta H_t \, \delta H_s^* + \eta_{21} \, \delta H_s \, \delta H_t^* + A_0 |\hat{n} . \nabla \times \delta \mathbf{H}|^2$$
$$+ A_1 |\nabla_{\!s} (\hat{n} . \nabla \times \delta \mathbf{H})|^2 + B_0 |\nabla_{\!s} . \delta \mathbf{H}|^2 + B_1 |\nabla_{\!s} (\nabla_{\!s} . \delta \mathbf{H})|^2 + \cdots \qquad (6.37)$$

and

$$I = \iint_S \left(\hat{n} . \nabla \times \{ \delta \mathbf{H}^* (A_0 \, \hat{n} . \nabla \times \delta \mathbf{H} - \nabla_{\!s} . [A_1 \nabla_{\!s} (\hat{n} . \nabla \times \delta \mathbf{H})] + \cdots) \} \right.$$
$$+ \nabla_{\!s} . \left\{ \delta \mathbf{H}^* (B_0 \nabla_{\!s} . \delta \mathbf{H} - \nabla_{\!s} . [B_1 \nabla_{\!s} (\nabla_{\!s} . \delta \mathbf{H})] + \cdots) \right.$$
$$\left. + \nabla_{\!s} . \delta \mathbf{H}^* B_1 \nabla_{\!s} (\nabla_{\!s} . \delta \mathbf{H}) + \hat{n} . \nabla \times \delta \mathbf{H}^* A_1 \nabla_{\!s} (\hat{n} . \nabla \times \delta \mathbf{H}) + \cdots \right\} \right) dS$$
$$(6.38)$$

If the A_m and B_m have the necessary continuity on S, the divergence theorem shows $I = 0$, and therefore

$$-\text{Re.} \iint_S P \, dS = \omega \iiint_V \left(\mu'' |\delta \mathbf{H}|^2 + \epsilon'' |\delta \mathbf{E}|^2 \right) dV \qquad (6.39)$$

(see (2.132)). The solution is then unique if

$$\text{Re. } A_m \geq 0, \qquad \text{Re. } B_m \geq 0 \qquad (6.40)$$
$$\text{Re. } \eta_{11} \geq 0, \qquad \text{Re. } \eta_{22} \geq 0 \qquad (6.41)$$

with

$$\eta_{21} = -\eta_{12}^* \qquad (6.42)$$

and we again note that these are only *sufficient* conditions for a passive surface.

If some or all of the A_m and B_m are discontinuous across a line C on S, an Mth order boundary condition must be supplemented with $\frac{1}{2}M$ additional constraints in the form of contact conditions at C. Since S is now equivalent to the union of two open surfaces that are joined at C, it is no longer possible to apply the divergence theorem to I, but using (5.93) and (5.94) instead gives

$$
\begin{aligned}
I = \oint_C \Big[&\hat{\ell}.\delta\mathbf{H}^* \left(A_0\, \hat{n}.\nabla \times \delta\mathbf{H} - \nabla_{\!s}.\left[A_1\nabla_{\!s}\left(\hat{n}.\nabla \times \delta\mathbf{H} \right) \right] \right) \\
&+ (\hat{\ell} \times \hat{n}).\delta\mathbf{H}^* \left(B_0\nabla_{\!s}.\delta\mathbf{H} - \nabla_{\!s}.\left[B_1\nabla_{\!s}\left(\nabla_{\!s}.\delta\mathbf{H} \right) \right] \right) \\
&+ (\hat{\ell} \times \hat{n}).\nabla_{\!s}(\nabla_{\!s}.\delta\mathbf{H})\, B_1\nabla_{\!s}.\delta\mathbf{H}^* \\
&+ (\hat{\ell} \times \hat{n}).\nabla_{\!s}(\hat{n}.\nabla \times \delta\mathbf{H})\, A_1\, \hat{n}.\nabla \times \delta\mathbf{H}^* \Big]_-^+ \, dC
\end{aligned}
\tag{6.43}
$$

where $[\ ]_-^+$ denotes the discontinuity across C. For brevity, we have shown only the terms corresponding to a fourth order boundary condition. In (6.43) the terms involving A_m vanish by virtue of the edge condition, and then $I = 0$ if

$$(\hat{\ell} \times \hat{n}).\mathbf{H} = 0 \qquad \text{on } C \tag{6.44}$$

or

$$[B_0\nabla_{\!s}.\mathbf{H} - \nabla_{\!s}.\{B_1\nabla_{\!s}(\nabla_{\!s}.\mathbf{H})\}]_-^+ = 0 \tag{6.45}$$

and

$$\nabla_{\!s}.\mathbf{H} = 0 \qquad \text{on } C \tag{6.46}$$

or

$$\left[B_1(\hat{\ell} \times \hat{n}).\nabla_{\!s}(\nabla_{\!s}.\mathbf{H}) \right]_-^+ = 0 \tag{6.47}$$

These are the required contact conditions.

6.3   Generalised Babinet principles

Babinet's principle is due to Jacques Babinet (1794–1872), who also discovered one of the three (neutral) points of zero polarisation in diffuse sky radiation. In a study of coronas, Babinet (1837) observed that the interference pattern of light behind a small opaque obstacle is the same as the light transmitted through an opening of the same size and shape, and explained this in terms of edge waves. In other words, given two complementary screens such that the openings in one correspond exactly to the opaque portions of the other, and vice versa, the complex fields at a point similarly placed behind the two structures are such that

$$U_1 + U_2 = U$$

(Born and Wolf, 1964) where U is the value when no screen is present.

Prior to World War II most books on optics either failed to mention the principle or discussed it in the context of Kirchhoff's theory of diffraction. Since this implies the concept of a "black" screen (Baker and Copson, 1950) and treats light as a scalar phenomenon, the formulation is not immediately applicable to electromagnetic fields. EPSTEIN (1915) appears to have been the first to derive the correct form (HUXLEY, 1947) of the electromagnetic analogue of Babinet's principle, although BOOKER (1946) is generally credited with this. In 1941, when comparing the fields scattered by thin slits and wires, Booker observed that in the complementary problem a different (complementary) incident field is required. He then proved (BOOKER, 1946) the principle in its new form, and an alternative proof based on integral equations was provided by COPSON (1946). However, the latter treatment was criticised because of its use of scalar potentials and its failure to recognise the role of edge conditions, and a proof that avoids these difficulties was given later (COPSON, 1950).

In its standard form Babinet's principle involves idealised screens and apertures, i.e. hard surfaces in acoustics and perfect electric conductors in electromagnetics. The extension to soft surfaces and perfect magnetic conductors is trivial and, as we shall now show, the generalised boundary conditions discussed in the previous section also lead to surfaces that are complementary in the sense of the principle.

6.3.1   Scalar fields

Babinet's principle is a consequence of the symmetry of fields radiated by planar distributions analogous to the single and double layer distributions in potential theory. This property is exemplified by the Rayleigh-Sommerfeld formulas for diffraction by apertures in soft and hard screens, and the principle relates the field scattered by an aperture in a soft (hard) screen to that scattered by the complementary hard (soft) disk (SENIOR, 1975a). Thus, soft and hard surfaces are complementary in the context of the principle.

There are also complementary surfaces associated with the generalised boundary condition (6.7). To see this, consider a thin slice of material about the plane $y = 0$ with (6.7) applied at both sides. Then

$$\frac{\partial U}{\partial y} = \pm LU \tag{6.48}$$

at $y = \pm 0$ where L is the differential operator shown in (6.10). Addition and subtraction of the two conditions give

$$\frac{\partial}{\partial y}U^+ + \frac{\partial}{\partial y}U^- - L[U]^+_- = 0 \tag{6.49}$$

$$\left[\frac{\partial U}{\partial y}\right]^+_- - L\left(U^+ + U^-\right) = 0 \tag{6.50}$$

where the signs refer to the upper (positive) and lower (negative) sides. In general, (6.49) and (6.50) imply discontinuities in both U and $\partial U/\partial y$ across the plane $y = 0$, but in special cases one or other is continuous. The resulting membranes are the surfaces of interest.

If U is continuous across $y = 0$ then

$$[U]_-^+ = 0 \qquad (6.51)$$

$$\left[\frac{\partial U}{\partial y}\right]_-^+ = 2LU \qquad (6.52)$$

These represent the situation in which the pressure is equal on the two sides and there is a related jump discontinuity in the normal component of the fluid velocity. They are the scalar analogues of the conditions at a generalised resistive sheet for which U is a tangential component of the electric field, and we shall term the resulting surface a (generalised) resistive membrane. When $M = 0$ and $\alpha = \infty$ the surface is simply a soft one and when $\alpha = 0$ the membrane ceases to exist. The conditions (6.51) and (6.52) therefore include an infinitesimally thin soft surface as a special case.

In the same way that a soft (or Dirichlet) boundary condition has a hard (Neumann) one as its complement, so the resistive conditions (6.51) and (6.52) have theirs, namely

$$\left[\frac{\partial U}{\partial y}\right]_-^+ = 0 \qquad (6.53)$$

$$L[U]_-^+ = 2\frac{\partial U}{\partial y} \qquad (6.54)$$

We refer to these as the conditions at a (generalised) conductive membrane, and for (6.51), (6.52) and (6.53), (6.54) to be complementary it is necessary that L (involving α and the coefficients $\beta_{ij}^{(m)}$) is identical in the two cases. The conditions imply that the normal component of the fluid velocity is continuous whereas the pressure is not, and the resulting membrane has the same difficulty of practical realisation as a conductive sheet has in electromagnetics. Nevertheless, (6.53) and (6.54) are permissible from a mathematical viewpoint, and are essential for the subsequent analysis. When $M = 0$ and $\alpha = 0$ the surface is hard, and when $\alpha = \infty$ it ceases to exist.

To demonstrate the Babinet equivalence of the two membranes, consider first a resistive membrane in the plane $y = 0$. Assuming sources entirely in $y < 0$ that generate a primary field $U_0(x, y, z)$, the total field everywhere is

$$U_1(x, y, z) = U_0(x, y, z) - U_0(x, \pm y, z) \mp 2 \iint_{-\infty}^{\infty} U_1(x', 0, z') \frac{\partial g}{\partial y} \, dx' \, dz' \quad (6.55)$$

(SENIOR, 1975a) where

$$g(\mathbf{r} \mid \mathbf{r}') = \frac{1}{4\pi} \frac{e^{-jk_0|\mathbf{r}-\mathbf{r}'|}}{|\mathbf{r} - \mathbf{r}'|} \qquad (6.56)$$

is the free space Green's function, and the upper and lower signs refer to the regions $y > 0$ and $y < 0$, respectively. Application of (6.52) then gives

$$\lim_{y \to 0} 2 \iint_{-\infty}^{\infty} U_1(x', 0, z') \frac{\partial^2 g}{\partial y^2} \, dx' \, dz' + \frac{\partial U_0}{\partial y} + LU_1(x, 0, z) = 0 \qquad (6.57)$$

for $-\infty < x, z < \infty$, and this is an integral equation from which to determine $U_1(x, 0, z)$. In the second (complementary) problem, the same primary field illuminates a conductive membrane in $y = 0$ and, if the total field is now $U_2(x, y, z)$,

$$U_2(x, y, z) = U_0(x, y, z) - \iint_{-\infty}^{\infty} [U_2(x', y', z')]_-^+ \frac{\partial g}{\partial y} \, dx' \, dz' \tag{6.58}$$

(SENIOR, 1975a). On applying (6.54) we obtain

$$\lim_{y \to 0} 2 \iint_{-\infty}^{\infty} [U_2(x', y', z')] \frac{\partial^2 g}{\partial y^2} \, dx' \, dz' - 2 \frac{\partial U_0}{\partial y} + L \, [U_2(x, y, z)]_-^+ = 0 \tag{6.59}$$

for $-\infty < x, z < \infty$, and provided the two boundary value problems are well-posed, comparison of (6.57) with (6.59) shows

$$[U_2(x, y, z)]_-^+ = -2U_1(x, y, z) \tag{6.60}$$

Hence, from (6.55) and (6.58)

$$U_2(x, y, z) \pm U_1(x, y, z) = U_0(x, \pm y, z) \tag{6.61}$$

at all points in space, where the upper (lower) signs apply for $y > 0$ $(y < 0)$. This shows the connection between the two fields. Since α and the $\beta_{ij}^{(m)}$ can be functions of position, the membranes include, for example, an aperture in a soft screen and its complementary hard disk as a special case, and (6.61) is consistent with Babinet's original observation.

6.3.2 Vector fields

For electromagnetic fields the analysis is similar to that above but the results are richer because of the effect of the polarisation. We again start by considering a surface $y = 0$ subject to the same GIBC on both sides. From (6.17) the boundary conditions on $y = \pm 0$ are

$$y \times \mathbf{E} = \pm \hat{y} \times (\hat{y} \times \overline{\overline{L}}.\mathbf{H})$$

where $\overline{\overline{L}}$ is the Mth order tensor impedance operator shown in (6.18). Addition and subtraction of the boundary conditions then give

$$\hat{y} \times \left(\mathbf{E}^+ + \mathbf{E}^- \right) = \hat{y} \times \left[\hat{y} \times \overline{\overline{L}}.\mathbf{H} \right]^+ \tag{6.62}$$

$$[\hat{y} \times \mathbf{E}]_-^+ = \hat{y} \times \left\{ \hat{y} \times \overline{\overline{L}}. \left(\mathbf{H}^+ + \mathbf{H}^- \right) \right\} \tag{6.63}$$

The surface is, of course, opaque, and supports electric and magnetic currents but, since the currents do not interact with one another, we can treat the surface as a superposition of electric and magnetic current sheets. For the electric current

$$\hat{y} \times \left(\mathbf{E}^+ + \mathbf{E}^-\right) = \overline{\overline{L}}.\hat{y} \times [\hat{y} \times \mathbf{H}]^+_- \tag{6.64}$$

and

$$[\hat{y} \times \mathbf{E}]^+_- = 0 \tag{6.65}$$

representing a generalised resistive sheet with (tensor) resistivity $\overline{\overline{R}}_e = \frac{1}{2}\overline{\overline{L}}$. When $\overline{\overline{L}} = 0$ the sheet is a perfect electric conductor, and when $\overline{\overline{L}} = \infty$ it no longer exists. Similarly, for the magnetic current

$$[\hat{y} \times \mathbf{E}]^+_- = \overline{\overline{L}}.\hat{y} \times \left\{\hat{y} \times \left(\mathbf{H}^+ + \mathbf{H}^-\right)\right\} \tag{6.66}$$

with

$$[\hat{y} \times \mathbf{H}]^+_- = 0 \tag{6.67}$$

and as evident from (2.68), these represent a generalised conductive sheet with (tensor) conductivity $\mathbf{R}_m = \frac{1}{2}\overline{\overline{L}}^T/\Delta_L$ where $\Delta_L = \det \overline{\overline{L}}$. In the special case when $\overline{\overline{L}} = \infty$ the sheet is a perfect magnetic conductor, and when $\overline{\overline{L}} = 0$ it ceases to exist. As we shall now show, the sheets whose transition conditions are given in (6.64), (6.65) and (6.66), (6.67) are complementary in the sense of Babinet's principle.

We consider first a generalised resistive sheet occupying the entire plane $y = 0$ and illuminated by a primary field $(\mathbf{E}^i = \mathbf{F},\ \mathbf{H}^i = \mathbf{G})$ resulting from sources in $y < 0$. The tensor resistivity is $\frac{1}{2}\overline{\overline{L}}$ and may vary over the surface in accordance with (6.18) and (6.19). To find the total field $(\mathbf{E}^{(1)}, \mathbf{H}^{(1)})$ we note that in the half space $y \geq 0$ the field can be obtained from the electric and magnetic Hertz vectors

$$\mathbf{\Pi}_e(\mathbf{r}) = 0, \qquad \mathbf{\Pi}_m(\mathbf{r}) = -\frac{2jY_0}{k_0} \iint_{-\infty}^{\infty} \mathbf{K}_m(\mathbf{r}')\, g(\mathbf{r} \mid \mathbf{r}')\, dx'\, dz'$$

where $g(\mathbf{r} \mid \mathbf{r}')$ is the free space Green's function (6.56) and

$$\mathbf{K}_m = -\hat{y} \times \mathbf{E}^{(1)}$$

is the magnetic current density on the surface $y = 0+$ of integration. In $y \geq 0$

$$\mathbf{E}^{(1)} = 2 \iint_{-\infty}^{\infty} \mathbf{K}_m \times \nabla g\, dx'\, dz'$$

$$\mathbf{H}^{(1)} = -2jk_0 Y_0 \iint_{-\infty}^{\infty} \left\{\mathbf{K}_m g + k_0^{-2}(\mathbf{K}_m.\nabla)\nabla g\right\} dx'\, dz'$$

where the differentiation is with respect to the unprimed coordinates of the observation point, and, since the tangential (normal) components of $\mathbf{E}^{(1)}$ $(\mathbf{H}^{(1)})$

are symmetric about the plane $y = 0$ with the other components antisymmetric, extension to the whole space gives

$$\mathbf{E}^{(1)} = \mathbf{F} + \left\{ \begin{matrix} -\mathbf{F} \\ \mathbf{F}^r \end{matrix} \right\} \pm 2 \iint_{-\infty}^{\infty} \mathbf{K_m} \times \nabla g \, dx' \, dz' \qquad (6.68)$$

$$\mathbf{H}^{(1)} = \mathbf{G} + \left\{ \begin{matrix} -\mathbf{G} \\ \mathbf{G}^r \end{matrix} \right\} \mp 2jk_0Y_0 \iint_{-\infty}^{\infty} \left\{ \mathbf{K_m} g + k_0^{-2}(\mathbf{K_m}.\nabla)\nabla g \right\} dx' \, dz' \qquad (6.69)$$

Here and throughout the rest of this section, upper and lower signs/alternatives apply to the regions $y > 0$ and $y < 0$, respectively, and $(\mathbf{F}^r, \mathbf{G}^r)$ is the field reflected by a pec plane $y = 0$ when the field $(\mathbf{F}, \mathbf{G})$ is incident upon it. Thus

$$\begin{matrix} F^r_{x,z}(\mathbf{r}) = -F_{x,z}(\tilde{\mathbf{r}}), & F^r_y(\mathbf{r}) = F_y(\tilde{\mathbf{r}}) \\ G^r_{x,z}(\mathbf{r}) = G_{x,z}(\tilde{\mathbf{r}}), & G^r_y(\mathbf{r}) = -G_y(\tilde{\mathbf{r}}) \end{matrix}$$

where $\tilde{\mathbf{r}}$ is the image of the point $\mathbf{r}$ in the plane $y = 0$. It only remains to enforce (6.64) and, when this is done, the following integral equation is obtained:

$$-\mathbf{K_m} = \overline{\overline{L}}. \left(\mathbf{G} + 2jk_0Y_0 \lim_{y \to 0} \iint_{-\infty}^{\infty} \left\{ \mathbf{K_m} g + k_0^{-2}(\mathbf{K_m}.\nabla)\nabla g \right\} dx' \, dz' \right) \qquad (6.70)$$

holding at all points of the plane $y = 0$.

There are two problems that are complementary to this. The first has the same field $(\mathbf{E}^i = \mathbf{F}, \ \mathbf{H}^i = \mathbf{G})$ incident on the dual of the resistive sheet, namely, a conductive sheet in the plane $y = 0$. Instead of proceeding as before, we now express the field scattered by the sheet in terms of the total current it supports. Since there is no electric current,

$$\mathbf{\Pi_e}(\mathbf{r}) = 0, \qquad \mathbf{\Pi_m}(\mathbf{r}) = -\frac{jY_0}{k_0} \iint_{-\infty}^{\infty} \mathbf{J}_{m(s)}(\mathbf{r}') g(\mathbf{r} \mid \mathbf{r}') \, dx' \, dz'$$

where

$$\mathbf{J}_{m(s)} = - [\hat{y} \times \mathbf{E}]_-^+$$

is the total magnetic sheet current, and the entire field at all points of space is then

$$\mathbf{E}^{(2)} = \mathbf{F} + \iint_{-\infty}^{\infty} \mathbf{J}_{m(s)} \times \nabla g \, dx' \, dz' \qquad (6.71)$$

$$\mathbf{H}^{(2)} = \mathbf{G} - jk_0Y_0 \iint_{-\infty}^{\infty} \left\{ \mathbf{J}_{m(s)} g + k_0^{-2}(\mathbf{J}_{m(s)}.\nabla)\nabla g \right\} dx' \, dz' \qquad (6.72)$$

On applying (6.66), an integral equation from which to determine $\mathbf{J}_{m(s)}$ is found to be

$$\tfrac{1}{2}\mathbf{J}_{m(s)} = \overline{\overline{L}}. \left(\mathbf{G} - jk_0Y_0 \lim_{y \to 0} \iint_{-\infty}^{\infty} \left\{ \mathbf{J}_{m(s)} g + k_0^{-2}(\mathbf{J}_{m(s)}.\nabla)\nabla g \right\} dx' \, dz' \right) \qquad (6.73)$$

and comparison with (6.70) shows

$$\mathbf{J}_{m(s)} = -2\mathbf{K}_m \tag{6.74}$$

Thus, if the resistivity and conductivity of the two sheets are such that

$$\overline{\overline{R}}_e.\overline{\overline{R}}_m = \tfrac{1}{2}\overline{\overline{L}}.\tfrac{1}{2}\overline{\overline{L}}^{-1} = \tfrac{1}{4}\overline{\overline{I}} \tag{6.75}$$

in accordance with (6.64) and (6.66), the two problems are also related, and from (6.68), (6.69), (6.71), and (6.72) the total fields satisfy

$$\mathbf{E}^{(2)} \pm \mathbf{E}^{(1)} = \left\{ \begin{matrix} \mathbf{F} \\ -\mathbf{F}^r \end{matrix} \right\}$$

$$\tag{6.76}$$

$$\mathbf{H}^{(2)} \pm \mathbf{H}^{(1)} = \left\{ \begin{matrix} \mathbf{G} \\ -\mathbf{G}^r \end{matrix} \right\}$$

A special case of this result is an aperture (where $\overline{\overline{R}}_e = \infty$) in a pec screen ($\overline{\overline{R}}_e = 0$), and the complementary structure for the same incident field is a pmc disk ($\overline{\overline{R}}_m = 0$) coincident with the aperture. This is the direct analogue of the standard form of Babinet's principle in acoustics.

Another problem that is also complementary to the first has a resistive sheet of (different) resistivity $\overline{\overline{R}}'_e = \tfrac{1}{2}\overline{\overline{L}}'$ (say) and a different incident field. If $\mathbf{E}^i = -Z_0\mathbf{G}$, $\mathbf{H}^i = Y_0\mathbf{F}$, we can express the field scattered by the sheet in terms of the Hertz vectors

$$\mathbf{\Pi}_e(\mathbf{r}) = -\frac{jZ_0}{k_0} \iint_{-\infty}^{\infty} \mathbf{J}_{e(s)}(\mathbf{r}') g(\mathbf{r} \mid \mathbf{r}') \, dx' \, dz', \qquad \mathbf{\Pi}_m(\mathbf{r}) = 0$$

where $\mathbf{J}_{e(s)}$ is the total electric sheet current, and the complete field is then

$$\mathbf{E}^{(3)} = -Z_0\mathbf{G} - jk_0 Z_0 \iint_{-\infty}^{\infty} \left\{ \mathbf{J}_{e(s)} g + k_0^{-2} (\mathbf{J}_{e(s)}.\nabla)\nabla g \right\} dx' \, dz' \tag{6.77}$$

$$\mathbf{H}^{(3)} = Y_0\mathbf{F} - \iint_{-\infty}^{\infty} \mathbf{J}_{e(s)} \times \nabla g \, dx' \, dz' \tag{6.78}$$

From (6.64) with $\overline{\overline{L}}$ replaced by $\overline{\overline{L}}'$, an integral equation for $\mathbf{J}_{e(s)}$ is

$$\tfrac{1}{2}\overline{\overline{L}}'.Y_0\mathbf{J}_{e(s)} = \mathbf{G} + jk_0 Y_0 \lim_{y \to 0} \iint_{-\infty}^{\infty} \left\{ \mathbf{J}_{e(s)} g + k_0^{-2} (\mathbf{J}_{e(s)}.\nabla)\nabla g \right\} dx' \, dz' \tag{6.79}$$

and if

$$\overline{\overline{L}}.\overline{\overline{L}}' = Z_0^2 \overline{\overline{I}} \tag{6.80}$$

(6.79) is identical to (6.70) with

$$\mathbf{J}_{e(s)} = 2Y_0\mathbf{K}_m \tag{6.81}$$

Since (6.80) can be written as

$$Y_0 \overline{\overline{R}}_e . Y_0 \overline{\overline{R}}_e' = \tfrac{1}{4} \overline{\overline{I}} \tag{6.82}$$

$\overline{\overline{R}}_e'$ bears a similar relationship to $\overline{\overline{R}}_e$ as does the conductivity $\overline{\overline{R}}_m$ in the second problem, but we have now avoided the introduction of a conductive sheet at the expense of a change in the incident field. With the identification (6.81), it follows from (6.68), (6.69), (6.77), and (6.78) that

$$Y \mathbf{E}^{(3)} \mp \mathbf{H}^{(1)} = \left\{ \begin{matrix} -\mathbf{G} \\ \mathbf{G}^{\mathbf{r}} \end{matrix} \right\}$$

$$ZH^{(3)} \pm \mathbf{E}^{(1)} = \left\{ \begin{matrix} \mathbf{F} \\ -\mathbf{F}^{\mathbf{r}} \end{matrix} \right\} \tag{6.83}$$

The above results are a generalisation of those previously derived for first order sheets by SENIOR (1975b) and SAMADDAR (1985), and the relationship (6.82) between the resistivity tensors is identical to that obtained by BAUM AND SIN-GURAJU (1974) using a combined field formulation. In contrast to the surfaces considered by LANG (1973) (but see also HARRINGTON AND MAUTZ (1974)), those forming our complementary structures must have resistivities and conductivities satisfying (6.75) or (6.82) at all pairs of corresponding points. Since $\overline{\overline{R}}' = \infty$ (0) when $\overline{\overline{R}} = 0$ (∞), the problem complementary to that of an open aperture in a pec screen is a pec disk coincident with the aperture. The second problem pairing therefore includes the standard electromagnetic form of Babinet's principle as a special case, and (6.83) provides the usual connection between the fields.

The above Babinet equivalences are the most general yet known. Admittedly, the complementary sheets could be difficult to realise if $M > 1$. Since $\overline{\overline{R}}_e$ is then a differential operator, the $\overline{\overline{R}}_m$ and $\overline{\overline{R}}_e'$ specified by (6.75) and (6.82), respectively, involve integral operations, but they do at least show what is mathematically possible.

Attempts to develop a Babinet principle for "absorbing" surfaces have failed (NEUGEBAUER, 1956), and no such principle is known for combination sheets, including GIBC sheets as a special case. Since the scattered fields no longer have the symmetry that characterises electric and magnetic sheets individually, the type of proof that we have used is no longer valid, and it is unlikely that a Babinet principle exists. Nevertheless, there is an interesting consequence of the Babinet equivalence (6.76) that relates immediately to a GIBC, and follows because the complementary sheets *together* form a GIBC sheet. Thus, if the field ($\mathbf{E}^i = \mathbf{F}$, $\mathbf{H}^i = \mathbf{G}$) is incident on the surface $y = 0$ subject to the GIBC (6.17), the field in $y < 0$ is the sum of ($\mathbf{E}^{(1)}, \mathbf{H}^{(1)}$) and ($\mathbf{E}^{(2)}, \mathbf{H}^{(2)}$) with $\overline{\overline{L}} = 2\overline{\overline{R}}_e$. Hence, from (6.76), the scattered field is twice that due to the generalised resistive sheet in the first problem, and this has been used (SENIOR

ET AL., 1990) to determine the scattering from an impedance insert in a pec plane.

6.4 Applications

It is, perhaps, not surprising that there have been few published applications of boundary conditions of higher order than the second. As noted in Section 5.6, BERNARD (1987) considered the two-dimensional problem of a plane wave incident on a wedge of arbitrary angle subject to GIBCs of arbitrary order, but did not complete the solution. RICOY AND VOLAKIS (1992) also employed boundary conditions of arbitrary order to model a stratified dielectric slab, and then studied the diffraction of a plane wave by the junction of two semi-infinite layers simulated in this manner. Instead of using contact conditions to ensure uniqueness, they specified the arbitrary constants by requiring the continuity of the *interior* fields and their derivatives at the junction. Indeed, one of the main difficulties with higher order conditions is the enforcement of continuity at abrupt changes in the surface properties, and there are only a few instances (for example, SENIOR (1993a,b)) where the contact conditions have been successfully applied.

The difficulty disappears if there are no surface discontinuities, and higher order GIBCs have proved effective in modelling coatings applied to curved bodies. Another application of growing importance is in connection with absorbing boundary conditions to limit the domain in a finite element or finite difference solution of a scattering problem, and these topics are discussed in Chapters 7 and 8, respectively. But even when there are discontinuities in the surface properties, it may be possible to satisfy the contact conditions automatically. An example is a numerical solution based on an integral equation formulation where the basis functions are chosen to ensure continuity, and SABETFAKHRI AND KATEHI (1994) and VAN DEVENTER AND KATEHI (1995) have used boundary conditions of high and, in principle, infinite order to simulate the effect of the substrate in the analysis of microstrip problems.

References

Babinet, J. (1837), "Memoires d'optique météorologique", *Compt. Rend. Acad. Sci., Paris*, **4**, pp. 638–648.

Baker, B. B. and Copson, E. T. (1950), *The Mathematical Theory of Huygens' Principle* (2nd edition), Clarendon Press, Oxford, pp. 166–168.

Baum, C. E. and Singuraju, B. K. (1974), "Generalization of Babinet's principle in terms of the combined field to include impedance loaded aperture antennas and scatterers", Interaction Note No. 217, Air Force Weapons Laboratory, Kirtland Air Force Base, NM.

Bernard, J.-M. L. (1987), "Diffraction by a metallic wedge covered with a dielectric material", *Wave Motion*, **9**, pp. 543–561.

Booker, H. G. (1946), "Slot aerials and their relation to complementary wire aerials (Babinet's principle)", *J. IEE, London*, **93A**, pp. 620–626.

Born, M. and Wolf, E. (1964), *Principles of Optics* (2nd edition), Pergamon Press, Oxford, p. 381. Note that Babinet's initial is erroneously given as A.

Copson, E. T. (1946), "An integral-equation method of solving plane diffraction problems", *Proc. Roy. Soc. London, Ser. A*, **186**, pp. 100–118.

Copson, E. T. (1950), "Diffraction by a plane screen", *Proc. Roy. Soc. London, Ser. A*, **202**, pp. 277–284.

Epstein, P. (1915), "Spezielle Beugunsprobleme", in *Encyklopädie der mathematischen Wissenschaften*, **5**, pt. 3, Leipzig, pp. 489–525.

Harrington, R. F. and Mautz, J. R. (1974), "Comments on 'Babinet's principle for a perfectly conducting screen with aperture covered by resistive sheet'", *IEEE Trans. Antennas Propagat.*, **AP-22**, p. 842.

Hoppe, D. J. and Rahmat-Samii, Y. (1994), "Higher order impedance boundary conditions revisited: application to chiral coatings", *J. Electromagn. Waves Applics.*, **8**, pp. 1303–1329.

Huxley, L. G. H. (1947), *Wave Guides*, University Press, Cambridge, pp. 283–290.

Lang, K. C. (1973), "Babinet's principle for a perfectly conducting screen with aperture covered by resistive sheet", *IEEE Trans. Antennas Propagat.*, **AP-21**, pp. 738–740.

Ljalinov, M. A. (1992), "Boundary-contact problems in electromagnetic scattering", Proc. URSI International Symposium on Electromagnetic Theory, Sydney, Australia, pp. 308–310.

Neugebauer, H. E. J. (1956), "Diffraction of electromagnetic waves caused by apertures in absorbing plane screens", *IRE Trans. Antennas Propagat.*, **AP-4**, pp. 115–119.

Ricoy, M. A. and Volakis, J. L. (1992), "Diffraction by a symmetric material junction simulated with generalized sheet transition conditions", *IEEE Trans. Antennas Propagat.*, **AP-40**, pp. 742–754.

Sabetfakhri, K. and Katehi, L. P. B. (1994), "An integral transform technique for analysis of planar dielectric structures", *IEEE Trans. Microwave Theory Tech.*, **MTT-42**, pp. 1052–1062.

Samaddar, S. N. (1985), "Babinet's principle for an anisotropic surface using different approaches", Memo Rep. 5648, Naval Research Laboratory, Washington, D.C.

Senior, T. B. A. (1975a), "Some extensions of Babinet's principle", *J. Acoust. Soc. Amer.*, **58**, pp. 501–503.

Senior, T. B. A. (1975b), "Some extensions of Babinet's principle in electromagnetic theory", *IEEE Trans. Antennas Propagat.*, **AP-25**, pp. 417–420.

Senior, T. B. A. (1993a), "Diffraction by a half plane with third-order transition conditions", *Radio Sci.*, **28**, pp. 273–279.

Senior, T. B. A. (1993b), "Diffraction by half plane junctions", University of Michigan Radiation Laboratory Report No. RL 892.

Senior, T. B. A., Sarabandi, K. and Natzke, J. R. (1990), "Scattering by a narrow gap", *IEEE Trans. Antennas Propagat.*, **AP-38**, pp. 1102–1110.

Senior, T. B. A. and Volakis, J. L. (1989), "Derivation and application of a class of generalized boundary conditions", *IEEE Trans. Antennas Propagat.*, **AP-37**, pp. 1566–1572.

van Deventer, T. E. and Katehi, L. P. B. (1995), "Generalized boundary conditions with applications to submillimeter and optical waveguides", to be published in *Radio Sci.*.

Chapter 7

GIBC applications to cylindrical bodies

The GIBC applications discussed so far have been to planar surfaces but, as evident from the general forms presented in the last chapter, the conditions are also applicable to curved surfaces. To illustrate this fact, we now consider their application to the singly curved surface of a cylinder. In the special case of a homogeneous circular cylinder, the exact solution of the scattering problem is known, and this can be used to construct approximate boundary conditions as well as to determine their accuracy. An important practical application is to coated metallic cylinders. This is examined in the second section, where we demonstrate the accuracy that can be achieved with a second order GIBC, and we then develop a high frequency solution.

7.1 Homogeneous circular cylinder

We consider a right circular cylinder composed of a homogeneous dielectric of complex refractive index $N = \sqrt{\epsilon_r \mu_r}$.

7.1.1 Exact eigenfunction solution

In terms of the cylindrical polar coordinates ρ, ϕ, z the cylinder is defined as $\rho = \rho_0$ and is illuminated by the H-polarised plane wave

$$\mathbf{H}^i = \hat{z}\, e^{jk_0 x} \tag{7.1}$$

incident in a plane perpendicular to the z axis of the cylinder as shown in Fig. 7–1. The only non-zero components of the total field are then H_z, E_ρ and E_ϕ, and they can be expressed as follows:

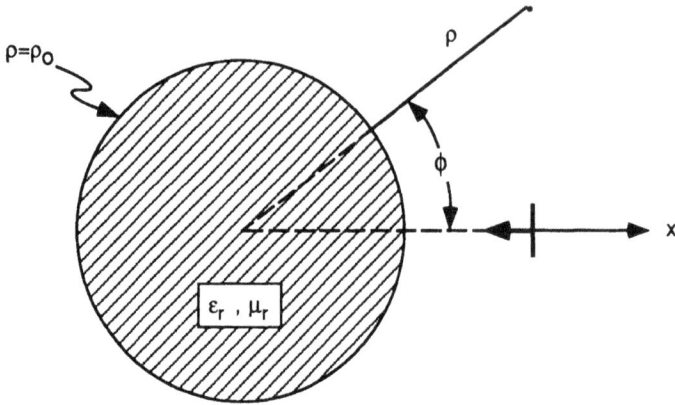

Figure 7–1: *Geometry for the homogeneous circular cylinder*

$\rho \geq \rho_0$

$$H_z = \sum_{m=-\infty}^{\infty} j^m \left\{ J_m(k_0\rho) + R_m^{\rm H} H_m^{(2)}(k_0\rho) \right\} e^{-jm\phi}$$

$$E_\rho = \frac{Z_0}{k_0\rho} \sum_{m=-\infty}^{\infty} j^m m \left\{ J_m(k_0\rho) + R_m^{\rm H} H_m^{(2)}(k_0\rho) \right\} e^{-jm\phi} \quad (7.2)$$

$$E_\phi = jZ_0 \sum_{m=-\infty}^{\infty} j^m \left\{ J_m'(k_0\rho) + R_m^{\rm H} H_m^{(2)\prime}(k_0\rho) \right\} e^{-jm\phi}$$

$\rho \leq \rho_0$

$$H_z = \sum_{m=-\infty}^{\infty} j^m a_m J_m(Nk_0\rho) e^{-jm\phi}$$

$$E_\rho = \frac{Z}{Nk_0\rho} \sum_{m=-\infty}^{\infty} j^m m a_m J_m(Nk_0\rho) e^{-jm\phi} \quad (7.3)$$

$$E_\phi = jZ \sum_{m=-\infty}^{\infty} j^m a_m J_m'(Nk_0\rho) e^{-jm\phi}$$

where Z is the intrinsic impedance of the dielectric, J_m is the Bessel function of order m, $H_m^{(2)}$ is the Hankel function of the second kind of order m, the prime denotes differentiation with respect to the entire argument, and $R_m^{\rm H}$ and a_m are coefficients to be determined. From the continuity of H_z and E_ϕ at $\rho = \rho_0$ we find

$$R_m^{\rm H} = -\frac{J_m'(k_0\rho_0) - jY_0 Q^{\rm H} J_m(k_0\rho_0)}{H_m^{(2)\prime}(k_0\rho_0) - jY_0 Q^{\rm H} H_m^{(2)}(k_0\rho_0)} \quad (7.4)$$

with

$$Q^{\rm H} = -jZ \frac{J_m'(Nk_0\rho_0)}{J_m(Nk_0\rho_0)} \quad (7.5)$$

and Q^H is a key quantity in the subsequent development. On the surface of the cylinder

$$H_z(\rho_0, \phi) = -\frac{2j}{\pi k_0 \rho_0} \sum_{m=-\infty}^{\infty} \left\{ H_m^{(2)\prime}(k_0\rho_0) - jY_0 Q^H H_m^{(2)}(k_0\rho_0) \right\}^{-1} e^{-jm(\phi-\pi/2)}$$

(7.6)

and in the far zone the scattered field is

$$H_z^s(\rho, \phi) \sim \sqrt{\frac{2\pi}{k_0\rho}} e^{-j(k_0\rho - \pi/4)} P^H(\phi)$$

where

$$P^H(\phi) = \frac{1}{\pi} \sum_{m=-\infty}^{\infty} (-1)^m R_m^H e^{-jm\phi}$$

(7.7)

with $\phi = 0$ in the backscattering direction.

Under the assumption that $|N|k_0\rho_0 \gg 1$ with Im. $N < 0$ (to prevent any penetration through the cylinder), Q can be expanded in an asymptotic series for large $x = Nk_0\rho_0$. Since

$$J_m(x) = \tfrac{1}{2}\left\{ H_m^{(1)}(x) + H_m^{(2)}(x) \right\}$$

it follows that (BOWMAN ET AL., 1987)

$$J_m(x) = \frac{e^{j(x - m\pi/2 - \pi/4)}}{\sqrt{2\pi x}} \sum_{\ell=0}^{\infty} \frac{(m,\ell)}{(-2jx)^\ell} =$$

$$\frac{e^{j(x - m\pi/2 - \pi/4)}}{\sqrt{2\pi x}} \left\{ 1 - \frac{4m^2 - 1}{8jx} - \frac{(4m^2 - 1)(4m^2 - 9)}{128x^2} \right.$$

$$+ \frac{(4m^2 - 1)(4m^2 - 9)(4m^2 - 25)}{128 \cdot 24 j x^3}$$

$$+ \frac{(4m^2 - 1)(4m^2 - 9)(4m^2 - 25)(4m^2 - 49)}{(128)^2 \cdot 6 x^4}$$

$$\left. - \frac{(4m^2 - 1)(4m^2 - 9)(4m^2 - 25)(4m^2 - 49)(4m^2 - 81)}{(128)^2 \cdot 240 j x^5} + O(x^{-6}) \right\}$$

Hence

$$J_m'(x) = j\frac{e^{j(x - m\pi/2 - \pi/4)}}{\sqrt{2\pi x}} \left\{ 1 - \frac{4m^2 + 3}{8jx} - \frac{(4m^2 - 1)(4m^2 + 15)}{128x^2} \right.$$

$$+ \frac{(4m^2 - 1)(4m^2 - 9)(4m^2 + 35)}{128 \cdot 24 j x^3}$$

$$+ \frac{(4m^2 - 1)(4m^2 - 9)(4m^2 - 25)(4m^2 + 63)}{(128)^2 \cdot 6 x^4}$$

$$\left. - \frac{(4m^2 - 1)(4m^2 - 9)(4m^2 - 25)(4m^2 - 49)(4m^2 + 99)}{(128)^2 \cdot 240 j x^5} + O(x^{-6}) \right\}$$

giving

$$Q^H = Z\left\{1 - \frac{1}{2jx} - \frac{4m^2 - 1}{8x^2} - \frac{4m^2 - 1}{8jx^3} - \frac{(4m^2 - 1)(4m^2 - 25)}{128x^4}\right.$$
$$\left. - \frac{(4m^2 - 1)(4m^2 - 13)}{32jx^5} + O(x^{-6})\right\}$$

and therefore

$$Q^H = Z\left\{1 - \frac{1}{2jx} + \frac{1}{8x^2} + \frac{1}{8jx^3} - \frac{25}{128x^4} - \frac{13}{32jx^5}\right.$$
$$- \frac{1}{2}\left(\frac{m}{x}\right)^2\left(1 + \frac{1}{jx} - \frac{13}{8x^2} - \frac{7}{2jx^3}\right)$$
$$\left. - \frac{1}{8}\left(\frac{m}{x}\right)^4\left(1 + \frac{4}{jx}\right) + O(x^{-6})\right\} \tag{7.8}$$

For E polarisation with

$$\mathbf{E}^i = \hat{z}\,e^{jk_0 z} \tag{7.9}$$

the analysis is similar. If the total electric field in $\rho \geq \rho_0$ is written as

$$E_z = \sum_{m=-\infty}^{\infty} j^m\left\{J_m(k_0\rho) + R_m^E\,H_m^{(2)}(k_0\rho)\right\}e^{jm\phi} \tag{7.10}$$

we have

$$R_m^E = -\frac{J_m'(k_0\rho_0) - jZ_0Q^E\,J_m(k_0\rho_0)}{H_m^{(2)'}(k_0\rho_0) - jZ_0Q^E\,H_m^{(2)}(k_0\rho_0)} \tag{7.11}$$

with

$$Q^E = -jY\frac{J_m'(Nk_0\rho_0)}{J_m(Nk_0\rho_0)} \tag{7.12}$$

and thus

$$ZQ^E = YQ^H \tag{7.13}$$

in accordance with duality.

7.1.2   GIBC simulations

An approximate boundary condition should reproduce R_m (and therefore Q) to a specified order in $1/x$. A general expression for a GIBC is given in (6.20). In the case of H-polarisation, $\nabla_s.\mathbf{H} = 0$ since there is no z dependence, and (6.20) reduces to

$$E_\phi = -\eta_{zz}H_z - \frac{1}{\rho_0}\frac{\partial}{\partial\phi}f(\hat{n}.\nabla \times \mathbf{H})$$

Also

$$f(\hat{n}.\nabla \times \mathbf{H}) = f(jk_0 Y_0 E_\rho)$$
$$= jk_0 Y_0 \left(A_0 - A_1 \frac{1}{\rho_0^2} \frac{\partial^2}{\partial\phi^2} + \cdots \right) E_\rho$$

where $A_0, A_1, \ldots$ are the arbitrary constants in (6.19), and the boundary condition is therefore

$$E_\phi = -\eta_{zz} H_z - \frac{jk_0 Y_0}{\rho_0} \frac{\partial}{\partial\phi} \left(A_0 - A_1 \frac{1}{\rho_0^2} \frac{\partial^2}{\partial\phi^2} + \cdots \right) E_\rho \qquad (7.14)$$

When the expressions (7.2) for the field components are inserted, we find

$$\frac{J_m'(k_0\rho_0) + R_m^{\mathrm{H}} H_m^{(2)\prime}(k_0\rho_0)}{J_m(k_0\rho_0) + R_m^{\mathrm{H}} H_m^{(2)}(k_0\rho_0)} = jY_0 \left\{ \eta_{zz} - \left(\frac{m}{\rho_0}\right)^2 A_0 - \left(\frac{m}{\rho_0}\right)^4 A_1 - \cdots \right\}$$

and the resulting formula for R_m^{H} is identical to that in (7.4) if

$$Q^{\mathrm{H}} = \eta_{zz} - \left(\frac{m}{\rho_0}\right)^2 A_0 - \left(\frac{m}{\rho_0}\right)^4 A_1 - \cdots \qquad (7.15)$$

The standard (first order) impedance boundary condition (2.27) has $\eta_{zz} = Z$ with $A_0 = A_1 = \cdots = 0$, and is accurate to the zeroth order in $1/x$, but a comparision of (7.8) and (7.15) shows that by choosing

$$\eta_{zz} = Z \left(1 - \frac{1}{2jx} \right) \qquad (7.16)$$

the SIBC is made accurate to the first order in $1/x$. To improve the accuracy still further it is necessary to go to a higher order boundary condition, and the additional terms in (7.15) enable us to simulate the different impedances seen by the individual modes. For a second order boundary condition, $A_1 = \cdots = 0$, and if

$$\eta_{zz} = Z \left(1 - \frac{1}{2jx} + \frac{1}{8x^2} + \frac{1}{8jx^3} \right) \qquad (7.17)$$

with

$$A_0 = \frac{Z}{2(Nk_0)^2} \left(1 + \frac{1}{jx} \right) \qquad (7.18)$$

the condition is accurate to the third order in $1/x$. Similarly, a third order GIBC with

$$\eta_{zz} = Z \left(1 - \frac{1}{2jx} + \frac{1}{8x^2} + \frac{1}{8jx^3} - \frac{25}{128x^4} - \frac{13}{32jx^5} \right) \qquad (7.19)$$

$$A_0 = \frac{Z}{2(Nk_0)^2} \left(1 + \frac{1}{jx} - \frac{13}{8x^2} - \frac{7}{2jx^3} \right) \qquad (7.20)$$

and

$$A_1 = \frac{Z}{8(Nk_0)^4}\left(1 + \frac{4}{jx}\right) \tag{7.21}$$

is accurate to the fifth order. In all of the above, $x = Nk_0\rho_0$. We observe that each increase in the order of the boundary condition produces two orders of improvement in accuracy, and the results are identical to those obtained using Rytov's technique (see Appendix A) where the formal expansion parameter is $|N|^{-1}$.

For E-polarisation $\hat{\rho}.\nabla \times \mathbf{H} = 0$ and the GIBC (6.20) becomes

$$E_z = \eta_{\phi\phi}H_\phi - \frac{1}{\rho_0}\frac{\partial}{\partial\phi}g(\nabla_{\mathrm{s}}.\mathbf{H})$$

with (see (6.19))

$$\begin{aligned}
g(\nabla_{\mathrm{s}}.\mathbf{H}) &= \left(B_0 - B_1\frac{1}{\rho_0^2}\frac{\partial^2}{\partial\phi^2} + \cdots\right)\nabla_{\mathrm{s}}.\mathbf{H} \\
&= \frac{1}{\rho_0}\frac{\partial}{\partial\phi}\left(B_0 - B_1\frac{1}{\rho_0^2}\frac{\partial^2}{\partial\phi^2} + \cdots\right)H_\phi
\end{aligned}$$

giving

$$E_z = \left(\eta_{\phi\phi} - B_0\frac{1}{\rho_0^2}\frac{\partial^2}{\partial\phi^2} + B_1\frac{1}{\rho_0^4}\frac{\partial^4}{\partial\phi^4} - \cdots\right)H_\phi \tag{7.22}$$

where $B_0, B_1, \ldots$ are constants. By inserting the eigenfunction expansions of the field components we obtain

$$\frac{1}{Q^{\mathrm{E}}} = \eta_{\phi\phi} + \left(\frac{m}{\rho_0}\right)^2 B_0 + \left(\frac{m}{\rho_0}\right)^4 B_1 + \cdots \tag{7.23}$$

and hence, from (7.8) and (7.13),

$$\eta_{\phi\phi} = Z\left(1 + \frac{1}{2jx} - \frac{3}{8x^2} - \frac{3}{8jx^3} + \frac{63}{128x^4} + \frac{27}{32jx^5}\right) \tag{7.24}$$

$$B_0 = \frac{Z}{2(Nk_0)^2}\left(1 + \frac{2}{jx} - \frac{39}{8x^2} - \frac{29}{4jx^3}\right) \tag{7.25}$$

and

$$B_1 = \frac{3}{8}\frac{Z}{(Nk_0)^4}\left(1 + \frac{4}{jx}\right) \tag{7.26}$$

accurate to the fifth order.

For simplicity we now confine attention to the second order boundary conditions for which $A_1 = B_1 = 0$. Then

$$\eta_{\phi\phi} = Z\left(1 + \frac{1}{2jx} - \frac{3}{8x^2} - \frac{3}{8jx^3}\right) \tag{7.27}$$

$$B_0 = \frac{Z}{2(Nk_0)^2}\left(1 + \frac{2}{jx}\right) \tag{7.28}$$

with η_{zz} and A_0 shown in (7.17) and (7.18), respectively, and the boundary conditions are

$$E_\phi = -Z\left\{1 - \frac{1}{2jx} + \frac{1}{2x^2}\left(1 + \frac{1}{jx}\right)\left(\frac{\partial^2}{\partial\phi^2} + \frac{1}{4}\right)\right\}H_z \qquad (7.29)$$

$$E_z = Z\left\{1 + \frac{1}{2jx} - \frac{3}{8x^2}\left(1 + \frac{1}{jx}\right) - \frac{1}{2x^2}\left(1 + \frac{2}{jx}\right)\frac{\partial^2}{\partial\phi^2}\right\}H_\phi \qquad (7.30)$$

accurate to $O(x^{-3})$. From (7.29) we obtain precisely

$$\frac{\partial H_z}{\partial\rho} = jk_0 Y_0 Z\left\{1 - \frac{1}{2jx} + \frac{1}{2x^2}\left(1 + \frac{1}{jx}\right)\left(\frac{\partial^2}{\partial\phi^2} + \frac{1}{4}\right)\right\}H_z \qquad (7.31)$$

on using Maxwell's equation, and similarly, from (7.30) by inverting the differential operator,

$$\frac{\partial E_z}{\partial\rho} = jk_0 Z_0 Y\left\{1 - \frac{1}{2jx} + \frac{1}{2x^2}\left(1 + \frac{1}{jx}\right)\left(\frac{\partial^2}{\partial\phi^2} + \frac{1}{4}\right)\right\}E_z \qquad (7.32)$$

with $x = Nk_0\rho_0$ as before. As expected, (7.32) is simply the dual of (7.31).

If the planar second order boundary condition (5.28) is applied locally at every point of the cylinder with the parameters chosen as indicated in (5.59) appropriate to a lossy half space, the result is equivalent to (7.31) and (7.32) with the terms in braces replaced by $\{1 + (1/2x^2)(\partial^2/\partial\phi^2)\}$. Thus, the additional terms (which are contributed by those other than the leading ones in the expressions for η_{zz}, $\eta_{\phi\phi}$, A_0 and B_0) represent the effect of the surface curvature. We also note that when (7.31) and (7.32) are specialised to the case of an absorbing boundary condition (see Chapter 8), it is customary to replace the factor $\left(1 + \frac{1}{jx}\right)$ by $\left(1 - \frac{1}{jx}\right)^{-1}$ but, as evident from the above analysis, this neither reproduces nor reduces the terms of higher order.

7.1.3 Accuracy

The exact eigenfunction expansion can be used to assess the validity of the approximate boundary conditions, and for this purpose it is sufficient to confine attention to H-polarisation. The most accurate first order boundary condition is then

$$E_\phi = -\eta_{zz}H_z \qquad (7.33)$$

with η_{zz} given in (7.16), and the corresponding second order GIBC is

$$E_\phi = -\eta_{zz}H_z - jk_0 Y_0 A_0\frac{1}{\rho}\frac{\partial E_\rho}{\partial\phi} \qquad (7.34)$$

where η_{zz} and A_0 are given in (7.17) and (7.18), respectively.

From physical considerations it is obvious that Im. N must be large enough to prevent field penetration through the cylinder, and without this, no local boundary condition can suffice. In Fig. 7-2 the backscattered far field amplitude $|P^H|$ is plotted as a function of $k_0\rho_0$ for $N = 1.6$ and $\mu_r = 1$, and neither boundary condition is effective even when $k_0\rho_0$ becomes large. For fixed Im. $N \neq 0$ the errors associated with each boundary condition decrease with increasing $k_0\rho_0$, but the GIBC attains a specified accuracy for smaller values of $k_0\rho_0$. In Fig. 7-3 where $N = 2 - j$ and $\mu_r = 1$ the GIBC provides an accuracy of 1% if $k_0\rho_0 > 2.8$, whereas the SIBC does not achieve this until $k_0\rho_0 > 8.7$. From a large number of comparison of this type it has been found that the GIBC is accurate to (about) 1% if

$$|\text{Im. } N|k_0\rho_0 \gtrsim 2.3 \qquad \text{for } |N| \gtrsim 4 \tag{7.35}$$

$$|\text{Im. } N|k_0\rho_0 \gtrsim 3.4 \qquad \text{for } 4 > |N| \gtrsim 2 \tag{7.36}$$

For large $|N|$ the agreement between (7.35) and the SIBC criterion (2.97) developed by WANG (1987) is due to the restriction on field penetration. Compared with the SIBC, the GIBC produces the most significant improvement when $|N|$ is relatively small and Re. N and Im. N are comparable in magnitude. The specified accuracy is then achieved for much smaller values of $k_0\rho_0$.

Similar calculations have been carried out for bistatic scattering and for the surface field, and these are illustrated in Fig. 7-4 and Fig. 7-5, respectively, where $k_0\rho_0 = 17$, $N = 2 - j$ and $\mu_r = 1$. The GIBC faithfully reproduces the detailed structure of the field, and its superiority to the SIBC is evident. For an accuracy of (about) 1% the requirement is now

$$|\text{Im. } N|k_0\rho_0 \gtrsim 2.3 \qquad \text{for } |N| \gtrsim 6 \tag{7.37}$$

but if somewhat larger errors at a few angles are acceptable, the restriction on $|N|$ can be relaxed. The phase accuracy is comparable and the results are not significantly affected if $\mu_r \neq 1$.

7.1.4   Skew incidence

A more general case is that of skew (oblique) incidence on the cylinder and, although a similar treatment is possible, the analysis is much more involved. Nevertheless, there are some subtle differences which warrant a brief discussion.

If the incident plane wave is such that

$$E_z^i = e_z e^{-jk_0 \hat{\imath}.\mathbf{r}}, \qquad H_z^i = h_z e^{-jk_0 \hat{\imath}.\mathbf{r}} \tag{7.38}$$

with

$$\hat{\imath} = -\hat{x}\sin\beta + \hat{z}\cos\beta \tag{7.39}$$

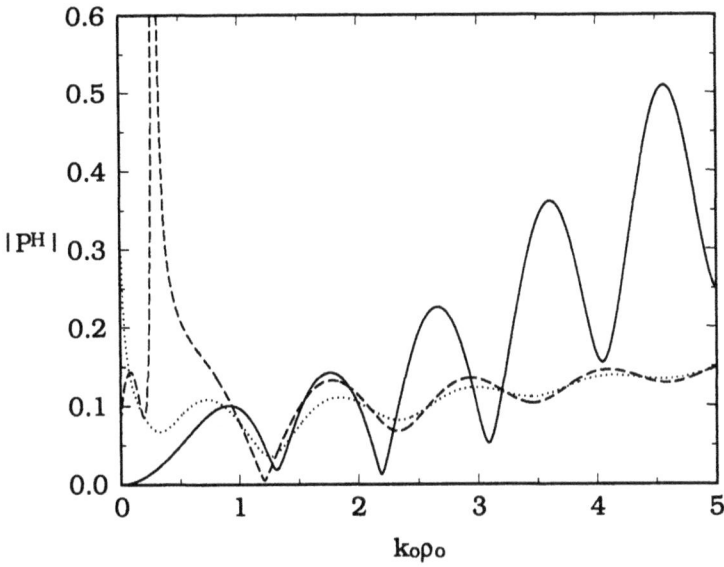

Figure 7–2: *Backscattered far field amplitude for a cylinder with $N = 1.6$ and $\mu_r = 1$: exact (——), GIBC (- - - -) and SIBC (·····)*

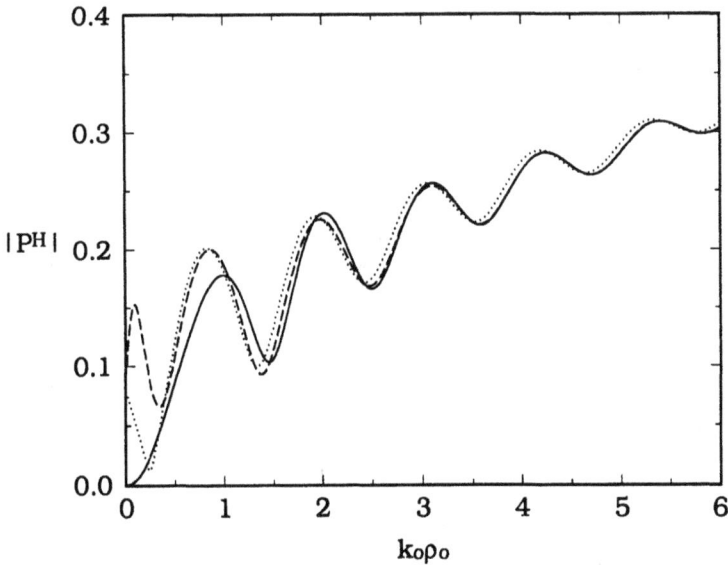

Figure 7–3: *Backscattered far field amplitude for a cylinder with $N = 2 - j$ and $\mu_r = 1$: exact (——), GIBC (- - - -) and SIBC (·····)*

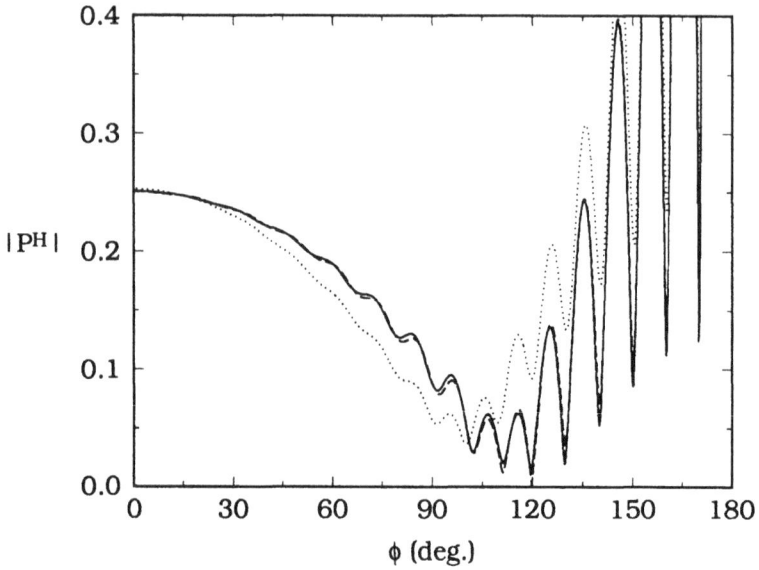

Figure 7–4: *Bistatic far field amplitude for a cylinder with $k_0\rho_0 = 17$, $N = 1.5 - j0.2$ and $\mu_r = 1$: exact (———), GIBC (- - - -) and SIBC (····)*

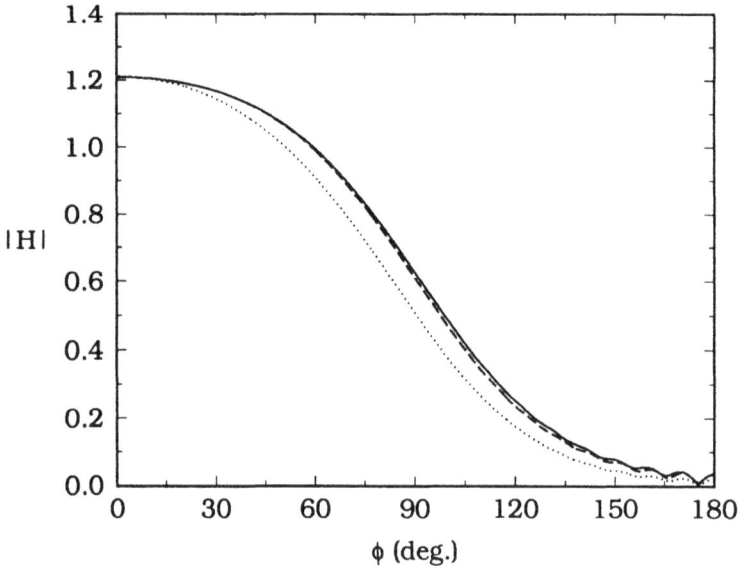

Figure 7–5: *Surface field magnitude for a cylinder with $k_0\rho_0 = 17$, $N = 1.5 - j0.2$ and $\mu_r = 1$: exact (———), GIBC (- - - -) and SIBC (····)*

all field components must display the same z dependence, and we can construct the other incident field components from (4.46). The total field components E_z and H_z in $\rho \geq \rho_0$ can be written as

$$
E_z = \sum_{m=-\infty}^{\infty} j^m \left\{ e_z[J_m(k_0\rho \sin \beta) + C_m^{\mathrm{E}} H_m^{(2)}(k_0\rho \sin \beta)] \right.
$$
$$
\left. - Z_0 h_z \bar{C}_m H_m^{(2)}(k_0\rho \sin \beta) \right\} e^{-jm\phi} e^{-jk_0 z \cos \beta}
$$

$$(7.40)$$

$$
H_z = \sum_{m=-\infty}^{\infty} j^m \left\{ h_z[J_m(k_0\rho \sin \beta) + C_m^{\mathrm{H}} H_m^{(2)}(k_0\rho \sin \beta)] \right.
$$
$$
\left. + Y_0 e_z \bar{C}_m H_m^{(2)}(k_0\rho \sin \beta) \right\} e^{-jm\phi} e^{-jk_0 z \cos \beta}
$$

and the expressions for the other field components can be obtained by substituting (7.40) into (4.46). There are analogous expressions for the field components in $\rho \leq \rho_0$ and, by enforcing the continuity of E_z, E_ϕ, H_z and H_ϕ at $\rho = \rho_0$, the coefficients C_m^{E}, C_m^{H} and $\bar{C}_m$ can be determined. The results are given in RUCK ET AL. (1970).

For a circular cylinder the second order GIBC (6.20) is

$$
\hat{\rho} \times \mathbf{E} = \hat{\rho} \times (\hat{\rho} \times \{\bar{\bar{\eta}}.\mathbf{H} - \nabla(B_0 \nabla_{\mathrm{s}}.\mathbf{H}) - \hat{\rho} \times \nabla(A_0 \hat{\rho}.\nabla \times \mathbf{H})\}) \qquad (7.41)
$$

and since

$$
\nabla_{\mathrm{s}}.\mathbf{H} = \frac{1}{\rho_0} \frac{\partial H_\phi}{\partial \phi}
$$

$$
\hat{\rho}.\nabla \times \mathbf{H} = \frac{1}{\rho_0} \frac{\partial H_z}{\partial \phi} + jk_0 \cos \beta \, H_\phi
$$

the boundary conditions are

$$
E_\phi = -\left(\eta_{zz} + k_0^2 B_0 \cos^2 \beta - \frac{A_0}{\rho_0^2} \frac{\partial^2}{\partial \phi^2} \right) H_z + \frac{jk_0 \cos \beta}{\rho_0} (A_0 - B_0) \frac{\partial H_\phi}{\partial \phi}
$$

$$(7.42)$$

$$
E_z = \left(\eta_{\phi\phi} + k_0^2 A_0 \cos^2 \beta - \frac{B_0}{\rho_0^2} \frac{\partial^2}{\partial \phi^2} \right) H_\phi - \frac{jk_0 \cos \beta}{\rho_0} (A_0 - B_0) \frac{\partial H_z}{\partial \phi}
$$

When the exact eigenfunction expansions for the field components are inserted into (7.42), four equations are obtained involving η_{zz}, $\eta_{\phi\phi}$, A_0 and B_0. The only Bessel and Hankel functions that appear are the ratios

$$
\frac{J_m'\left(k_0\rho_0\sqrt{N^2 - \cos^2 \beta}\right)}{J_m\left(k_0\rho_0\sqrt{N^2 - \cos^2 \beta}\right)} \qquad \text{and} \qquad \frac{H_m^{(2)'}(k_0\rho_0 \sin \beta)}{H_m^{(2)}(k_0\rho_0 \sin \beta)}
$$

which are then expanded under the assumption that

$$
|k_0\rho_0\sqrt{N^2 - \cos^2 \beta}|, \qquad k_0\rho_0 \sin \beta \gg 1
$$

with Im. $N < 0$. If, in addition, it is assumed that $|N| \gg 1$ so that $(N^2 - \cos^2 \beta)^\alpha$ can be replaced by $N^{2\alpha}\{1 - (\alpha/N^2)\cos^2 \beta\}$ wherever it appears, we recover the expressions for η_{zz}, $\eta_{\phi\phi}$, A_0 and B_0 given in (7.17), (7.27), (7.18) and (7.28), respectively. In essence, the requirement is for a lossy cylinder of large radius with incidence which is not too oblique, and the fact that the parameters η_{zz} etc. are the same as for normal incidence $(\beta = \pi/2)$ is consistent with Rytov's analysis where the illumination is unspecified.

7.2   Coated metallic cylinder

Approximate boundary conditions have proved very useful for modelling the effect of a coating applied to a metallic body. In the case of a perfectly conducting plane with a layer of either a low or higher contrast dielectric, the second order GIBC was given in Section 5.2 with the parameters identified in (5.60) and (5.63), respectively. We now employ the known solution for a uniformly coated right circular cylinder to determine the role played by the surface curvature, and then consider the accuracy of the resulting scattered field.

7.2.1   Low and high contrast coatings

A perfectly conducting cylinder of radius ρ_1 is covered with a layer of thickness $t = \rho_0 - \rho_1$ composed of a homogeneous dielectric whose complex refractive index is N. The body is illuminated by the H-polarised plane wave (7.1) or the E-polarised plane wave (7.9) incident in the direction of the negative x axis (see Fig. 7-6), and we seek to model the structure using second order boundary conditions applied at the outer surface $\rho = \rho_0$.

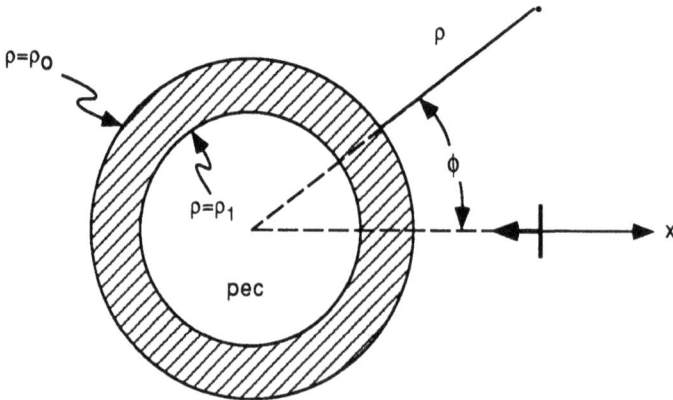

Figure 7–6: *Geometry for the coated circular cylinder*

Once again the exact solution of the scattering problem is known in the form of an eigenfunction expansion, and as shown by RUCK ET AL. (1970)

$$Q^H = -jZ\frac{H_m^{(1)'}(x_0) - \Gamma^H H_m^{(2)'}(x_0)}{H_m^{(1)}(x_0) - \Gamma^H H_m^{(2)}(x_0)} \tag{7.43}$$

$$Q^E = -jY\frac{H_m^{(1)'}(x_0) - \Gamma^E H_m^{(2)'}(x_0)}{H_m^{(1)}(x_0) - \Gamma^E H_m^{(2)}(x_0)} \tag{7.44}$$

where

$$\Gamma^H = \frac{H_m^{(1)'}(x_1)}{H_m^{(2)'}(x_1)}, \qquad \Gamma^E = \frac{H_m^{(1)}(x_1)}{H_m^{(2)}(x_1)} \tag{7.45}$$

with $x_0 = Nk_0\rho_0$ and $x_1 = Nk_0\rho_1$. Thus $t = (x_0 - x_1)/(Nk_0)$. For a second order GIBC, (7.15) and (7.23) become

$$Q^H = \eta_{zz} - \left(\frac{m}{\rho_0}\right)^2 A_0 \tag{7.46}$$

$$\frac{1}{Q^E} = \eta_{\phi\phi} + \left(\frac{m}{\rho_0}\right)^2 B_0 \tag{7.47}$$

There are several different approximations that can be made to cast the expressions for Q^H and Q^E into the forms required by (7.46) and (7.47), and we will describe two of them. If $x_0 - x_1 \ll x_0$ implying $t \ll \rho_0$ we have

$$H_m^{(1)}(x_1) = H_m^{(1)}(x_0) - Nk_0tH_m^{(1)'}(x_0) + O\left\{(Nk_0t)^2\right\}$$

and similarly for the other functions involved in (7.45). Then

$$Q^H = jZNk_0t\left\{1 - \left(\frac{m}{x_0}\right)^2\right\}\{1 + O(Nk_0t)\}$$

$$Q^E = -j\frac{Y}{Nk_0t}\{1 + O(Nk_0t)\}$$

and to the leading order in Nk_0t

$$\eta_{zz} = \eta_{\phi\phi} = jZ_0\mu_r k_0 t$$
$$A_0 = jZ_0\mu_r k_0t(Nk_0)^{-2}, \qquad B_0 = 0 \tag{7.48}$$

From (7.14) and (7.22) the resulting boundary conditions are

$$E_\phi = -jZ_0\mu_r k_0t\left\{1 + \frac{1}{(Nk_0)^2}\frac{1}{\rho_0^2}\frac{\partial^2}{\partial\phi^2}\right\}H_z \tag{7.49}$$
$$E_z = jZ_0\mu_r k_0t$$

and are identical to those obtained by the local application of the planar boundary condition (5.28) at each point of the cylindrical surface, with the parameters given the values (5.60) appropriate to a low contrast dielectric. To this order of approximation the curvature of the surface has no effect, and it is necessary to include terms $O\{(Nk_0t)^2\}$ to produce a curvature dependence.

If $|N| \gg 1$ the low contrast boundary conditions (7.49) are accurate only if k_0t is very small, and an alternative approximation is to replace each Hankel function in (7.43) and (7.44) by its asymptotic expansion for large argument. In the case of Q^H we have

$$
Q^H = jZ\left\{1 - \frac{4m^2 - 3}{8x_0^2} + \frac{4m^2 + 3}{16x_0^2} \cdot \frac{t}{\rho_0} + O\left(x_0^{-3}\right)\right\}
$$

$$
\cdot \left\{\frac{T' - \dfrac{4m^2 + 3}{8x_0^2} \cdot \dfrac{t}{\rho_0}}{1 + \dfrac{1}{2x_0}\left(1 + \dfrac{4m^2 + 3}{4x_0} \cdot \dfrac{t}{\rho_0}\right)T'} + O\left(x_0^{-3}\right)\right\}
$$

where $T' = \tan(Nk_0t)$, and thus to the leading order

$$
\eta_{zz} = jZT'\left(1 - \frac{T'}{2Nk_0\rho_0}\right)
$$

$$
A_0 = jZT'\frac{1}{2(Nk_0)^2}\left\{1 + \left(T' + \frac{1}{T'}\right)Nk_0t - \frac{t}{2\rho_0}\right\}
$$

(7.50)

Similarly

$$
\frac{1}{Q^E} = jZ\left\{1 + \frac{4m^2 - 1}{8x_0^2} - \frac{4m^2 - 1}{16x_0^2} \cdot \frac{t}{\rho_0} + O\left(x_0^{-3}\right)\right\}
$$

$$
\cdot \left\{\frac{T' - \dfrac{4m^2 - 1}{8x_0^2} \cdot \dfrac{t}{\rho_0}}{1 - \dfrac{1}{2x_0}\left(1 - \dfrac{4m^2 - 1}{4x_0} \cdot \dfrac{t}{\rho_0}\right)T'} + O\left(x_0^{-3}\right)\right\}
$$

giving

$$
\eta_{\phi\phi} = jZT'\left(1 + \frac{T'}{2Nk_0\rho_0}\right)
$$

$$
B_0 = jZT'\frac{1}{2(Nk_0)^2}\left\{1 - \left(T' + \frac{1}{T'}\right)Nk_0t - \frac{t}{2\rho_0}\right\}
$$

(7.51)

and the boundary conditions (7.14) and (7.22) are then

$E_\phi = -jZT'$

$$\cdot \left\{ 1 - \frac{T'}{2Nk_0\rho_0} + \frac{1}{2(Nk_0)^2} \left[1 + \left(T' + \frac{1}{T'} \right) Nk_0 t - \frac{t}{2\rho_0} \right] \frac{1}{\rho_0^2} \frac{\partial^2}{\partial\phi^2} \right\} H_z$$

(7.52)

$E_z = jZT'$

$$\cdot \left\{ 1 + \frac{T'}{2Nk_0\rho_0} - \frac{1}{2(Nk_0)^2} \left[1 - \left(T' + \frac{1}{T'} \right) Nk_0 t - \frac{t}{2\rho_0} \right] \frac{1}{\rho_0^2} \frac{\partial^2}{\partial\phi^2} \right\} H_\phi$$

As $k_0 t \to 0$, $(T' + 1/T')Nk_0 t \to 1$. To the leading order in $k_0 t$ the high contrast conditions (7.52) then reduce to the simpler low contrast ones (7.49), showing that (7.52) include (7.49) as a special case. A rough criterion for when (7.49) is adequate is $(T' + 1/T')Nk_0 \lesssim 1.25$, i.e. $|N|k_0 t \lesssim 0.5$.

For a planar surface the analogous boundary condition is (5.28) with the coefficients given in (5.63). If these high contrast parameters are approximated under the assumption that $Nk_0 t$ is of zeroth order with terms $O(N^{-2})$ negligible by comparison, and the boundary condition is then applied locally at each point of the cylindrical surface, the result is identical to (7.52) apart from the terms $T'/(2Nk_0\rho_0)$ and $t/(2\rho_0)$, which therefore represent the curvature corrections. For most coatings of practical interest, the corrections are only a few percent and can be neglected.

7.2.2 Accuracy

Since the scattering from a coated perfectly conducting cylinder has a complicated behaviour, it is convenient to start by looking at the exact solution. In Fig. 7–7(a) the modulus of the backscattered far field amplitude $P^H(0)$ is shown as a function of $k_0\rho_0$ and N for a lossless coating with $t = 0.05\lambda_0$. The dark regions correspond to field maxima and the light ones to field minima. The most prominent features are the "ridges" at approximately $N = 5$ and $N = 15$. In general, ridges are centred at

$$\text{Re. } N \simeq (2n - 1)\frac{\lambda_0}{4t} \qquad (n = 1, 2, 3, \ldots) \qquad (7.53)$$

implying that the separation (in Re. N) between successive ridges decreases with increasing t. As $k_0\rho_0$ increases, the first ridge (corresponding to $n = 1$) shifts to somewhat smaller Re. N, but the shift is much less pronounced for $n > 1$. Below the first ridge there is a gradual transition from the damped sinusoidal oscillation as a function of $k_0\rho_0$ characteristic of a conducting cylinder ($N \simeq 1$) to the sharp oscillations that typify the ridge, and we also observe the curling effect produced by the growing displacement of the maxima and minima as the ridge is approached from below. This occurs with all of the ridges, but as n increases, the effects are confined to a narrower range of Re. N, and the ridges become more compact. Another feature is the series of sharp resonances which emerge from the decaying oscillations above the ridges. These

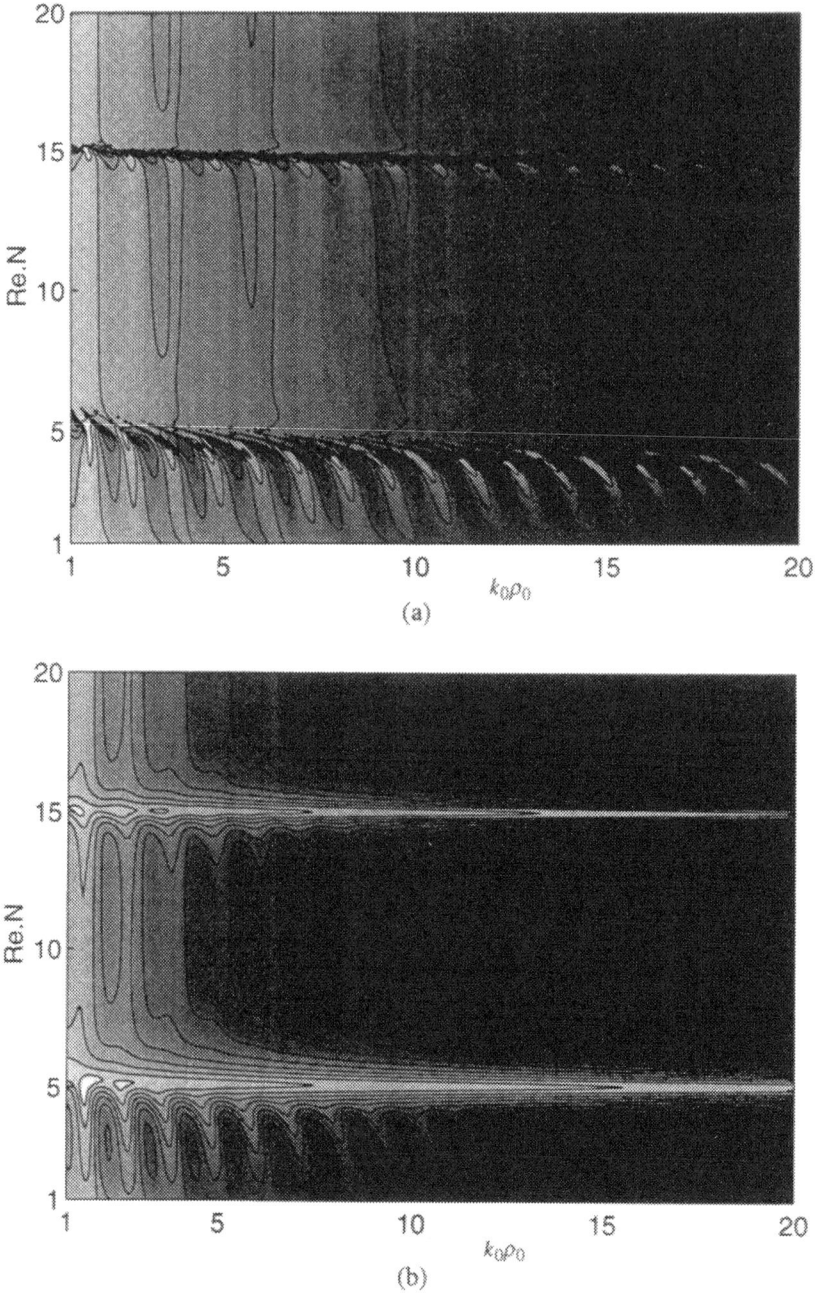

Figure 7–7: *Backscattered far field amplitude* $|P^{\mathrm{H}}(0)|$ *for a coated circular cylinder with* $t = 0.05\lambda_0$ *and (a) N real, (b) Im. N = −0.4. The dark regions correspond to field maxima and the light ones to field minima*

show up as black dots in Fig. 7–7(a). The resonances associated with a given sequence become ever sharper away from the ridge and are therefore increasingly difficult to detect numerically. Beyond these, i.e. for larger Re. N, the field is almost monotonic as a function of $k_0\rho_0$, but as Re. N increases towards the next ridge, the field behaviour described above repeats itself. For Re. $N \simeq \lambda_0/2t$, midway between the ridges, the field is similar to that for a perfectly conducting cylinder of radius $\rho_1 = \rho_0 + t$. The analogous results for a lossy coating are illustrated in Fig. 7–7(b) where Im. $N = -0.4$. We observe that the loss serves to suppress the sharp resonances and to diffuse the ridge-like structure.

Because of the complicated field behaviour, it is almost impossible to quantify the accuracy of the approximate boundary conditions developed in Section 7.2.1, and we have chosen instead to rely on a presentation similar to that in Fig. 7–7. Thus, Fig. 7–8(a) corresponds to Fig. 7–7(a) and shows the percentage error in $|P^H(0)|$ associated with the second order GIBC (7.52). Black denotes an error of 5 percent or more, and the successive shades are in steps of one percent down to white where the error is less than 1 percent. Not surprisingly, the GIBC fails in the immediate vicinity of the ridges, but otherwise the boundary condition is remarkably accurate even for values of $k_0\rho_0$ as small as unity. For a lossy coating (see Fig. 7–8(b)) the performance is even better, particularly for the larger $k_0\rho_0$, and if $k_0\rho_0 \gtrsim 9$ the GIBC provides an accuracy of better than 5 percent for all Re. N. The results are similar for other $t \lesssim 0.25\lambda_0$.

The planar version of the boundary condition is obtained by suppressing the terms in T'/ρ_0 and t/ρ_0 in (7.52), and the error plots analogous to those in Fig. 7–8 are shown in Fig. 7–9. As expected the errors are larger, especially for small $k_0\rho_0$ and/or Re. N, but above the first ridge the GIBC is accurate to within a few percent for a wide range of $k_0\rho_0$. The performance is certainly superior to that of the first order boundary condition obtained by eliminating the terms $O(N^{-2})$ in (7.52), and this is evident from Fig. 7–10. For the bistatic field the errors are similar to those shown in Figs. 7–8 to 7–10 except at the angles ϕ corresponding to deep minima in the field, and this is also true for the surface field.

7.3   High frequency solution

The geometrical theory of diffraction (GTD) and its uniform (UTD) version are powerful techniques for calculating the field scattered by a body whose radii of curvature are electrically large. We now consider the UTD solution for a circular cylinder subject to a second order GIBC imposed at its surface, and indicate the extension to an arbitrary convex cylinder. We follow the treatment given by SYED AND VOLAKIS (1991).

If the incident field u^i is either the H-polarised plane wave (7.1) or the

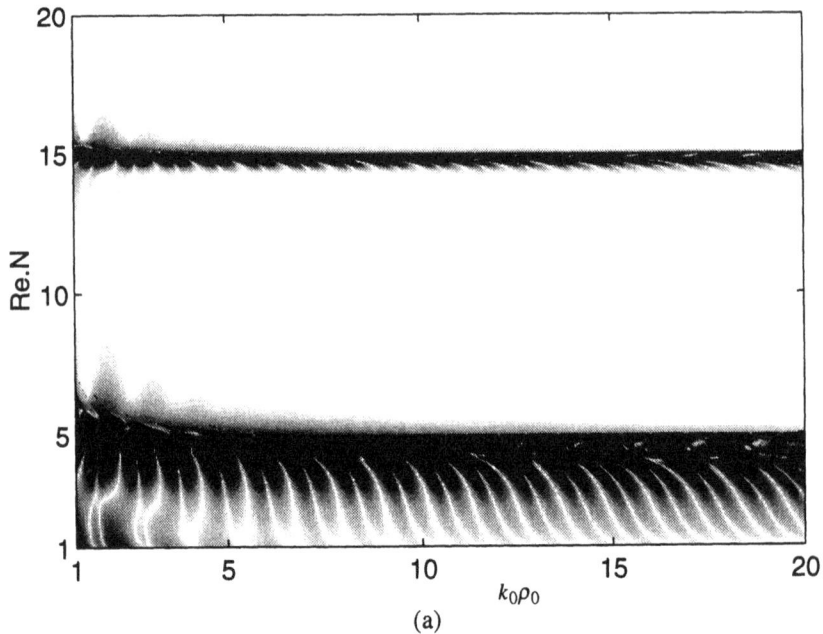

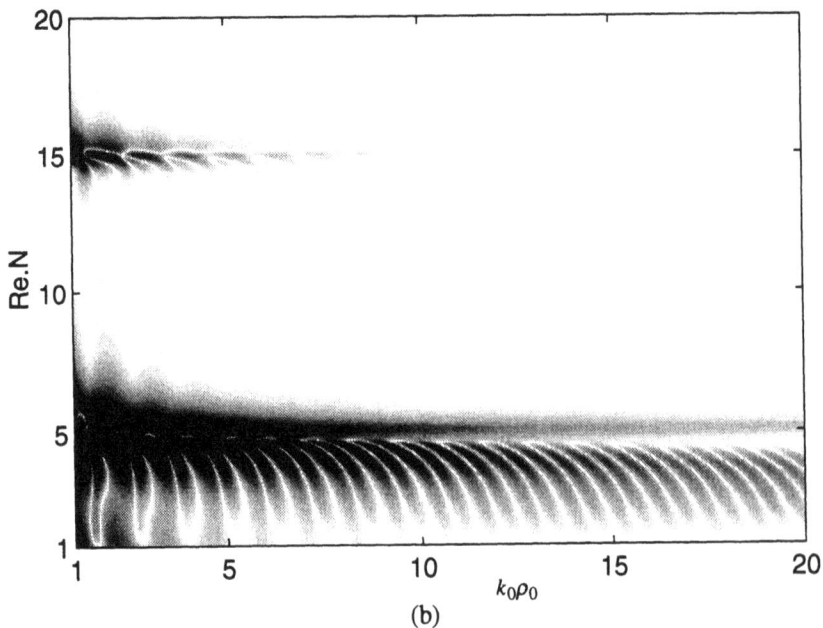

Figure 7–8: *Percentage error in the backscattered far field amplitude* $|P^{\mathrm{H}}(0)|$ *for the second order GIBC (7.53) when* $t = 0.05\lambda_0$ *and (a) N real, (b) Im. N =* -0.4. *The errors range from 5% or more (black) to 1% or less (white)*

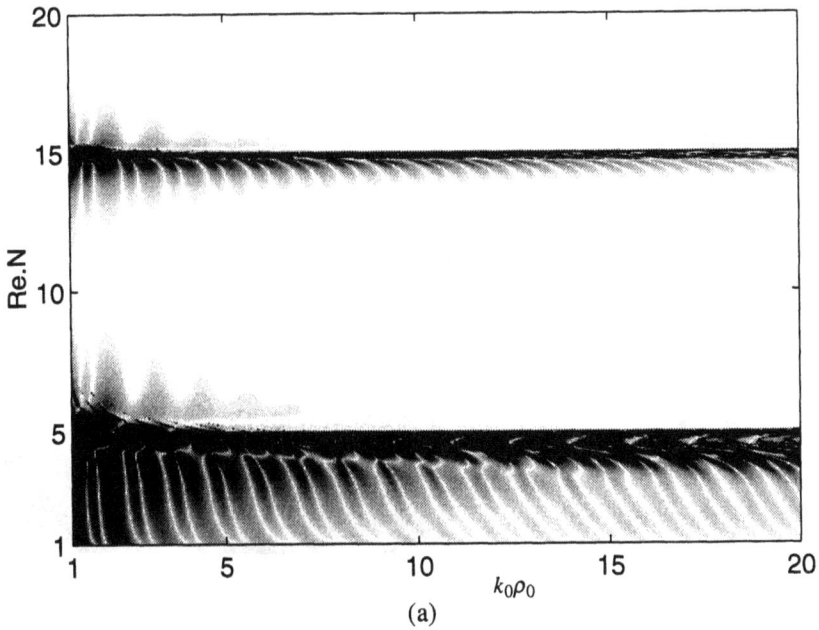

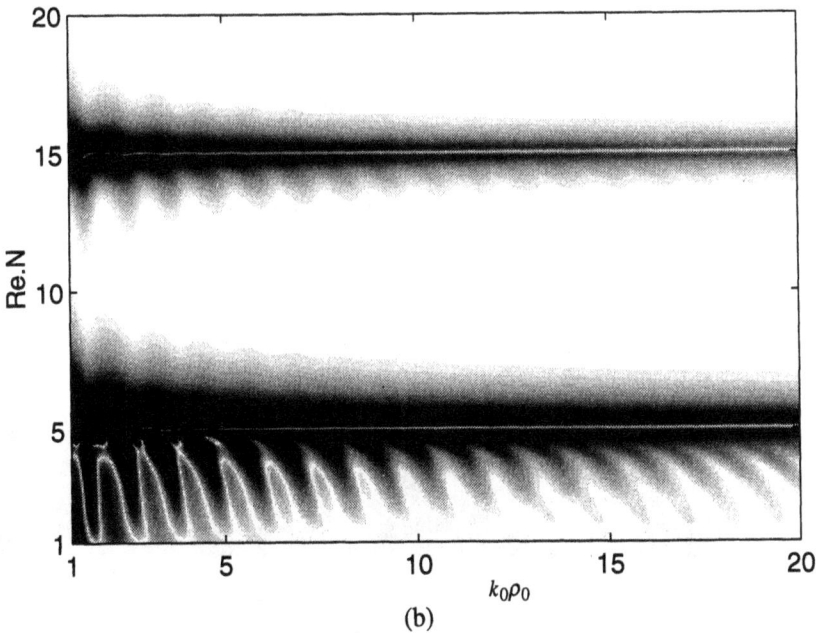

Figure 7-9: *Percentage error in the backscattered far field amplitude $|P^H(0)|$ for the planar version of (7.53) with the terms in T'/ρ_0 and t/ρ_0 eliminated. The coating thickness is $t = 0.05\lambda_0$ with (a) N real, (b) Im. $N = -0.4$. The errors range from 5% or more (black) to 1% or less (white)*

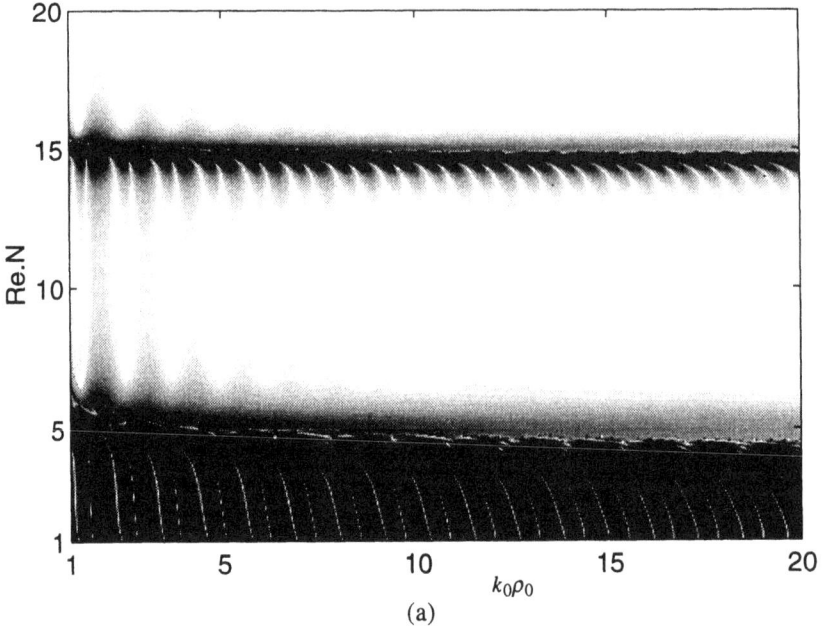

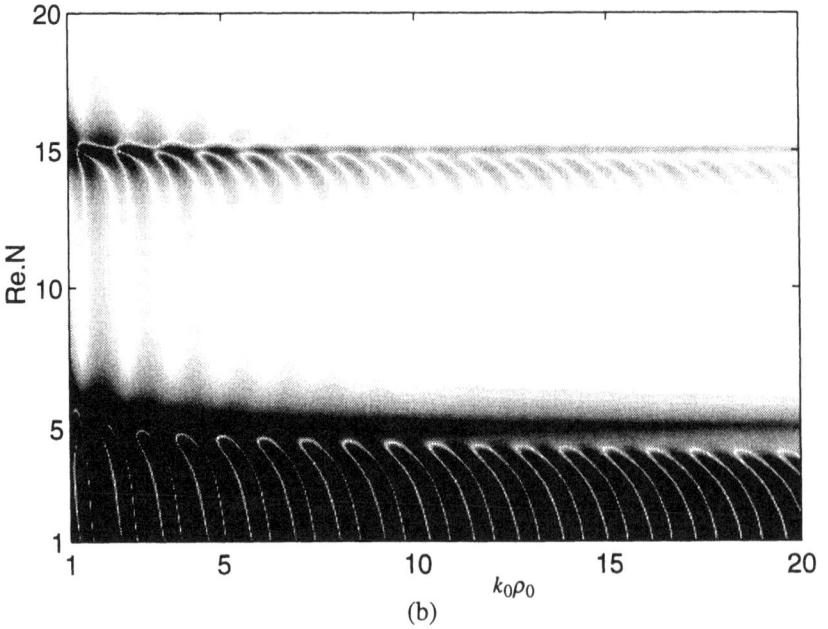

Figure 7–10: *Percentage error in the backscattered far field amplitude $|P^{\mathrm{H}}(0)|$ for the SIBC, i.e. (7.53) with terms $O(N^{-2})$ eliminated. The coating thickness is $t = 0.05\lambda_0$ with (a) N real, (b) Im. $N = -0.4$. The errors range from 5% or more (black) to 1% or less (white)*

E-polarised wave (7.9), the total field u ($= H_z$ or E_z) in $\rho \geq \rho_0$ is given in (7.2) or (7.10), respectively, and this can be written as

$$u = \sum_{m=-\infty}^{\infty} j^m \left\{ J_m(k_0\rho) - \frac{J'_m(k_0\rho_0) + S(m) J_m(k_0\rho_0)}{H_m^{(2)'}(k_0\rho_0) + S(m) H_m^{(2)}(k_0\rho_0)} H_m^{(2)}(k_0\rho) \right\} e^{-jm\phi}$$

(7.54)

where

$$S(m) = \begin{cases} -jY_0 Q^{\mathrm{H}} & \text{for H-polarisation} \\ -jZ_0 Q^{\mathrm{E}} & \text{for E-polarisation} \end{cases}$$

is specified by the boundary condition at $\rho = \rho_0$. For a second order GIBC

$$S(m) = -j \left\{ \frac{a_0 + a_2}{a_1} - \frac{a_0}{a_1} \left(\frac{m}{k_0\rho_0} \right)^2 \right\}$$

for H-polarisation and

$$S(m) = -j \left\{ \frac{a'_0 + a'_2}{a'_1} - \frac{a'_0}{a'_1} \left(\frac{m}{k_0\rho_0} \right)^2 \right\}$$

for E-polarisation. For an SIBC $a_0 = a'_0 = 0$. In the case of a coated metallic cylinder, the parameters a_i and a'_i, $i = 0, 1, 2$, can be deduced from (7.46), (7.47), (7.50) and (7.51) and, if the corresponding planar boundary condition is imposed locally, the parameters are those given in (5.63).

By application of a Watson transformation (BOWMAN ET AL., 1987) to (7.54) we obtain

$$u = u_1 + u_2$$

(7.55)

with

$$u_1 = \int_{-\infty}^{\infty} \left\{ H_\nu^{(1)}(k_0\rho) - \frac{H_\nu^{(1)'}(k_0\rho_0) + S(\nu) H_\nu^{(1)}(k_0\rho_0)}{H_\nu^{(2)'}(k_0\rho_0) + S(\nu) H_\nu^{(2)}(k_0\rho_0)} H_\nu^{(2)}(k_0\rho) \right\} e^{-j\nu\psi} \, d\nu$$

(7.56)

$$u_2 = \int_{-\infty}^{\infty} \left\{ J_\nu(k_0\rho) - \frac{J'_\nu(k_0\rho_0) + S(\nu) J_\nu(k_0\rho_0)}{H_\nu^{(2)'}(k_0\rho_0) + S(\nu) H_\nu^{(2)}(k_0\rho_0)} H_\nu^{(2)}(k_0\rho) \right\}$$
$$\cdot \left\{ e^{-j\nu(2\pi+\phi)} + e^{-j\nu(2\pi-\phi)} \right\} \frac{e^{j\nu\pi/2}}{1 - e^{-j2\pi\nu}} \, d\nu$$

where

$$\psi = |\phi| - \frac{\pi}{2}$$

The partial field u_1 includes the geometrical optics and dominant surface wave contributions, and u_2 is the contribution of those creeping waves which have made at least one circuit of the cylinder. For most purposes u_2 is negligible, and to evaluate u_1 it is necessary to consider separately the three regions shown in Fig. 7–11.

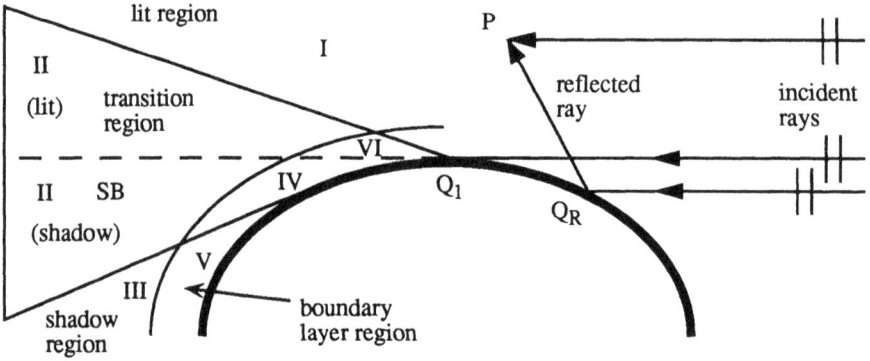

Figure 7–11: *Different spatial regions for plane wave scattering from a smooth convex cylinder*

7.3.1   Lit and deep shadow regions

In the lit region I the incident and reflected rays of geometrical optics provide an accurate first order approximation the total field and, from an asymptotic evaluation of (7.56),

$$u(P) \simeq u^i(P) + u^i(Q_{\mathrm{R}}) R \sqrt{\frac{\tilde{\rho}}{\tilde{\rho} + \ell}} \, e^{-jk_0\ell} \qquad (7.57)$$

(FELSEN AND MARCUVITZ, 1973), where R is the reflection coefficient given in (5.4) or (5.5) with $\phi = \pi/2 - \theta^i$. The angle θ^i is shown in Fig. 7–12, ℓ is the distance from the reflection point Q_{R} to the observation point P, assumed

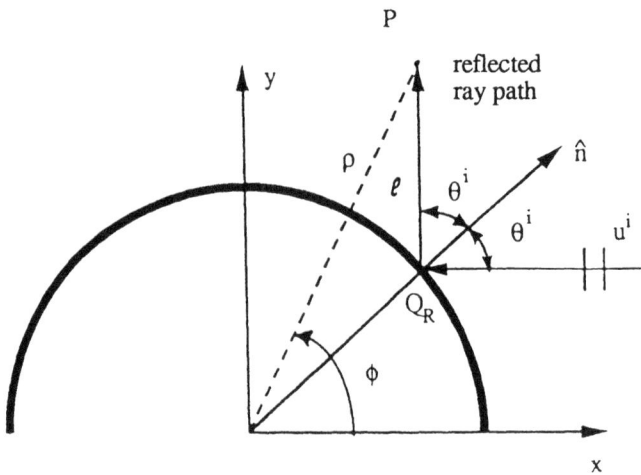

Figure 7–12: *Reflected ray path*

large, and

$$\tilde{\rho} = \tfrac{1}{2}\rho_0 \cos \theta^{\mathrm{i}}$$

is the distance from Q_{R} to the caustic of the reflected rays. For a general convex cylinder, ρ_0 must be replaced by the radius of curvature $\rho_{\mathrm{s}}(Q_{\mathrm{R}})$ of the surface at Q_{R}.

When the observation point is in the deep shadow region III, a residue series evaluation of (7.56) gives

$$u_1 = -\frac{4}{k_0\rho_0} \sum_{n=1}^{\infty} \frac{H_{\nu_n}^{(2)}(k_0\rho)}{H_{\nu_n}^{(2)}(k_0\rho_0)} \left\{ \frac{\partial}{\partial \nu} \left[H_{\nu}^{(2)'}(k_0\rho_0) + S(\nu)\, H_{\nu}^{(2)}(k_0\rho_0) \right]_{\nu=\nu_n} \right\}^{-1}$$
$$\cdot e^{-j\nu_n(\phi - \pi/2)} \qquad (7.58)$$

where ν_n are the zeros of the transcendental equation

$$H_{\nu_n}^{(2)'}(k_0\rho_0) + S(\nu_n)\, H_{\nu_n}^{(2)}(k_0\rho_0) = 0$$

A ray theory description now follows by inserting the Debye approximations to the Hankel functions in (7.58), and is

$$u_1(P) \simeq u^{\mathrm{i}}(Q_1)\, T \frac{e^{-jk_0 s}}{\sqrt{s}} \qquad (7.59)$$

where

$$T = -\sum_{n=1}^{\infty} \mathcal{D}_n(Q_1)\, e^{-j\nu_n \theta}\, \mathcal{D}_n(Q_2)$$

is the diffraction coefficient for a circular cylinder,

$$\mathcal{D}(Q_1) = \mathcal{D}(Q_2) =$$
$$e^{j\pi/4} \left\{ H_{\nu_n}^{(2)}(k_0\rho_0) \frac{\partial}{\partial \nu} \left[H_{\nu_n}^{(2)'}(k_0\rho_0) + S(\nu)\, H_{\nu_n}^{(2)}(k_0\rho_0) \right]_{\nu=\nu_n} \right\}^{-1} \qquad (7.60)$$

and θ is the angle shown in Fig. 7-13. For a general convex cylinder the required substitutions are

$$\rho_0 \rightarrow \rho_{\mathrm{s}}(Q_{1,2})$$

and

$$e^{-j\nu_n \theta} \rightarrow \exp\left\{ -j \int_{Q_1}^{Q_2} \frac{\nu_n(t')}{\rho_{\mathrm{s}}(t')} \, dt' \right\}$$

where t' is the distance along the surface of the cylinder. As a result of these changes, the attachment coefficient $\mathcal{D}_n(Q_1)$ and the launching coefficient $\mathcal{D}_n(Q_2)$ may differ.

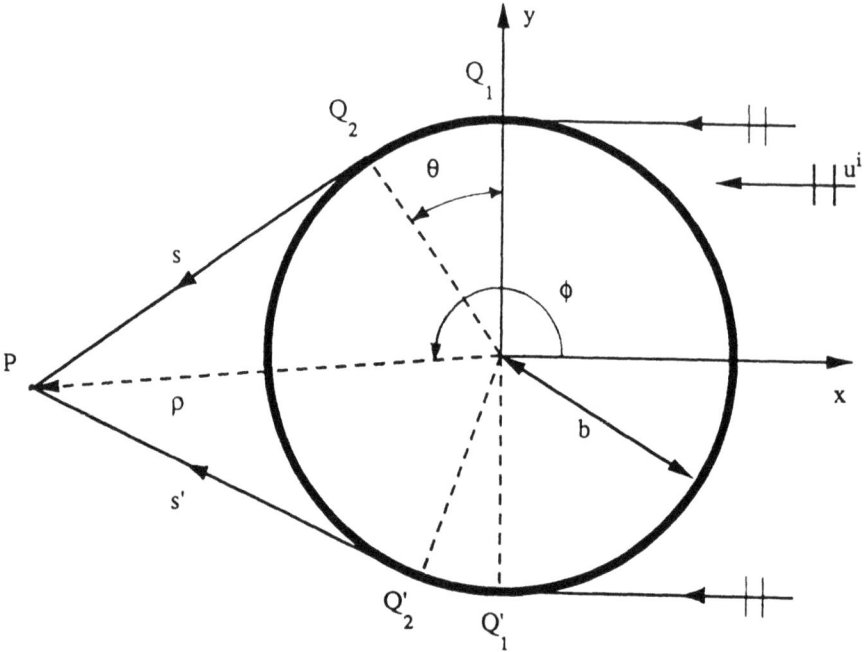

Figure 7-13: *Propagation ray paths in the shadow region*

7.3.2   Transition region

In the transition region II about the shadow boundary the field changes rapidly as a function of angle, and the geometrical optics and creeping wave solutions (7.57) and (7.59) are not sufficient. Alternative (uniform) expressions are necessary here, and these can be deduced from the results developed by PATHAK (1979) for a perfectly conducting convex cylinder.

In the lit portion of the region

$$u_1(P) \simeq u^i(P) + u^i(Q_R)\, R \sqrt{\frac{\tilde{\rho}}{\tilde{\rho}+\ell}}\, e^{-jk_0\ell} \qquad (7.61)$$

where R is now

$$R = -\sqrt{\frac{4}{x}} \exp\left(-j\frac{x^3}{12}\right) \left\{ \frac{e^{-j\pi/4}}{2x\sqrt{\pi}} \left[1 - F_{\mathrm{KP}}(2k_0\ell \cos^2\theta^i)\right] + G(x,q) \right\}$$

with

$$x = -2m(Q_R)\cos\theta^i$$

and

$$m(Q_R) = \left\{ \tfrac{1}{2} k_0\, \rho_s(Q_R) \right\}^{1/3}$$

The function F_{KP} is the UTD transition function defined in (3.60) and

$$G(x, q) = \frac{e^{-j\pi/4}}{\sqrt{\pi}} \int_{-\infty}^{\infty} \frac{V'(\tau) - q(\tau) V(\tau)}{W_2'(\tau) - q(\tau) W_2(\tau)} e^{-jx\tau} \, d\tau$$

where $W_{1,2}(\tau)$ are the Fock-type Airy functions

$$W_{1,2}(\tau) = \frac{1}{\sqrt{\pi}} \int_{\Gamma_{1,2}} e^{\tau u - u^3/3} \, du$$

$$V(\tau) = \frac{1}{2j} \{W_1(\tau) - W_2(\tau)\}$$

and the prime denotes the derivative. The contour Γ_1 runs from $\infty e^{-j2\pi/3}$ to $\infty - j\epsilon$, and Γ_2 is the complex conjugate of Γ_1. We remark that, in the case of an SIBC, $q(\tau)$ is a constant.

In the shadow portion of the region

$$u_1(P) \simeq u^i(Q_1) T \frac{e^{-jk_0 s}}{\sqrt{s}} \tag{7.62}$$

where the diffraction coefficient T is now defined as

$$T = -\sqrt{\frac{2}{k_0}} \, m(Q_1) \, m(Q_2) \, e^{-jk_0 t} \left\{ \frac{e^{-j\pi/4}}{2x\sqrt{\pi}} [1 - F_{\mathrm{KP}}(k_0 s a)] + G(x, q) \right\}$$

with

$$t = \int_{Q_1}^{Q_2} dt'$$

$$x = \int_{Q_1}^{Q_2} \frac{m(t')}{\rho_s(t')} \, dt'$$

and

$$a = x^2 \{2 \, m(Q_1) \, m(Q_2)\}^{-1}$$

All of these results were first derived for a circular cylinder and then generalised to an arbitrary smooth convex cylinder.

7.3.3   Near-surface field

If the observation point is close to the surface of the cylinder as shown in Fig. 7-14, ℓ is not large and the previous results are no longer valid. To treat this case, the Bessel and Hankel functions in (7.56) are expressed in terms of Airy functions which are then expanded in Taylor series. By a process similar to that used by PATHAK (1979), it can be shown that for P_N in the lit region (see Fig. 7-14)

$$u_1(P) \simeq u^i(P) \left\{ e^{-jhz} - \sum_{n=0}^{5} \frac{(-1)^n}{n!} (jhz)^n + e^{-jz^3/3} [\Lambda_1(z) - \Lambda_2(z)] \right\} \tag{7.63}$$

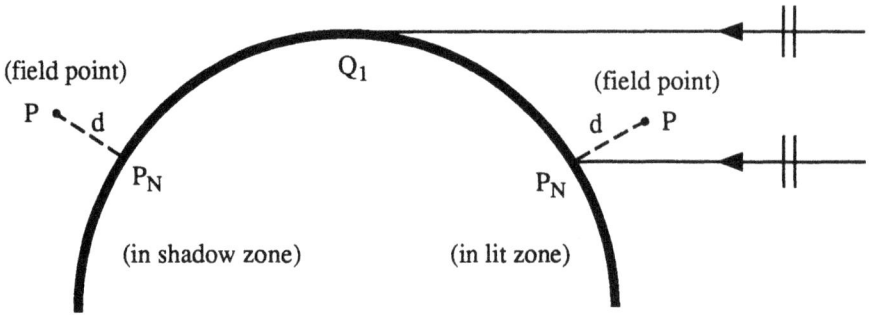

Figure 7–14: *Projection of the field point P in the direction of the normal to the surface at P_N*

with

$$\Lambda_1(z) = g_1(z) + j\frac{h^2}{2}\,g_1'(z) - \frac{h^3}{3!}\,g_1(z) - \frac{h^4}{4!}\,g_1''(z) - 4j\frac{h^5}{5!}\,g_1'(z)$$

$$\Lambda_2(z) = h\,g_2(z) + j\frac{h^3}{3}\,g_2'(z) - 2\frac{h^4}{4!}\,g_2(z) - \frac{h^5}{5!}\,g_2''(z)$$

$$g_1(z) = \frac{1}{\sqrt{\pi}}\int_{-\infty}^{\infty}\frac{e^{-jz\tau}}{W_2'(\tau) - q(\tau)\,W_2(\tau)}\,d\tau$$

$$g_2(z) = \frac{1}{\sqrt{\pi}}\int_{-\infty}^{\infty}\frac{q(\tau)\,e^{-jz\tau}}{W_2'(\tau) - q(\tau)\,W_2(\tau)}\,d\tau$$

$$h = \frac{k_0 d}{m(P_n)} = \frac{k_0}{m(P_N)}\{\rho - \rho_s(P_N)\}$$

and

$$z = -m(P_n)\cos\theta^i$$

When P_N is in the shadow region

$$u_1(P) \simeq u^i(Q_1)\,e^{-jk_0 t}\left\{\frac{\rho_s(P_N)}{\rho_s(Q_1)}\right\}^{-1/6}[\Lambda_1(x) - \Lambda_2(x)] \qquad (7.64)$$

with

$$t = \int_{Q_1}^{P_N} dt'$$

and

$$x = \int_{Q_1}^{P_N}\frac{m(t')}{\rho_s(t')}\,dt'$$

The functions g_1, g_2 and G are Fock-type integrals which can be computed using the method developed by PEARSON (1987). In the particular case when the boundary condition is an SIBC, $g_2 = qg_1$ since q is a constant.

If the creeping and/or surface waves are only slowly attenuating, it may be necessary to improve the accuracy of (7.63). This can be done by adding the appropriate form of (7.64) representing the contribution of the creeping wave which has travelled the least distance to reach P_N. For even more accuracy, the contribution u_2 of those waves which have travelled at least once around the cylinder can also be included. In Fig. 7–15 the UTD expressions for the total

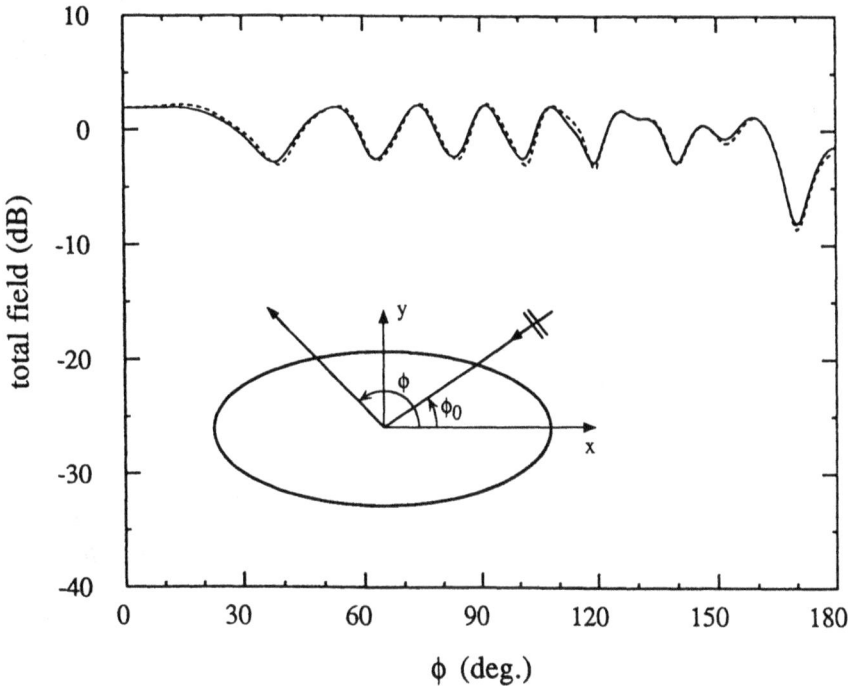

Figure 7–15: *Elliptic cylinder (semi axes $2\lambda_0$ and λ_0) subject to a second order GIBC simulating a coating with $t = 0.2\lambda_0$, $\epsilon_r = 8$ and $\mu_r = 1$, illuminated by the H-polarised plane wave (7.1). The total field is computed at $\rho = 5\lambda_0$ using a moment method (——) and the UTD expressions (- - - -)*

field of an elliptic cylinder computed at a distance $\rho = 5\lambda_0$ from its centre are compared with a moment method solution. In both instances a GIBC was used simulating a coating $0.2\lambda_0$ thick with $\epsilon_r = 8$ and $\mu_r = 1$, and the agreement is excellent. For the same coating applied to a circular cylinder with $\rho_0 = 3\lambda_0$, the UTD expressions (7.63) and (7.64) for the total field computed at a distance $0.05\lambda_0$ above the surface are compared with the eigenfunction solution in Fig. 7–16. The same GIBC was used and, from the studies described in Section 7.2.2, it is known that this is accurate to about 2 percent. Although the

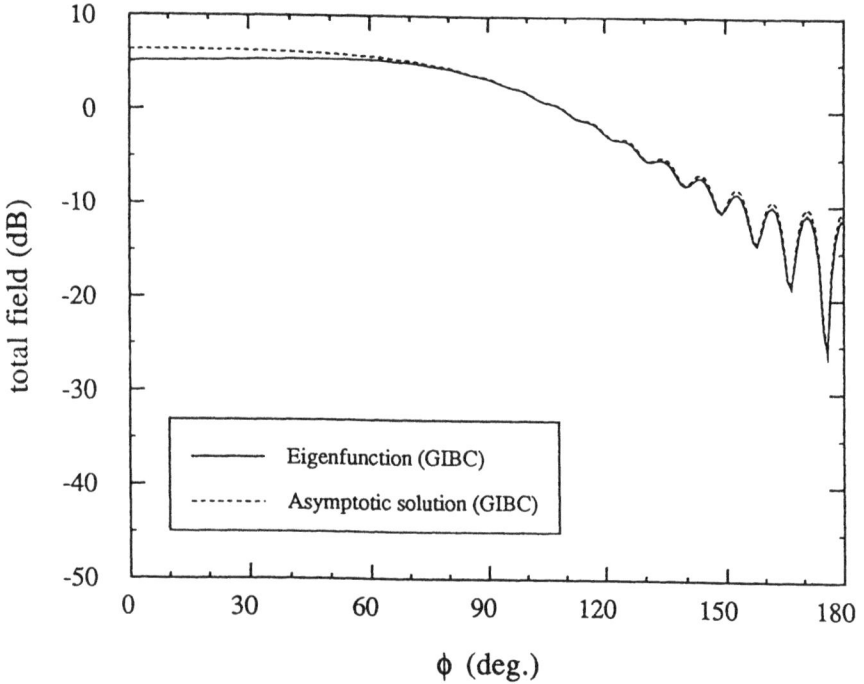

Figure 7–16: *Circular cylinder with $\rho_0 = 3\lambda_0$ subject to a second order GIBC simulating a coating with $t = 0.2\lambda_0$, $\epsilon_r = 8$ and $\mu_r = 1$, illuminated by the H-polarised plane wave (7.1). The total field is computed at $\rho = 3.05\lambda_0$ using the eigenfunction expansion (——) and the UTD expressions (- - - -)*

discrepancies in Fig. 7–16 are larger than this, particularly in the lit region, the agreement is still good. Additional comparisons of the UTD and eigenfunction solutions have been given by SYED AND VOLAKIS (1991).

7.4   Edges and junctions

We close this chapter by commenting on the more complicated problem of edges and junctions where the task is to determine the field diffracted by the edge of (a segment of) a curved cylindrical impedance surface, or by a discontinuity in the impedance of a curved surface. The geometries are illustrated in Fig. 7–17.

Assuming that the impedance is modelled by an SIBC, the solution is possible in certain cases, but requires the use of more sophisticated analytical techniques including the Bessel/Hankel transform. For the cylindrical edge of

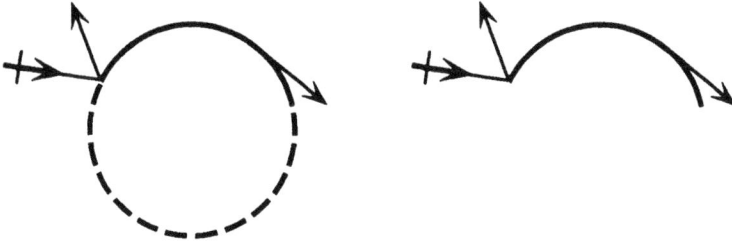

Figure 7–17: *Impedance junctions and edges of a cylindrical surface*

a perfectly conducting surface, IDEMEN AND FELSEN (1981) reduced the mixed boundary value problem to a matrix Hilbert equation, which was then factored into upper and lower matrices to determine the angular spectra. A generalisation of this method, namely the Wiener-Hopf-Hilbert method (HURD, 1976), was later used by BÜYÜKAKSOY AND UZGÖREN (1988) to factor the matrix equation for an impedance surface, and thereby obtain the diffraction coefficients for a cylindrical impedance edge. These same authors also derived (BÜYÜKAKSOY AND UZGÖREN, 1987) the diffraction coefficients for an impedance junction on a cylindrical surface, and MOLINET (1995) has recently found the solution for an impedance surface with a discontinuity in curvature. Not surprisingly, the diffraction coefficients for all these types of discontinuity are quite complicated, and they have yet to be used effectively in problems of practical interest.

References

Bowman, J. J., Senior, T. B. A. and Uslenghi, P. L. E. (1987), *Electromagnetic and Acoustic Scattering by Simple Shapes*, Hemisphere Pub. Corp., New York, pp. 34–36, 53.

Büyükaksoy, A. and Uzgören, G. (1987), "High-frequency scattering from the impedance discontinuity on a cylindrically curved surface", *IEEE Trans. Antennas Propagat.*, **AP-35**, pp. 234–236.

Büyükaksoy, A. and Uzgören, G. (1988), "Diffraction of high-frequency waves by a cylindrically curved surface with different face impedances", *IEEE Trans. Antennas Propagat.*, **AP-36**, pp. 690–695.

Felsen, L. B. and Marcuvitz, N. (1973), *Radiation and Scattering of Waves*, Prentice-Hall, Inc., Englewood Cliffs NJ, pp. 693–697.

Hurd, R. A. (1976), "The Wiener-Hopf-Hilbert method for diffraction problems", *Can. J. Phys.*, **54**, pp. 775–780.

Idemen, M. and Felsen, L. B. (1981), "Diffraction of a whispering gallery mode by the edge of a thin cylindrically curved surface", *IEEE Trans. Antennas Propagat.*, **AP-29**, pp. 571–579.

Molinet, F. (1995), "Uniform asymptotic solution for the diffraction by a discontinuity in curvature", *Ann. des Telecomm*.

Pathak, P. H. (1979), "An asymptotic analysis of the scattering of plane waves by a smooth convex cylinder", *Radio Sci.*, **14**, pp. 419–435.

Pearson, L. W. (1987), "A scheme for automatic computation of Fock-type integrals", *IEEE Trans. Antennas Propagat.*, **AP-35**, pp. 1111–1118.

Ruck, G. T., Barrick, D. E., Stuart, W. D. and Krichbaum, C. K. (1970), *Radar Cross Section Handbook*, Plenum Press, New York, pp. 239 *et seq.*, pp. 273 *et seq.*

Syed, H. H. and Volakis, J. L. (1991), "High-frequency scattering by a smooth coated cylinder simulated with generalized impedance boundary conditions", *Radio Sci.*, **26**, pp. 1305–1314.

Wang, D.-S. (1987), "Limits and validity of the impedance boundary condition on penetrable surfaces", *IEEE Trans. Antennas Propagat.*, **AP-35**, pp. 453–457.

Chapter 8

Absorbing boundary conditions

8.1 Introduction

The GIBCs discussed in the preceding chapters were designed to simulate the surface properties of a scatterer, thereby eliminating the need to consider fields interior to the body. There is, however, another purpose for approximate boundary conditions, and this is to create a boundary which does not perturb a field incident upon it—in effect, to simulate a surface which is actually not there. The resulting conditions can be regarded as GIBCs for non-reflecting surfaces, and are generally referred to as absorbing boundary conditions (ABCs). They are of growing importance in numerical work where they are used to terminate the computational domain in a finite element (SILVESTER AND FERRARI, 1990) or finite difference (KUNZ AND LUEBBERS, 1993) solution of the wave equation (see Fig. 8–1). In considering their two- and three-dimensional forms, emphasis will be placed on second order ABCs because of their extensive use in scattering and radiation problems.

Two general methods for deriving ABCs have been described in the literature, along with a number of variations of each. Regardless of the method, the goal is to construct a local differential operator which minimises the reflection coefficient for any wave impinging on the surface where the ABC is applied. After a survey of the various methods, we then present a third one based on Rytov's derivation of a GIBC described in Appendix A. In this case the general form of a second order ABC for two-dimensional fields is

$$\frac{\partial U}{\partial n} = \alpha U - \frac{\partial}{\partial s}\left(\beta \frac{\partial U}{\partial s}\right) \tag{8.1}$$

(see (5.74)) where s and n are shown in Fig. 8–1, and α and β may be functions of s. The aim now is to specify α and β such that the reflection coefficient (5.66) is minimised. Similarly, for three-dimensional fields the general form is

$$\hat{n} \times \mathbf{E} = \hat{n} \times (\hat{n} \times \{\bar{\bar{\eta}}.\mathbf{H} - \nabla(A\nabla_s.\mathbf{H}) - \hat{n} \times \nabla(B\hat{n}.\nabla \times \mathbf{H})\}) \tag{8.2}$$

(see (5.83)) and its dual. Here also the coefficients $\bar{\bar{\eta}}$, A and B may be functions of position on the surface, and they must be determined to minimise the

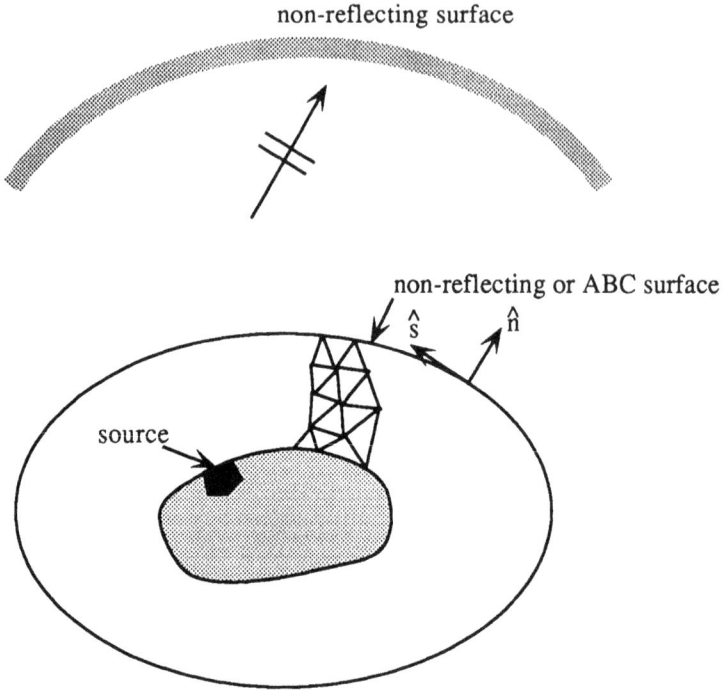

Figure 8–1: *Illustration of a non-reflecting surface and its use for terminating a finite element mesh*

reflection.

In the one-way wave equation method the coefficients are chosen so that (8.1) or (8.2) is satisfied only by those waves propagating beyond a boundary such as that illustrated in Fig. 8–2. Ideally, enforcement of an ABC allows these waves to pass through the surface without reflection, but this can only

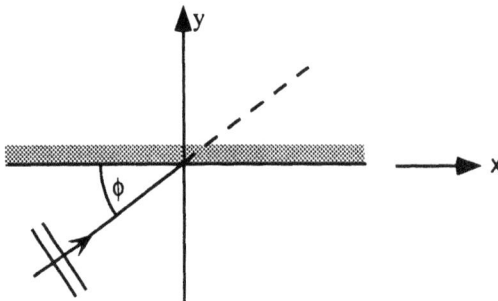

Figure 8–2: *Wave impinging on an ABC boundary $y = constant$*

be achieved precisely with an ABC of infinite order. Since this is impractical, the order must be finite, and this leads to approximations in simulating
a non-reflecting body. The first useful ABCs of second (and higher) order
were developed by ENGQUIST AND MAJDA (1977), and their well-posedness was
later examined by TREFETHEN AND HALPERN (1986). However, it appears that
LINDMAN (1975) was the first to use this approach for deriving ABCs, and
the concept of a one-way differential equation was introduced as early as 1970
(CLAERBOUT, 1985) for seismic wave analysis.

   The second general method is based on an assumed representation of the
field in terms of outgoing waves in which the field is expanded in inverse
powers of a large parameter and the coefficients are then determined by substitution back into the wave equation. In two dimensions the procedure was
first employed by BAYLISS AND TURKEL (1980). In three dimensions, WEBB
AND KANELLOPOULOS (1989) used a similar procedure to derive a second order
ABC for spherical surfaces, and for a general curvilinear surface the analogous
condition was developed by CHATTERJEE AND VOLAKIS (1993).

   In the following the various ABCs are presented and then considered in the
context of Rytov's GIBC.

8.2   Two-dimensional ABCs

8.2.1   One-way wave equation method

The essence of this method is to factor the wave equation into a product of
two operators each of which is associated with one of the solutions of the wave
equation.

Cartesian coordinates

For simplicity we consider first the planar surface $y = 0$ shown in Fig. 8–2.
The ABC (8.1) then becomes

$$\frac{\partial U}{\partial y} = \alpha U - \beta \frac{\partial^2 U}{\partial x^2} \tag{8.3}$$

and, assuming no z dependence, the wave equation

$$\left(\frac{\partial^2}{\partial x^2} + \frac{\partial^2}{\partial y^2} + k_0^2 \right) U = 0$$

can be factored as

$$\left(\frac{\partial}{\partial y} - L \right) \left(\frac{\partial}{\partial y} + L \right) U = 0 \tag{8.4}$$

where

$$L = jk_0 \sqrt{1 + \frac{1}{k_0^2} \frac{\partial^2}{\partial x^2}} \tag{8.5}$$

is a pseudodifferential operator. Clearly (8.4) implies

$$\left(\frac{\partial}{\partial y} - L\right) U = 0 \quad \text{or} \quad \left(\frac{\partial}{\partial y} + L\right) U = 0$$

and the second of these is satisfied by waves propagating in the positive y direction alone. Consequently, the condition

$$\left(\frac{\partial}{\partial y} + L\right) U = 0 \tag{8.6}$$

perfectly absorbs all waves incident on the surface from below.

The key to deriving a finite order ABC from (8.6) is the approximation of (8.5) by a finite series. The simplest approximation is the two-term Taylor series expansion

$$\sqrt{1 + \frac{1}{k_0^2}\frac{\partial^2}{\partial x^2}} \simeq 1 + \frac{1}{2k_0^2}\frac{\partial^2}{\partial x^2} \tag{8.7}$$

which leads to the classic ENGQUIST AND MAJDA (1977) second order ABC

$$\left(\frac{\partial}{\partial y} + jk_0 + \frac{j}{2k_0}\frac{\partial^2}{\partial x^2}\right) U = 0 \tag{8.8}$$

Comparison with (8.3) shows

$$\alpha = -jk_0, \qquad \beta = \frac{j}{2k_0} \tag{8.9}$$

and the corresponding plane wave reflection coefficient is

$$R = -\left(\frac{1 - \sin\phi}{1 + \sin\phi}\right)^2 \tag{8.10}$$

As evident from the derivation, the form (8.8) is that appropriate for a surface $y = $ constant and, if the ABC is to be enforced on $x = $ constant instead, x and y must be interchanged (see Fig. 8–3).

Higher order approximations to the radical in (8.7) were developed by ENGQUIST AND MAJDA (1977) and later by HALPERN AND TREFETHEN (1988). Among them, the Padé approximants of L are quite popular. In this case

$$\sqrt{1 - w^2} = \frac{p_m(w)}{q_n(w)} \tag{8.11}$$

where $w = (j/k_0)(\partial/\partial x)$ and $p_m(w)$ and $q_n(w)$ are polynomials of degree m and n, respectively. The rational function on the right hand side of (8.11) is referred to as the (m, n) Padé approximant. As an example, the $(2, 0)$ Padé approximant is

$$\sqrt{1 - w^2} = 1 - \tfrac{1}{2}w^2 \tag{8.12}$$

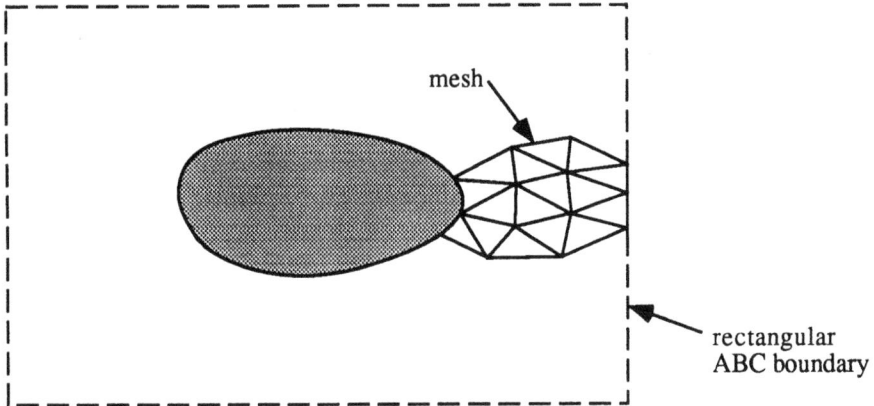

Figure 8–3: *Typical boundary for enforcing the ABC (8.8)*

and recovers the two-term Taylor series expansion (8.7). Other approximations of the radical can be obtained using interpolation techniques such as
the Chebyshev method and the method of least squares, and the $(2,2)$ Padé
approximant (MOORE ET AL., 1988) is

$$\sqrt{1 - w^2} = \frac{1 - \frac{3}{4}w^2}{1 - \frac{1}{4}w^2} \tag{8.13}$$

This leads to the third order ABC

$$\left(\frac{1}{4k_0^2} \frac{\partial^3}{\partial y\, \partial x^2} + j\frac{3}{4k_0} \frac{\partial^2}{\partial x^2} + \frac{\partial}{\partial y} + jk_0 \right) U = 0 \tag{8.14}$$

which can be cast into the form (6.1).
    A more general version of (8.13) is

$$\sqrt{1 - w^2} = \frac{1 - cw^2}{1 - \left(c - \frac{1}{2}\right)w^2} \tag{8.15}$$

valid for any c. When $c = \frac{1}{2}$ we recover (8.12) but, for other values of c, (8.15)
leads to the third order ABC

$$\left(\frac{c - \frac{1}{2}}{k_0^2} \frac{\partial^3}{\partial y\, \partial x^2} + j\frac{c}{k_0} \frac{\partial^2}{\partial x^2} + \frac{\partial}{\partial y} + jk_0 \right) U = 0 \tag{8.16}$$

This is a generalisation of (8.14) and reduces to it when $c = 0.75$. The reflection
coefficient implied by (8.16) is

$$R = -\left(\frac{1 - \sin\phi}{1 + \sin\phi} \right)^2 \left(\frac{\frac{c-1}{c-0.5} + \sin\phi}{\frac{c-1}{c-0.5} - \sin\phi} \right) \tag{8.17}$$

and we observe that for $0 \leq \phi \leq \pi$ (see Fig. 8–2), $|R| \leq 1$ only if $0.5 \leq c \leq 1$. Thus, in the context of a GIBC, the surface is no longer passive for other values of c. The reflection coefficient (8.17) differs from (8.10) in having an additional zero at the angle $\phi = \sin^{-1}\left(\frac{1-c}{c-0.5}\right)$, and this is a real angle if $0.75 \leq c \leq 1$. By choosing c appropriately in this range we can therefore place the zero at any angle we desire, and this is illustrated in Fig. 8–4. As c increases, the zero

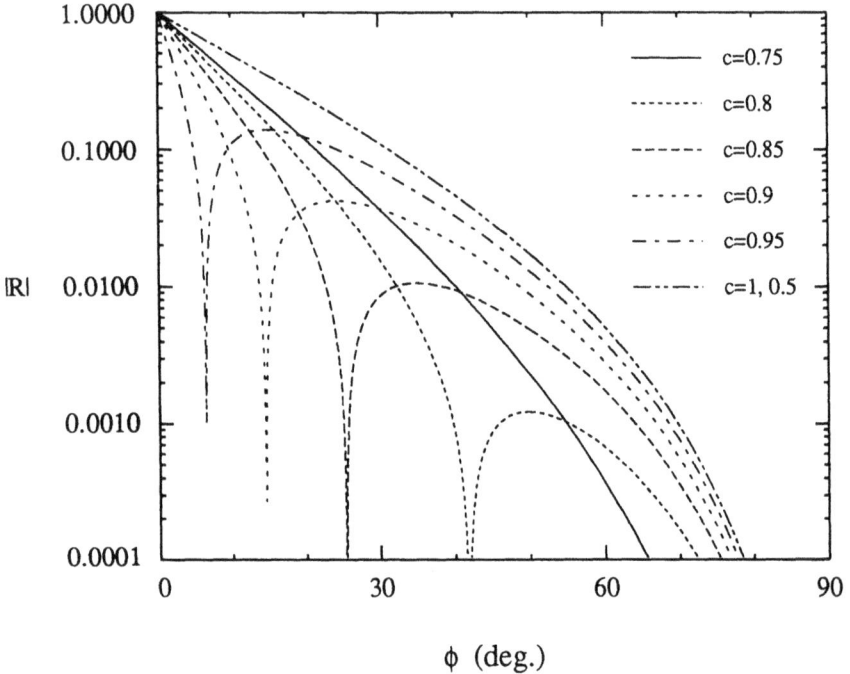

Figure 8–4: *The reflection coefficient (8.17) as a function of ϕ*

approaches $\phi = 0$, but when $c = 1$ the reflection coefficient reverts to (8.10). However, because of the narrowness of the null, the overall best performance is achieved when $c = 0.75$ corresponding to the $(2, 2)$ Padé approximant, and for this

$$R = -\left(\frac{1 - \sin\phi}{1 + \sin\phi}\right)^{3} \tag{8.18}$$

To improve the performance still further, it is logical to seek higher order Padé approximants, and McINTURFF AND SIMON (1993) have developed closed form expressions for the Padé approximants of any order. Using the Lindman series representation

$$\sqrt{1 - w^{2}} = 1 - \sum_{m=1}^{M} \frac{a_{m} w^{2}}{1 - b_{m} w^{2}} \tag{8.19}$$

they showed that

$$a_m = \frac{1}{2M+1}\left\{1 + \cos\frac{(2m-1)\pi}{2M+1}\right\}$$

$$b_m = \frac{1}{2}\left\{1 - \cos\frac{(2m-1)\pi}{2M+1}\right\}$$

(8.20)

for $m = 1, 2, \ldots, M$. The ABC implied by (8.17) has order $2M + 1$ and is therefore always an odd order condition (if $M \geq 1$). The corresponding reflection coefficient is

$$R = -\left(\frac{1 - \sin\phi}{1 + \sin\phi}\right)^{2M+1}$$

(8.21)

and this is shown in Fig. 8–5 for $M = 0.5$, 1 and 2. It is obvious that increasing

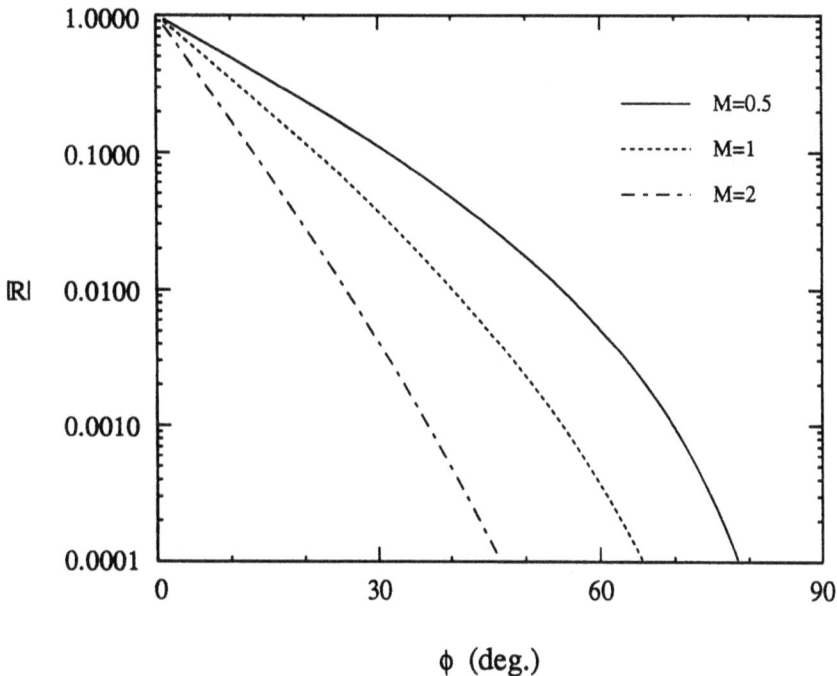

Figure 8–5: *The reflection coefficients (8.10) and (8.21) for $M = 1$ and 2*

M reduces R at the expense of a higher order boundary condition.

An alternative method was considered by HIGDON (1992) who chose to write the ABC in the form (6.6) originally proposed by KARP AND KARAL (1965).

He required that (8.6) be satisfied for M plane waves incident at the angles $\phi = \phi_m$, $m = 1, 2, \ldots, M$, and since, for each of the plane waves,

$$L = jk_0 \sqrt{1 + \frac{1}{k_0^2} \frac{\partial^2}{\partial x^2}} = jk_0 \sin \phi_m$$

the ABC

$$\prod_{m=1}^{M} \left(\frac{\partial}{\partial y} + jk_0 \sin \phi_m \right) = 0 \tag{8.22}$$

ensures that all of the plane waves are perfectly absorbed. Unfortunately, the numerical implementation of (8.22) leads to instabilities and, to remedy these, (8.22) is replaced by (FANG, 1993)

$$\prod_{m=1}^{M} \left(\frac{\partial}{\partial y} + jk_0 \sin \phi_m + \epsilon_m \right) = 0 \tag{8.23}$$

The ϵ_m are referred to as damping factors and are generally chosen such that $0.01/\Delta y < \epsilon_m < 0.5/\Delta y$ where Δy is the element size in the computational domain (ESWARAPPA AND HOEFER, 1994).

Alternative higher order ABCs have been presented by LIAO ET AL. (1984) and these must also be stabilised (MOGHADDAM AND CHEW, 1991; LUEBBERS AND PENNEY, 1994) prior to their application.

Cylindrical coordinates

In circular cylindrical coordinates the wave equation is

$$\left(\frac{\partial^2}{\partial \rho^2} + \frac{1}{\rho} \frac{\partial}{\partial \rho} + \frac{1}{\rho^2} \frac{\partial^2}{\partial \phi^2} + k_0^2 \right) U = 0 \tag{8.24}$$

where we have again assumed that there is no z dependence, and this can be written as (LEE ET AL., 1990)

$$\left\{ \left(\frac{\partial}{\partial \rho} + \frac{1}{2\rho} \right)^2 + \frac{1}{4\rho^2} + \frac{1}{\rho^2} \frac{\partial^2}{\partial \phi^2} + k_0^2 \right\} U = 0 \tag{8.25}$$

In a similar manner to (8.5) we factor this as

$$\left(\frac{\partial}{\partial \rho} + \frac{1}{2\rho} - L \right) \left(\frac{\partial}{\partial \rho} + \frac{1}{2\rho} + L \right) U = 0 \tag{8.26}$$

where L is again a pseudodifferential operator. On equating (8.25) and (8.26) we find

$$\frac{1}{\rho^2} \frac{\partial^2}{\partial \phi^2} + k_0^2 + \frac{1}{4\rho^2} = \frac{\partial L}{\partial \rho} - (L)^2 \tag{8.27}$$

If L is expanded as (ENGQUIST AND MAJDA, 1977)

$$L = L^{(1)} + L^{(0)} + L^{(-1)} + L^{(-2)} + \cdots \qquad (8.28)$$

each of the $L^{(n)}$ can be determined recursively, with $L^{(1)} = O(1)$, $L^{(0)} = O(\rho^{-3})$, and so on. To obtain $L^{(1)}$ we equate the left hand side of (8.27) to terms which are $O(1)$. Assuming that $\partial L^{(1)}/\partial \rho = O(\rho^{-3})$, to be confirmed later, we set

$$\frac{1}{\rho^2} \frac{\partial^2}{\partial \phi^2} + k_0^2 + \frac{1}{4\rho^2} = -(L^{(1)})^2 + O(\rho^{-3})$$

and thus

$$L^{(1)} = jk_0 \sqrt{1 + \frac{1}{k_0^2 \rho^2} \frac{\partial^2}{\partial \phi^2} + \frac{1}{4k_0^2 \rho^2}} + O(\rho^{-3})$$

giving

$$L^{(1)} = jk_0 \left\{ 1 + \frac{1}{2(k_0\rho)^2} \left(\frac{\partial^2}{\partial \phi^2} + \frac{1}{4} \right) \right\} \qquad (8.29)$$

$L^{(0)}$ is specified by the terms $O(\rho^{-3})$ in (8.27) and, when these are collected, we find

$$\frac{\partial L^{(1)}}{\partial \rho} - L^{(1)}L^{(0)} - L^{(0)}L^{(1)} = 0$$

Assuming $L^{(0)}$ and $L^{(1)}$ commute,

$$L^{(0)} = \frac{1}{2L^{(1)}} \frac{\partial L^{(1)}}{\partial \rho}$$

and, retaining only the leading term in the denominator, we have

$$L^{(0)} = \frac{1}{2k_0^2 \rho^3} \left(\frac{\partial^2}{\partial \phi^2} + \frac{1}{4} \right) \qquad (8.30)$$

Hence

$$L \simeq L^{(1)} + L^{(0)} \simeq jk_0 + \frac{j}{2k_0\rho^2} \left(1 + \frac{1}{jk_0\rho} \right) \left(\frac{\partial^2}{\partial \phi^2} + \frac{1}{4} \right) \qquad (8.31)$$

accurate to $O(\rho^{-3})$, and from (8.26) the corresponding ABC is the second order one

$$\left\{ \frac{\partial}{\partial \rho} + \frac{1}{2\rho} + jk_0 + \frac{j}{2k_0\rho^2} \left(1 - \frac{1}{jk_0\rho} \right) \left(\frac{\partial^2}{\partial \phi^2} + \frac{1}{4} \right) \right\} U = 0 \qquad (8.32)$$

A more commonly used form is

$$\left\{ \frac{\partial}{\partial \rho} + \frac{1}{2\rho} + jk_0 + \frac{j}{2k_0\rho^2} \frac{\frac{\partial^2}{\partial \phi^2} + \frac{1}{4}}{1 + \frac{1}{jk_0\rho}} \right\} U = 0 \qquad (8.33)$$

which can be obtained from (8.32) by the approximation

$$1 - \frac{1}{jk_0\rho} = \left(1 + \frac{1}{jk_0\rho}\right)^{-1} + O\left\{(k_0\rho)^{-2}\right\} \qquad (8.34)$$

The second order ABC (8.33) is due to LEE ET AL. (1990) and is more accurate than the simpler form derived by ENGQUIST AND MAJDA (1977). The latter did not contain the term $\frac{1}{4}$ added to $\partial^2/\partial\phi^2$, and this has proved to be important for numerical purposes.

The ABCs given above must be enforced on the cylindrical boundary $\rho =$ constant shown in Fig. 8-6, and this leads to large computational domains

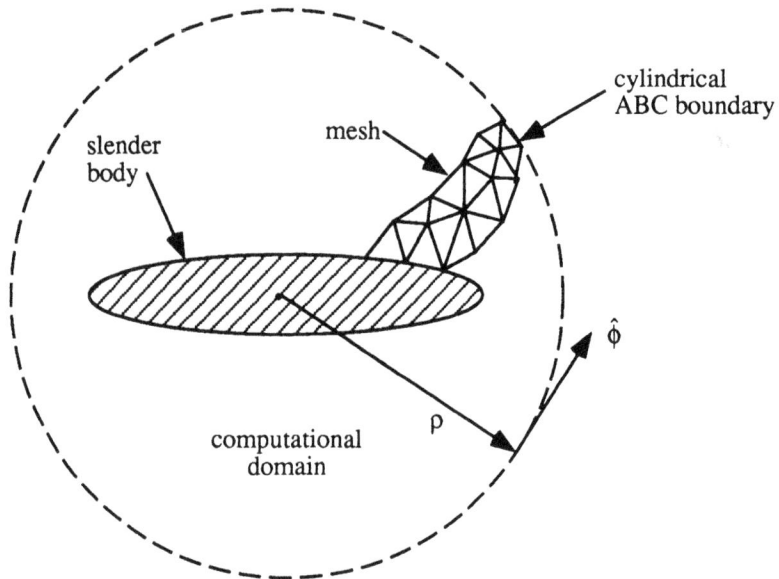

Figure 8–6: *A circular cylindrical boundary enclosing a slender body*

when modelling the field around slender bodies. To reduce the domain size, LEE ET AL. (1990) also developed the ABC analogous to (8.33) for the boundary of an elliptic cylinder, and we present this next.

Elliptical coordinates

In terms of the elliptic cylinder coordinates u, v, z where

$$x = \frac{d}{2}\cosh u \cos v, \qquad y = \frac{d}{2}\sinh u \sin v$$

with $0 \le u < \infty$ and $0 \le v < 2\pi$, and assuming no z dependence, the wave equation can be written as

$$\left\{ \frac{\partial^2}{\partial n^2} + \frac{2 \sinh u \cosh u}{d(\cosh^2 u - \cos^2 v)^{3/2}} \frac{\partial}{\partial n} + \frac{4}{d^2(\cosh^2 u - \cos^2 v)} \frac{\partial^2}{\partial v^2} + k_0^2 \right\} U = 0$$

(8.35)

where d is the interfocal distance of the family of ellipses $u =$ constant, n is the outward normal to the surface, and

$$\frac{\partial}{\partial n} = \frac{2}{d \left(\cosh^2 u - \cos^2 v \right)^{1/2}} \frac{\partial}{\partial u}$$

The coordinate system is shown in Fig. 8–7 and we remark that

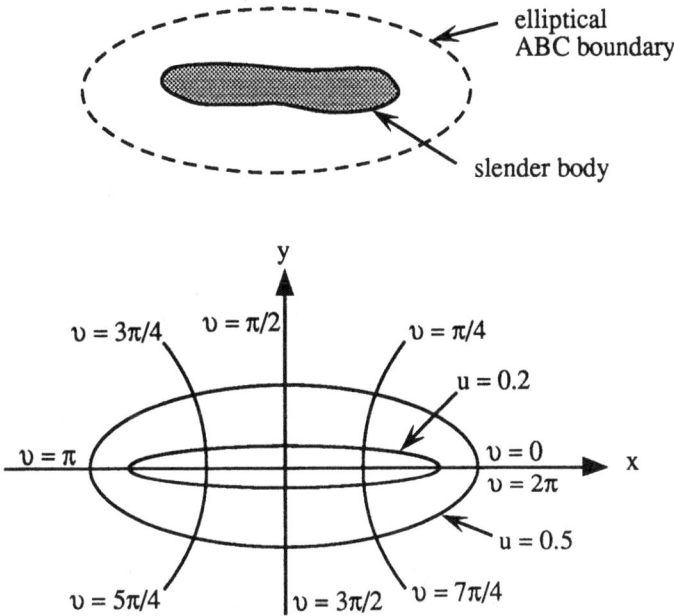

Figure 8–7: *Elliptical ABC boundary and the associated coordinates*

$$\frac{d}{2} \frac{\left(\cosh^2 u - \cos^2 v \right)^{3/2}}{\sinh u \cosh u} = \rho_c$$

(8.36)

where ρ_c is the radius of curvature of the ellipse. In the limit $u = \infty$ the ellipse becomes a circle.

Following the procedure used for the circular cylinder, we seek a pseudo-differential operator L such that (8.35) can be written as

$$\left(\frac{\partial}{\partial n} + \frac{\sinh u \cosh u}{d(\cosh^2 u - \cos^2 v)^{3/2}} - L \right)$$

$$\cdot \left(\frac{\partial}{\partial n} + \frac{\sinh u \cosh u}{d(\cosh^2 u - \cos^2 v)^{3/2}} + L \right) U = 0 \qquad (8.37)$$

and this leads to the second order ABC

$$\left\{ \frac{\partial}{\partial n} + \frac{1}{2\rho_c} + jk_0 + \frac{j}{2k_0\rho_c^2} \frac{\dfrac{\cosh^2 u - \cos^2 v}{\sinh^2 u \cosh^2 u} \dfrac{\partial^2}{\partial v^2} + \dfrac{1}{4}}{1 + \dfrac{1}{jk_0\rho_c}} \right\} U = 0 \qquad (8.38)$$

(LEE ET AL., 1990), which clearly reduces to (8.33) as $u \to \infty$. Although this is only applicable on an elliptical boundary, it suggests the possibility of a similar condition in arbitrary (body fitted) coordinates. This is addressed in Section 8.2.3.

8.2.2   Mode annihilation method

The method is based on an assumed expansion of the field in a series of terms each representing an outward propagating wave. The series must, of course, be convergent and recover the far zone field behaviour, and to see where the terminology comes from, we note that in the Sommerfeld radiation condition

$$\sqrt{\rho} \left(\frac{\partial U}{\partial \rho} + jk_0 U \right) \to 0 \qquad (8.39)$$

the operator annihilates the outward travelling cylindrical "mode" $e^{-jk_0\rho}/\sqrt{\rho}$ to $O(\rho^{-3/2})$.

The general method for deriving higher order ABCs using mode annihilating operators was introduced by BAYLISS AND TURKEL (1980) and BAYLISS ET AL. (1982). They began with the Wilcox-type expansion

$$U = \frac{e^{-jk_0\rho}}{\sqrt{\rho}} \sum_{m=0}^{\infty} \frac{a_m(\phi)}{\rho^m} \qquad (8.40)$$

(WILCOX, 1956; KARP, 1961) where the coefficients a_m are independent of ρ. By substituting this expansion into the wave equation (8.24), it can be shown that a_m satisfy the recursion relationship (MITTRA ET AL., 1989)

$$-2jk_0(m+1)a_{m+1} = \left(m + \tfrac{1}{2} \right)^2 a_m + \frac{\partial^2}{\partial \phi^2} a_m \qquad (8.41)$$

From (8.40)

$$\begin{aligned}
\frac{\partial U}{\partial \rho} &= -jk_0 U - \frac{e^{-jk_0\rho}}{\rho^{3/2}} \sum_{m=0}^{\infty} \frac{\left(m + \tfrac{1}{2} \right) a_m}{\rho^m} \\
&= -jk_0 U - \frac{e^{-jk_0\rho}}{2\rho^{3/2}} \sum_{m=0}^{\infty} \frac{a_m}{\rho^m} - \frac{e^{-jk_0\rho}}{\rho^{3/2}} \sum_{m=0}^{\infty} \frac{m a_m}{\rho^m} \\
&= \left(-jk_0 - \frac{1}{2\rho} \right) U - \frac{e^{-jk_0\rho}}{\rho^{5/2}} \sum_{m=1}^{\infty} \frac{m a_m}{\rho^{m-1}} \qquad (8.42)
\end{aligned}$$

and by using the recursion relationship (8.41) twice, we find

$$\left\{\frac{\partial}{\partial\rho} + \frac{1}{2\rho} + jk_0 + \frac{j}{2k_0\rho^2}\left(1 - \frac{1}{jk_0\rho}\right)\left(\frac{\partial^2}{\partial\phi^2} + \frac{1}{4}\right)\right\}U = 0 + O(\rho^{-9/2}) \quad (8.43)$$

Apart from the order term, this is identical to the second order ABC (8.32) obtained by the one-way wave equation method.

BAYLISS AND TURKEL (1980) also provided an alternative derivation based on the recursion relation

$$B_m U = \left(\frac{\partial}{\partial\rho} + jk_0 + \frac{4m-3}{2\rho}\right)B_{m-1}U \quad (8.44)$$

where the B_m are referred to as annihilating differential operators with

$$B_1 U = \left(\frac{\partial}{\partial\rho} + jk_0 + \frac{1}{2\rho}\right)U \quad (8.45)$$

It can be shown that (8.44) is satisfied by the expansion (8.40), and based on (8.40)

$$B_1 U = 0 + O(\rho^{-5/2}) \quad (8.46)$$

This is a first order ABC and is more accurate than the Sommerfeld radiation condition.

Higher order ABCs can be obtained using (8.44) in conjunction with the wave equation. By applying the operator B_m to (8.40) it is found that

$$B_m U = 0 + O(\rho^{-2m-1/2}) \quad (8.47)$$

showing that B_m annihilates the first m terms of expansion (8.40). From (8.44) and (8.45) we have

$$B_2 U = \left\{\frac{\partial}{\partial\rho} + \frac{1}{2\rho} + jk_0 + \frac{j}{2k_0\rho^2}\frac{\frac{\partial^2}{\partial\phi^2} + \frac{1}{4}}{1 + \frac{1}{jk_0\rho}}\right\}U = 0 + O(\rho^{-9/2}) \quad (8.48)$$

and this is often referred to as the second order Bayliss-Turkel ABC. It is the same as (8.33), and a comparison with the generic form (8.1) shows

$$\alpha = -jk_0\left\{1 + \frac{3}{2jk_0\rho} - \frac{3}{8(k_0\rho)^2}\right\}\left(1 + \frac{1}{jk_0\rho}\right)^{-1}$$

$$\beta = \frac{j}{2k_0}\left(1 + \frac{1}{jk_0\rho}\right)^{-1} \quad (8.49)$$

with $\partial^2/\partial s^2$ replaced by $\rho^{-2}\,\partial^2/\partial\phi^2$.

One of the first applications of (8.48) to a scattering problem was reported by PETERSON AND CASTILLO (1989). It has been demonstrated that (8.48) is more accurate than the Engquist-Majda ABC (8.8), which must be enforced on a rectangular boundary as illustrated in Fig. 8–3.

8.2.3   Method of successive approximations

Even the Bayliss-Turkel ABC (8.48) has the disadvantage that it can only be enforced on a circular cylindrical boundary as illustrated in Fig. 8–6, but MA (1991) has shown that it is possible to develop analogous conditions in curvilinear coordinates. The method used has some similarities to the annihilating operator method, but is more accurately described as a method of successive approximations.

We begin by writing the wave equation in a general Dupin coordinate system (TAI, 1992). Denoting the orthogonal coordinates as v_1, v_2, v_3 and the associated vectors as $\hat{v}_1$, $\hat{v}_2$, $\hat{v}_3$,

$$\nabla^2 U = \frac{1}{h_1 h_2 h_3} \sum_{i=1}^{3} \frac{\partial}{\partial v_i} \left(\frac{h_1 h_2 h_3}{h_i^2} \frac{\partial U}{\partial v_i} \right)$$

where h_1, h_2, h_3 are the corresponding metric coefficients. By analogy with the coordinate system α, β, γ used in Appendix A where γ was normal to the boundary, we now choose v_3 in this direction. Then $\hat{v}_3 = \hat{n}$ and, in accordance with the Dupin system, $h_3 = 1$. If v_2 is the axial coordinate so that $h_2 = 1$ with $\partial/\partial v_2 = 0$, the wave equation becomes

$$\frac{\partial}{\partial n} \left(h_1 \frac{\partial U}{\partial n} \right) + \frac{\partial}{\partial v_1} \left(\frac{1}{h_1} \frac{\partial U}{\partial v_1} \right) + h_1 k_0^2 U = 0 \tag{8.50}$$

From Fig. 8–7 we observe that $h_1 = n + a$ and

$$\frac{1}{h_1} \frac{\partial}{\partial v_1} = \frac{\partial}{\partial s} = \frac{1}{1 + n/a} \frac{\partial}{\partial \ell}$$

where a is the local radius of curvature and may be a function of s (see Fig. 8–8). With these substitutions, (8.50) can be written as

$$\frac{\partial}{\partial n} \left\{ \left(1 + \frac{n}{a} \right) \frac{\partial U}{\partial n} \right\} + \frac{\partial}{\partial \ell} \left\{ \frac{1}{1 + n/a} \frac{\partial U}{\partial \ell} \right\} + k_0^2 \left(1 + \frac{n}{a} \right) U = 0 \tag{8.51}$$

and this is the starting point for Ma's derivation.

To proceed further it is assumed that the field propagating away from the scattering surface has the form

$$U(s, n) = A_0 e^{-jk_0 n} f(s, n) \tag{8.52}$$

This implies that the phase front of the scattered field conforms to the surface of the scatterer, and we note the similarity of this assumption to that made by Rytov (see Appendix A). When (8.52) is inserted into (8.51), we obtain

$$\left\{ 2jk_0 - \frac{1}{2(n+a)} \right\} \left\{ \frac{\partial f}{\partial n} + \frac{f}{2(n+a)} \right\} =$$
$$\left\{ \frac{a}{n+a} \frac{\partial}{\partial \ell} \left(\frac{a}{n+a} \frac{\partial f}{\partial \ell} \right) + \frac{f}{4(n+a)^2} \right\} + \frac{\partial}{\partial n} \left\{ \frac{\partial f}{\partial n} + \frac{f}{2(n+a)} \right\} \tag{8.53}$$

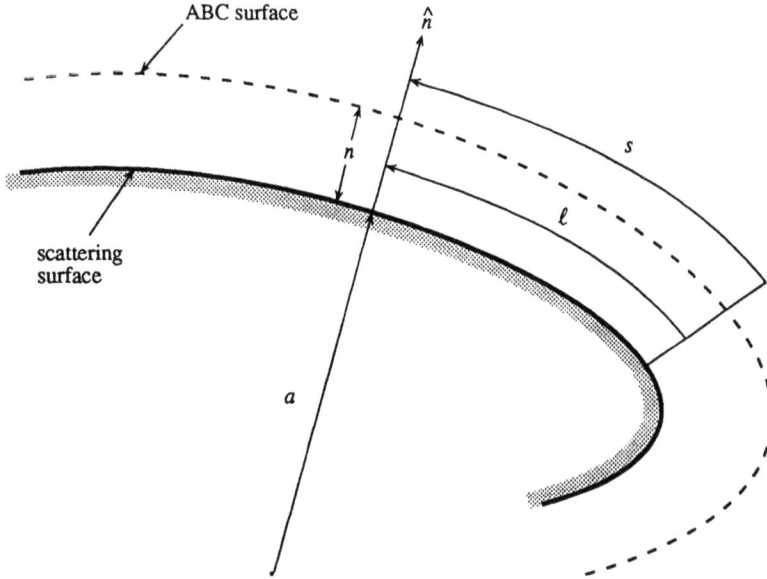

Figure 8–8: *Illustration of body-fitted coordinates*

and, provided f can be expanded in inverse powers of $(n + a)^{1/2}$, the right hand side is of higher order than $n^{-1/2}$ for $n \gg 1$. Consequently, (8.53) gives

$$\frac{\partial f}{\partial n} + \frac{f}{2(n + a)} = 0 + O\left\{(n + a)^{-5/2}\right\} \qquad (8.54)$$

and the corresponding equation for U is

$$B_1 U = \left(\frac{\partial}{\partial n} + jk_0 + \frac{1}{2\rho_c}\right) U = 0 \qquad (8.55)$$

where $\rho_c = n + a$ is the radius of curvature of the ABC boundary. This is the generalised version of (8.45).

To derive a second order ABC we note that to the next higher order in $(n + a)^{-1/2}$, (8.53) is

$$\frac{\partial f}{\partial n} + \frac{f}{2(n + a)} =$$
$$\frac{1}{2jk_0}\left\{\frac{a}{n + a}\frac{\partial}{\partial \ell}\left(\frac{a}{n + a}\frac{\partial f}{\partial \ell}\right) + \frac{f}{4(n + a)^2}\right\} + O\left\{(n + a)^{-7/2}\right\} \qquad (8.56)$$

In terms of the coordinate s we have

$$\frac{\partial f}{\partial n} + \frac{f}{2(n + a)} = \frac{1}{2jk_0}\left\{\frac{\partial^2 f}{\partial s^2} + \frac{f}{4(n + a)^2}\right\} + O\left\{(n + a)^{-7/2}\right\}$$

and the corresponding equation for U is

$$\tilde{B}_2 U = \left\{ \frac{\partial}{\partial n} + \frac{1}{2\rho_c} + jk_0 + \frac{j}{2k_0\rho_c^2}\left(\rho_c^2\frac{\partial^2}{\partial s^2} + \frac{1}{4}\right)\right\} U = 0 \qquad (8.57)$$

In the limit $\rho_c \to \rho$ we recover (8.48) apart from the factor $1 + 1/(jk_0\rho)$ in the denominator.

We can improve the accuracy of (8.57) by a process of successive approximation in which the boundary conditions of lower order are used to approximate those terms in (8.53) which were omitted from (8.56). When moved to the right hand side of (8.53) those terms are

$$\Delta F = \left\{ \frac{\partial}{\partial n} + \frac{1}{2(n+a)}\right\}\left\{ \frac{\partial f}{\partial n} + \frac{f}{2(n+a)}\right\} \qquad (8.58)$$

and hence, from (8.56),

$$2jk_0\Delta F = \left\{ \frac{\partial}{\partial n} + \frac{1}{2(n+a)}\right\}\left\{ \frac{a}{n+a}\frac{\partial}{\partial \ell}\left(\frac{a}{n+a}\frac{\partial f}{\partial \ell}\right) + \frac{f}{4(n+a)^2}\right\}$$

On carrying out the differentiation with respect to n and using the first order condition (8.54) to eliminate $\partial f/\partial n$, we obtain

$$\begin{aligned} 2jk_0\Delta F &= -\frac{2}{n+a}\left\{ \frac{a}{n+a}\frac{\partial}{\partial \ell}\left(\frac{a}{n+a}\frac{\partial f}{\partial \ell}\right) + \frac{f}{4(n+a)^2}\right\} \\ &\quad - \frac{3}{2}\frac{a}{n+a}\frac{\partial}{\partial \ell}\left(\frac{1}{n+a}\right)\frac{\partial}{\partial \ell}\left(\frac{a}{n+a}f\right) \end{aligned} \qquad (8.59)$$

The last term on the right hand side involves the rate of change of curvature and, if this is neglected, the inclusion of the remaining term in (8.56) gives

$$\begin{aligned} \frac{\partial f}{\partial n} + \frac{f}{2(n+a)} &= \frac{1}{2jk_0}\left\{1 - \frac{1}{jk_0(n+a)}\right\}\left\{ \frac{a}{n+a}\frac{\partial}{\partial \ell}\left(\frac{a}{n+a}\frac{\partial f}{\partial \ell}\right)\right. \\ &\quad \left. + \frac{f}{4(n+a)^2}\right\} + O\left\{(n+a)^{-9/2}\right\} \end{aligned} \qquad (8.60)$$

The corresponding equation for U is

$$B_2 U = \left[\frac{\partial}{\partial n} + \frac{1}{2\rho_c} + jk_0 + \frac{j}{2k_0\rho_c^2}\left(1 - \frac{1}{jk_0\rho_c}\right)\left(\rho_c^2\frac{\partial^2}{\partial s^2} + \frac{1}{4}\right)\right] U = 0 \quad (8.61)$$

and since this is (virtually) identical to (8.48) when $\rho_c = \rho$, it represents the simplest generalisation of the second order Bayliss-Turkel ABC to a non-circular boundary.

The process can be continued by using (8.60) in conjunction with (8.54) to construct ABCs of still higher order (MA, 1991). JONES (1992) has also used a method of successive approximation to develop a second order ABC, and his result is identical to (8.61) when (8.34) is employed.

8.2.4 ABCs from GIBCs

In the preceding sections we have described several methods for deriving two-dimensional ABCs, culminating in the second order condition (8.61) for a non-circular cylinder. The conditions were obtained without any reference to the GIBCs which are the subject of this book, but it is obvious that the two types of condition are related. Our goal now is to establish that connection and to demonstrate how the techniques described in Appendix A can also be used to derive ABCs.

In Appendix A it was shown how Rytov's method can be employed to develop GIBCs of any desired order applicable at the convex surface of a (possibly inhomogeneous) dielectric body. A second order boundary condition is given in (A.52) in terms of a curvilinear coordinate system (α, β, γ) with γ in the direction of the outward normal to the surface, and for a homogeneous dielectric whose intrinsic impedance and complex refractive index are Z and N, respectively, the result is

$$\hat{\gamma} \times \mathbf{E} \;=\; \hat{\gamma} \times \left(\hat{\gamma} \times \left\{ \overline{\overline{\eta}}.\mathbf{H} - \frac{Z}{2(k_0 N)^2 h_\gamma} \nabla \left[h_\gamma \nabla_{\!s}.\mathbf{H} \right] \right.\right.$$
$$\left.\left. - \frac{Z}{2(k_0 N)^2 h_\gamma} \nabla \left[h_\gamma \hat{\gamma}.\nabla_{\!s} \times \mathbf{H} \right] \right\} \right) \qquad (8.62)$$

where $\overline{\overline{\eta}}$ is given in (A.25) with (see (A.47) and (A.49))

$$\eta_{\beta\beta}^{\alpha\alpha} \;=\; Z \left\{ 1 \pm \frac{1}{2jk_0 N h_\gamma} \frac{\partial}{\partial \gamma} \left(\ln \frac{h_\alpha}{h_\beta} \right) \right.$$
$$\pm \frac{1}{(2k_0 N h_\gamma)^2} \left[h_\gamma \frac{\partial}{\partial \gamma} \left\{ \frac{1}{h_\gamma} \frac{\partial}{\partial \gamma} \left(\ln \frac{h_\alpha}{h_\beta} \right) \right\} \right.$$
$$\left.\left. \mp \frac{1}{2} \left\{ \frac{\partial}{\partial \gamma} \left(\ln \frac{h_\alpha}{h_\beta} \right) \right\}^2 \right] \right\} \qquad (8.63)$$

accurate to $O(|N|^{-2})$. In the special case of the two-dimensional problem for a cylindrical body whose generators are parallel to the β axis we have $\partial/\partial\beta = 0$ and $h_\beta = 1$, and (8.62) reduces to (A.85). Consistent with the coordinates in Figs. 8–8 and 8–9(a),

$$\frac{1}{h_\gamma} \frac{\partial}{\partial \gamma} = \frac{\partial}{\partial n}, \qquad \frac{1}{h_\alpha} \frac{\partial}{\partial \alpha} = \frac{\partial}{\partial s}$$

and (A.85) then becomes

$$\left\{ Z_0 Y \frac{\partial}{\partial n} - jk_0 + \frac{1}{2N\rho_c} - \frac{j}{8k_0(N\rho_c)^2} \left(2\frac{\partial \rho_c}{\partial n} - 1 \right) \right.$$
$$\left. - \frac{j}{2k_0 N^2} \frac{1}{h_\gamma} \frac{\partial}{\partial s} \left(h_\gamma \frac{\partial}{\partial s} \right) \right\} U \;=\; 0 \qquad (8.64)$$

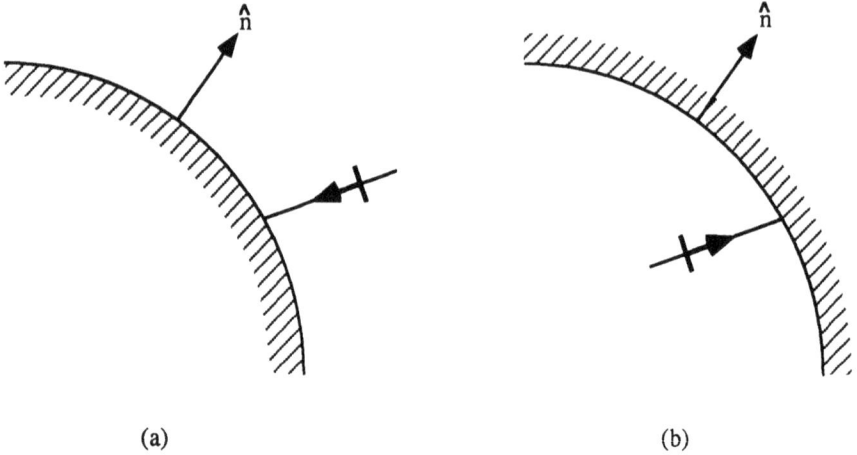

(a)   (b)

Figure 8–9: *Geometry for a GIBC (a) compared with that for an ABC (b)*

where $U = H_\beta$. The analogous result for E_β can be obtained using duality.

In contrast to (8.64), which was developed for a *convex* surface, an ABC is applicable to a *concave* surface as illustrated in Fig. 8–9(b). In Rytov's method the change requires us to reverse $\hat{\gamma}$ in Section A–2 (or, more precisely, identify $\hat{s}$ of Appendix A as $\hat{\gamma}$ instead of $-\hat{\gamma}$). This amounts to a reversal in the direction of the incident field, and the same effect can be achieved by replacing j by $-j$, equivalent to a time reversal. In addition, a non-reflecting surface requires an impedance match, and it is therefore necessary that $Z = Z_0$ and $N = 1$. With these changes, (8.64) now represents a second order ABC.

We have described the Rytov solution as an expansion in inverse powers of $|N|$ and since, for a homogeneous body, a GIBC is accurate only for a lossy dielectric (see, for example, Section 7.1.3), it is not immediately evident that the choice $Z = Z_0$ and $N = 1$ can be justified. In reality, however, the method is based on an expansion of the interior field close to the surface in inverse powers of $|k|\ell$ when $k = N k_0$ is the interior wave number and ℓ is a length associated with the boundary, and it is therefore valid for $N = 1$ provided $k_0 \ell \gg 1$ and the interior field is still a local one. In the case of a convex surface, an impedance match maximises the field transmitted into the body, and the resulting boundary condition will fail because the local field is no longer the incident field alone. On the other hand, for a concave surface the match *minimises* the field reflected back into the interior, and this is indeed the objective of an ABC.

With the changes indicated above, the ABC analogous to (8.64) is

$$\left\{ \frac{\partial}{\partial n} + jk_0 + \frac{1}{2\rho_c} + \frac{j}{2k_0}\left[\frac{1}{h_\gamma}\frac{\partial}{\partial s}\left(h_\gamma \frac{\partial}{\partial s}\right) + \frac{1}{4\rho_c^2}\left(2\frac{\partial \rho_c}{\partial n} - 1\right) \right] \right\} U = 0 \quad (8.65)$$

and, on putting $\partial \rho_c / \partial n = 1$ and $h_\gamma = 1$, appropriate to a Dupin coordinate system, the boundary condition becomes identical to the ABC (8.57) developed by MA (1991). To the next higher order, i.e. to $O\{(k_0\rho_c)^{-3}\}$, the GIBC is given in (A.83) and, if we neglect the operator L associated with the tangential derivatives of the metric coefficients, the analogous ABC is

$$\left\{ \frac{\partial}{\partial n} + jk_0 + \frac{1}{2\rho_c} + \frac{j}{2k_0} \left[\left(1 - \frac{1}{jk_0\rho_c} \right) \frac{1}{h_\gamma} \frac{\partial}{\partial s} \left(h_\gamma \frac{\partial}{\partial s} \right) \right. \right.$$
$$\left. \left. + \frac{1}{4} \left(1 + \frac{1}{2jk_0} \frac{\partial}{\partial n} \right) \frac{1}{\rho_c^2} \left(2 \frac{\partial \rho_c}{\partial n} - 1 \right) \right] \right\} U \;=\; 0 \qquad (8.66)$$

When expressed in the Dupin coordinate system, this is identical to Ma's ABC (8.61).

We have now established the connection between GIBCs and ABCs and, in particular, have shown how Rytov's method provides a systematic procedure for rigorously developing ABCs of any desired order. In two dimensions, the results agree with the most accurate ones obtained by other methods, and show the influence of tangential derivatives of the curvature. The method is also applicable to three-dimensional geometries.

8.3 Three-dimensional ABCs

As before the goal is to develop ABCs which minimise the field reflected from the surface where the ABC is imposed, thereby allowing us to truncate the computational volume in a numerical solution. For a three-dimensional scattering or radiation problem, the ABC surface is box-like, spherical or doubly-curved as illustrated in Figs. 8–10 to 8–12, and, if the surface is perfectly absorbing, the enclosed scatterer or radiator appears to be situated in a free space environment.

Paralleling the development of two-dimensional ABCs, we first derive the boundary condition from the one-way wave equation. The mode annihilation method is then used to derive ABCs applicable to spherical and body-conforming surfaces. This is based on an *a priori* convergent expansion of the field, and has proved successful in generating a rather general class of ABCs. Finally, the ABCs and GIBCs are compared, and some numerical results are presented.

8.3.1 One-way wave equation method

For simplicity we will confine attention to Cartesian coordinates for which the analysis is similar to that in Section 8.2.1. The starting point is the wave equation

$$\left(\frac{\partial^2}{\partial x^2} + \frac{\partial^2}{\partial y^2} + \frac{\partial^2}{\partial z^2} + k_0^2 \right) U = 0 \qquad (8.67)$$

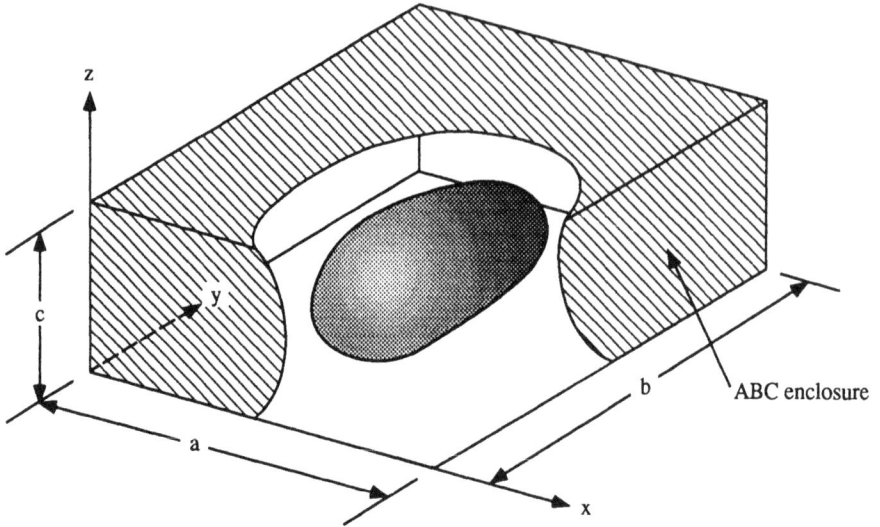

Figure 8–10: *Scatterer enclosed in a rectangular ABC surface*

where U represents a Cartesian component of $\mathbf{E}$ or $\mathbf{H}$. This is rewritten as

$$\left(\frac{\partial}{\partial y} - L\right)\left(\frac{\partial}{\partial y} + L\right)U = 0 \tag{8.68}$$

where

$$L = jk_0\sqrt{1 + \frac{1}{k_0^2}\left(\frac{\partial^2}{\partial x^2} + \frac{\partial^2}{\partial z^2}\right)} \tag{8.69}$$

and the factorisation is that appropriate for the faces $y = $ constant of the box-like surface in Fig. 8–10. The condition to be enforced at $y = b$ is then

$$\left(\frac{\partial}{\partial y} + L\right)U = 0$$

since this absorbs all waves travelling in the positive y direction, and the analogous condition at $y = 0$ is

$$\left(\frac{\partial}{\partial y} - L\right)U = 0$$

Practical ABCs can be obtained by approximating the radical in (8.69) using a finite series. Thus, from the two-term Taylor series approximation analogous to (8.7), we get the second order ABC

$$\left\{\frac{\partial}{\partial y} \pm jk_0 \pm \frac{j}{2k_0}\left(\frac{\partial^2}{\partial x^2} + \frac{\partial^2}{\partial z^2}\right)\right\}U = 0 \tag{8.70}$$

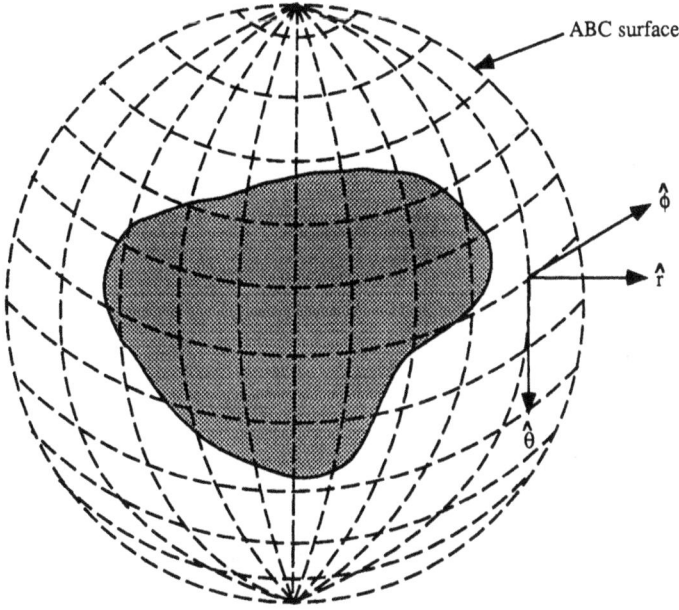

Figure 8–11: *Scatterer enclosed in a spherical ABC surface*

where the upper and lower signs are for waves propagating in the positive and negative y directions, respectively. This is the ABC derived by MUR (1981) and is the obvious analogue of the two-dimensional condition (8.8) of ENGQUIST AND MAJDA (1977). It can be used in connection with a finite difference solution of the wave equation when the uniform grid is terminated on the box-like surface in Fig. 8–10. This was done first by MUR (1981) and the condition has since been widely used (TAFLOVE AND UMASHANKAR, 1987; KUNZ AND LUEBBERS, 1992). Only the tangential field components are required, and it is desirable to place the ABC surface a wavelength or so from the body. However, it is often placed as close as $\lambda_0/2$ to (some parts of) the outer surface of the body. On the surfaces $x = $ constant of the box, the analogous ABCs are

$$\left\{ \frac{\partial}{\partial x} \pm jk_0 \pm \frac{j}{2k_0} \left(\frac{\partial^2}{\partial y^2} + \frac{\partial^2}{\partial z^2} \right) \right\} U = 0$$

with a similar change for the surfaces $z = $ constant.

Higher order ABCs can be derived by using higher order approximations to the pseudodifferential operator L. As an example, from the expansion (8.15) we obtain the third order ABC

$$\left\{ \frac{c - \frac{1}{2}}{k_0^2} \frac{\partial}{\partial y} \left(\frac{\partial^2}{\partial x^2} + \frac{\partial^2}{\partial z^2} \right) + \frac{jc}{k_0} \left(\frac{\partial^2}{\partial x^2} + \frac{\partial^2}{\partial z^2} \right) + \frac{\partial}{\partial y} + jk_0 \right\} U = 0 \qquad (8.71)$$

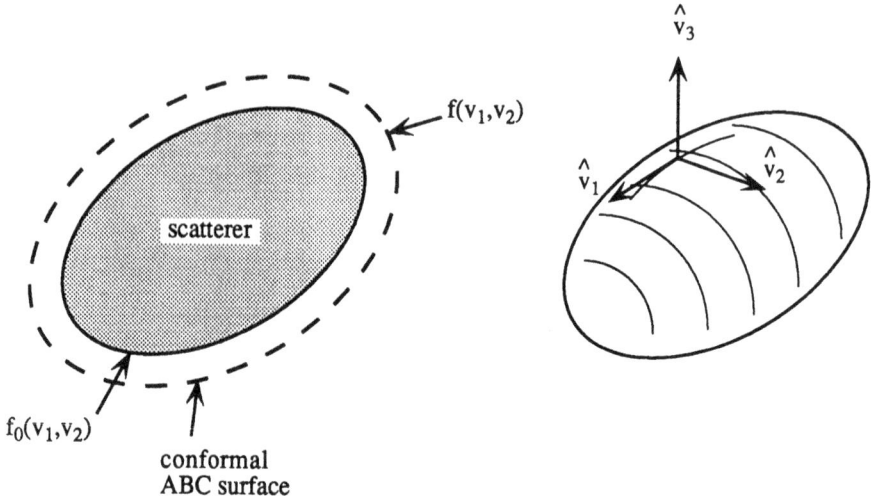

Figure 8–12: *Scatterer enclosed in a conformal ABC surface*

where the form shown is appropriate for the surface $y = b$ in Fig. 8–10. Such higher order ABCs have only recently been employed in numerical solutions, but, in spite of the improved absorptivity, they often lead to poorly conditioned matrix systems or slowly convergent temporal solutions.

8.3.2   Mode annihilation method

It is now assumed that the ABC surface which truncates the computational domain is a sphere (see Fig. 8–11), and we seek a boundary conditon which can be enforced on the surface to absorb (and not reflect) waves coming from the interior. The derivation is based on the Wilcox expansion

$$\mathbf{E} = \frac{e^{-jk_0 r}}{r} \sum_{m=0}^{\infty} \frac{\mathbf{E}_m(\theta, \phi)}{r^m} \tag{8.72}$$

for the field on the ABC surface $r = $ constant where r, θ, ϕ are spherical polar coordinates. The coefficients or partial fields $\mathbf{E}_m$ are functions only of the tangential variables and $\mathbf{E}_m.\hat{r} = E_{mr}$ is allowed to be non-zero except in the case $m = 0$.

The procedure is similar to that described in Section 8.2.2 in connection with the Bayliss-Turkel ABC, and consists of substituting the expansion (8.72) into Maxwell's equations and then constructing differential (curl) operators which annihilate the first M terms of the series. To derive the first order ABC we consider the operation $\hat{r} \times (\nabla \times \mathbf{E})$. Since

$$\nabla \times \left(\frac{e^{-jk_0 r}}{r^{m+1}} \mathbf{E}_m \right) = \nabla \left(\frac{e^{-jk_0 r}}{r^{m+1}} \right) \times \mathbf{E}_m + \frac{e^{-jk_0 r}}{r^{m+1}} \nabla \times \mathbf{E}_m$$

it follows that

$$\hat{r} \times \nabla \times \left(\frac{e^{-jk_0 r}}{r^{m+1}} \mathbf{E}_m \right) = \left[jk_0 \frac{e^{-jk_0 r}}{r^{m+1}} + \frac{(m+1)e^{-jk_0 r}}{r^{m+2}} \right] \mathbf{E}_{ms} + \frac{e^{-jk_0 r}}{r^{m+1}} \hat{r} \times (\nabla \times \mathbf{E}_m)$$

(8.73)

where $\mathbf{E}_{ms} = \mathbf{E}_m - \hat{r}\, E_{mr}$ denotes the field tangential to the ABC surface. To evaluate the second term on the right hand side of (8.73) we note that (VAN BLADEL, 1985)

$$\nabla \times \mathbf{E}_m = \nabla_s \times \mathbf{E}_m + \hat{r} \times \frac{\partial \mathbf{E}_m}{\partial r}$$

where the subscript "s" indicates the surface operator, and

$$\nabla_s \times \mathbf{E}_m = -\hat{r} \times \nabla E_{mr} + \frac{1}{r} \left(\hat{\phi} E_{m\theta} - \hat{\theta} E_{m\phi} \right) + \hat{r} \nabla \cdot (\mathbf{E}_m \times \hat{r})$$

Since $\mathbf{E}_m$ is independent of r, we now have

$$\hat{r} \times \nabla \times \mathbf{E}_m = \nabla_s E_{mr} - \frac{\mathbf{E}_{ms}}{r}$$

(8.74)

where

$$\nabla_s f = -\hat{r} \times \hat{r} \times \nabla f$$

is the surface gradient and, when (8.74) is substituted into (8.73), we obtain

$$\hat{r} \times \nabla \times \left(\frac{e^{-jk_0 r}}{r^{m+1}} \mathbf{E}_m \right) = \frac{e^{-jk_0 r}}{r} \left[jk_0 \frac{\mathbf{E}_{ms}}{r^m} + \frac{r\nabla_s E_{mr} + m\mathbf{E}_{ms}}{r^{m+1}} \right]$$

Thus, the field $\mathbf{E}$ represented by the Wilcox expansion (8.72) satisfies the relationship

$$\hat{r} \times \nabla \times \mathbf{E} - jk_0 \mathbf{E}_s = \frac{e^{-jk_0 r}}{r} \sum_{m=0}^{\infty} \left(\frac{r\nabla_s E_{mr} + m\mathbf{E}_{ms}}{r^{m+1}} \right)$$

(8.75)

from which we conclude

$$\hat{r} \times \nabla \times \mathbf{E} - jk_0 \mathbf{E}_s = 0 + O(r^{-3})$$

(8.76)

since $E_{mr} = 0$ for $m = 0$. This is the first order ABC for a spherical surface, and is identical to the Sommerfeld or Silver-Müller radiation condition (2.12).

To derive the second order ABC the operator $\hat{r} \times \nabla \times$ is applied to (8.75), and on using the same identities as before we find

$$\hat{r} \times \nabla \times [\hat{r} \times \nabla \times \mathbf{E} - jk_0 \mathbf{E}_s] =$$

$$\frac{e^{-jk_0 r}}{r} \sum_{m=0}^{\infty} \left[\left(jk_0 + \frac{m+1}{r} \right) \frac{r\nabla_s E_{mr} + m\mathbf{E}_{ms}}{r^{m+1}} \right]$$

(8.77)

This can be simplified by employing (8.75) to express the right hand side in terms of differential operators. By combining (8.75) and (8.77) in this manner, we obtain

$$\left\{\hat{r} \times (\nabla \times) \;-jk_0 - \frac{2}{r}\right\} (\hat{r} \times \nabla \times \mathbf{E} - jk_0 \mathbf{E}_s) =$$
$$\frac{e^{-jk_0 r}}{r} \sum_{m=1}^{\infty} \left(\frac{m-1}{r}\right) \left(\frac{r\nabla_s E_{mr} + mE_{ms}}{r^{m+1}}\right)$$

and hence

$$\left\{\hat{r} \times (\nabla \times) \;-jk_0 - \frac{2}{r}\right\} (\hat{r} \times \nabla \times \mathbf{E} - jk_0 \mathbf{E}_s) = 0 + O(r^{-5}) \qquad (8.78)$$

which is a second order ABC. With the aid of the identities

$$\hat{r} \times \nabla \times (\hat{r} \times \nabla \times \mathbf{E}) \;=\; \nabla \times [\hat{r}(\hat{r}.\nabla \times \mathbf{E})] - k_0^2 \mathbf{E}_s \qquad (8.79)$$

$$\hat{r} \times \nabla \times \mathbf{E}_s \;=\; \hat{r} \times \nabla \times \mathbf{E} - \nabla_s E_r \qquad (8.80)$$

(VAN BLADEL, J., 1985), (8.78) can be written in a form which is more convenient for numerical implementation, namely

$$\hat{r} \times \nabla \times \mathbf{E} = jk_0 E_s + \left\{2\left(jk_0 + \frac{1}{r}\right)\right\}^{-1} \{\nabla \times [\hat{r}(\hat{r}.\nabla \times \mathbf{E})] + jk_0 \nabla_s E_r\} \quad (8.81)$$

and this is the second order ABC derived by PETERSON (1988).

Because the last term on the right hand side of (8.81) involves only a single del operator, it is expected that (8.81) will lead to an asymmetric matrix in a finite element implementation when Galerkin's method is used to generate the discrete system. Since this results in larger storage and solution times, an ABC which leads to a symmetric matrix is desirable, and WEBB AND KANEL-LOPOULOS (1989) have shown that to the same order of approximation (8.81) can be written as

$$\hat{r} \times \nabla \times \mathbf{E} = jk_0 \mathbf{E}_s + \left\{2\left(jk_0 + \frac{1}{r}\right)\right\}^{-1} \{\nabla \times [\hat{r}(\hat{r}.\nabla \times \mathbf{E})] + \nabla_s(\nabla \cdot \mathbf{E}_s)\}$$
$$(8.82)$$

This second order ABC produces a symmetric matrix and its form is similar to the generic condition (8.2). We note in passing that a rather simple way to show the equivalence of (8.81) and (8.82) is through the relationship

$$\hat{r} \times \nabla \times \nabla_s E_r = jk_0 \nabla_s E_r + \frac{2}{r}\nabla_s E_r + e^{-jk_0 r} \sum_{m=2}^{\infty} \left(\frac{m-1}{r}\right) \frac{\nabla_s E_{mr}}{r^{m+1}}$$

satisfied by the field (8.72). It then follows that

$$\hat{r} \times \nabla \times \nabla_s E_r = jk_0 \nabla_s E_r + \frac{2}{r}\nabla_s E_r + O(r^{-5}) \qquad (8.83)$$

and from the identity

$$\hat{r} \times \nabla \times \nabla_s E_r = \nabla_s (\nabla \cdot \mathbf{E}_s) + \frac{2}{r} \nabla_s E_r \qquad (8.84)$$

we have

$$\nabla_s (\nabla \cdot \mathbf{E}_s) = jk_0 \nabla_s E_r + O(r^{-5}) \qquad (8.85)$$

Compared with (5.87) the difference in sign is because of the difference in the direction of propagation.

A third order ABC can be obtained by applying the operator $\hat{r} \times \nabla \times$ to the second order condition (8.78) and this can be repeated as often as desired to generate ABCs of any order. The conversion process is similar to that used in deriving the Bayliss-Turkel ABC (8.48). We note that for the expansion (8.72) the relationship

$$\prod_{m=2}^{M} \left\{ \hat{r} \times (\nabla \times) \ - jk_0 - \frac{2(m-1)}{r} \right\} (\hat{r} \times \nabla \times \mathbf{E} - jk_0 \mathbf{E}_s) = 0 \qquad (8.86)$$

holds and, since this is accurate to $O(r^{-2M-1})$, it can be regarded as the Mth order ABC for a spherical surface.

8.3.3 Curvilinear coordinates

For the ABCs we have just developed, the need to impose them on a spherical surface may lead to a large computational domain, particularly when the enclosed body is elongated or planar. It is therefore important to seek conditions which can be imposed on a surface that conforms to the body. In this case, the expansion (8.72) is no longer convenient. Many terms would be necessary to provide an acceptable characterisation of the field on the ABC surface, and it is more appropriate to express the outgoing (scattered or radiated) field in a form whose wavefront conforms to the surface of the body. With this in mind we write (CHATTERJEE AND VOLAKIS, 1993)

$$\mathbf{E}(v_1, v_2, n) = \frac{e^{-jk_0 n}}{\sqrt{R_1 R_2}} \sum_{m=0}^{\infty} \frac{\mathbf{E}_m(v_1, v_2)}{\left(\sqrt{R_1 R_2}\right)^m} \qquad (8.87)$$

where $R_i = \rho_i + n$, $i = 1, 2$, with the ρ_i denoting the two principal radii of curvature of the ABC surface. As in Section 8.2.3, the coordinates v_1, v_2, v_3 ($= n$) of the Dupin system are associated with the unit vectors $\hat{v}_1, \hat{v}_2, \hat{v}_3$ ($= \hat{n}$) and the metric coefficients h_1, h_2, h_3 ($= 1$). The ABC surface and the outgoing wavefront are defined by

$$f(v_1, v_2) = n\hat{n} + f_0(v_1, v_2)$$

where $f_0(v_1, v_2)$ is the reference surface for the wavefront (see Fig. 8–12), and this may coincide with the body's surface. If this is so, the first term of the expansion (8.87) is simply the geometrical optics field measured at the body's surface $f_0(v_1, v_2)$. Thus, ABCs developed on the basis of (8.87) should allow perfect absorption of the geometrical optics field, and this is typically the dominant field in a scattering analysis. Since (8.87) can be cast in the form of the Wilcox expansion (8.72) by using a binomial series to expand $\sqrt{R_1 R_2}$, it will be referred to as a modified Wilcox expansion.

The derivation of the ABCs based on (8.87) is similar to that for spherical surfaces, but the analysis is more involved because the curl and gradient operations must be carried out in the Dupin system. To construct the first order ABC we consider the operation $\hat{n} \times \nabla \times \mathbf{E}$. Since

$$
\nabla \times \left[\frac{e^{-jk_0 n}}{\left(\sqrt{R_1 R_2}\right)^{m+1}} \mathbf{E}_m(v_1, v_2) \right] =
$$

$$
\nabla \left[\frac{e^{-jk_0 n}}{\left(\sqrt{R_1 R_2}\right)^{m+1}} \right] \times \mathbf{E}_m + \frac{e^{-jk_0 n}}{\left(\sqrt{R_1 R_2}\right)^{m+1}} \nabla \times \mathbf{E}_m \qquad (8.88)
$$

it follows that

$$
\hat{n} \times \nabla \times \left[\frac{e^{-jk_0 n}}{\left(\sqrt{R_1 R_2}\right)^{m+1}} \mathbf{E}_m \right] =
$$

$$
\left[jk_0 \frac{e^{-jk_0 n}}{\left(\sqrt{R_1 R_2}\right)^{m+1}} + (m+1) \frac{R_1 + R_2}{2(R_1 R_2)} \frac{e^{-jk_0 n}}{\left(\sqrt{R_1 R_2}\right)^{m+1}} \right] \mathbf{E}_{ms}
$$

$$
+ \frac{e^{-jk_0 n}}{\left(\sqrt{R_1 R_2}\right)^{m+1}} \hat{n} \times (\nabla \times \mathbf{E}_m) \qquad (8.89)
$$

where $E_{ms} = -\hat{n} \times \hat{n} \times \mathbf{E}_m$. Also (VAN BLADEL, 1985)

$$
\nabla \times \mathbf{E}_m = \nabla_s \times \mathbf{E}_m + \hat{n} \times \frac{\partial \mathbf{E}_m}{\partial n} = \nabla_s \times \mathbf{E}_m \qquad (8.90)
$$

with

$$
\nabla_s \times \mathbf{E} = -\hat{n} \times \nabla E_n + \hat{v}_2 \frac{E_{v1}}{R_1} - \hat{v}_1 \frac{E_{v2}}{R_2} + \hat{n} \nabla \cdot (\mathbf{E} \times \hat{n}) \qquad (8.91)
$$

and thus

$$
\hat{n} \times \nabla \times \mathbf{E}_m = \nabla_s E_{mn} - \overline{\overline{K}} \cdot \mathbf{E}_{ms} \qquad (8.92)
$$

where

$$
\overline{\overline{K}} = K_1 \hat{v}_1 \hat{v}_1 + K_2 \hat{v}_2 \hat{v}_2 = \frac{1}{R_1} \hat{v}_1 \hat{v}_1 + \frac{1}{R_2} \hat{v}_2 \hat{v}_2 \qquad (8.93)
$$

and

$$
\nabla_s g = \hat{v}_1 \frac{1}{h_1} \frac{\partial(h_1 g)}{\partial v_1} + \hat{v}_2 \frac{1}{h_2} \frac{\partial(h_2 g)}{\partial v_2} = -\hat{n} \times \hat{n} \times \nabla g
$$

is the surface gradient. When (8.92) is substituted into (8.89) we have

$$
\hat{n} \times \nabla \times \left[\frac{e^{-jk_0 n}}{\left(\sqrt{R_1 R_2}\right)^{m+1}} \mathbf{E}_m \right] =
$$

$$
\frac{e^{-jk_0 n}}{\sqrt{R_1 R_2}} \left[\left(jk_0 + (m+1)K_\mu - \overline{\overline{K}} \cdot \right) \mathbf{E}_{ms} + \nabla_s E_{mn} \right] \frac{1}{\left(\sqrt{R_1 R_2}\right)^m} \quad (8.94)
$$

where

$$
K_\mu = \frac{K_1 + K_2}{2} \quad (8.95)
$$

is the mean curvature. Thus, from (8.87),

$$
\hat{n} \times \nabla \times \mathbf{E} - \left(jk_0 + K_\mu - \overline{\overline{K}} \cdot \right) \mathbf{E}_s = \frac{e^{-jk_0 n}}{\sqrt{R_1 R_2}} \sum_{m=0}^{\infty} \frac{\nabla_s E_{mn} + m K_\mu \mathbf{E}_{ms}}{\left(\sqrt{R_1 R_2}\right)^m} \quad (8.96)
$$

implying

$$
\hat{n} \times \nabla \times \mathbf{E} - \left(jk_0 + K_\mu - \overline{\overline{K}} \cdot \right) \mathbf{E}_s = 0 + O\left\{ \left(\sqrt{R_1 R_2}\right)^{-3} \right\} \quad (8.97)
$$

and this is the first order ABC for a curvilinear surface. It reduces to (8.76) for a spherical surface (when $R_1 = R_2 = r$), and to the two-dimensional ABC (8.55) as $R_1 \to \infty$. It is therefore the most general first order ABC we have presented so far.

A higher order ABC is necessary to reduce the error in (8.97) and thereby improve the accuracy. By applying the operator $\hat{n} \times \nabla \times$ to (8.96) and then using (8.89) and (8.90)–(8.92) we obtain

$$
\hat{n} \times \nabla \times \left[\hat{n} \times \nabla \times \mathbf{E} - \left(jk_0 + K_\mu - \overline{\overline{K}} \cdot \right) \mathbf{E}_s \right] =
$$

$$
\frac{e^{-jk_0 n}}{\sqrt{R_1 R_2}} \sum_{m=1}^{\infty} \left[\left(jk_0 + (m+3)K_\mu - \frac{K_g}{K_\mu} \right) \left(\nabla_s E_{mn} + m K_\mu \mathbf{E}_{ms} \right) \right.
$$

$$
\left. - \left(2K_\mu - \frac{K_g}{K_\mu} \right) \nabla_s E_{mn} - m K_\mu \overline{\overline{K}} \cdot \mathbf{E}_{ms} \right] \frac{1}{\left(\sqrt{R_1 R_2}\right)^m} \quad (8.98)
$$

where $K_g = K_1 K_2$ is the Gaussian curvature, and (8.97) allows us to write the right hand side as

$$
\frac{e^{-jk_0 n}}{\sqrt{R_1 R_2}} \sum_{m=1}^{\infty} m K_\mu \frac{\nabla_s E_{mn} + m K_\mu \mathbf{E}_{ms}}{\left(\sqrt{R_1 R_2}\right)^m}
$$

$$
+ \left(jk_0 + 3K_\mu - \frac{K_g}{K_\mu} - \overline{\overline{K}} \cdot \right) \left[\hat{n} \times \nabla \times \mathbf{E} - \left(jk_0 + K_\mu - \overline{\overline{K}} \cdot \right) \mathbf{E}_s \right]
$$

Hence, from (8.98),

$$\left[\hat{n} \times \nabla \times -\left(jk_0 + 4K_\mu - \frac{K_g}{K_\mu} - \overline{\overline{K}}\cdot\right)\right]\left[\hat{n} \times \nabla \times \mathbf{E} - \left(jk_0 + K_\mu - \overline{\overline{K}}\cdot\right)\mathbf{E}_s\right]$$
$$+ \left(2K_\mu - \frac{K_g}{K_\mu} - \overline{\overline{K}}\cdot\right)\nabla_s E_n = \frac{e^{-jk_0 n}}{\sqrt{R_1 R_2}}\sum_{m=1}^{\infty}(m-1)\frac{\nabla_s(E_m)_n + mK_\mu \mathbf{E}_{ms}}{\left(\sqrt{R_1 R_2}\right)^m} \tag{8.99}$$

and, since the right hand side of this is $O\left\{\left(\sqrt{R_1 R_2}\right)^{-5}\right\}$, a second order ABC is

$$\left[\hat{n} \times \nabla \times -\left(jk_0 + 4K_\mu - \frac{K_g}{K_\mu} - \overline{\overline{K}}\cdot\right)\right]\left[\hat{n} \times \nabla \times \mathbf{E} - \left(jk_0 + K_\mu - \overline{\overline{K}}\cdot\right)\mathbf{E}_s\right]$$
$$+ \left(2K_\mu - \frac{K_g}{K_\mu} - \overline{\overline{K}}\cdot\right)\nabla_s E_n = 0 + O\left\{\left(\sqrt{R_1 R_2}\right)^{-5}\right\} \tag{8.100}$$

This can be simplified using the identities

$$\hat{n} \times \nabla \times (\hat{n} \times \nabla \times \mathbf{E}) = \nabla \times [\hat{n}(\nabla \times \mathbf{E} \cdot \hat{n})] - k_0^2 \mathbf{E}_s$$
$$- (K_1 - K_2)\{\hat{v}_1(\nabla \times \mathbf{E}).\hat{v}_2 - \hat{v}_2(\nabla \times \mathbf{E}).\hat{v}_1\} \tag{8.101}$$

$$\hat{n} \times \nabla \times \mathbf{E}_s = \hat{n} \times \nabla \times \mathbf{E} - \nabla_s E_n \tag{8.102}$$

These are generalisations of (8.79) and (8.80), and (8.101) is a corrected version of the one employed by CHATTERJEE AND VOLAKIS (1994). After some manipulation, (8.100) then becomes

$$(D - 2K_\mu)\hat{n} \times \nabla \times \mathbf{E} =$$
$$\left[4(K_\mu)^2 - K_g + D\left(jk_0 - \overline{\overline{K}}\cdot\right) + K_\mu \Delta K \overline{\overline{T}} \cdot + \overline{\overline{K}} \cdot \overline{\overline{K}}\cdot\right]\mathbf{E}_s$$
$$+ \nabla \times [\hat{n}(\nabla \times \mathbf{E}) \cdot \hat{n}] + \left(jk_0 + 3K_\mu - \frac{K_g}{K_\mu} - 2\overline{\overline{K}}\cdot\right)\nabla_s E_n \tag{8.103}$$

where

$$D = 2jk_0 + 5K_\mu - \frac{K_g}{K_\mu} \tag{8.104}$$

$$\Delta K = K_1 - K_2 = \frac{1}{R_1} - \frac{1}{R_2} \tag{8.105}$$

and

$$\overline{\overline{T}} = \hat{v}_1\hat{v}_1 - \hat{v}_2\hat{v}_2 \tag{8.106}$$

One final modification is required. As noted earlier, a single del operation leads to an asymmetric matrix even when $R_1 = R_2$. This is undesirable and, in addition, (8.103) is not in the form (8.2) consistent with a uniqueness proof (see Section 5.4.2). The process for remedying this is similar to that employed in deriving the symmetric ABC (8.82) for a sphere. From (8.87)

$$
\begin{aligned}
\hat{n} \times \nabla \times \nabla_s E_n & \\
&= \frac{e^{-jk_0 n}}{\sqrt{R_1 R_2}} \sum_{m=1}^{\infty} \left[jk_0 + (m+1)K_\mu \right] \frac{\nabla_s E_{mn}}{\left(\sqrt{R_1 R_2}\right)^m} \\
&= jk_0 \nabla_s E_n + 2K_\mu \nabla_s E_n + \frac{e^{-jk_0 n}}{\sqrt{R_1 R_2}} \sum_{m=2}^{\infty} (m-1)K_\mu \frac{\nabla_s E_{mn}}{\left(\sqrt{R_1 R_2}\right)^m}
\end{aligned}
$$

and since the generalisation of (8.84) is

$$
\hat{n} \times \nabla \times \nabla_s E_n = \nabla_s (\nabla \cdot \mathbf{E}_s) + 2K_\mu \nabla_s E_n \tag{8.107}
$$

we have

$$
\hat{n} \times \nabla \times \nabla_s E_n = jk_0 \nabla_s E_n + 2K_\mu \nabla_s E_n + O\left\{ \left(\sqrt{R_1 R_2}\right)^{-5} \right\}
$$

which is the obvious generalisation of (8.83). When substituted into (8.107) we conclude that

$$
\nabla_s (\nabla \cdot \mathbf{E}_s) = jk_0 \nabla_s E_n + O\left\{ \left(\sqrt{R_1 R_2}\right)^{-5} \right\} \tag{8.108}
$$

allowing (8.103) to be written as

$$
\hat{n} \times \nabla \times \mathbf{E} = \bar{\bar{\alpha}} \cdot \mathbf{E}_s + \beta \nabla \times [\hat{n}(\nabla \times \mathbf{E} \cdot \hat{n})] + \bar{\bar{\gamma}} \cdot \nabla_s (\nabla \cdot \mathbf{E}_s) \tag{8.109}
$$

where

$$
\begin{aligned}
\bar{\bar{\alpha}} &= \hat{v}_1 \hat{v}_1 \left[jk_0 + \tfrac{1}{4}\Delta K \{ (\Delta K + 4K_\mu)\Delta - 2 \} \right] \\
&\quad + \hat{v}_2 \hat{v}_2 \left[jk_0 + \tfrac{1}{4}\Delta K \{ (\Delta K - 4K_\mu)\Delta + 2 \} \right] \tag{8.110} \\
\beta &= \Delta \tag{8.111} \\
\bar{\bar{\gamma}} &= \hat{v}_1 \hat{v}_1 \left(jk_0 + K_\mu - \frac{K_g}{K_\mu} - \Delta K \right) \frac{\Delta}{jk_0} \\
&\quad + \hat{v}_2 \hat{v}_2 \left(jk_0 + K_\mu - \frac{K_g}{K_\mu} + \Delta K \right) \frac{\Delta}{jk_0} \tag{8.112}
\end{aligned}
$$

with

$$
\Delta = (D - 2K_\mu)^{-1} = \left(2jk_0 + 3K_\mu - \frac{K_g}{K_\mu} \right)^{-1} \tag{8.113}
$$

Apart from the corrections resulting from the (revised) identity (8.101), this is the ABC derived by CHATTERJEE AND VOLAKIS (1993).

The second order ABC (8.109) has the general form (8.2) provided the co-efficients are treated as constants, i.e. the derivatives of curvature are ignored. This is within the order of approximation of the boundary condition and was, in fact, assumed in going from (8.100) to (8.103). When $R_1 = R_2 = r$ implying a spherical surface,

$$K_\mu = r^{-1}, \qquad K_g = r^{-2}, \qquad \Delta K = 0, \qquad \Delta = (2jk_0 + 2/r)^{-1}$$

and (8.109) reduces to (8.82), as it should. Similarly, if $K_2 = 1/R_2 = 0$ implying a cylindrical surface, we have

$$K_\mu = (2R_1)^{-1}, \qquad K_g = 0, \qquad \Delta K = R_1^{-1}$$

with $R_1 = \rho_c$. On setting $\hat{v}_1 = \hat{\phi}$ and $\hat{v}_2 = \hat{z}$ the coefficients (8.110) and (8.112) become

$$\overline{\overline{\alpha}} = \left\{ jk_0\hat{\phi}\hat{\phi} + \left(jk_0 + \frac{1}{\rho_c} \right) \hat{z}\hat{z} \right\} \left(2jk_0 + \frac{1}{2\rho_c} \right) \Delta \qquad (8.114)$$

$$\overline{\overline{\gamma}} = \left\{ \left(jk_0 - \frac{1}{2\rho_c} \right) \hat{\phi}\hat{\phi} + \left(jk_0 + \frac{3}{2\rho_c} \right) \hat{z}\hat{z} \right\} \frac{\Delta}{jk_0} \qquad (8.115)$$

with

$$\Delta = \left(2jk_0 + \frac{3}{2\rho_c} \right)^{-1} \qquad (8.116)$$

In the special case of E-polarisation, (8.109) is then identical to (8.61) apart from having $[1 + (3/4jk_0\rho_c)]^{-1}$ in place of $[1 - (1/jk_0\rho_c)]$.

The ABC (8.109) has been implemented in a finite element solution of the wave equation and found to yield accurate results even when placed as close as $0.35\lambda_0$ to the body. As an example, Fig. 8–13 shows the backscatter-ing cross-section of a circular metallic cylinder which is open at the top and closed at the bottom. The calculations were carried out by CHATTERJEE ET AL. (1994) using a finite element code whose mesh was terminated on a conformal cylindrical surface with the ABC (8.109). The results are compared with the measured data of SCHUH ET AL. (1993) and with the values obtained from a body of revolution code (GLISSON AND WILTON, 1982). Another example is the backscattering from a rectangular patch antenna whose cavity is recessed into a cylinder as illustrated in Fig. 8–14. The finite element calculations were carried out with the mesh terminated on a cylinder, and Fig. 8–15 shows the effect of moving the ABC boundary away from the antenna (KEMPEL AND VO-LAKIS, 1995). The results demonstrate that the conformal ABC (8.109) can be applied quite close to the body and still yield an accuracy which is acceptable for most purposes.

We close this section by noting that STUPFEL (1994) has recently developed a conformal three-dimensional vector ABC whose form is the same as (8.109), but, because of the different expansions that were used, the higher order terms

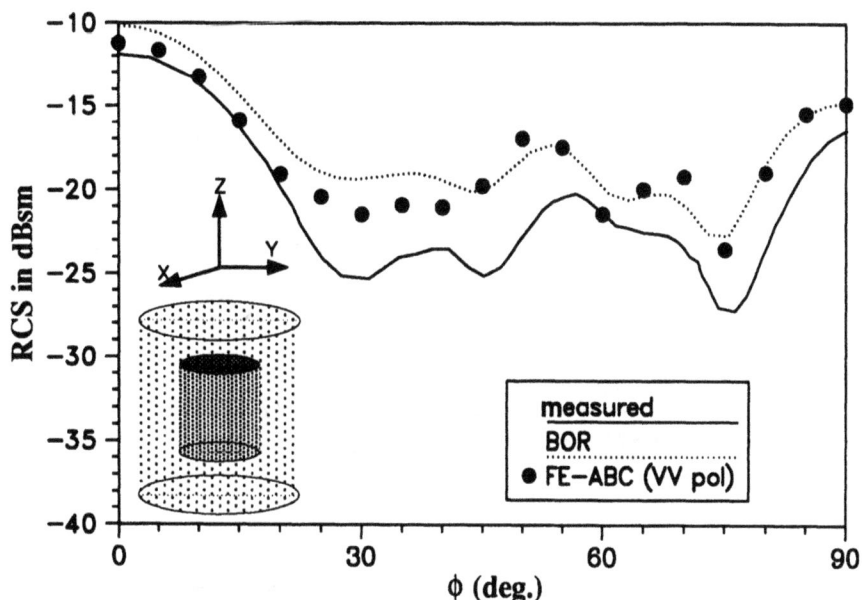

Figure 8–13: *Backscattering cross-section of a cylindrical inlet of height* $1.875\lambda_0$ *and diameter* $0.625\lambda_0$. *The corresponding dimensions of the conformal mesh termination are* $2.775\lambda_0$ *and* $1.075\lambda_0$

in the expressions for the coefficients $\bar{\bar{\alpha}}$, β and $\bar{\bar{\gamma}}$ differ from those in (8.110)–(8.112). Alternative ABCs valid for restricted classes of surface have also been presented by MAHADEVAN AND MITTRA (1993) and SUN AND BALANIS (1994).

8.3.4 Derivation from GIBCs

In Section 8.2.4 we showed that all of the two-dimensional ABCs could be deduced from the second order GIBCs derived in Appendix A, and a similar situation exists in three dimensions. From (A.52) we have, for a homogeneous body,

$$\hat{\gamma} \times \mathbf{E} = \hat{\gamma} \times \left(\hat{\gamma} \times \left\{ \bar{\bar{\eta}}.\mathbf{H} - \frac{Z}{2(k_0 N)^2 h_\gamma} \nabla[h_\gamma \nabla_s.\mathbf{H}] \right\} \right. $$
$$\left. - \frac{Z}{2(k_0 N)^2 h_\gamma} \nabla[h_\gamma \hat{\gamma}.\nabla \times \mathbf{H}] \right) \qquad (8.117)$$

where $\bar{\bar{\eta}}$ is defined in (A.47) and (A.49), and we note that (A.44) and (A.45) can be used to replace $\nabla_s.\mathbf{H}$ by $-jk_0 Y Z_0 H_\gamma$.

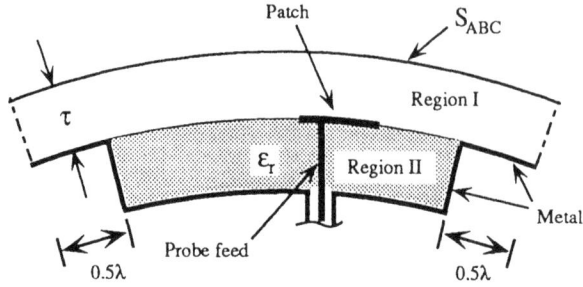

Figure 8–14: *Geometry for a cavity-backed patch antenna. The patch dimensions are 2 cm × 3 cm (long side along the axis), the cavity dimensions are 5 cm × 6 cm × 0.07874 cm, and $\epsilon_r = 2.7$*

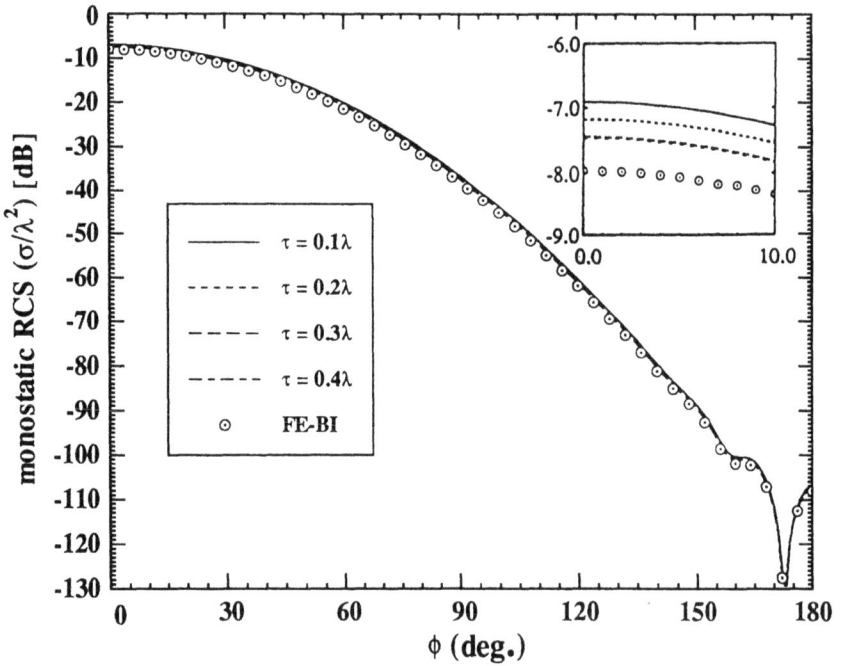

Figure 8–15: *Monostatic RCS of the cavity-backed patch antenna for E-polarisation as a function of the separation τ of the ABC surface, compared with data from a finite element–boundary integral formulation. In the inset the curves for $\tau = 0.3\lambda$ and 0.4λ are indistinguishable*

In the special case of a sphere, $\hat{\gamma} = \hat{r}$ and $h_\gamma = 1$, and (8.117) becomes

$$\hat{r} \times \mathbf{E} = Z\hat{r} \times \left(\hat{r} \times \left\{ \mathbf{H} + \frac{jYZ_0}{2k_0 N^2} \nabla_{\!s} H_r \right\} - \frac{1}{2(k_0 N)^2} \nabla[\hat{r}.\nabla \times \mathbf{H}] \right) \quad (8.118)$$

Since

$$\hat{r} \times \nabla[\hat{r}.\nabla \times \mathbf{H}] = -\nabla \times [\hat{r}(\hat{r}.\nabla \times \mathbf{H})]$$

Maxwell's equations can be used to write (8.118) as

$$\hat{r} \times (\nabla \times \mathbf{H}) = -jk_0 Y_0 Z \mathbf{H}_s - \frac{1}{2jk_0 N^2} \left\{ Y_0 Z \nabla \times [\hat{r}(\hat{r}.\nabla \times \mathbf{H})] - jk_0 \nabla_{\!s} H_r \right\}$$

whose dual is

$$\hat{r} \times (\nabla \times \mathbf{E}) = -jk_0 Z_0 Y \mathbf{E}_s - \frac{1}{2jk_0 N^2} \left\{ Z_0 Y \nabla \times [\hat{r}(\hat{r}.\nabla \times \mathbf{E})] - jk_0 \nabla_{\!s} E_r \right\}$$

If we now convert this to an ABC in the manner described in Section 8.2.4, we obtain

$$\hat{r} \times (\nabla \times \mathbf{E}) = jk_0 \mathbf{E}_s + \frac{1}{2jk_0} \left\{ \nabla \times [\hat{r}(\hat{r}.\nabla \times \mathbf{E})] + jk_0 \nabla_{\!s} E_r \right\} \quad (8.119)$$

which is identical to (8.81) if $1 + 1/(jk_0 r)$ is replaced by unity. As in the case of a cylinder, a third order Rytov expansion accurate to $O(|k_0 N|^{-3})$ is necessary to reproduce the term $1/(jk_0 r)$.

For a general curvilinear body it is convenient to express (8.117) in the Dupin coordinate system of the previous section. Then

$$\hat{n} \times \mathbf{E} = Z\hat{n} \times \left(\hat{n} \times \left\{ Y\overline{\overline{\eta}}.\mathbf{H} - \frac{1}{2(k_0 N)^2} \nabla(\nabla_{\!s}.\mathbf{H}) \right\} - \frac{1}{2(k_0 N)^2} \nabla[\hat{n}.\nabla \times \mathbf{H}] \right)$$

with

$$Y\overline{\overline{\eta}} = \left\{ 1 - \frac{1}{8k_0^2}(\Delta K)^2 \right\} \overline{\overline{I}} + \frac{1}{2jk_0} \left\{ \Delta K - \frac{1}{2jk_0} \frac{\partial}{\partial n} \Delta K \right\} \overline{\overline{T}}$$

where

$$\overline{\overline{I}} = \hat{v}_1 \hat{v}_1 + \hat{v}_2 \hat{v}_2$$

and $\overline{\overline{T}}$ is given in (8.106) and by a process analogous to that used above we obtain

$$\hat{n} \times (\nabla \times \mathbf{E}) = -jk_0 Z_0 Y (Y\overline{\overline{\eta}}.\mathbf{E}_s) - \frac{Z_0 Y}{2jk_0 N^2} \{ \nabla \times [\hat{n}(\hat{n}.\nabla \times \mathbf{E})]$$
$$+ \nabla_{\!s}(\nabla_{\!s}.\mathbf{E}) \} \quad (8.120)$$

The corresponding ABC is

$$\hat{n} \times (\nabla \times \mathbf{E}) = jk_0 \overline{\overline{\alpha}}_0.\mathbf{E}_s + \frac{1}{2jk_0} \{ \nabla \times [\hat{n}(\hat{n}.\nabla \times \mathbf{E})] + \nabla_{\!s}(\nabla_{\!s}.\mathbf{E}) \} \quad (8.121)$$

294   *Absorbing boundary conditions*

where

$$\bar{\bar{\alpha}}_0 = \left\{1 - \frac{1}{8k_0^2}(\Delta K)^2\right\}\bar{\bar{I}} - \frac{1}{2jk_0}\left\{\Delta K - \frac{1}{2jk_0}\frac{\partial}{\partial n}\Delta K\right\}\bar{\bar{T}} \qquad (8.122)$$

But

$$\frac{\partial}{\partial n}\Delta K = \frac{\partial}{\partial n}\left(\frac{1}{R_1} - \frac{1}{R_2}\right) = -\left(\frac{1}{R_1^2} - \frac{1}{R_2^2}\right)$$

from (8.103) and (8.87), so that

$$\frac{\partial}{\partial n}\Delta K = -(K_1 + K_2)(K_1 K_2) = -2K_\mu \Delta K$$

Hence

$$\bar{\bar{\alpha}}_0 = \left\{1 - \frac{1}{8k_0^2}(\Delta K)^2\right\}\bar{\bar{I}} - \frac{\Delta K}{2jk_0}\left\{1 - \frac{K_\mu}{jk_0}\right\}\bar{\bar{T}} \qquad (8.123)$$

and (8.121) agrees precisely with (8.109) if terms $O(k_0^{-3})$ are neglected.

References

Bayliss, A., Gunzburger, M. and Turkel, E. (1982), "Boundary conditions for the numerical solution of elliptic equations in exterior regions", *SIAM J. Appl. Math.*, **42**, pp. 430–451.

Bayliss, A. and Turkel, E. (1980), "Radiation boundary conditions for wave-like equations", *Commun. Pure Appl. Math.*, **23**, pp. 707–725.

Chatterjee, A. and Volakis, J. L. (1993), "Conformal absorbing boundary conditions for the vector wave equation", *Microwave Opt. Tech. Lett.*, **6**, pp. 886–888.

Chatterjee, A., Volakis, J. L. and Windheiser, D. (1994), "Parallel computation of 3D electromagnetic scattering using finite elements", *Int. J. Num. Modelling: Electr. Net. Dev. and Fields*, **7**, pp. 329–342.

Claerbout, J. F. (1985), *Imaging the Earth's Interior*, Blackwell Publ. Co.

Engquist, B. and Majda, A. (1977), "Absorbing boundary conditions for the numerical simulation of waves", *Math. Comput.*, **31**, pp. 629–651.

Eswarappa, C. and Hoefer, W. (1994), "One-way equation absorbing boundary conditions for 3-D TLM analysis of planar and quasi-planar structures", *IEEE Trans. Microwave Theory Tech.*, **MTT-42**, pp. 1669–1677.

Fang, J. (1993), "Absorbing boundary treatments in the simulation of wave propagation in microwave integrated-circuits", *Proc. 9th Annual Review of Progress in Applied Computational Electromagnetics*, Monterey CA, pp. 322–329.

Glisson, A. W. and Wilton, D. R. (1982), "Simple and efficient numerical techniques for treating bodies of revolution", University of Mississippi Technical Report No. 105 (revised).

Halpern, L. and Trefethen, L. N. (1988), "Wide-angle one-way wave equations", *J. Acoust. Soc. Amer.*, **84**, pp. 1397–1404.

Higdon, R. L. (1992), "Absorbing boundary conditions for acoustic and elastic waves in stratified media", *J. Comp. Phys.*, **101**, pp. 386–418.

Jones, D. S. (1992), "An improved surface radiation condition", *IMA J. Appl. Math.*, **48**, pp. 163–193.

Karp, S. N. and Karal, F. C., Jr. (1965), "Generalized impedance boundary conditions with applications to surface wave structures", in *Electromagnetic Wave Theory*, Pt. 1 (J. Brown, Ed.), Pergamon Press, New York, pp. 479–483.

Karp, S. N. (1961), "A convergent far-field expansion for two-dimensional radiation functions", *Commun. Pure Appl. Math.*, **14**, pp. 427–434.

Kempel, L. C. and Volakis, J. L. (1995), "Applying new vector ABCs to patch antennas on a circular cylinder", to be published.

Kunz, K. S. and Luebbers, R. J. (1993), *Finite Difference - Time Domain Method in Electromagnetics*, CRC Press, Boca Raton FL.

Lee, C. F., Shin, R. T., Kong, J. A. and McCartin, B. J. (1990), "Absorbing boundary conditions on circular and elliptical boundaries", *J. Electromagn. Waves Applics*, **4**, pp. 945–962.

Liao, Z. P., Wong, H. L., Yang, B. P. and Yuan, Y. F. (1984), "A transmitting boundary for transient wave analysis", *Scientia Sinica*, **28**, pp. 1063–1076.

Lindman, E. L. (1975), " 'Free-space' boundary conditions for the time dependent wave equation", *J. Comput. Phys.*, **18**, pp. 66–78.

Luebbers, R. J. and Penney, C. (1994), "Scattering from apertures in infinite ground planes using FDTD", *IEEE Trans. Antennas Propagat.*, **AP-42**, pp. 731–736.

Ma, Y.-C. (1991), "A note on the radiation boundary conditions for the Helmholtz equation", *IEEE Trans. Antennas Propagat.*, **AP-39**, pp. 1526–1530.

Mahadevan, K. and Mittra, R. (1993), "Radar cross section computation of inhomogeneous scatterers using edge-based finite element methods in frequency and time domains", *Radio Sci.*, **28**, pp. 1181–1193.

McInturff, K. and Simon, P. S. (1993), "Closed-form expressions for coefficients used in FD-TD high-order boundary conditions", *IEEE Microwave Guided Wave Lett.*, **3**, pp. 222–223.

Mei, K. K. and Fang, J. (1992), "Superabsorption: A method to improve absorbing boundary conditions", *IEEE Trans. Antennas Propagat.*, **AP-40**, pp. 1001–1010.

Mittra, R., Ramahi, O., Khebir, A., Gordon, R. and Kouki, A. (1989), "A review of absorbing boundary conditions for two and three-dimensional electromagnetic scattering problems", *IEEE Trans. Magnetics*, **25**, pp. 3034–3039.

Moghaddam, M. and Chew, W. C. (1991), "Stabilizing Liao's absorbing boundary conditions using single-precision arithmetic", *Proc. IEEE AP-S Int. Symp.*, London, Canada, pp. 430–433.

Moore, T. G., Blaschak, J. G., Taflove, A. and Kriegsmann, G. A. (1988), "Theory and application of radiation boundary operators", *IEEE Trans. Antennas Propagat.*, **AP-36**, pp. 1797–1811.

Mur, G. (1981), "Absorbing boundary conditions for the finite difference approximation of the time-domain electromagnetic-field equations", *IEEE Trans. Electromagn. Compat.*, **EMC-23**, pp. 377–382.

Peterson, A. F. (1988), "Absorbing boundary conditions for the vector wave equation", *Microwave Opt. Tech. Lett.*, **1**, pp. 62–64.

Peterson, A. F. and Castillo, S. P. (1989), "A frequency-domain differential equation formulation for electromagnetic scattering from inhomogeneous cylinders", *IEEE Trans. Antennas Propagat.*, **AP-37**, pp. 601–607.

Schuh, M. J., Woo, A., Sanders, M. and Wang, H. T. G. (1993), "Radar cross-section measurement data of four small cavities", NASA Tech. Memo. No. 108782.

Silvester, P. P. and Ferrari, R. L. (1990), *Finite Elements for Electrical Engineers* (2nd edn.), Cambridge University Press, Cambridge, U.K.

Stupfel, B. (1994), "Absorbing boundary conditions on arbitrary boundaries for the scalar and vector wave equations", *IEEE Trans. Antennas Propagat.*, **AP-42**, pp. 773–780.

Sun, W. and Balanis, C. A. (1994), "Vector one-way wave absorbing boundary conditions for FEM applications", *IEEE Trans. Antennas Propagat.*, **AP-42**, pp. 872–878.

Taflove, A. and Umashankar, K. R. (1987), "The finite difference time-domain (FD-TD) method for electromagnetic scattering and interaction problems", *J. Electromagn. Waves Applics*, 1, pp. 243–267.

Tai, C.-T. (1992), *Generalized Vector and Dyadic Analysis*, IEEE Press, Piscataway, NJ.

Trefethen, L. N. and Halpern, L. (1986), "Well-posedness of one-way wave equations and absorbing boundary conditions", *Math. Comput.*, **47**, pp. 421–435.

Van Bladel, J. (1985), *Electromagnetic Fields*, Hemisphere Pub. Co., New York, p. 503.

Webb, J. P. and Kanellopoulos, V. N. (1989), "Absorbing boundary conditions for the finite element solution of the vector wave equation", *Microwave Opt. Tech. Lett.*, **2**, pp. 370–372.

Wilcox, C. H. (1956), "An expansion theorem for electromagnetic fields", *Commun. Pure Appl. Math.*, **9**, pp. 115–132.

Appendix A

Generalised Rytov analysis

One of the first rigorous derivations of an approximate boundary condition was by RYTOV (1940), who developed conditions through the second order applicable at the curved surface of a highly conducting body. The results are important in showing the effect of surface curvature and material variations, and we will now apply the same method to the formulation used by LEONTOVICH (1948). The boundary conditions obtained (SENIOR, 1990) are more general than those given by either author, and reveal some errors in the expressions quoted by Leontovich.

A.1 Formulation

A lossy body composed of a material whose complex permittivity ϵ and complex permeability μ may vary as functions of position is immersed in free space and illuminated by an electromagnetic field. On the assumption that the external field varies slowly over the surface S, we seek a boundary condition that can be applied at S to simulate the effect of the material.

Inside the body Maxwell's equations are

$$\nabla \times \mathbf{E}' = -j\omega\mu\mathbf{H}', \qquad \nabla \times \mathbf{H}' = j\omega\epsilon\mathbf{E}'$$

where the prime denotes the interior field. Since

$$\nabla \times (\sqrt{\epsilon}\,\mathbf{E}') = \nabla(\sqrt{\epsilon}) \times \mathbf{E}' + \sqrt{\epsilon}\,\nabla \times \mathbf{E}'$$

we have

$$\nabla \times (\sqrt{\epsilon}\,\mathbf{E}') + \sqrt{\epsilon}\,\mathbf{E}' \times \nabla(\ln\sqrt{\epsilon}) = -jk_0 N\sqrt{\mu}\,\mathbf{H}' \qquad (\text{A.1})$$

where $N = \sqrt{\epsilon\mu/\epsilon_0\mu_0}$ is the complex refractive index of the material, and similarly

$$\nabla \times (\sqrt{\mu}\,\mathbf{H}') + \sqrt{\mu}\,\mathbf{H}' \times \nabla(\ln\sqrt{\mu}) = jk_0 N\sqrt{\epsilon}\,\mathbf{E}' \qquad (\text{A.2})$$

Assume $|N|$ is large everywhere inside the body, and on this basis write

$$N = \frac{w}{q} \qquad (\text{A.3})$$

where w is a function of position and q is a small parameter. With geometrical optics as a guide, let

$$\sqrt{\epsilon}\,\mathbf{E}' = \mathbf{A}e^{-jk_0\psi/q}, \qquad \sqrt{\mu}\,\mathbf{H}' = \mathbf{B}e^{-jk_0\psi/q}$$

Then

$$\nabla \times (\sqrt{\epsilon}\,\mathbf{E}') = \left(\nabla \times \mathbf{A} - \frac{jk_0}{q}\nabla\psi \times \mathbf{A}\right)e^{-jk_0\psi/q}$$

with a similar equation for $\nabla \times (\sqrt{\mu}\,\mathbf{H}')$, and (A.1) and (A.2) now become

$$w\mathbf{A} + \nabla\psi \times \mathbf{B} = \frac{q}{jk_0}\sqrt{\mu}\,\nabla \times \frac{\mathbf{B}}{\sqrt{\mu}}$$

(A.4)

$$w\mathbf{B} - \nabla\psi \times \mathbf{A} = -\frac{q}{jk_0}\sqrt{\epsilon}\,\nabla \times \frac{\mathbf{A}}{\sqrt{\epsilon}}$$

We seek a solution for $\mathbf{A}$ and $\mathbf{B}$ in the form of power series in q, *viz.*

$$\mathbf{A} = \mathbf{A}_0 + q\mathbf{A}_1 + q^2\mathbf{A}_2 + \cdots$$

(A.5)

$$\mathbf{B} = \mathbf{B}_0 + q\mathbf{B}_1 + q^2\mathbf{B}_2 + \cdots$$

and for this purpose we introduce the orthogonal curvilinear coordinates α, β, γ with metric coefficients h_α, h_β, h_γ such that S is the surface $\gamma = $ constant with $\hat{\gamma}$ in the direction of the outward normal.

A.2   Zeroth order solution

Inserting (A.5) into (A.4) and retaining only terms which are independent of q, we find

$$w\mathbf{A}_0 + \nabla\psi \times \mathbf{B}_0 = 0$$

(A.6)

$$w\mathbf{B}_0 - \nabla\psi \times \mathbf{A}_0 = 0$$

showing that $\mathbf{A}_0$, $\mathbf{B}_0$ and $\nabla\psi$ are mutually perpendicular. By eliminating (say) $\mathbf{B}_0$,

$$(w^2 - |\nabla\psi|^2)\mathbf{A}_0 = 0$$

implying

$$|\nabla\psi|^2 = w^2$$

analogous to the eikonal equation of geometrical optics. Hence

$$\nabla\psi = w\hat{s}$$

where $\hat{s}$ is a unit vector in the direction of propagation in the body.

The tangential components of the electric and magnetic fields are continuous across the surface and, as already noted, the fields outside are slowly varying over S. Hence, the fields inside must also vary slowly, and this is only possible if ψ is constant on S. It follows that $\nabla\psi$ (and therefore $\hat{s}$) are normal to S and, consistent with propagation into the body,

$$\nabla\psi = -w\hat{\gamma} \tag{A.7}$$

From (A.6) and (A.7)

$$\mathbf{A}_0 = \hat{\gamma} \times \mathbf{B}_0, \qquad \mathbf{B}_0 = -\hat{\gamma} \times \mathbf{A}_0 \tag{A.8}$$

and to this order of approximation the local field inside the body looks like a plane wave propagating in the direction of the inward normal to S. In particular, just inside the surface,

$$A_{0\alpha} = -B_{0\beta}, \qquad A_{0\beta} = B_{0\alpha}, \qquad A_{0\gamma} = B_{0\gamma} = 0 \tag{A.9}$$

From the continuity of the tangential field components across S

$$\mathbf{A}_0 = \sqrt{\epsilon}\,\mathbf{E}, \qquad \mathbf{B}_0 = \sqrt{\mu}\,\mathbf{H}$$

where $\mathbf{E}, \mathbf{H}$ are the external fields, and thus, to the zeroth order,

$$E_\alpha = -ZH_\beta, \qquad E_\beta = ZH_\alpha \tag{A.10}$$

on S, with

$$E_\gamma = H_\gamma = 0 \tag{A.11}$$

where

$$Z = \sqrt{\frac{\mu}{\epsilon}} \tag{A.12}$$

is the intrinsic impedance of the material at the surface. In vector form,

$$\hat{\gamma} \times \mathbf{E} = Z\,\hat{\gamma} \times (\hat{\gamma} \times \mathbf{H}) \tag{A.13}$$

and this is the standard impedance boundary condition, customarily derived on the basis of a homogeneous half space (see Section 2.2).

A.3 First order solution

Equating the coefficients of q on both sides of (A.4) we obtain

$$\mathbf{A}_1 - \hat{\gamma} \times \mathbf{B}_1 = \frac{1}{jk_0 w}\sqrt{\mu}\,\nabla \times \frac{\mathbf{B}_0}{\sqrt{\mu}} \tag{A.14}$$

$$\mathbf{B}_1 + \hat{\gamma} \times \mathbf{A}_1 = -\frac{1}{jk_0 w}\sqrt{\epsilon}\,\nabla \times \frac{\mathbf{A}_0}{\sqrt{\epsilon}} \tag{A.15}$$

In terms of the chosen coordinates, the components of (A.14) are

$$A_{1\alpha} + B_{1\beta} = -\frac{1}{jk_0 w}\frac{\sqrt{\mu}}{h_\beta h_\gamma}\frac{\partial}{\partial\gamma}\left(h_\beta\frac{B_{0\beta}}{\sqrt{\mu}}\right)$$

$$A_{1\beta} - B_{1\alpha} = \frac{1}{jk_0 w}\frac{\sqrt{\mu}}{h_\alpha h_\gamma}\frac{\partial}{\partial\gamma}\left(h_\alpha\frac{B_{0\alpha}}{\sqrt{\mu}}\right)$$

$$A_{1\gamma} = \frac{1}{jk_0 w}\frac{\sqrt{\mu}}{h_\alpha h_\beta}\left\{\frac{\partial}{\partial\alpha}\left(h_\beta\frac{B_{0\beta}}{\sqrt{\mu}}\right) - \frac{\partial}{\partial\beta}\left(h_\alpha\frac{B_{0\alpha}}{\sqrt{\mu}}\right)\right\} \quad \text{(A.16)}$$

and in view of (A.9), the components of (A.15) are

$$B_{1\alpha} - A_{1\beta} = \frac{1}{jk_0 w}\frac{\sqrt{\epsilon}}{h_\beta h_\gamma}\frac{\partial}{\partial\gamma}\left(h_\beta\frac{B_{0\alpha}}{\sqrt{\epsilon}}\right)$$

$$B_{1\beta} + A_{1\alpha} = \frac{1}{jk_0 w}\frac{\sqrt{\epsilon}}{h_\alpha h_\gamma}\frac{\partial}{\partial\gamma}\left(h_\alpha\frac{B_{0\beta}}{\sqrt{\epsilon}}\right)$$

$$B_{1\gamma} = -\frac{1}{jk_0 w}\frac{\sqrt{\epsilon}}{h_\alpha h_\beta}\left\{\frac{\partial}{\partial\alpha}\left(h_\beta\frac{B_{0\alpha}}{\sqrt{\epsilon}}\right) + \frac{\partial}{\partial\beta}\left(h_\alpha\frac{B_{0\beta}}{\sqrt{\epsilon}}\right)\right\} \quad \text{(A.17)}$$

The expressions for $A_{1\alpha} + B_{1\beta}$ are identical if

$$\frac{\sqrt{\mu}}{h_\beta}\frac{\partial}{\partial\gamma}\left(h_\beta\frac{B_{0\beta}}{\sqrt{\mu}}\right) + \frac{\sqrt{\epsilon}}{h_\alpha}\frac{\partial}{\partial\gamma}\left(h_\alpha\frac{B_{0\beta}}{\sqrt{\epsilon}}\right) = 0$$

implying

$$\frac{\partial B_{0\beta}}{\partial\gamma} = \frac{1}{2}B_{0\beta}\frac{\partial}{\partial\gamma}\left(\ln\frac{\sqrt{\epsilon\mu}}{h_\alpha h_\beta}\right) \quad \text{(A.18)}$$

and therefore

$$\frac{\partial}{\partial\gamma}\left(h_\beta\frac{B_{0\beta}}{\sqrt{\mu}}\right) = -\frac{1}{2}\frac{h_\beta}{\sqrt{\mu}}B_{0\beta}\frac{\partial}{\partial\gamma}\left(\ln\frac{h_\alpha}{h_\beta}Z\right)$$

Similarly, the expressions for $A_{1\beta} - B_{1\alpha}$ are the same if

$$\frac{\partial B_{0\alpha}}{\partial\gamma} = \frac{1}{2}B_{0\alpha}\frac{\partial}{\partial\gamma}\left(\ln\frac{\sqrt{\epsilon\mu}}{h_\alpha h_\beta}\right) \quad \text{(A.19)}$$

giving

$$\frac{\partial}{\partial\gamma}\left(h_\alpha\frac{B_{0\alpha}}{\sqrt{\mu}}\right) = -\frac{1}{2}\frac{h_\alpha}{\sqrt{\mu}}B_{0\alpha}\frac{\partial}{\partial\gamma}\left(\ln\frac{h_\beta}{h_\alpha}Z\right)$$

Hence

$$A_{1\alpha} = -B_{1\beta} + \frac{B_{0\beta}}{2jk_0 w h_\gamma}\frac{\partial}{\partial\gamma}\left(\ln\frac{h_\alpha}{h_\beta}Z\right) \quad \text{(A.20)}$$

$$A_{1\beta} = B_{1\alpha} - \frac{B_{0\alpha}}{2jk_0 w h_\gamma}\frac{\partial}{\partial\gamma}\left(\ln\frac{h_\beta}{h_\alpha}Z\right) \quad \text{(A.21)}$$

with $A_{1\gamma}$ and $B_{1\gamma}$ as shown in (A.16) and (A.17), respectively.

It is now a trivial matter to construct approximate boundary conditions for the tangential components of the external field on S. From (A.9) and (A.20)

$$
\begin{aligned}
A_\alpha &= A_{0\alpha} + qA_{1\alpha} + O(q^2) \\
&= -B_{0\beta} - qB_{1\beta} + \frac{qB_{0\beta}}{2jk_0wh_\gamma} \frac{\partial}{\partial\gamma}\left(\ln\frac{h_\alpha}{h_\beta}Z\right) + O(q^2) \\
&= -B_\beta\left\{1 - \frac{q}{2jk_0wh_\gamma}\frac{\partial}{\partial\gamma}\left(\ln\frac{h_\alpha}{h_\beta}Z\right)\right\} + O(q^2)
\end{aligned}
$$

and thus, to the first order in q,

$$
A_\alpha = -B_\beta\left\{1 - \frac{1}{2jk_0Nh_\gamma}\frac{\partial}{\partial\gamma}\left(\ln\frac{h_\alpha}{h_\beta}Z\right)\right\} \tag{A.22}
$$

Similarly, from (A.9) and (A.21),

$$
A_\beta = B_\alpha\left\{1 - \frac{1}{2jk_0Nh_\gamma}\frac{\partial}{\partial\gamma}\left(\ln\frac{h_\beta}{h_\alpha}Z\right)\right\} \tag{A.23}
$$

and, to the first order in $|N|^{-1}$, the boundary conditions on S are

$$
\begin{aligned}
E_\alpha &= -ZH_\beta\left\{1 - \frac{1}{2jk_0Nh_\gamma}\frac{\partial}{\partial\gamma}\left(\ln\frac{h_\alpha}{h_\beta}Z\right)\right\} \\
E_\beta &= ZH_\alpha\left\{1 - \frac{1}{2jk_0Nh_\gamma}\frac{\partial}{\partial\gamma}\left(\ln\frac{h_\beta}{h_\alpha}Z\right)\right\}
\end{aligned} \tag{A.24}
$$

where all quantities are evaluated at the surface. The factor $\frac{1}{2}$ is missing from the formulas quoted by LEONTOVICH (1948).

To the order shown, only the normal variation of the impedance has any effect. This is consistent with the interpretation of the surface impedance as the local impedance looking in, and provides justification for applying the SIBC (A.13) at each point of a surface even when the properties of the material vary laterally. If $h_\alpha \neq h_\beta$ the effective surface impedance implied by (A.24) is a tensor:

$$
\bar{\bar{\eta}} = \eta_{\alpha\alpha}\hat{\alpha}\hat{\alpha} + \eta_{\beta\beta}\hat{\beta}\hat{\beta} \tag{A.25}
$$

with

$$
\eta_{\alpha\alpha} = Z\left\{1 - \frac{1}{2jk_0Nh_\gamma}\frac{\partial}{\partial\gamma}\left(\ln\frac{h_\beta}{h_\alpha}Z\right)\right\} \tag{A.26}
$$

$$
\eta_{\beta\beta} = Z\left\{1 - \frac{1}{2jk_0Nh_\gamma}\frac{\partial}{\partial\gamma}\left(\ln\frac{h_\alpha}{h_\beta}Z\right)\right\}
$$

and in terms of $\bar{\bar{\eta}}$

$$
\hat{\gamma} \times \mathbf{E} = \bar{\bar{\eta}}.\hat{\gamma} \times (\hat{\gamma} \times \mathbf{H}) \tag{A.27}
$$

Since $N \to N$ and $Z \to 1/Z$ under the duality transformation, it can be verified that (A.27) satisfies duality to the first order in $|N|^{-1}$. The anisotropy is a consequence of the curvature of the surface. In the special case when the coordinates coincide with the directions of the principal curvatures at every point of S,

$$\frac{1}{h_\gamma} \frac{\partial}{\partial \gamma} \left(\ln \frac{h_\alpha}{h_\beta} \right) = \frac{1}{R_\alpha} - \frac{1}{R_\beta} \tag{A.28}$$

where R_α and R_β are the principal radii of curvature, and, if $R_\alpha = R_\beta$ (including a planar surface as a particular example), the impedance becomes a scalar. Thus, for a planar surface $y =$ constant ($\alpha = z$, $\beta = x$, $\gamma = y$ where x, y, z are Cartesian coordinates, implying $h_\alpha = h_\beta = h_\gamma = 1$)

$$\overline{\overline{\eta}} = Z \left(1 - \frac{1}{2jk_0 N} \frac{\partial}{\partial y} (\ln Z) \right) \overline{\overline{I}} \tag{A.29}$$

where $\overline{\overline{I}}$ is the identity tensor in the α, β coordinates. Likewise, for a spherical surface $r =$ constant ($\alpha = \theta$, $\beta = \phi$, $\gamma = r$ where r, θ, ϕ are spherical polar coordinates, implying $h_\alpha = r$, $h_\beta = r \sin \theta$, $h_\gamma = 1$)

$$\overline{\overline{\eta}} = Z \left(1 - \frac{1}{2jk_0 N} \frac{\partial}{\partial r} (\ln Z) \right) \overline{\overline{I}} \tag{A.30}$$

but for a circular cylindrical surface $\rho =$ constant ($\alpha = \phi$, $\beta = z$, $\gamma = \rho$ where ρ, ϕ, z are cylindrical polar coordinates, implying $h_\alpha = \rho$, $h_\beta = h_\gamma = 1$)

$$\overline{\overline{\eta}} = Z \left(1 - \frac{1}{2jk_0 N} \frac{\partial}{\partial \rho} (\ln Z) \right) \overline{\overline{I}} - \frac{Z}{2jk_0 N \rho} (\hat{\alpha}\hat{\alpha} - \hat{\beta}\hat{\beta}) \tag{A.31}$$

In all of these results the derivative is evaluated at the surface.

A.4   Second order solution

For the terms involving q^2 the analysis is more tedious. From (A.4)

$$\mathbf{A}_2 - \hat{\gamma} \times \mathbf{B}_2 = \frac{1}{jk_0 w} \sqrt{\mu} \, \nabla \times \frac{\mathbf{B}_1}{\sqrt{\mu}} \tag{A.32}$$

$$\mathbf{B}_2 + \hat{\gamma} \times \mathbf{A}_2 = -\frac{1}{jk_0 w} \sqrt{\epsilon} \, \nabla \times \frac{\mathbf{A}_1}{\sqrt{\epsilon}} \tag{A.33}$$

the components of which are

$$A_{2\alpha} + B_{2\beta} = \frac{\sqrt{\mu}}{jk_0 w h_\beta h_\gamma} \left\{ \frac{\partial}{\partial \beta} \left(\frac{h_\gamma B_{1\gamma}}{\sqrt{\mu}} \right) - \frac{\partial}{\partial \gamma} \left(\frac{h_\beta B_{1\beta}}{\sqrt{\mu}} \right) \right\} \tag{A.34}$$

$$A_{2\beta} - B_{2\alpha} = \frac{\sqrt{\mu}}{jk_0 w h_\alpha h_\gamma} \left\{ \frac{\partial}{\partial \gamma} \left(\frac{h_\alpha B_{1\alpha}}{\sqrt{\mu}} \right) - \frac{\partial}{\partial \alpha} \left(\frac{h_\gamma B_{1\gamma}}{\sqrt{\mu}} \right) \right\} \tag{A.35}$$

$$A_{2\gamma} = \frac{1}{jk_0 w} \frac{\sqrt{\mu}}{h_\alpha h_\beta} \left\{ \frac{\partial}{\partial \alpha} \left(\frac{h_\beta B_{1\beta}}{\sqrt{\mu}} \right) - \frac{\partial}{\partial \beta} \left(\frac{h_\alpha B_{1\alpha}}{\sqrt{\mu}} \right) \right\} \qquad (A.36)$$

$$B_{2\alpha} - A_{2\beta} = -\frac{\sqrt{\epsilon}}{jk_0 w h_\beta h_\gamma} \left\{ \frac{\partial}{\partial \beta} \left(\frac{h_\gamma A_{1\gamma}}{\sqrt{\epsilon}} \right) - \frac{\partial}{\partial \gamma} \left(\frac{h_\beta A_{1\beta}}{\sqrt{\epsilon}} \right) \right\} \qquad (A.37)$$

$$B_{2\beta} + A_{2\alpha} = -\frac{\sqrt{\epsilon}}{jk_0 w h_\alpha h_\gamma} \left\{ \frac{\partial}{\partial \gamma} \left(\frac{h_\alpha A_{1\alpha}}{\sqrt{\epsilon}} \right) - \frac{\partial}{\partial \alpha} \left(\frac{h_\gamma A_{1\gamma}}{\sqrt{\epsilon}} \right) \right\} \qquad (A.38)$$

$$B_{2\gamma} = -\frac{1}{jk_0 w} \frac{\sqrt{\epsilon}}{h_\alpha h_\beta} \left\{ \frac{\partial}{\partial \alpha} \left(\frac{h_\beta A_{1\beta}}{\sqrt{\epsilon}} \right) - \frac{\partial}{\partial \beta} \left(\frac{h_\alpha A_{1\alpha}}{\sqrt{\epsilon}} \right) \right\} \qquad (A.39)$$

The expressions for $A_{2\alpha} + B_{2\beta}$ are the same if

$$\frac{\partial B_{1\beta}}{\partial \gamma} = \frac{\partial A_{1\alpha}}{\partial \gamma} + A_{1\alpha} \frac{\partial}{\partial \gamma} \left(\ln \frac{h_\alpha}{\sqrt{\epsilon}} \right) - B_{1\beta} \frac{\partial}{\partial \gamma} \left(\ln \frac{h_\beta}{\sqrt{\mu}} \right)$$

$$- \frac{\sqrt{\epsilon}}{h_\alpha} \frac{\partial}{\partial \alpha} \left(\frac{h_\gamma A_{1\gamma}}{\sqrt{\epsilon}} \right) + \frac{\sqrt{\mu}}{h_\beta} \frac{\partial}{\partial \beta} \left(\frac{h_\gamma B_{1\gamma}}{\sqrt{\mu}} \right)$$

and, in view of (A.20), this serves to determine $\partial B_{1\beta}/\partial \gamma$ as

$$\frac{\partial B_{1\beta}}{\partial \gamma} = -\frac{1}{2} B_{1\beta} \frac{\partial}{\partial \gamma} \left(\ln \frac{h_\alpha h_\beta}{\sqrt{\epsilon \mu}} \right) - \frac{\sqrt{\epsilon}}{2 h_\alpha} \frac{\partial}{\partial \alpha} \left(\frac{h_\gamma A_{1\gamma}}{\sqrt{\epsilon}} \right)$$

$$+ \frac{\sqrt{\mu}}{2 h_\beta} \frac{\partial}{\partial \beta} \left(\frac{h_\gamma B_{1\gamma}}{\sqrt{\mu}} \right) + \frac{B_{0\beta}}{4 j k_0 w h_\gamma} \frac{\partial}{\partial \gamma} \left(\ln \frac{h_\alpha}{h_\beta} Z \right) \frac{\partial}{\partial \gamma} \left(\ln \frac{h_\alpha}{\sqrt{\epsilon}} \right)$$

$$+ \frac{1}{4 j k_0} \frac{\partial}{\partial \gamma} \left\{ \frac{B_{0\beta}}{w h_\gamma} \frac{\partial}{\partial \gamma} \left(\ln \frac{h_\alpha}{h_\beta} Z \right) \right\} \qquad (A.40)$$

Substitution into (say) (A.34) then provides a unique expression for $A_{2\alpha} + B_{2\beta}$, and using (A.16), (A.17) and (A.20) we obtain (SENIOR, 1990)

$$A_{2\alpha} = -B_{2\beta} + \frac{1}{2 j k_0 w h_\gamma} \left\{ B_{1\beta} \frac{\partial}{\partial \gamma} \left(\ln \frac{h_\alpha}{h_\beta} Z \right) \right.$$

$$- \frac{B_{0\beta}}{2 j k_0} \left[\frac{1}{2 w h_\gamma} \left\{ \frac{\partial}{\partial \gamma} \left(\ln \frac{h_\alpha}{h_\beta} Z \right) \right\}^2 + \frac{\partial}{\partial \gamma} \left\{ \frac{1}{w h_\gamma} \frac{\partial}{\partial \gamma} \left(\ln \frac{h_\alpha}{h_\beta} Z \right) \right\} \right]$$

$$+ \frac{\sqrt{\epsilon}}{j k_0 h_\alpha} \frac{\partial}{\partial \alpha} \left[\frac{Z}{w} \frac{h_\gamma}{h_\alpha h_\beta} \left\{ \frac{\partial}{\partial \alpha} \left(\frac{h_\beta B_{0\beta}}{\sqrt{\mu}} \right) - \frac{\partial}{\partial \beta} \left(\frac{h_\alpha B_{0\alpha}}{\sqrt{\mu}} \right) \right\} \right]$$

$$\left. - \frac{\sqrt{\mu}}{j k_0 h_\beta} \frac{\partial}{\partial \beta} \left[\frac{1}{Z w} \frac{h_\gamma}{h_\alpha h_\beta} \left\{ \frac{\partial}{\partial \alpha} \left(\frac{h_\beta B_{0\alpha}}{\sqrt{\epsilon}} \right) + \frac{\partial}{\partial \beta} \left(\frac{h_\alpha B_{0\beta}}{\sqrt{\epsilon}} \right) \right\} \right] \right\} \qquad (A.41)$$

Hence

$$A_\alpha = A_{0\alpha} + q A_{1\alpha} + q^2 A_{2\alpha} + O(q^3)$$

and when the expressions for $A_{0\alpha}$, $A_{1\alpha}$ and $A_{2\alpha}$ are inserted, the result is

$$A_\alpha = -\left\{ 1 - \frac{q}{2 j k_0 w h_\gamma} \frac{\partial}{\partial \gamma} \left(\ln \frac{h_\alpha}{h_\beta} Z \right) - \frac{q^2}{(2 k_0 w h_\gamma)^2} \left[\frac{\partial^2}{\partial \gamma^2} \left(\ln \frac{h_\alpha}{h_\beta} Z \right) \right. \right.$$

$$+ \frac{1}{2} \left\{ \frac{\partial}{\partial \gamma} \left(\ln \frac{h_\alpha}{h_\beta} Z \right) \right\}^2 - \frac{\partial}{\partial \gamma} \left(\ln \frac{h_\alpha}{h_\beta} Z \right) \frac{\partial}{\partial \gamma} (\ln w h_\gamma) \right] \right\} B_\beta$$

$$- \frac{q^2 \sqrt{\epsilon}}{2 k_0^2 w h_\alpha h_\gamma} \frac{\partial}{\partial \alpha} \left[\frac{Z}{w} \frac{h_\gamma}{h_\alpha h_\beta} \left\{ \frac{\partial}{\partial \alpha} \left(\frac{h_\beta B_\beta}{\sqrt{\mu}} \right) - \frac{\partial}{\partial \beta} \left(\frac{h_\alpha B_\alpha}{\sqrt{\mu}} \right) \right\} \right]$$

$$+ \frac{q^2 \sqrt{\mu}}{2 k_0^2 w h_\beta h_\gamma} \frac{\partial}{\partial \beta} \left[\frac{1}{Zw} \frac{h_\gamma}{h_\alpha h_\beta} \left\{ \frac{\partial}{\partial \alpha} \left(\frac{h_\beta B_\alpha}{\sqrt{\epsilon}} \right) + \frac{\partial}{\partial \beta} \left(\frac{h_\alpha B_\beta}{\sqrt{\epsilon}} \right) \right\} \right] \quad \text{(A.42)}$$

accurate to the second order in q.

In terms of the external fields

$$\frac{1}{h_\alpha h_\beta} \left\{ \frac{\partial}{\partial \alpha} \left(\frac{h_\beta B_\beta}{\sqrt{\mu}} \right) - \frac{\partial}{\partial \beta} \left(\frac{h_\alpha B_\alpha}{\sqrt{\mu}} \right) \right\} = \hat{\gamma}. \nabla \times \mathbf{H} = jk_0 Y_0 E_\gamma \quad \text{(A.43)}$$

where the substitution is valid because ψ is constant on S and the derivatives are tangential ones. Similarly

$$\frac{1}{h_\alpha h_\beta} \left\{ \frac{\partial}{\partial \alpha} \left(\frac{h_\beta B_\alpha}{\sqrt{\epsilon}} \right) + \frac{\partial}{\partial \beta} \left(\frac{h_\alpha B_\beta}{\sqrt{\epsilon}} \right) \right\} = \nabla_{\mathbf{s}}.(Z\mathbf{H}) \quad \text{(A.44)}$$

where $\nabla_{\mathbf{s}}.$ is the surface divergence or, by using (A.9),

$$\frac{1}{h_\alpha h_\beta} \left\{ \frac{\partial}{\partial \alpha} \left(\frac{h_\beta B_\alpha}{\sqrt{\epsilon}} \right) + \frac{\partial}{\partial \beta} \left(\frac{h_\alpha B_\beta}{\sqrt{\epsilon}} \right) \right\} = \hat{\gamma}. \nabla \times \mathbf{E} = -jk_0 Z_0 H_\gamma \quad \text{(A.45)}$$

Since $w = Nq$ implying

$$\frac{\partial}{\partial \gamma} (\ln w) = \frac{\partial}{\partial \gamma} (\ln N)$$

(A.42) can now be written as

$$E_\alpha - \frac{1}{2jk_0 Nh_\gamma} \hat{\alpha}. \nabla \left(\frac{\epsilon_0}{\epsilon} h_\gamma E_\gamma \right) = -\eta_{\beta\beta} \left\{ H_\beta - \frac{1}{2jk_0 Nh_\gamma} \hat{\beta}. \nabla \left(\frac{\mu_0}{\mu} h_\gamma H_\gamma \right) \right\} \quad \text{(A.46)}$$

where

$$\eta_{\beta\beta} = Z \left\{ 1 - \frac{1}{2jk_0 Nh_\gamma} \frac{\partial}{\partial \gamma} \left(\ln \frac{h_\alpha}{h_\beta} Z \right) - \frac{1}{(2k_0 Nh_\gamma)^2} \left[\frac{\partial^2}{\partial \gamma^2} \left(\ln \frac{h_\alpha}{h_\beta} Z \right) \right. \right.$$

$$\left. \left. + \frac{1}{2} \left\{ \frac{\partial}{\partial \gamma} \left(\ln \frac{h_\alpha}{h_\beta} Z \right) \right\}^2 - \frac{\partial}{\partial \gamma} \left(\ln \frac{h_\alpha}{h_\beta} Z \right) \frac{\partial}{\partial \gamma} (\ln h_\gamma N) \right] \right\} \quad \text{(A.47)}$$

A similar analysis starting with (A.35) and (A.37) gives (SENIOR, 1990)

$$E_\beta - \frac{1}{2jk_0 Nh_\gamma} \hat{\beta}. \nabla \left(\frac{\epsilon_0}{\epsilon} h_\gamma E_\gamma \right) = \eta_{\alpha\alpha} \left\{ H_\alpha - \frac{1}{2jk_0 Nh_\gamma} \hat{\alpha}. \nabla \left(\frac{\mu_0}{\mu} h_\gamma H_\gamma \right) \right\} \quad \text{(A.48)}$$

where

$$
\eta_{\alpha\alpha} = Z\left\{1 - \frac{1}{2jk_0Nh_\gamma}\frac{\partial}{\partial\gamma}\left(\ln\frac{h_\beta}{h_\alpha}Z\right) - \frac{1}{(2k_0Nh_\gamma)^2}\left[\frac{\partial^2}{\partial\gamma^2}\left(\ln\frac{h_\beta}{h_\alpha}Z\right)\right.\right.
$$
$$
\left.\left.+ \frac{1}{2}\left\{\frac{\partial}{\partial\gamma}\left(\ln\frac{h_\beta}{h_\alpha}Z\right)\right\}^2 - \frac{\partial}{\partial\gamma}\left(\ln\frac{h_\beta}{h_\alpha}Z\right)\frac{\partial}{\partial\gamma}(\ln h_\gamma N)\right]\right\} \quad (A.49)
$$

This differs from $\eta_{\beta\beta}$ only in having h_α and h_β interchanged, and to the required order

$$
\eta_{\alpha\alpha} = Z\left(1 - \frac{1}{2jk_0Nh_\gamma}\frac{\partial}{\partial\gamma}\right)\exp\left\{-\frac{1}{2jk_0Nh_\gamma}\frac{\partial}{\partial\gamma}\left(\ln\frac{h_\beta}{h_\alpha}Z\right)\right\}
$$

$$
(A.50)
$$

$$
\eta_{\beta\beta} = Z\left(1 - \frac{1}{2jk_0Nh_\gamma}\frac{\partial}{\partial\gamma}\right)\exp\left\{-\frac{1}{2jk_0Nh_\gamma}\frac{\partial}{\partial\gamma}\left(\ln\frac{h_\alpha}{h_\beta}Z\right)\right\}
$$

The boundary conditions are accurate to the second order in $|N|^{-1}$ and in vector form

$$
\hat{\gamma}\times\left\{\mathbf{E} - \frac{1}{2jk_0Nh_\gamma}\nabla\left(\frac{\epsilon_0}{\epsilon}h_\gamma E_\gamma\right)\right\}
$$

$$
= \bar{\bar{\eta}}.\hat{\gamma}\times\left(\hat{\gamma}\times\left\{\mathbf{H} - \frac{1}{2jk_0Nh_\gamma}\nabla\left(\frac{\mu_0}{\mu}h_\gamma H_\gamma\right)\right\}\right) \quad (A.51)
$$

where $\bar{\bar{\eta}}$ is the tensor surface impedance shown in (A.25). It can be verified that duality is satisfied.

A.5   Special cases

The second order boundary condition (A.51) is remarkably compact and most of the complication is embedded in the expression for $\bar{\bar{\eta}}$. It has the same form regardless of any variation in the material properties of the surface, and in this respect it is similar to the zeroth and first order conditions. Any lateral variation in the properties is taken care of by the gradient operations, and a change in the normal direction affects only the surface impedance $\bar{\bar{\eta}}$. However, as noted in Section 5.2, the form is not convenient for most applications, and it is more desirable to have $\hat{\gamma}\times\mathbf{E}$ expressed in terms of the tangential magnetic field components and their tangential derivatives. Recognising that (A.51) is actually a second order expansion in powers of q, the transformation is easy to make, and, using (A.43) and (A.44), the boundary condition can be written as

$$
\hat{\gamma}\times\mathbf{E} = \hat{\gamma}\times\left(\hat{\gamma}\times\left\{\bar{\bar{\eta}}.\mathbf{H} - \frac{Y_0Z}{2k_0^2Nh_\gamma}\nabla\left[\frac{\mu_0}{\mu}h_\gamma\nabla_s.(Z\mathbf{H})\right]\right\}\right.
$$

$$
\left. - \frac{Z_0}{2k_0^2Nh_\gamma}\nabla\left[\frac{\epsilon_0}{\epsilon}h_\gamma\hat{\gamma}.\nabla\times\mathbf{H}\right]\right) \quad (A.52)
$$

This is analogous to (5.25).

For simplicity we now assume that the material is homogeneous. In the special case of a planar surface $y = \text{constant}$, $\bar{\bar{\eta}} = Z\bar{\bar{I}}$ and (A.51) reduces to

$$\hat{y} \times \left\{ \mathbf{E} - \frac{1}{2jk_0 N\epsilon_r} \nabla E_y \right\} = Z\,\hat{y} \times \left(\hat{y} \times \left\{ \mathbf{H} - \frac{1}{2jk_0 N\mu_r} \nabla H_y \right\} \right)$$

where $\epsilon_r = \epsilon/\epsilon_0$ and $\mu_r = \mu/\mu_0$. The components of this are

$$E_x - \frac{1}{2jk_0 N\epsilon_r}\frac{\partial E_y}{\partial x} = Z\left\{ H_z - \frac{1}{2jk_0 N\mu_r}\frac{\partial H_y}{\partial z} \right\}$$

$$\tag{A.53}$$

$$E_z - \frac{1}{2jk_0 N\epsilon_r}\frac{\partial E_y}{\partial z} = -Z\left\{ H_x - \frac{1}{2jk_0 N\mu_r}\frac{\partial H_y}{\partial x} \right\}$$

and therefore

$$E_x = Z\left\{ H_z + \frac{1}{2(k_0 N)^2}\left(\frac{\partial^2 H_z}{\partial x^2} - \frac{\partial^2 H_x}{\partial x\,\partial z} \right) - \frac{Y}{2(k_0 N)^2}\left(\frac{\partial^2 E_x}{\partial z^2} - \frac{\partial^2 E_z}{\partial x\,\partial z} \right) \right\}$$

$$E_z = -Z\left\{ H_x + \frac{1}{2(k_0 N)^2}\left(\frac{\partial^2 H_x}{\partial z^2} - \frac{\partial^2 H_z}{\partial x\,\partial z} \right) - \frac{Y}{2(k_0 N)^2}\left(\frac{\partial^2 E_x}{\partial x\,\partial z} - \frac{\partial^2 E_z}{\partial x^2} \right) \right\}$$

On inserting the zeroth order approximations to the electric field components on the right hand side (a process that can be justified by starting with (A.52) instead), we obtain

$$E_x = Z\left\{ H_z + \frac{1}{2(k_0 N)^2}\left[\left(\frac{\partial^2}{\partial x^2} - \frac{\partial^2}{\partial z^2} \right) H_z - 2\frac{\partial^2 H_x}{\partial x\,\partial z} \right] \right\}$$

$$\tag{A.54}$$

$$E_z = -Z\left\{ H_x + \frac{1}{2(k_0 N)^2}\left[\left(\frac{\partial^2}{\partial z^2} - \frac{\partial^2}{\partial x^2} \right) H_x - 2\frac{\partial^2 H_z}{\partial x\,\partial z} \right] \right\}$$

in agreement with (5.26) and (5.53). The analogous results cited by LEON-TOVICH (1948) have a factor $j/2$ multiplying the $O(|N|^{-2})$ terms, and it appears that the error was made in extending the correct but specialised formulas of RYTOV (1940). Also, by differentiating (A.53) tangentially,

$$\left\{ \frac{\partial^2}{\partial y^2} - 2jk_0 N\epsilon_r\frac{\partial}{\partial y} - k_0^2(2N^2 - 1) \right\} E_y = 0$$

$$\tag{A.55}$$

$$\left\{ \frac{\partial^2}{\partial y^2} - 2jk_0 N\mu_r\frac{\partial}{\partial y} - k_0^2(2N^2 - 1) \right\} H_y = 0$$

which are the same as the boundary conditions derived from the Fresnel reflection coefficients in Section 5.3.2.

For a cylindrical surface $\rho = $ constant, the surface impedance is a tensor and (A.51) gives

$$E_\phi - \frac{1}{2jk_0 N \epsilon_r} \frac{1}{\rho} \frac{\partial E_\rho}{\partial \phi} = -\eta_{zz} \left\{ H_z - \frac{1}{2jk_0 N \mu_r} \frac{\partial H_\rho}{\partial z} \right\}$$

$$E_z - \frac{1}{2jk_0 N \epsilon_r} \frac{\partial E_\rho}{\partial z} = \eta_{\phi\phi} \left\{ H_\phi - \frac{1}{2jk_0 N \mu_r} \frac{1}{\rho} \frac{\partial H_\rho}{\partial \phi} \right\}$$

(A.56)

where

$$\eta_{zz} = Z \left\{ 1 - \frac{1}{2jk_0 N \rho} + \frac{1}{8(k_0 N \rho)^2} \right\}$$

$$\eta_{\phi\phi} = Z \left\{ 1 + \frac{1}{2jk_0 N \rho} - \frac{3}{8(k_0 N \rho)^2} \right\}$$

(A.57)

and (A.56) can be written as

$$E_\phi = -\eta_{zz} \left\{ \left[1 - \frac{1}{2(k_0 N)^2} \left(\frac{\partial^2}{\partial z^2} - \frac{1}{\rho^2} \frac{\partial^2}{\partial \phi^2} \right) \right] H_z - \frac{1}{(k_0 N)^2 \rho} \frac{\partial^2 H_\phi}{\partial \phi \, \partial z} \right\}$$ (A.58)

$$E_z = \eta_{\phi\phi} \left\{ \left[1 + \frac{1}{2(k_0 N)^2} \left(\frac{\partial^2}{\partial z^2} - \frac{1}{\rho^2} \frac{\partial^2}{\partial \phi^2} \right) \right] H_\phi - \frac{1}{(k_0 N)^2 \rho} \frac{\partial^2 H_z}{\partial \phi \, \partial z} \right\}$$ (A.59)

In the special case of a two-dimensional problem for H-polarisation ($\mathbf{H} = \hat{z} \, H_z$ with no z dependence), (A.59) disappears and (A.58) becomes

$$\frac{\partial H_z}{\partial \rho} = jk_0 Y_0 Z \left\{ 1 - \frac{1}{2jk_0 N \rho} + \frac{1}{2(k_0 N \rho)^2} \left(\frac{\partial^2}{\partial \phi^2} + \frac{1}{4} \right) \right\} H_z$$

(A.60)

Similarly, for E-polarisation (A.58) is null and (A.59) gives

$$E_z = \eta_{\phi\phi} \left\{ 1 - \frac{1}{2(k_0 N \rho)^2} \frac{\partial^2}{\partial \phi^2} \right\} \frac{1}{jk_0 Z_0} \frac{\partial E_z}{\partial \rho}$$

from which we obtain

$$\frac{\partial E_z}{\partial \rho} = jk_0 Z_0 Y \left\{ 1 - \frac{1}{2jk_0 N \rho} + \frac{1}{2(k_0 N \rho)^2} \left(\frac{\partial^2}{\partial \phi^2} + \frac{1}{4} \right) \right\} E_z$$

(A.61)

to the second order in $|N|^{-1}$.

Finally, for a spherical surface $r = $ constant, the surface impedance reverts to a scalar ($\overline{\overline{\eta}} = Z\overline{\overline{I}}$) and (A.51) can be written as

$$\hat{r} \times \mathbf{E} = Z\hat{r} \times \left(\hat{r} \times \left\{ \mathbf{H} - \frac{1}{2jk_0 N \mu_r} \nabla H_r \right\} \right) + \frac{1}{2(k_0 N)^2} \nabla \times \{ \hat{r} \, (\hat{r}.\nabla \times \mathbf{H}) \}$$

(A.62)

and the boundary conditions analogous to (A.58) and (A.59) are

$$
\begin{aligned}
E_\theta \;=\; & -Z\Bigg\{\left[1+\frac{1}{2(k_0 N r)^2}\left(\frac{\partial^2}{\partial\theta^2}+\frac{\cos\theta}{\sin\theta}\frac{\partial}{\partial\theta}-\frac{1}{\sin^2\theta}-\frac{1}{\sin^2\theta}\frac{\partial^2}{\partial\phi^2}\right)\right]H_\phi \\
& -\frac{1}{(k_0 N r)^2\sin\theta}\frac{\partial^2 H_\theta}{\partial\theta\,\partial\phi}\Bigg\}
\end{aligned}
\tag{A.63}
$$

$$
\begin{aligned}
E_\phi \;=\; & Z\Bigg\{\left[1-\frac{1}{2(k_0 N r)^2}\left(\frac{\partial^2}{\partial\theta^2}+\frac{\cos\theta}{\sin\theta}\frac{\partial}{\partial\theta}-\frac{1}{\sin^2\theta}-\frac{1}{\sin^2\theta}\frac{\partial^2}{\partial\phi^2}\right)\right]H_\theta \\
& -\frac{1}{(k_0 N r)^2\sin\theta}\frac{\partial^2 H_\phi}{\partial\theta\,\partial\phi}\Bigg\}
\end{aligned}
\tag{A.64}
$$

All of these results are accurate to the second order in $|N|^{-1}$ and, just as the zeroth and first order expansions both lead to a first order boundary condition, so the second and third order expansions give rise to a second order condition. Since the inclusion of the third order terms does not increase the order of the boundary condition, it is logical to seek them, but for a general inhomogeneous body the resulting expressions are too complicated to be usable. However, for the two-dimensional problem of a homogeneous cylinder the formulas are relatively simple, and we now consider this case.

A.6   Homogeneous cylinder

If the generators of the cylinder are parallel to the $\beta\ (=z)$ axis, then $h_\beta = 1$ and, for the two-dimensional problem of an H- or E-polarised field, there is no β dependence, implying $\partial/\partial\beta = 0$. This leads to a considerable simplification of the results in Sections A.2 through A.4. The relationships in (A.9) for the zeroth order fields are unaffected, and (A.20) and (A.21) reduce to

$$
A_{1\alpha} = -B_{1\beta}+\frac{B_{0\beta}}{2jk_0\rho_c w}, \qquad A_{1\beta} = B_{1\alpha}+\frac{B_{0\alpha}}{2jk_0\rho_c w}
\tag{A.65}
$$

where

$$
\rho_c = \left\{\frac{1}{h_\gamma}\frac{\partial}{\partial\gamma}(\ln h_\alpha)\right\}^{-1}
\tag{A.66}
$$

is the local radius of curvature. Also, from (A.16) and (A.17),

$$
A_{1\gamma} = \frac{1}{jk_0 w h_\alpha}\frac{\partial B_{0\beta}}{\partial\alpha}, \qquad B_{1\gamma} = -\frac{1}{jk_0 w h_\alpha}\frac{\partial B_{0\alpha}}{\partial\alpha}
\tag{A.67}
$$

and from (A.18) and (A.19)

$$
\frac{\partial B_{0\beta}}{\partial\gamma} = -\frac{h_\gamma}{2\rho_c}B_{0\beta}, \qquad \frac{\partial B_{0\alpha}}{\partial\alpha} = -\frac{h_\gamma}{2\rho_c}B_{0\alpha}
\tag{A.68}
$$

For the second order fields (A.41) gives

$$A_{2\alpha} = -B_{2\beta} + \frac{1}{2jk_0\rho_c w} B_{1\beta} + \frac{1}{4(k_0 w)^2} Q B_{0\beta} - \frac{1}{2(k_0 w)^2} \frac{1}{h_\alpha h_\gamma} \frac{\partial}{\partial \alpha} \left(\frac{h_\gamma}{h_\alpha} \frac{\partial B_{0\beta}}{\partial \alpha} \right) \tag{A.69}$$

where

$$Q = \frac{1}{\rho_c^2} \left(\frac{1}{2} - \frac{1}{h_\gamma} \frac{\partial \rho_c}{\partial \gamma} \right) \tag{A.70}$$

and similarly

$$A_{2\beta} = B_{2\alpha} + \frac{1}{2jk_0\rho_c w} B_{1\alpha} - \frac{1}{4(k_0 w)^2} \tilde{Q} B_{0\alpha} - \frac{1}{2(k_0 w)^2} \frac{1}{h_\alpha h_\gamma} \frac{\partial}{\partial \alpha} \left(\frac{h_\gamma}{h_\alpha} \frac{\partial B_{0\alpha}}{\partial \alpha} \right) \tag{A.71}$$

where

$$\tilde{Q} = \frac{1}{\rho_c^2} \left(\frac{1}{2} + \frac{1}{h_\gamma} \frac{\partial \rho_c}{\partial \gamma} \right) \tag{A.72}$$

Also, from (A.36) and (A.39)

$$A_{2\gamma} = \frac{1}{jk_0 w h_\alpha} \frac{\partial B_{1\beta}}{\partial \alpha}, \qquad B_{2\gamma} = -\frac{1}{jk_0 w h_\alpha} \frac{\partial A_{1\beta}}{\partial \alpha} \tag{A.73}$$

and from (A.40), (A.67) and (A.68)

$$\frac{\partial B_{1\beta}}{\partial \gamma} = -\frac{h_\gamma}{2\rho_c} B_{1\beta} + \frac{h_\gamma}{4jk_0 w} Q B_{0\beta} - \frac{1}{2jk_0 w h_\alpha} \frac{\partial}{\partial \alpha} \left(\frac{h_\gamma}{h_\alpha} \frac{\partial B_{0\beta}}{\partial \alpha} \right) \tag{A.74}$$

Similarly

$$\frac{\partial B_{1\alpha}}{\partial \gamma} = -\frac{h_\gamma}{2\rho_c} B_{1\alpha} + \frac{h_\gamma}{4jk_0 w} \tilde{Q} B_{0\alpha} - \frac{1}{2jk_0 w h_\alpha} \frac{\partial}{\partial \alpha} \left(\frac{h_\gamma}{h_\alpha} \frac{\partial B_{0\alpha}}{\partial \alpha} \right) \tag{A.75}$$

We now consider the third order fields. From (A.4) the terms involving q^3 are

$$\mathbf{A}_3 - \hat{\gamma} \times \mathbf{B}_3 = \frac{1}{jk_0 w} \nabla \times \mathbf{B}_1$$

$$\mathbf{B}_3 + \hat{\gamma} \times \mathbf{A}_3 = -\frac{1}{jk_0 w} \nabla \times \mathbf{A}_1$$

the transverse components of which are

$$A_{3\alpha} + B_{3\beta} = -\frac{1}{jk_0 w h_\gamma} \frac{\partial B_{2\beta}}{\partial \gamma} \tag{A.76}$$

$$A_{3\beta} - B_{3\alpha} = \frac{1}{jk_0 w h_\alpha h_\gamma} \left\{ \frac{\partial}{\partial \gamma} (h_\alpha B_{2\alpha}) - \frac{\partial}{\partial \alpha} (h_\gamma B_{2\gamma}) \right\} \tag{A.77}$$

$$B_{3\alpha} - A_{3\beta} = \frac{1}{jk_0 w h_\gamma} \frac{\partial A_{2\beta}}{\partial \gamma} \tag{A.78}$$

$$B_{3\beta} + A_{3\alpha} = -\frac{1}{jk_0 w h_\alpha h_\gamma} \left\{ \frac{\partial}{\partial \gamma} (h_\alpha A_{2\alpha}) - \frac{\partial}{\partial \alpha} (h_\gamma A_{2\gamma}) \right\} \tag{A.79}$$

Since $\partial B_{2\beta}/\partial\gamma$ is not required for the third order results, a unique expression for $A_{3\alpha} + B_{3\beta}$ can be obtained by simply averaging (A.76) and (A.79), giving

$$A_{3\beta} + B_{3\alpha} = \frac{1}{jk_0 w h_\gamma}\left\{\frac{1}{h_\alpha}\frac{\partial}{\partial\alpha}(h_\gamma A_{2\gamma}) - \frac{h_\gamma}{\rho_c}A_{2\alpha} - \frac{\partial}{\partial\gamma}(A_{2\alpha} + B_{2\beta})\right\}$$

and by substituting the expressions for $A_{2\alpha} + B_{2\beta}$ and $A_{2\gamma}$ from (A.69) and (A.73), a straightforward but tedious analysis shows

$$\begin{aligned}
A_{3\alpha} = {} & -B_{3\beta} + \frac{1}{2jk_0\rho_c w}B_{2\beta} + \frac{1}{4(k_0 w)^2}QB_{1\beta} \\
& - \frac{1}{2(k_0 w)^2}\frac{1}{h_\alpha h_\gamma}\frac{\partial}{\partial\alpha}\left(\frac{h_\gamma}{h_\alpha}\frac{\partial B_{1\beta}}{\partial\alpha}\right) \\
& - \frac{1}{2j(k_0 w)^3\rho_c}\frac{1}{h_\alpha h_\gamma}\frac{\partial}{\partial\alpha}\left(\frac{h_\gamma}{h_\alpha}\frac{\partial B_{0\beta}}{\partial\alpha}\right) \\
& - \frac{1}{4j(k_0 w)^3}\left(L + \frac{1}{2h_\gamma}\frac{\partial Q}{\partial\gamma}\right)B_{0\beta}
\end{aligned} \qquad (A.80)$$

where L is the operator

$$L = \frac{1}{h_\alpha^2 h_\gamma}\left\{\frac{\partial^2}{\partial\alpha\,\partial\gamma}\left(\ln\frac{h_\alpha^2}{h_\gamma}\right)\frac{\partial}{\partial\alpha} + \frac{1}{2}\left[\frac{\partial^2}{\partial\alpha^2}\left(\frac{h_\gamma}{\rho_c}\right) + \frac{\partial}{\partial\alpha}\left(\frac{h_\gamma}{\rho_c}\right)\frac{\partial}{\partial\alpha}\left(\ln\frac{h_\gamma}{h_\alpha}\right)\right]\right\}$$
$$(A.81)$$

We observe that L vanishes if the tangential derivatives of the metric coefficients are zero. Hence

$$A_\alpha = A_{0\alpha} + qA_{1\alpha} + q^2 A_{2\alpha} + q^3 A_{3\alpha} + O(q^4)$$

and when the expressions for $A_{0\alpha}$, $A_{1\alpha}$, $A_{2\alpha}$ and $A_{3\alpha}$ are inserted, the result is

$$\begin{aligned}
E_\alpha = {} & -Z\left\{1 - \frac{1}{2jNk_0\rho_c} + \frac{1}{2(Nk_0)^2}\left(1 + \frac{1}{jNk_0\rho_c}\right)\frac{1}{h_\alpha h_\gamma}\frac{\partial}{\partial\alpha}\left(\frac{h_\gamma}{h_\alpha}\frac{\partial}{\partial\alpha}\right)\right. \\
& \left. - \frac{1}{4(Nk_0)^2}\left(Q - \frac{1}{2jNk_0 h_\gamma}\frac{\partial Q}{\partial\gamma}\right) + \frac{1}{4j(Nk_0)^3}L\right\}H_\beta
\end{aligned} \qquad (A.82)$$

accurate to the third order in $|N|^{-1}$, where we have used the fact that $E_\alpha/H_\beta = ZA_\alpha/B_\beta$ at the surface. Since

$$E_\alpha = -\frac{Z_0}{jk_0}\frac{1}{h_\gamma}\frac{\partial H_\beta}{\partial\gamma}$$

(A.82) can be expressed as a scalar boundary condition for the component H_β ($= H_z$) alone, namely

$$\begin{aligned}
\frac{1}{h_\gamma}\frac{\partial H_z}{\partial\gamma} = {} & jk_0 Y_0 Z\left\{1 - \frac{1}{2jNk_0\rho_c}\right. \\
& + \frac{1}{2(Nk_0)^2}\left(1 + \frac{1}{jNk_0\rho_c}\right)\frac{1}{h_\alpha h_\gamma}\frac{\partial}{\partial\alpha}\left(\frac{h_\gamma}{h_\alpha}\frac{\partial}{\partial\alpha}\right) \\
& \left. - \frac{1}{4(Nk_0)^2}\left(Q - \frac{1}{2jNk_0 h_\gamma}\frac{\partial Q}{\partial\gamma}\right) + \frac{1}{4j(Nk_0)^3}L\right\}H_z \qquad (A.83)
\end{aligned}$$

Similarly,

$$
E_\beta = Z\left\{1 + \frac{1}{2jNk_0\rho_c} - \frac{1}{2(Nk_0)^2}\left(1 + \frac{3}{2jNk_0\rho_c}\right)\frac{1}{h_\alpha h_\gamma}\frac{\partial}{\partial\alpha}\left(\frac{h_\gamma}{h_\alpha}\frac{\partial}{\partial\alpha}\right)\right.
$$
$$
\left. - \frac{1}{4(Nk_0)^2}\left(\tilde{Q} - \frac{1}{2jNk_0 h_\gamma}\frac{\partial\tilde{Q}}{\partial\gamma}\right) - \frac{1}{4j(Nk_0)^3}L\right\}H_\alpha \qquad (A.84)
$$

and by inverting the differential operator we obtain a boundary condition for E_z which differs from (A.83) only in having YZ_0 in place of Y_0Z in accordance with duality. If we retain only the terms $O(|N|^{-2})$, (A.83) becomes

$$
\frac{1}{h_\gamma}\frac{\partial H_z}{\partial\gamma} = jk_0Y_0Z\left\{1 - \frac{1}{2jNk_0\rho_c} + \frac{1}{2(Nk_0)^2}\frac{1}{h_\alpha h_\gamma}\frac{\partial}{\partial\alpha}\left(\frac{h_\gamma}{h_\alpha}\frac{\partial}{\partial\alpha}\right)\right.
$$
$$
\left. + \frac{1}{8(Nk_0\rho_c)^2}\left(\frac{2}{h_\gamma}\frac{\partial\rho_c}{\partial\gamma} - 1\right)\right\}H_z \qquad (A.85)
$$

In the special case of a circular cylinder of radius ρ, we have $\alpha = \phi$ and $\gamma = \rho$, implying $h_\alpha = \rho$ and $h_\gamma = 1$. Then

$$
\rho_c = \rho, \qquad Q = -\frac{1}{2\rho^2}, \qquad \frac{\partial Q}{\partial\gamma} = \frac{1}{\rho^3}, \qquad L = 0
$$

and (A.83) reduces to

$$
\frac{\partial H_z}{\partial\rho} = jk_0Y_0Z\left\{1 - \frac{1}{2jNk_0\rho} + \frac{1}{2(Nk_0\rho)^2}\left(1 + \frac{1}{jNk_0\rho}\right)\left(\frac{\partial^2}{\partial\phi^2} + \frac{1}{4}\right)\right\}H_z
$$
$$
(A.86)
$$

in agreement with (7.29). We observe that the only effect of including terms $O(|N|^{-3})$ is to produce the factor $1 + 1/(jNk_0\rho)$ in the last term. For a noncircular cylinder, a similar factor with ρ replaced by the variable radius ρ_c multiplies the derivative term in (A.83), but it does not extend to the terms which are analogous to $\frac{1}{4}$ in (A.86).

References

Leontovich, M. A. (1948), *Investigations on Radiowave Propagation, Part II*, Printing House of the Academy of Sciences, Moscow, pp. 5–12.

Rytov, S. M. (1940), "Calcul du skin-effet par la méthode des perturbations", *J. Phys. USSR*, **2**, pp. 233–242. The paper was republished in Russian in *Zhur. Eksp. i Teoret. Fiz.*, **10**, pp. 180–189 (1940). An English translation of the latter has been made by V. Kerdemelidis and K. M. Mitzner, Northrop Navair, Hawthorne, CA 90250.

Senior, T. B. A. (1990), "Approximate boundary conditions, part 1", University of Michigan Radiation Laboratory Report RL-861, Ann Arbor, MI.

Appendix B

Special functions

B.1 Numerical Wiener-Hopf factorisation

A crucial step in the solution of Wiener-Hopf or dual integral equations is the factorisation (or splitting) of an even function $F(\lambda)$ into a product of two functions such that

$$F(\lambda) = F_+(\lambda)\, F_-(\lambda) \qquad (B.1)$$

where $\lambda = \sigma + j\tau$, $F_+(\lambda)$ is free of singularities and zeros in the upper ($\tau > \tau_-$) half of the complex λ-plane, and $F_-(\lambda)$ has similar properties in the lower half $\tau < \tau_+$, with $\tau_+ > \tau_-$. For the factorisation to be possible, $F(\lambda)$ must be analytic in the strip $\tau_- < \tau < \tau_+$ where $\tau_\pm$ may be vanishingly small. If we further demand that $F(\lambda) \to 1$ uniformly as $|\sigma| \to \infty$ in the strip, then (MITTRA AND LEE, 1971)

$$F_+(\lambda) = F_-(-\lambda) = \exp\{H(\lambda)\} \qquad (\text{Im. } \lambda > 0) \qquad (B.2)$$

where

$$H(\lambda) = \frac{1}{2\pi j} \int_{C_1} \frac{\ln F(\beta)}{\beta - \lambda}\, d\beta \qquad (B.3)$$

and C_1 is the path shown in Fig. B–1. The last of the above conditions is not a restriction since we can always modify $F(\lambda)$ to produce this behaviour. In

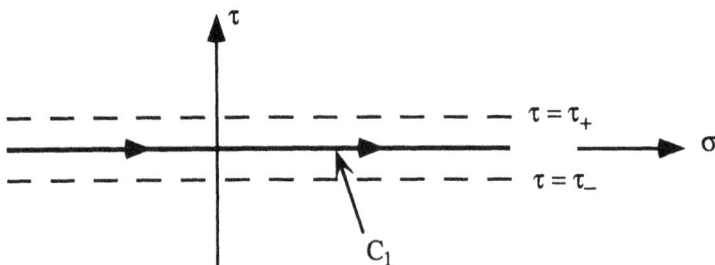

Figure B–1: *The path of integration C_1*

addition, because $F(\lambda)$ is an even function, we can set $\tau_- = -\tau_+$, ensuring that the path C_1 (for which $\tau = 0$) remains within the strip as τ_+ approaches zero. The integral (B.3) is not convenient for numerical evaluation, and we describe below a procedure originally presented by RICOY AND VOLAKIS (1989) to simplify its evaluation.

Apart from the infinite limits of integration, the integrand may be singular in the vicinity of the path, but it is usually possible to alleviate these problems. Most functions $F(\beta)$ that occur in diffraction theory are also analytic in the angular sector $\{\beta(t) = te^{j\gamma},\ 0 < \gamma < \gamma_0,\ 0 < t < \infty\}$ for some γ_0, with F approaching unity as $t \to \infty$ in the sector. Since $F(\beta)$ is even, it then has the same properties in $\{\beta(t) = te^{j\gamma},\ \pi < \gamma < \pi + \gamma_0,\ 0 < t < \infty\}$ and, under this assumption, we can rotate the path C_1 through an angle γ counterclockwise about the origin into the path C_2 shown in Fig. B-2. This gives

$$H(\lambda) = h(\gamma - \gamma_\lambda) \ln F(\lambda) + \frac{1}{2\pi j} \int_{C_2} \frac{\ln F(\beta)}{\beta - \lambda}\, d\beta \qquad (B.4)$$

where

$$h(\gamma - \gamma_\lambda) = \begin{cases} 0 & \text{for } \gamma < \gamma_\lambda \\ 1 & \text{for } \gamma > \gamma_\lambda \end{cases}$$

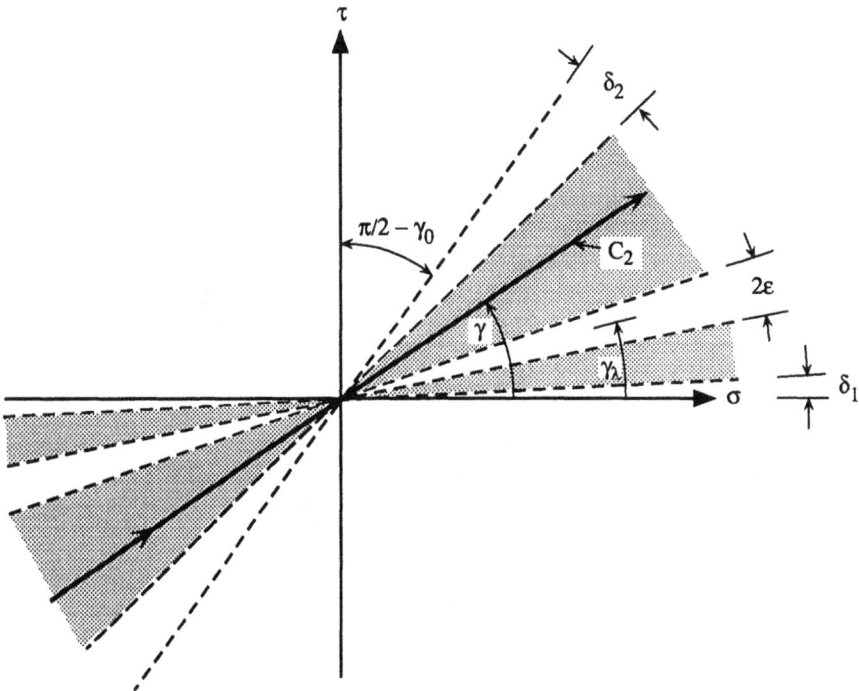

Figure B–2: *Illustration of the path C_2 and the permitted values of γ*

and

$$\gamma_\lambda = \arctan \frac{\text{Im. } \lambda}{\text{Re. } \lambda}$$

The next task is to address the singularities of the integrand in (B.4). It is clear that the numerator is infinite at the zeros $\beta = \beta_n$, $n = 1, 2, \ldots, N$, of $F(\beta)$. Although these are excluded from the angular sector by virtue of the stipulated behaviour of $F(\beta)$, one or more of the β_n may be very close to C_2 as γ approaches 0 or γ_λ. This is often the case and is true, for example, as γ approaches zero if $\tau_\pm \to 0$, but, since the resulting singularity is logarithmic, it is only necessary to keep C_2 a small distance away to eliminate the problem. The obvious solution is to restrict γ to the angular sector $\delta_1 < \gamma < \gamma_0 - \delta_2$ where $\delta_{1,2}$ are small angles which are determined empiricially, and this works provided none of the β_n is close to the origin. The integrand of (B.4) also has a pole at the zero $\beta = \lambda$ of the denominator and, since the only restriction on λ is that Im. $\lambda > 0$, the pole may lie close to or upon C_2. The simplest way to prevent this is to exclude γ from the range $(\gamma_\lambda - \epsilon, \ \gamma_\lambda + \epsilon)$ where ϵ is also determined empirically. Because the singularity is now a pole, it is clear that $\epsilon \gg \delta$ and, if $0 < \gamma_\lambda < \gamma_0$, the restriction on γ could be severe. This is undesirable since the convergence of $\ln F(\beta)$ as $\beta \to \infty$ may be a function of γ, and we would like to be able to choose C_2 to maximise the convergence. The possibility that the optimum path lies within the range $(\gamma_\lambda - \epsilon, \ \gamma_\lambda + \epsilon)$ is reduced if we can shrink this range, and we can do so as follows. By making the substitution $\beta = te^{j\gamma}$ and using the fact that $F(\beta)$ is an even function, (B.4) becomes

$$H(\lambda) = h(\gamma - \gamma_\lambda) \ln F(\lambda) + \frac{\lambda e^{j\gamma}}{j\pi} \int_0^\infty \frac{\ln F(te^{j\gamma})}{t^2 e^{2j\gamma} - \lambda^2} \, dt \qquad (\text{B.5})$$

which presents a difficulty when the pole $t = \lambda e^{-j\gamma}$ is near the real axis. However, by adding and subtracting its value at this pole from the numerator, and then evaluating the added term analytically, we obtain

$$H(\lambda) = \tfrac{1}{2} \ln F(\lambda) + \frac{\lambda e^{j\gamma}}{j\pi} \int_0^\infty \frac{\ln F(te^{j\gamma}) - \ln F(\lambda)}{t^2 e^{2j\gamma} - \lambda^2} \, dt$$

and the integrand is now analytic at $t = \lambda e^{-j\gamma}$. This effectively reduces ϵ to the same order as δ, and addresses the concern noted above.

The final obstacle is the infinite range of integration, but, with the change of integration variable from t to v where $v = \frac{2}{\pi} \arctan t$, we have

$$H(\lambda) = \tfrac{1}{2} \ln F(\lambda) + \frac{\lambda e^{j\gamma}}{2j} \int_0^1 \frac{\ln F\left(e^{j\gamma} \tan \frac{\pi v}{2}\right) - \ln F(\lambda)}{e^{2j\gamma} \sin^2 \frac{\pi v}{2} - \lambda^2 \cos^2 \frac{\pi v}{2}} \, dv \qquad (\text{B.6})$$

and this is the desired result. In conjunction with (B.2), (B.6) provides a factorisation of an even function of λ, analytic in the strip $|\tau| < \tau_+$ and in the angular sector $0 < \gamma < \gamma_0$. The integral is over a convenient finite interval and is numerically tractable for $\delta_1 < \gamma < \gamma_0 - \delta_2$ with the proviso that if $0 < \gamma_\lambda < \gamma_0$, γ does not lie in the range $(\gamma_\lambda - \epsilon, \gamma_\lambda + \epsilon)$, and any zeros of $F(\beta)$ are not too close to the origin. This allows γ to be chosen for rapid convergence of the integral. In (B.6), care must be taken in defining the branch of the logarithm so that $F(e^{j\gamma} \tan \frac{\pi v}{2})$ is continuous on the path of integration.

B.2 Half-plane split function computation

The following FORTRAN programs were written by J. L. Volakis of the University of Michigan and compute the Maliuzhinets half-plane function $\psi_\pi(\text{CANG})$ and the upper half-plane function $K_+(\text{ETA},\text{COS(ANG)})$.

```
      COMPLEX FUNCTION PSIPI(CANG,ICOD)
C     Computes the Half-Plane Maliuzhinets function
C     CANG : Complex Argument of the function
C     Uses identities to get values when REAL(CANG)>PI/2
C     ICOD=0 Exact evaluation via 6-point Gaussian Integr.
C         =1 Approximate non-integral evaluation
      DIMENSION X(6),W(6)
      COMPLEX CANG,CANG1,YI,CINT,CF1,YR,COEF,XP1,XP2
      COMPLEX FINT,FINT1,FINT3,FINT4
      DATA X/.2386192,-.2386192,.6612094,-.6612094,.9324695,-.9324695/
      DATA W/.4679139,.4679139,.3607616,.3607616,.1713245,.1713245/
      DATA PI,EGPI,SRT2/3.141592654,25.13274123,1.41421356/
      DATA PI2,PSISQ,FACTR/1.5707963,.9324274,.68280233/
      FINT3=(0.,0.)
      FINT4=(0.,0.)
      CINT=(0.,0.)
      AR=REAL(CANG)
      AI=AIMAG(CANG)
      ITTT=0
      IF(AR.GT.0)GO TO 30
       AR=-AR
       ITTT=1
30    ITT=0
      IF(AI.GT.0)GO TO 40
       ITT=1
       AI=-AI
40    IT=0
      CANG1=CMPLX(AR,AI)
      IF(AR.LT.PI2)GO TO 90
       AR=AR-PI
       COEF=CCOS(.25*(CANG1-PI2))*PSISQ
       IT=1
      IF(AR.LT.PI2)GO TO 90
       AR=AR-PI
       COEF=CCOS(.25*(CANG1-PI2))/CCOS(.25*(CANG1-1.5*PI))
       IT=2
      IF(AR.LT.PI2)GO TO 90
```

```
         AR=AR-PI
         COEF=PSISQ*CCOS(.25*CANG1-.125*PI)*CCOS(.25*CANG1-.625*PI)
         COEF=COEF/CCOS(.25*CANG1-.375*PI)
         IT=3
  90     CONTINUE
C        IF(ABS(AR).GT.PI2+.01)PRINT *,'AR > PI2 ',AR
         IF(ICOD.NE.0)GO TO 110
         DO 100 I=1,6
         YR=.5*AR*(X(I)+1.)+(0.,1.)*AI
         FINT1=(PI*CSIN(YR)-2.*SRT2*PI*CSIN(.5*YR)+2.*YR)/CCOS(YR)
         YI=.5*AI*(X(I)+1.)*(0.,1.)
         FINT=(PI*CSIN(YI)-2.*SRT2*PI*CSIN(.5*YI)+2.*YI)/CCOS(YI)
         FINT3=.5*W(I)*FINT1*AR+FINT3
         FINT4=(0.,1.)*.5*W(I)*AI*FINT
         CINT=CINT+.5*W(I)*(AR*FINT1+(0.,1.)*AI*FINT)
 100     CONTINUE
         CINT=-CINT/EGPI
         PSIPI=CEXP(CINT)
         GO TO 200
 110     CONTINUE
         CANG1=CMPLX(AR,AI)
         IF(AI.GT.4.6)ICOD=3
         IF(ICOD.NE.1)GO TO 120
          PSIPI=1.-.0139*CANG1*CANG1
          GO TO 200
 120     CONTINUE
         IF(ICOD.LT.2)GO TO 200
          XP1=SRT2*CEXP((0.,1.)*.25*CANG1)
          XP2=CEXP(-(0.,1.)*.25*CANG1)
          PSIPI=FACTR*CSQRT(XP1+XP2)
C         IF(ICOD.EQ.4)GO TO 200
          XP2=CEXP(-(0.,1.)*CANG1)
          IF(ICOD.EQ.3)XP1=(0.,1.)*CANG1
          IF(ICOD.EQ.2)XP1=(0.,1.)*CANG1-1.
          XP1=CEXP(XP1/(2.*PI*XP2))
          PSIPI=PSIPI*XP1
 200     CONTINUE
         IF(IT.EQ.1)PSIPI=COEF/PSIPI
         IF(IT.EQ.2)PSIPI=COEF*PSIPI
         IF(IT.EQ.3)PSIPI=COEF/PSIPI
         IF(ITT.EQ.1)PSIPI=CONJG(PSIPI)
         IF(ITTT.EQ.1)PSIPI=CONJG(PSIPI)
         RETURN
         END
CIII
         COMPLEX FUNCTION KPLUSM(ANG,ETAOR,IUD,SB0)
C        Complex Split Function K+(ETAOR,cos(ANG))
C        ETAOR=Complex, Normalized Impedance
C        ANG=Angle in rad; May be Complex
C        SB0=sin(B0), B0=Skewness Angle, SB0=1 for normal incidence
C        IUD=0 Computes K+(1/(ETAOR*SB0),cos(ANG))
C        IUD=1 Computes K+(ETAOR/SB0,cos(ANG))
```

```
      COMPLEX*8 HEE,C1,C2,C3,C4,C5,ETA,ETAOR,CJ,ANG,HEEK,PSIPI,ETAI
      DATA SRT2,FPI,CJ/1.414213562,12.56637061,(0.,1.)/
      DATA PSIPI2,PI/.9656228,3.14159265/
      ETA=ETAOR
      IF(REAL(ETAOR).LT.0.)ETA=-ETAOR
      IF(IUD.EQ.0)GO TO 10
      ETAI=SBO/ETA
C     Compute arccos(ETA/SBO)
      HEEK=HEE(ETA,1,SBO)
      GO TO 20
10    ETAI=SBO*ETA
C     Compute arccos(1/CETA*SBO))
      HEEK=HEE(ETA,0,SBO)
20    CONTINUE
      IF(REAL(ETAOR).LT.0.)HEEK=-HEEK
      C1=2.*CCOS(HEEK)*(1.-CCOS(ANG))
      C1PH=BTAN2(AIMAG(C1),REAL(C1))
      C1=SQRT(CABS(C1))*CEXP(.5*CJ*C1PH)
      C2=SRT2*CSIN(.5*(ANG-HEEK))+1
      C3=SRT2*CSIN(.5*(ANG+HEEK))+1
      C4=PSIPI(PI-ANG+HEEK,1)/PSIPI2
      C5=PSIPI(PI-ANG-HEEK,1)/PSIPI2
      KPLUSM=(2.*C1*C4*C4*C5*C5)/(C2*C3)
      RETURN
      END
CIII
      COMPLEX FUNCTION HEE(ETA,IUD,SBO)
C     COMPUTES THE INVERSE COSINE/SINE OF COMPLEX NUMBER
      COMPLEX ETA,ETA1,CJ
      DOUBLE PRECISION RE,AE,REP,REM,AA,BB,SGN,RAA
      DATA SRT2,FPI,CJ/1.414213562,12.56637061,(0.,1.)/
      DATA PSIPI2,PI/.9656228,3.14159265/
      ETA1=1./(ETA*SBO)
      IF(IUD.EQ.1)ETA1=ETA/SBO
      RE=REAL(ETA1)
      AE=AIMAG(ETA1)
      REP=RE+1.
      REM=RE-1.
      AA=.5*(DSQRT(REP*REP+AE*AE)+DSQRT(REM*REM+AE*AE))
      BB=.5*(DSQRT(REP*REP+AE*AE)-DSQRT(REM*REM+AE*AE))
      IF(AE.NE.0.DO)THEN
        SGN=AE/DABS(AE)
      ELSE
        SGN=1.DO
      ENDIF
      RAA=AA*AA-1.
      IF(RAA.LT.1.E-6)RAA=0.
C see page 80 of Ambramowitz and Stegun
C DASIN=double precision arcsin
      HEE=DASIN(BB)+CJ*DLOG(AA+DSQRT(RAA))*SGN
C uncomment the line to get inverse cosine of eta1
      HEE=.5*PI-HEE
```

```
300   RETURN
      END
CC
      REAL FUNCTION BTAN2(Y,X)
      DATA PI/3.1415926/
      IF(ABS(X).GT.1.E-6) GO TO 20
      IF(ABS(Y).GT.1.E-6) GO TO 10
      BTAN2=0.
      RETURN
 10      BTAN2=.5*PI
      IF(Y.LT.0.)BTAN2=-BTAN2
      RETURN
 20      BTAN2=ATAN2(Y,X)
      RETURN
      END
```

B.3   Transition function computation

The following FORTRAN programs compute the Clemmow transition function given in (3.55), and the Kouyoumjian-Pathak transition function (3.60) for $z > 0$. The first was written by M. I. Herman at the University of Michigan, and the second by W. D. Burnside, R. J. Marhefka and N. Wang of The Ohio State University.

```
      COMPLEX FUNCTION FFCL(XF)
C**********************************************************************
C
CTHIS ROUTINE COMPUTES THE CLEMMOW TRANSITION FUNCTION,COMPLEX ARG.
C
      COMPLEX CJ,XF,UJ,CN1,CN2,CN3,CN4,CSQ
      DATA CJ,UJ/(0.,1.),(1.,0.)/
      DATA PI/3.14159265/
CFLAG CORRECTS FOR ERROR OF BRANCH IN COMPLEX SQRT OPERATION(ARG>PI)
      R1=REAL(XF)
      R2=AIMAG(XF)
      IF(R1.EQ.0.)GO TO 4
      R3=R2/R1
      R3=ATAN(R3)
      IF((R1.LT.0.).AND.(R2.LE.0.))R3=R3-PI
      IF((R1.LT.0.).AND.(R2.GT.0.))R3=R3+PI
      IF((R3.GE.-PI).AND.(R3.LT.(-3.*PI/4.)))R3=R3+PI*2.
      GO TO 3
    4 R3=PI/2.
      IF(R2.LT.0.)R3=-R3
    3 PHASE=R3
      X=CABS(XF)
      IF(X.GE.4.0)GO TO 1
      CN1=XF*CEXP(CJ*PI/4.)
      IF(ABS(REAL(CN1)).GE.3.)GO TO 1
```

```
      IF(X.GE.0.3)GO TO 10
C     SMALL ARGUMENT FORM
      CN1=XF+XF*XF*XF*(CJ*.66667-.26667*XF*XF)
      FFCL=.5*SQRT(PI)*CEXP(-CJ*PI/4.)*CEXP(CJ*XF*XF)-CN1
      GO TO 20
C     APPROXIMATION OF THE EXACT SOLUTION
   10 CN3=CMPLX(.7071067,.7071067)
      CN2=XF*CN3
      X1=REAL(CN2)
      SIGN2=1.
      IF(X1.GE.0.)GO TO 5
      SIGN2=-1.
      X1=-X1
    5 Y1=AIMAG(CN2)*SIGN2
CFORM THE COMP. ERROR FUNC.USING 1-ERF (ABRAM.& STEGUN 7.1.29)
C||| ERFC(Z)=1-;ERF(X)+EXP(-X*X)]   (/2PIX+ HIGHER ORDERS:*SIGN
C            ERF         CN1              CN3
C     ERF(X)=1-(A1*T+A2*T*T+A3*T*T*T)EXP(-X*X)
      T=1./(1.+.47047*X1)
      R3=EXP(-X1*X1)
      ERF=1.-(.3480242*T-.0958798*T*T+.7478556*T*T*T)*R3
CNOW FORM THE NEXT SERIES OF TERMS IN 7.1.29
      IF(X1.EQ.0.)GO TO 15
      R2=1.-COS(2.*X1*Y1)
      R4=SIN(2.*X1*Y1)
      CN1=CMPLX(R2,R4)
      CN1=(R3/(2.*PI*X1))*CN1
      GO TO 17
   15 CN1=(CJ*(Y1/PI))
CCALCULATE SECOND ORDER TERM
   17 R1=2.*X1
      CN3=(0.,0.)
      DO 18 N=1,20
      FN=FLOAT(N)
      XG=R1*COSH(Y1*FN)*SIN(R1*Y1)+FN*SINH(Y1*FN)*COS(R1*Y1)
      XF1=R1-R1*COSH(Y1*FN)*COS(R1*Y1)+FN*SINH(Y1*FN)*SIN(R1*Y1)
      CN2=CMPLX(XF1,XG)
      R3=((2.*EXP(-.25*FN*FN-X1*X1))/(FN*FN+4.*X1*X1))/PI
      CN2=R3*CN2
      CN3=CN3+CN2
   18 CONTINUE
      CN1=UJ-(ERF*UJ+CN3+CN1)*(SIGN2*UJ)
CCOMP. ERROR FUNC. IS IN CN1
CTHIS NEXT SEQUENCE CALCULATES THE FRESNEL FUNCTION FOR COMPARISON
CWITH TABLE VALUES IN ABRAMOWITZ AND STEGUN,UPDATE CHECK FOR Fc
C     CN3=CN1*CEXP(XF*CJ)
C     CN2=CSQ
C     CN4=(-.7071067,.7071067)
C     CN2=CN2*CN4
C     R1=REAL(CN2)
C     X1=AIMAG(CN2)
C     R2=REAL(CN3)
```

```
C     X2=AIMAG(CN3)
C     WRITE(6,560)R1,X1,R2,X2
C 560 FORMAT(' ','X+jY=',E11.5,'+j',E11.5,5X,'TABLE ERFC=',
C     1E11.5,'+j',E11.5)
      CN3=CMPLX(.7071067,-.7071607)
      CN2=CEXP(XF*XF*CJ)
      CN1=CN1*CN2*CN3
      R1=.5*SQRT(3.14159256)
      FFCL=CN1*(R1*UJ)
      GO TO 20
CLARGE ARGUMENT FORM
    1 FLAG=0.
      CN1=UJ/(-2.*CJ*XF*XF)
      FFCL=(UJ/(2.*CJ*XF))*(UJ+CN1+3.*CN1*CN1)
      IF((PHASE.LT.(5.*PI/4.)).AND.(PHASE.GT.(PI/4.)))FLAG=1.
      IF(FLAG.EQ.1.)FFCL=FFCL+SQRT(PI)*CEXP(-CJ*PI/4.)*CEXP(CJ*XF*XF)
   20 CONTINUE
      RETURN
      END




      COMPLEX FUNCTION FFCT(XF)
C     THIS ROUTINE COMPUTES THE WEDGE TRANSITION FUNCTION.
      COMPLEX FXX(8),FX(8),CJ
      DIMENSION XX(8)
      DATA XX/.3,.5,.7,1.,1.5,2.3,4.,5.5/
      DATA CJ/(0.,1.)/
      DATA FX/(0.5729,0.2677),(0.6768,0.2682),(0.7439,0.2549),
     1(0.8095,0.2322),(0.873,0.1982),(0.9240,0.1577),(0.9658,0.1073),
     2(0.9797,0.0828)/
      DATA FXX/(0.,0.),(0.5195,0.0025),(0.3355,-0.0665),
     1(0.2187,-0.0757),(0.127,-0.068),(0.0638,-0.0506),
     2(0.0246,-0.0296),(0.0093,-0.0163)/
      X=ABS(XF)
      IF(X.GT.5.5)GO TO 1
      IF(X.GT.0.3)GO TO 10
C     SMALL ARGUMENT FORM
      FFCT=((1.253,1.253)*SQRT(X)-(0.,2.)*X-0.6667*X*X)*CEXP(CJ*X)
      GO TO 20
C     LINEAR INTERPOLATION REGION
   10 DO 11 N=2,7
   11 IF(X.LT.XX(N))GO TO 12
   12 FFCT=FXX(N)*(X-XX(N))+FX(N)
      GO TO 20
C     LARGE ARGUMENT FORM
    1 FFCT=1.+CMPLX(-0.75/X,0.5)/X
   20 IF(XF.GE.0.) RETURN
      FFCT=CONJG(FFCT)
      RETURN
      END
```

B.4 Maliuzhinets function

To solve (4.13) and the difference equation (4.14), MALIUZHINETS (1951) introduced the meromorphic function $\psi_\Phi(\alpha)$ such that

$$\sin\frac{1}{2}\left(\alpha+\frac{\pi}{2}\right)\psi_\Phi\left(\alpha+2\Phi\right)=\cos\frac{1}{2}\left(\alpha+\frac{\pi}{2}\right)\psi_\Phi\left(\alpha-2\Phi\right) \qquad (B.7)$$

and since

$$\sin(\alpha\pm\Phi)+\sin\theta \;=\; 2\sin\tfrac{1}{2}(\alpha\pm\Phi+\theta)\,\sin\tfrac{1}{2}(\alpha\pm\Phi-\theta+\pi)$$
$$\sin(\alpha\pm\Phi)-\sin\theta \;=\; -2\cos\tfrac{1}{2}(\alpha\pm\Phi+\theta)\,\cos\tfrac{1}{2}(\alpha\pm\Phi-\theta+\pi)$$

it follows from (4.14) that

$$\begin{aligned}
\Psi(\alpha) \;=\;& \psi_\Phi\left(\alpha+\Phi+\theta_+-\frac{\pi}{2}\right)\psi_\Phi\left(\alpha+\Phi-\theta_++\frac{\pi}{2}\right)\\
& \cdot\,\psi_\Phi\left(\alpha-\Phi+\theta_--\frac{\pi}{2}\right)\psi_\Phi\left(\alpha-\Phi-\theta_-+\frac{\pi}{2}\right) \qquad (B.8)
\end{aligned}$$

This shows the connection between $\Psi(\alpha)$ and the Maliuzhinets function $\psi_\Phi(\alpha)$.
A particular solution of (B.7) having the desired analyticity is

$$\psi_\Phi(\alpha)=\exp\left\{\frac{j}{8\Phi}\int_0^\alpha\int_{-j\infty}^{j\infty}\frac{\tan\frac{\pi u}{4\Phi}}{\cos(u-t)}\,du\,dt\right\}$$

and when the outer integration is carried out (ZAVADSKII AND SAKHAROVA,
1967), we obtain

$$\psi_\Phi(\alpha)=\exp\left\{-\frac{1}{2}\int_0^\infty\frac{\cosh\alpha t-1}{t\cosh\frac{\pi t}{2}\sinh 2\Phi t}\,dt\right\} \qquad (B.9)$$

In special cases the integral can be evaluated in closed form. Thus, for $\Phi=\pi/4$
or $3\pi/4$ corresponding to a right-angled exterior or interior wedge,

$$\psi_{\pi/4}(\alpha)=\cos\frac{\alpha}{2},\qquad \psi_{3\pi/4}(\alpha)=\frac{4}{3}\frac{\cos\frac{\alpha-\pi}{6}\cos\frac{\alpha+\pi}{6}}{\cos\frac{\alpha}{6}} \qquad (B.10)$$

and for $\Phi=\pi/2$ or π corresponding to a full or half-plane, $\psi_\Phi(\alpha)$ can be
expressed as a finite integral, *viz.*

$$\psi_{\pi/2}(\alpha) \;=\; \exp\left\{\frac{1}{4\pi}\int_0^\alpha\frac{2u-\pi\sin u}{\cos u}\,du\right\} \qquad (B.11)$$

$$\psi_\pi(\alpha) \;=\; \exp\left\{-\frac{1}{8\pi}\int_0^\alpha\frac{\pi\sin u-2\sqrt{2}\,\pi\sin\frac{u}{2}+2u}{\cos u}\,du\right\} \qquad (B.12)$$

It should be noted that the expression for $\psi_\pi(\alpha)$ given by MALIUZHINETS (1958)
has sign errors which were corrected later.

Some of the identities satisfied by $\psi_\Phi(\alpha)$ are as follows:

$$\psi_\Phi(-\alpha) = \psi_\Phi(\alpha), \qquad \psi_\Phi(\alpha^*) = \psi_\Phi^*(\alpha) \qquad \text{(B.13)}$$

where the asterisk denotes the complex conjugate, showing that $\psi_\Phi(\alpha)$ is a real, even function of α:

$$\psi_\Phi\left(\alpha + \frac{\pi}{2}\right) \psi_\Phi\left(\alpha - \frac{\pi}{2}\right) = \left\{\psi_\Phi\left(\frac{\pi}{2}\right)\right\}^2 \cos\frac{\pi\alpha}{4\Phi} \qquad \text{(B.14)}$$

$$\psi_\Phi(\alpha + \Phi)\,\psi_\Phi(\alpha - \Phi) = \left\{\psi_\Phi(\Phi)\right\}^2 \psi_{\Phi/2}(\alpha) \qquad \text{(B.15)}$$

$$\psi_\Phi(\alpha + 2\Phi) = \cot\frac{1}{2}\left(\alpha + \frac{\pi}{2}\right) \psi_\Phi(\alpha - 2\Phi) \qquad \text{(B.16)}$$

$$\psi_\Phi\left(\alpha \pm 2\Phi \pm \frac{3\pi}{2}\right) = \pm\frac{\sin\frac{\pi}{4\Phi}(\pi \pm \alpha)}{\sin\frac{\pi\alpha}{4\Phi}} \psi_\Phi\left(\alpha \pm 2\Phi \mp \frac{\pi}{2}\right) \quad \text{(B.17)}$$

and

$$\psi_\Phi(\alpha) = O\left\{\exp\left(\frac{\pi}{8\Phi}|\text{Im. }\alpha|\right)\right\} \qquad \text{(B.18)}$$

as $|\text{Im. }\alpha| \to \infty$. Additional results for the case $\Phi = \pi$ have been given by BOWMAN (1967). The function $\psi_\Phi(\alpha)$ has a denumerably infinite set of poles and zeros, and the poles and zeros nearest to the origin are at $\alpha = \pm(3\pi/2 + 2\Phi)$ and $\pm(\pi/2 + 2\Phi)$, respectively.

A direct numerical evaluation of (B.9) can be difficult, but approximate analytical expressions for $\psi_\Phi(\alpha)$ are available, and these may be accurate enough. HERMAN ET AL. (1987) showed that a combination of the small argument approximation

$$\psi_\Phi(\alpha) \simeq 1 - \alpha^2\delta/\Phi^2 \qquad \text{(B.19)}$$

with

$$\delta = 0.04626 + 0.054\,\Phi - 0.0078\,\Phi^2$$

and the large argument approximation

$$\psi_\Phi(\alpha) \simeq \sqrt{\cos\frac{\pi\alpha}{4\Phi}}\, \exp\left(-\frac{\gamma}{\pi}\right) \qquad \text{(B.20)}$$

with

$$\gamma = 2.556343\,\Phi - 3.259678\,\Phi^2 + 1.659306\,\Phi^3 \\ - 0.3883548\,\Phi^4 + 0.03473964\,\Phi^5$$

is accurate to better than 2 percent in amplitude and 2° in phase for all values of α provided $0 \leq \text{Re. }\alpha < \pi/2$ or (B.14) is first used to place α in this strip. Two percent equi-error curves for these approximations are given in Fig. B–3 from which we conclude that for all Φ (B.20) can be used if $\text{Im. }\alpha > 4$ and (B.19) otherwise. For greater accuracy it may be necessary to resort to

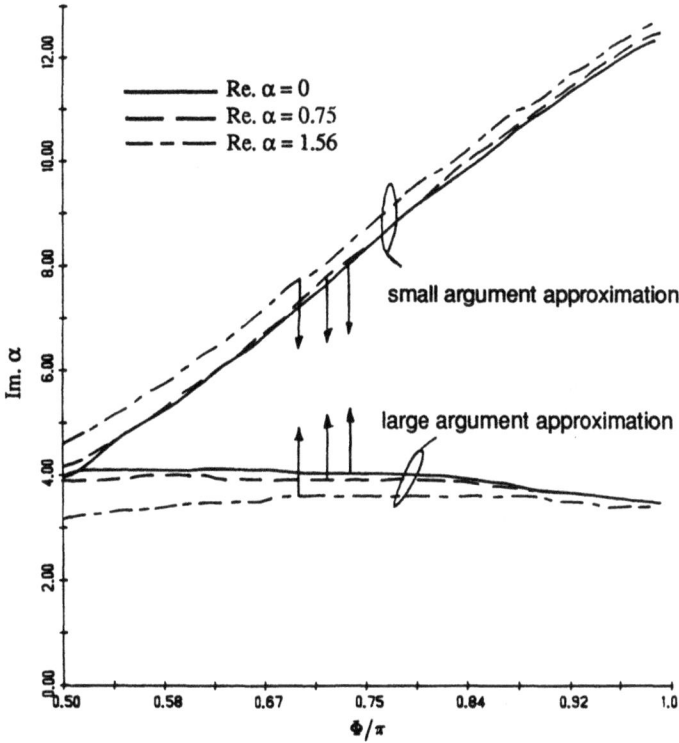

Figure B–3: *Two percentile equi-error amplitude curves for the small and large argument approximations. The direction of the arrows indicates the region where the error is less than 2 percent*

a numerical evaluation of the integral expression (B.9). Since the integrand decreases rapidly with increasing t, it is possible to truncate the range of integration, and, if Im. $\alpha < 11$, the range $(0, 1.5)$ is sufficient for an accuracy of better than 0.5 percent and 0.5°. A simple 5-point midpoint approximation in (B.9) then gives (HERMAN ET AL., 1987)

$$\psi_\Phi(\alpha) \simeq \exp\left\{-0.15 \sum_{n=1}^{5} \frac{\cosh \alpha t_n - 1}{t_n \cosh \dfrac{\pi t_n}{2} \sinh 2\Phi t_n}\right\} \tag{B.21}$$

with

$$t_n = 0.3n - 0.15.$$

If Im. $\alpha > 11$ the large argument approximation (B.20) provides this accuracy and can be used instead.

These approximations have been implemented in a FORTRAN program for the computation of $\psi_\Phi(\alpha)$. By examining the real and imaginary parts of α, the program automatically selects the appropriate approximation, and the subroutine PSIPHI incorporates the identity (B.14). A program listing follows.

```
      COMPLEX FUNCTION PSIPHI(CANG,PHI)
C     COMPUTES THE MALIUZHINETS FUNCTION OF ARGUMENT CANG
C     CANG=Complex Argument of the Function
C     2*PHI=External Angle of the Impedance Wedge
C     ROUTINE IS BASED ON THE PAPER BY HERMAN,VOLAKIS AND SENIOR(1987)
C     AND WAS WRITTEN BY M. HERMAN AT THE UNIVERSITY OF MICHIGAN
C     IF IMAG(CARG)<10, THE RIEMANN SUM IS PERFORMED,
C     OTHERWISE THE LARGE ARGUMNET APPROXIMATION IS USED
C
      COMPLEX CANG,CANG1,COEF
      COMPLEX CN1,CN2,CN3,CN4,CN5
      COMPLEX PSISQ,U,UJ
      DATA PI,PI2/3.141592654,1.5707963/
      DATA UJ/(1.,0.)/
C     MIDPOINT INTEGR. METHOD USING 5 POINTS (INTERVAL 0,1.5)
      UR=PI/2.
      UI=0.
      SUM=0.
      SUM1=0.
      FH=1.5/5.
      FH2=FH/2.
      DO 10 I=1,5
      S=FLOAT(I-1)*FH+FH2
      FS=(COSH(UR*S)*COS(UI*S)-1.)
      DENOM=S*COSH(PI*S/2.)*SINH(PHI*2.*S)
      FS=FS*FH/DENOM
      FS1=(SINH(UR*S)*SIN(UI*S))
      FS1=FS1*FH/DENOM
      SUM=SUM+FS
      SUM1=SUM1+FS1
   10 CONTINUE
      CN1=-.5*CMPLX(SUM,SUM1)
      CN1=CEXP(CN1)
      PSISQ=CN1*CN1
      AR=REAL(CANG)
      AI=AIMAG(CANG)
      ITTT=0
      IF(AR.GT.0)GO TO 30
       AR=-AR
       ITTT=1
   30 ITT=0
      IF(AI.GT.0)GO TO 40
       ITT=1
       AI=-AI
   40 IT=0
      CANG1=CMPLX(AR,AI)
      IF(AR.LT.PI2)GO TO 90
       AR=AR-PI
```

```
      CN5=UJ*(PI*PI/(8.*PHI))
      CN4=UJ*PI
      R5=PI/(4.*PHI)
      COEF=PSISQ*CCOS(CANG1*R5-CN5)
      IT=1
     IF(AR.LT.PI2)GO TO 90
      AR=AR-PI
      COEF=(CCOS(CANG1*R5-CN5))/CCOS((CANG1-CN4)*R5-CN5)
      IT=2
     IF(AR.LT.PI2)GO TO 90
      AR=AR-PI
      COEF=PSISQ*CCOS(CANG1*R5-CN5)*CCOS((CANG1-2.*CN4)*R5-CN5)
      COEF=COEF/CCOS((CANG1-CN4)*R5-CN5)
      IT=3
     IF(AR.LT.PI2)GO TO 90
     AR=AR-PI
     COEF=CCOS((CANG1-2.*CN4)*R5-CN5)*CCOS(CANG1*R5-CN5)
     COEF=COEF/(CCOS((CANG1-3.*CN4)*R5-CN5)*CCOS((CANG1-CN4)*R5-CN5))
     IT=2
     IF(AR.LT.PI2)GO TO 90
     AR=AR-PI
     COEF=CCOS((CANG1-2.*CN4)*R5-CN5)*CCOS(CANG1*R5-CN5)
     COEF=COEF/(CCOS((CANG1-3.*CN4)*R5-CN5)*CCOS((CANG-CN4)*R5-CN5))
     COEF=PSISQ*COEF*CCOS((CANG1-4.*CN4)*R5-CN5)
     IT=3
  90 CONTINUE
     IF(ABS(AR).GT.PI2)PRINT *,'AR > PI2 ',AR
     U=CMPLX(AR,AI)
     UR=AR
     UI=AI
     IF(UI.LE.10.)THEN
C    SMALL ARGUMENT APPROXIMATION USING
C    MIDPOINT INTEGR. METHOD--5 POINTS (INTERVAL 0,1.5)
     SUM=0.
     SUM1=0.
     FH=1.5/5.
     FH2=FH
     DO 100 I=1,5
     S=FLOAT(I-1)*FH+FH2
     FS=(COSH(UR*S)*COS(UI*S)-1.)
     DENOM=S*COSH(PI*S/2.)*SINH(PHI*2.*S)
     FS=FS*FH/DENOM
     FS1=(SINH(UR*S)*SIN(UI*S))
     FS1=FS1*FH/DENOM
     SUM=SUM+FS
     SUM1=SUM1+FS1
 100 CONTINUE
     CN1=-.5*CMPLX(SUM,SUM1)
     PSIPHI=CEXP(CN1)
     ELSE
C    LARGE ARGUMENT APPROXIMATION
     CN1=U*PI/(4.*PHI)
```

```
CN2=CCOS(CN1)
AMP=CABS(CN2)
AMP=SQRT(AMP)
R1=REAL(CN2)
R2=AIMAG(CN2)
PH=ATAN2(R2,R1)
IF(PH.LT.0.)PH=2.*PI+PH
PH=PH/2.
R1=AMP*COS(PH)
R2=AMP*SIN(PH)
CN1=CMPLX(R1,R2)
 B=2.556343
 C=-3.259678
 D=1.659306
 E=-.3883548
 F=.03473964
 PSIPHI=CN1*EXP(-(B*PHI+C*PHI**2+D*PHI**3+E*PHI**4+
1F*PHI**5)/PI)
 IF(REAL(PSIPHI).LT.0.)PSIPHI=-PSIPHI
 END IF
 IF(IT.EQ.1)PSIPHI=COEF/PSIPHI
 IF(IT.EQ.2)PSIPHI=COEF*PSIPHI
 IF(IT.EQ.3)PSIPHI=COEF/PSIPHI
 IF(ITT.EQ.1)PSIPHI=CONJG(PSIPHI)
 IF(ITTT.EQ.1)PSIPHI=CONJG(PSIPHI)
 RETURN
 END
```

References

Abramowitz, M. and Stegun, I. A. (1964), *Handbook of Mathematical Functions*, National Bureau of Standards, Washington DC, p. 255.

Bowman, J. J. (1967), "High-frequency backscattering from an absorbing infinite strip with arbitrary face impedances", *Can. J. Phys.*, **45**, pp. 2409–2430.

Herman, M. I., Volakis, J. L. and Senior, T. B. A. (1987), "Analytic expressions for a function occurring in diffraction theory", *IEEE Trans. Antennas Propagat.*, **AP-35**, pp. 1083–1086.

Maliuzhinets, G. D. (1951), "Some generalizations of the method of reflections in the theory of sinusoidal wave diffraction", Doctoral Dissertation, Fiz. Inst. Lebedev, Acad. Nauk. SSR (in Russian).

Maliuzhinets, G. D. (1958), "Excitation, reflection and emission of surface waves from a wedge with given face impedances", *Sov. Phys. Doklady*, **3**, pp. 752–755.

Mittra, R. and Lee, S.-W. (1971), *Analytical Techniques in the Theory of Guided Waves*, MacMillan Co., New York, pp. 4–11.

Ricoy, M. A. and Volakis, J. L. (1989), "E-polarization diffraction by a thick metal-dielectric join", *J. Electromagn. Waves Applics*, **3**, pp. 383–407.

Zavadskii, V. and Sakharova, M. (1967), "Application of the special function in problems of wave diffraction in wedge-shaped regions", *Sov. Phys. Acoust.*, **13**, pp. 48–54.

Appendix C

Steepest descent method

An integral which is important in diffraction theory is

$$I_C(\kappa) = \int_C g(\alpha)\, e^{\kappa\, f(\alpha)}\, d\alpha \qquad (C.1)$$

where κ is a positive real quantity, C is some contour in the complex $\alpha = \xi + j\eta$ plane, and

$$f(\alpha) = u(\xi, \eta) + jv(\xi, \eta)$$

is an analytic function with non-zero imaginary part. Since κ is typically large, the exponential portion of the integrand is highly oscillatory and, to facilitate the evaluation of the integral, C is deformed into the path C_{SDP} shown in Fig. C–1. The new path is such that on C_{SDP}

$$
\begin{aligned}
v(\xi, \eta) &= v(\xi_s, \eta_s) \\[4pt]
u(\xi, \eta) &\le u(\xi_s, \eta_s)
\end{aligned}
\qquad (C.2)
$$

where $\alpha_s = \xi_s + j\eta_s$ is a point still to be determined and, because v is constant, the integrand is no longer oscillatory. From Cauchy's residue theorem we then have

$$I_C(\kappa) = 2\pi j \sum \mathrm{Res} + I_b(\kappa) + I(\kappa) \qquad (C.3)$$

where the residues are those of any poles which are captured, and $I_b(\kappa)$ is the contribution from any branch cuts crossed in the deformation of C into C_{SDP}. We will henceforth omit $I_b(\kappa)$. The remaining integral in (C.3) is

$$I(\kappa) = \int_{C_{\text{SDP}}} g(\alpha)\, e^{\kappa\, f(\alpha)}\, d\alpha = e^{\kappa\, f(\alpha_s)} \int_{C_{\text{SDP}}} g(\alpha)\, e^{\kappa\{f(\alpha) - f(\alpha_s)\}}\, d\alpha \qquad (C.4)$$

and in view of (C.2),

$$f(\alpha) - f(\alpha_s) = -\{u(\xi_s, \eta_s) - u(\xi, \eta)\} = -\mu^2 < 0 \qquad (C.5)$$

Since μ is clearly real, (C.4) can be written as

$$I(\kappa) = e^{\kappa\, f(\alpha_s)} \int_{-\infty}^{\infty} g(\alpha)\, \frac{d\alpha}{d\mu}\, e^{-\kappa\mu^2}\, d\mu \qquad (C.6)$$

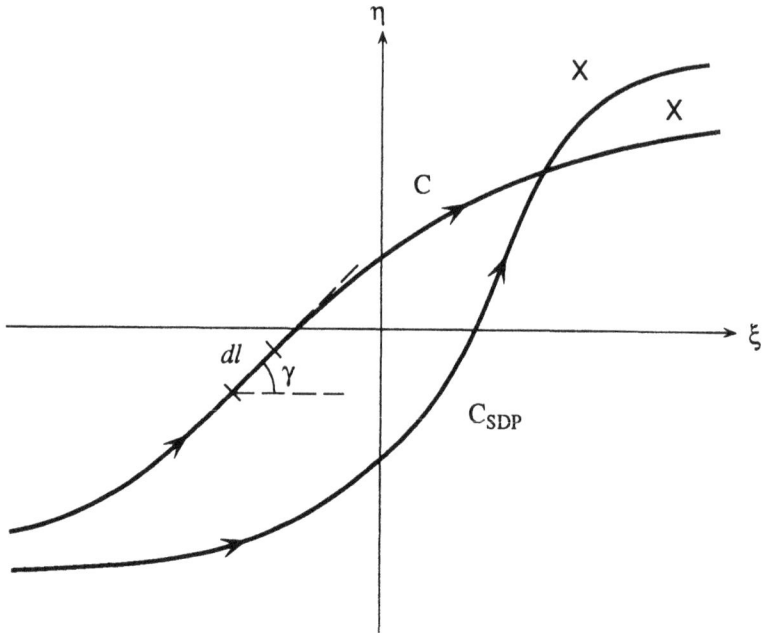

Figure C-1: *The path of integration C and the steepest descent path C_{SDP}, showing typical pole locations (X)*

and for $\kappa \gg 1$ the exponential decreases rapidly away from $\mu = 0$. The only significant contribution to the integral is from the immediate vicinity of α_s, and we will exploit this fact in the evaluation of $I(\kappa)$.

To better understand the character of C_{SDP} and, in particular, to determine the point α_s, we now consider the topology of $f(\alpha)$. If

$$\left. \frac{df}{d\alpha} \right|_{\alpha=\alpha_s} = f'(\alpha_s) = 0 \tag{C.7}$$

the Cauchy-Riemann equations (CHURCHILL, 1960)

$$\frac{\partial u}{\partial \xi} = \frac{\partial v}{\partial \eta}, \qquad \frac{\partial u}{\partial \eta} = -\frac{\partial v}{\partial \xi} \tag{C.8}$$

show

$$\frac{\partial u}{\partial \xi} = \frac{\partial u}{\partial \eta} = \frac{\partial v}{\partial \xi} = \frac{\partial v}{\partial \eta} = 0$$

at $\alpha = \alpha_s$. The α_s specified by (C.7) is therefore a local extremum of u and v, but we note that, because these are harmonic functions, they have no absolute maxima or minima in the complex plane. As a consequence of (C.8), the curves $u(\xi, \eta) = \mathrm{constant}$ and $v(\xi, \eta) = \mathrm{constant}$ are perpendicular to each other, and

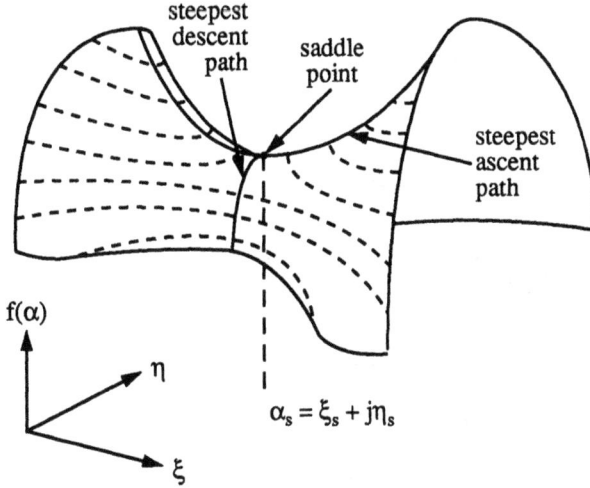

Figure C-2: *Topology of $f(\alpha) = u + jv$. The curves $u = $ const. (- - - -) and $v = $ const. (———) are orthogonal to each other*

this is illustrated in Fig. C-2. In addition, $v(\xi, \eta)$ is constant on those paths for which $u(\xi, \eta)$ changes (either increases or decreases) most rapidly. To show this it is necessary to consider $\partial^2 u / \partial \ell^2$ where ℓ is the arclength illustrated in Fig. C-1. For a maximum rate of change of $\partial u / \partial \ell$

$$\frac{\partial^2 u}{\partial \ell^2} = 0$$

or equivalently

$$
\begin{aligned}
\frac{\partial}{\partial \gamma}\left(\frac{\partial u}{\partial \ell}\right) &= \frac{\partial}{\partial \gamma}\left(\frac{\partial u}{\partial \xi} \cos \gamma + \frac{\partial u}{\partial \eta} \sin \gamma\right) \\
&= -\frac{\partial u}{\partial \xi} \sin \gamma + \frac{\partial u}{\partial \eta} \cos \gamma = 0
\end{aligned}
$$

where γ is the angle between the ξ axis and the path. This specifies the angle γ of the path for the maximum rate of change of $\partial u / \partial \ell$, and from the Cauchy-Riemann equations

$$
\begin{aligned}
\frac{\partial}{\partial \gamma}\left(\frac{\partial u}{\partial \ell}\right) &= -\left(\frac{\partial v}{\partial \eta} \sin \gamma + \frac{\partial v}{\partial \xi} \cos \gamma\right) \\
&= -\left(\frac{\partial v}{\partial \eta}\frac{\partial \eta}{\partial \ell} + \frac{\partial v}{\partial \xi}\frac{\partial \xi}{\partial \ell}\right) \\
&= -\frac{\partial v}{\partial \ell} = 0
\end{aligned}
$$

showing that v is constant on this path.

The path C_{SDP} defined by (C.2) is the *steepest descent path* (or SDP) through the point (ξ_s, η_s) and, because of the topology illustrated in Fig. C–2, this point is referred to as the *saddle point*. The fact that the exponential portion of the integrand in (C.4) decays most rapidly on the SDP simplifies the evaluation of $I(\kappa)$, and is the primary reason for the path deformation from C to C_{SDP}. We now consider the approximate closed form evaluation of the integral and show how the results depend on the algebraic properties of $g(\alpha)$ near to the SDP.

Case A: $g(\alpha)$ free of poles near the SDP

Because of the exponential decay of the integrand of (C.6), it is sufficient if we have a good knowledge of $g(\alpha)$ in the vicinity of the saddle point, and, provided $g(\alpha)$ is free of singularities (including poles) in this region, it can be expanded in a Taylor series about $\mu = 0$. Setting

$$g(\alpha)\frac{d\alpha}{d\mu} = G(\mu) = \sum_{m=0}^{\infty} \frac{G^{(m)}(0)}{m!}\mu^m \tag{C.9}$$

where

$$G^{(m)}(0) = \frac{d^m}{d\mu^m}G(\mu)\bigg|_{\mu=0} \tag{C.10}$$

we have

$$I(\kappa) = e^{\kappa f(\alpha_s)}\sum_{m=0}^{\infty}\frac{G^{(m)}(0)}{m!}\int_{-\infty}^{\infty}\mu^m e^{-\kappa\mu^2}\,d\mu$$

But

$$\int_{-\infty}^{\infty}\mu^m e^{-\kappa\mu^2}\,d\mu = \begin{cases} 0 & \text{if } m \text{ is odd} \\ \frac{1}{\kappa^{(m+1)/2}}\Gamma\left(\frac{m}{2}+\frac{1}{2}\right) & \text{if } m \text{ is even} \end{cases} \tag{C.11}$$

where $\Gamma(\,\cdot\,)$ is the Gamma function (ABRAMOWITZ AND STEGUN, 1964), and hence

$$I(\kappa) = \frac{e^{\kappa f(\alpha_s)}}{\sqrt{\kappa}}\sum_{m=0}^{\infty}\frac{G^{(2m)}(0)}{(2m)!}\frac{\Gamma\left(m+\frac{1}{2}\right)}{\kappa^m} \tag{C.12}$$

For large κ the first term alone provides a good approximation to $I(\kappa)$, and the result is

$$\begin{aligned} I(\kappa) &= e^{\kappa f(\alpha_s)}\sqrt{\frac{\pi}{\kappa}}\,G(0)\left\{1+O(\kappa^{-1})\right\} \\ &= e^{\kappa f(\alpha_s)}\sqrt{\frac{\pi}{\kappa}}\,g(\alpha_s)\frac{d\alpha}{d\mu}\bigg|_{\alpha=\alpha_s} + O(\kappa^{-3/2}) \end{aligned} \tag{C.13}$$

To evaluate $d\alpha/d\mu$ at $\alpha = \alpha_s$, we note that from (C.5)

$$f'(\alpha)\frac{d\alpha}{d\mu} = -2\mu$$

and

$$f''(\alpha)\left(\frac{d\alpha}{d\mu}\right)^2 + f'(\alpha)\frac{d^2\alpha}{d\mu^2} = -2$$

where the primes denote derivatives. Since $f'(\alpha_s) = 0$, it then follows that

$$\left.\frac{d\alpha}{d\mu}\right|_{\mu=0} = \sqrt{\frac{-2}{f''(\alpha_s)}}$$

provided α_s is a *first order* saddle point so that $f''(\alpha_s) \neq 0$. On substituting this into (C.13), we obtain

$$I_C(\kappa) = e^{\kappa f(\alpha_s)}\sqrt{\frac{-2\pi}{\kappa f''(\alpha_s)}}\,g(\alpha_s) + O(\kappa^{-3/2})$$

which can be written alternatively as

$$I_C(\kappa) \sim e^{\kappa f(\alpha_s)}\left|\frac{2\pi}{\kappa f''(\alpha_s)}\right|^{1/2} e^{j\phi_s}\,g(\alpha_s) \tag{C.14}$$

where ϕ_s is the angle between the ξ axis and the path at the saddle point as illustrated in Fig. C-3. This is the classic expression for the steepest descent evaluation of the integral (C.1). It is valid if α_s is a first order saddle point and $g(\alpha)$ is slowly varying in its vicinity—requiring, in particular, that $g(\alpha)$ be free of poles in this region. The computation of additional terms in (C.12) is tedious because of having to express $d^m\alpha/d\mu^m$ in terms of the higher derivatives of $f(\alpha)$, but $G^{(2)}(0)$ and $G^{(4)}(0)$ have been determined by SHAFER AND KOUYOUMJIAN (1967).

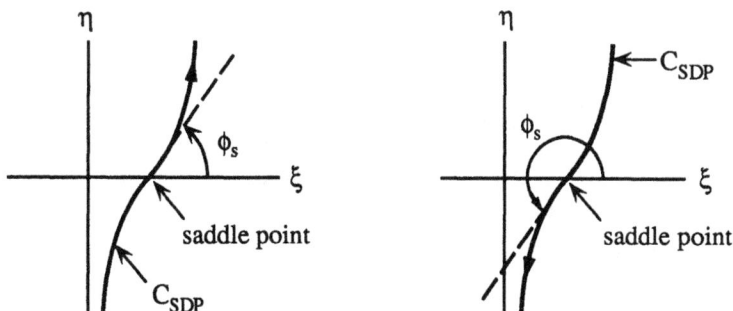

Figure C-3: *Definition of the angle ϕ_s in (C.14)*

Case B: $g(\alpha)$ has a simple pole at $\alpha = \alpha_p$

Multiplicative approach

When $g(\alpha)$ has a pole at $\alpha = \alpha_p$ which may be in the vicinity of the saddle point, $G(\mu)$ may not be analytic in this region, and cannot automatically be expanded in a Taylor series. To overcome the difficulty we introduce the function

$$G_1(\mu) = (\mu - \mu_p)\, G(\mu)$$

where

$$\mu_p^2 = f(\alpha_s) - f(\alpha_p) = -ja \qquad (C.15)$$

and a is a measure of the distance between α_s and α_p. Then

$$I(\kappa) = e^{\kappa f(\alpha_s)} \int_{-\infty}^{\infty} G_1(\mu) \frac{e^{-\kappa\mu^2}}{\mu - \mu_p}\, d\mu$$

and since $G_1(\mu)$ is analytic near $\mu = 0$, it can be expanded in a Taylor series to give

$$
\begin{aligned}
I(\kappa) &= e^{\kappa f(\alpha_s)} \sum_{m=0}^{\infty} \frac{G_1^{(m)}(0)}{m!} \int_{-\infty}^{\infty} \frac{\mu^m}{\mu - \mu_p} e^{-\kappa\mu^2}\, d\mu \\
&= e^{\kappa f(\alpha_s)} \sum_{m=0}^{\infty} \frac{G_1^{(m)}(0)}{m!} \int_{-\infty}^{\infty} \frac{\mu^m(\mu + \mu_p)}{\mu^2 + ja} e^{-\kappa\mu^2}\, d\mu \qquad (C.16)
\end{aligned}
$$

on using (C.15). From GRADSHTEYN AND RYZHIK (1965)

$$
\begin{aligned}
\int_{-\infty}^{\infty} \frac{\mu + \mu_p}{\mu^2 + ja} e^{-\kappa\mu^2}\, d\mu &= \int_{-\infty}^{\infty} \frac{\mu_p}{\mu^2 + ja} e^{-\kappa\mu^2}\, d\mu \\
&= 2e^{j\kappa a} \mu_p \sqrt{\frac{\pi}{a}} \int_{\sqrt{\kappa a}}^{\infty} e^{-j\tau^2}\, d\tau \\
&= \mu_p \sqrt{\frac{\pi}{\kappa}} \frac{F_{KP}(\kappa a)}{ja} \qquad (C.17)
\end{aligned}
$$

where

$$F_{KP}(\kappa a) = 2j\sqrt{\kappa a}\, e^{j\kappa a} \int_{\sqrt{\kappa a}}^{\infty} e^{-j\tau^2}\, d\tau$$

is the Kouyoumjian-Pathak transition function introduced in Section 3.3.2. It is related to the Clemmow transition function

$$F_C\left(\sqrt{\kappa a}\right) = e^{j\kappa a} \int_{\sqrt{\kappa a}}^{\infty} e^{-j\tau^2}\, d\tau$$

by the equation

$$F_{KP}(\kappa a) = \pm 2j\sqrt{\kappa a}\, F_C\left(\pm\sqrt{\kappa a}\right) \qquad (C.18)$$

where the lower sign applies if $\frac{\pi}{4} < \arg \sqrt{\kappa a} < \frac{5\pi}{4}$ and the upper sign otherwise. The Stokes lines $\arg \sqrt{\kappa a} = \frac{\pi}{4}, \frac{5\pi}{4}$ correspond to the pole crossing the real μ axis or, equivalently, the SDP in the α plane, and, when this occurs, the pole goes from being captured to not captured (or vice versa) in the deformation of C into C_{SDP}. Appearances to the contrary, this does not produce a discontinuity in the resulting expression for $I_C(\kappa)$ and, because of the identity (3.56), the discontinuity caused by the pole residue in (C.3) is balanced by the discontinuity in $F_C(\sqrt{\kappa a})$.

Substituting (C.17) into (C.16) and retaining only the first term in the series gives

$$I(\kappa) \simeq G_1(0) \, e^{\kappa f(\alpha_s)} \mu_{\text{p}} \sqrt{\frac{\pi}{\kappa}} \, \frac{F_{\text{KP}}(\kappa a)}{ja}$$

and since

$$G_1(0) = -\mu_{\text{p}} \, G(0) = -\mu_{\text{p}} \sqrt{\frac{-2}{f''(\alpha_s)}} \, g(\alpha_s)$$

we have

$$I(\kappa) \simeq g(\alpha_s) \, e^{\kappa f(\alpha_s)} \sqrt{\frac{-2\pi}{\kappa f''(\alpha_s)}} \, F_{\text{KP}}(\kappa a) \qquad (C.19)$$

or

$$I(\kappa) \simeq \pm 2j\sqrt{\kappa a} \, g(\alpha_s) \, e^{\kappa f(\alpha_s)} \sqrt{\frac{-2\pi}{\kappa f''(\alpha_s)}} \, F_C\left(\pm\sqrt{\kappa a}\right)$$

This is referred to as a *uniform* asymptotic approximation because $I(\kappa)$ is well-behaved as $\alpha_{\text{p}} \to \alpha_s$. In this case $g(\alpha_s) \to \infty$ but $F_{\text{KP}}(\kappa a) \to 0$. Thus, the uniformity is achieved by multiplying a large quantity by a small one. As indicated in Section 3.3.2, the multiplicative approach is also the modified Pauli-Clemmow method for the asymptotic evaluation of diffraction integrals. Note that for $\kappa a \gg 1$ and $-\frac{3\pi}{4} < \arg \kappa a < \frac{\pi}{4}$, $F_{\text{KP}}(\kappa a) \simeq 1$ and (C.19) reduces to the non-uniform result (C.14).

To evaluate the original integral $I_C(\kappa)$ we now return to (C.3) and consider the residue of the captured pole. Provided α_{p} crosses the SDP at α_s,

$$\begin{aligned} \text{Res} = R_{\text{p}} &= \lim_{\mu \to \mu_{\text{p}}} \left\{ (\mu - \mu_{\text{p}}) \, G(\mu) \, e^{\kappa f(\alpha_{\text{p}})} \right\} \\ &= \lim_{\alpha \to \alpha_{\text{p}}} \left\{ (\alpha - \alpha_{\text{p}}) \, g(\alpha) \, e^{\kappa f(\alpha_{\text{p}})} \right\} \end{aligned} \qquad (C.20)$$

and by using (3.56) it follows that

$$\begin{aligned} I_C(\kappa) &= 2\pi j R_{\text{p}} + I(\kappa) \\ &= -2j\sqrt{\kappa a} \, g(\alpha_s) \, e^{\kappa f(\alpha_s)} \sqrt{\frac{-2\pi}{\kappa f''(\alpha_s)}} \, F_C\left(-\sqrt{\kappa a}\right) \end{aligned} \qquad (C.21)$$

if the pole is captured when $-\frac{3\pi}{4} < \arg \sqrt{\kappa a} < \frac{\pi}{4}$. Alternatively

$$I_C(\kappa) = 2j\sqrt{\kappa a} \, g(\alpha_s) \, e^{\kappa f(\alpha_s)} \sqrt{\frac{-2\pi}{\kappa f''(\alpha_s)}} \, F_C\left(\sqrt{\kappa a}\right) \qquad (C.22)$$

if the pole is captured when $\frac{\pi}{4} < \arg\sqrt{\kappa a} < \frac{5\pi}{4}$.

If the pole crosses the SDP at a point other than the saddle point α_s, the discontinuity associated with the residue R_p is not precisely cancelled by the discontinuity in F_C, and (C.19) and (C.21) or (C.22) are no longer uniform approximations. This is not a fundamental limitation of the multiplicative approach, and continuity is restored by including the remaining terms in the series (C.16) (GENNARELLI AND PALUMBO, 1984), but the expression is no longer simple. As we will now show, there is an alternative approach which remedies the situation without requiring the other terms in the Taylor series.

Additive approach

This was developed by VAN DER WAERDEN (1951) and was applied to diffraction problems by CLEMMOW (1966) (see also FELSEN AND MARCUVITZ (1973)). It was later generalised to the case of multiple poles by VOLAKIS AND HERMAN (1986). The method starts by writing

$$G(\mu) = \frac{\tilde{R}_p}{\mu - \mu_p} + \tilde{G}(\mu) \tag{C.23}$$

where

$$R_p = \tilde{R}_p \exp\{\kappa\, f(\alpha_p)\} \tag{C.24}$$

is the residue of the pole α_p given in (C.20). Since $\tilde{G}(\mu)$ is well behaved as $\mu \to \mu_p$, it can be expanded in a Taylor series about $\mu = 0$, leading to the expression

$$I(\kappa) = e^{\kappa f(\alpha_s)}\left\{ \tilde{R}_p\mu_p \int_{-\infty}^{\infty} \frac{e^{-\kappa\mu^2}}{\mu^2 + ja}\, d\mu + \sum_{m=0}^{\infty} \frac{\tilde{G}^{(m)}(0)}{m!}\int_{-\infty}^{\infty} e^{-\kappa\mu^2}\, d\mu \right\}$$

where we have used (C.15) and (C.23). From (C.11) and (C.17) we then obtain

$$I(\kappa) = \frac{e^{\kappa f(\alpha_s)}}{\sqrt{\kappa}}\left\{ \tilde{R}_p\mu_p\sqrt{\pi}\, \frac{F_{\mathrm{KP}}(\kappa a)}{ja} + \sum_{m=0}^{\infty}\frac{\tilde{G}^{(2m)}(0)}{(2m)!}\frac{\Gamma(m+\frac{1}{2})}{\kappa^m} \right\}$$

(ROJAS, 1987) and, if we retain only the first term of the series,

$$I(\kappa) = e^{\kappa f(\alpha_s)}\sqrt{\frac{\pi}{\kappa}}\left\{ -\frac{\tilde{R}_p}{\mu_p}F_{\mathrm{KP}}(\kappa a) + \tilde{G}(0) \right\}$$

Finally, since

$$\tilde{G}(0) = \frac{\tilde{R}_p}{\mu_p} + G(0) = \frac{\tilde{R}_p}{\mu_p} + \sqrt{\frac{-2}{f''(\alpha_s)}}\, g(\alpha_s)$$

we have

$$I(\kappa) = e^{\kappa f(\alpha_s)}\sqrt{\frac{\pi}{\kappa}}\frac{\tilde{R}_p}{\mu_p}\{1 - F_{\mathrm{KP}}(\kappa a)\} + g(\alpha_s)\, e^{\kappa f(\alpha_s)}\sqrt{\frac{-2\pi}{\kappa f''(\alpha_s)}} \tag{C.25}$$

and this should be compared with (C.19) and (C.14).

In contrast to (C.19), (C.25) makes $I_C(\kappa)$ continuous as α_p crosses the SDP at *any* point. To show this, we consider the term

$$T(\kappa) = -e^{\kappa f(\alpha_s)} \sqrt{\frac{\pi}{\kappa} \frac{\tilde{R}_p}{\mu_p}} F_{KP}(\kappa a)$$

as the pole passes into the region enclosed by the paths C and C_{SDP} in Fig. C–1. From (C.18)

$$T(\kappa) = \mp 2 j e^{\kappa f(\alpha_s) + j\pi/4} \sqrt{\pi} \, \tilde{R}_p \, F_C \left(\pm \sqrt{\kappa a}\right)$$

and if the path deformation captures the pole (with $-\frac{3\pi}{4} < \arg \sqrt{\kappa a} < \frac{\pi}{4}$)

$$I_C(\kappa) = 2\pi j R_p - 2 j e^{\kappa f(\alpha_s) + j\pi/4} \sqrt{\pi} \, \tilde{R}_p \, F_C \left(\sqrt{\kappa a}\right) + \text{cont. terms} \qquad (C.26)$$

but if the pole lies outside the region

$$I_C(\kappa) = 2 j e^{\kappa f(\alpha_s) + j\pi/4} \sqrt{\pi} \, \tilde{R}_p \, F_C \left(-\sqrt{\kappa a}\right) + \text{cont. terms} \qquad (C.27)$$

The explicit terms in (C.26) and (C.27) are identical by virtue of the definition (C.24) and the identity (3.56). In (C.25) the transition function multiplies the residue of the pole, and it is this fact which makes $I_C(\kappa)$ continuous regardless of the pole's location. It is important to note that (C.25) remains valid as $\alpha_p \to \alpha_s$ in spite of the fact that $g(\alpha_s) \to \infty$. Since $\mu \to 0$ as $\alpha_p \to \alpha_s$, it follows from (C.9), (C.15) and (C.20) that

$$\lim_{\alpha_p = \alpha_s} \left\{ \frac{R_p}{\mu_p} + g(\alpha_s) \, e^{\kappa f(\alpha_s)} \sqrt{\frac{-2}{f''(\alpha_s)}} \right\} = 0$$

showing that the infinity of $g(\alpha_s)$ is cancelled by the first term on the right hand side of (C.25).

The additive method is easily generalised to the case when $g(\alpha)$ has multiple first order poles which may coalesce and cross the path of integration. If $g(\alpha)$ has N poles α_{pi}, $i = 1, 2, \ldots, N$, (C.23) is replaced by

$$G(\mu) = \sum_{i=1}^{N} \frac{\tilde{R}_{pi}}{\mu - \mu_{pi}} + \tilde{G}(\mu) \qquad (C.28)$$

where

$$\mu_{pi}^2 = f(\alpha_s) - f(\alpha_{pi}) = -j a_i$$

and

$$R_{pi} = \tilde{R}_{pi} \exp \left\{ \kappa f(\alpha_{pi}) \right\}$$

is the residue of the ith pole. Substituting (C.28) into (C.9) and (C.6) and then expanding $\tilde{G}(\mu)$ in a Taylor series, we find

$$I(\kappa) = \frac{e^{\kappa f(\alpha_s)}}{\sqrt{\kappa}} \left\{ \sum_{i=1}^{N} \tilde{R}_{p_i} \mu_{p_i} \sqrt{\pi} \frac{F_{\mathrm{KP}}(\kappa a_i)}{j a_i} + \sum_{m=0}^{\infty} \frac{\tilde{G}^{(2m)}(0)}{(2m)!} \frac{\Gamma\left(m + \frac{1}{2}\right)}{\kappa^m} \right\}$$

Since

$$\tilde{G}(0) = \sum_{i=1}^{N} \frac{\tilde{R}_{p_i}}{\mu_{p_i}} + G(0)$$

the first term in the Taylor series gives

$$I(\kappa) = e^{\kappa f(\alpha_s)} \sqrt{\frac{\pi}{\kappa}} \sum_{i=1}^{N} \frac{\tilde{R}_{p_i}}{\mu_{p_i}} \{1 - F_{\mathrm{KP}}(\kappa a_i)\} + g(\alpha_s) e^{\kappa f(\alpha_s)} \sqrt{\frac{-2\pi}{\kappa f''(\alpha_s)}} \quad \text{(C.29)}$$

This was obtained by VOLAKIS AND HERMAN (1986) and later by ROJAS (1987) (see also YIP AND CHIAVETTA (1987)), who showed the equivalence of the multiplicative and additive approaches when all terms in the Taylor series are retained.

Example

Consider the integral

$$I_C(k\rho) = \int_C \sec\left(\frac{\alpha - \alpha_0}{2n}\right) e^{-jk\rho \cos(\alpha - \phi)} \, d\alpha \quad \text{(C.30)}$$

with $0 < \phi < \pi$. Comparison with (C.1) and (C.4) leads to the identification

$$f(\alpha) = -j\cos(\alpha - \phi), \qquad\qquad \kappa = k\rho$$

$$g(\alpha) = \sec\left(\frac{\alpha - \alpha_0}{2n}\right), \qquad\qquad \alpha_s = \phi$$

and we observe that $g(\alpha)$ has a pole at $\alpha = \alpha_p$ where $\alpha_p = n\pi + \alpha_0$. Also

$$f''(\alpha_s) = j, \qquad g(\alpha_s) = \sec\left(\frac{\phi - \alpha_0}{2n}\right)$$

and

$$a = -j\{f(\alpha_p) - f(\alpha_s)\} = 2\sin^2\left(\frac{n\pi + \alpha_0 - \phi}{2}\right)$$

From (C.3) it then follows that

$$I_C(k\rho) = \left\{ \begin{array}{c} -4j\pi n e^{-jk\rho\cos(n\pi + \alpha_0 - \phi)} \\ 0 \end{array} \right\} + I(k\rho) \qquad (\phi \lessgtr n\pi + \alpha_0)$$

with

$$I(k\rho) = \int_{C_{\mathrm{SDP}}} \sec\left(\frac{\alpha - \alpha_0}{2n}\right) e^{-jk\rho\cos(\alpha-\phi)}\, d\alpha \qquad (\mathrm{C.31})$$

The remaining task is to specify the SDP and from (C.2) this is defined by

$$\cos(\xi - \phi)\cosh\eta = 1 \qquad (\mathrm{C.32})$$

with

$$\sin(\xi - \phi)\sinh\eta > 0 \qquad (\mathrm{C.33})$$

where (C.33) serves to restrict the solution domain of (C.32). We observe that the inequality is satisfied in the strips $\{2\pi N + \phi < \xi < 2\pi N + \pi + \phi,\ \eta > 0\}$ and $\{2\pi N - \pi + \phi < \xi < 2\pi N + \phi,\ \eta < 0\}$ for any integer N including zero, and since the strips corresponding to $N = 0$ are closest to the integration path, the SDP must be chosen to traverse these. From (C.32) the path is such that

$$\xi = \phi \pm \cos^{-1}\left(\frac{1}{\cosh\eta}\right) \qquad (\eta \gtrless 0)$$

which can be written as

$$\xi = \phi + \mathrm{gd}(\eta) \qquad (\mathrm{C.34})$$

where $\mathrm{gd}(\eta)$ is the Gudermann function

$$\mathrm{gd}(\eta) = \cos^{-1}\left(\frac{1}{\cosh\eta}\right)\operatorname{sgn}\eta \qquad (\mathrm{C.35})$$

The corresponding path is illustrated in Fig. C–4.

The evaluation of $I(k\rho)$ can now be carried out in accordance with the previous analysis. From (C.19)

$$I(k\rho) = \sec\left(\frac{\phi - \alpha_0}{2n}\right) e^{-jk\rho}e^{j\pi/4}\sqrt{\frac{2\pi}{k\rho}}\, F_{\mathrm{KP}}\left(2k\rho\sin^2\frac{n\pi + \alpha_0 - \phi}{2}\right)$$

and since the pole is captured when $-\frac{3\pi}{4} < \arg\sqrt{a} < \frac{\pi}{4}$, (C.21) can be used to show

$$I_C(k\rho) =$$

$$4\sqrt{\pi}\, e^{-jk\rho - j\pi/4}\, \frac{\sin\left(\dfrac{n\pi + \alpha_0 - \phi}{2}\right)}{\cos\left(\dfrac{\phi - \alpha_0}{2n}\right)}\, F_C\left(-\sqrt{2k\rho}\,\sin\frac{n\pi + \alpha_0 - \phi}{2}\right)$$

$$(\mathrm{C.36})$$

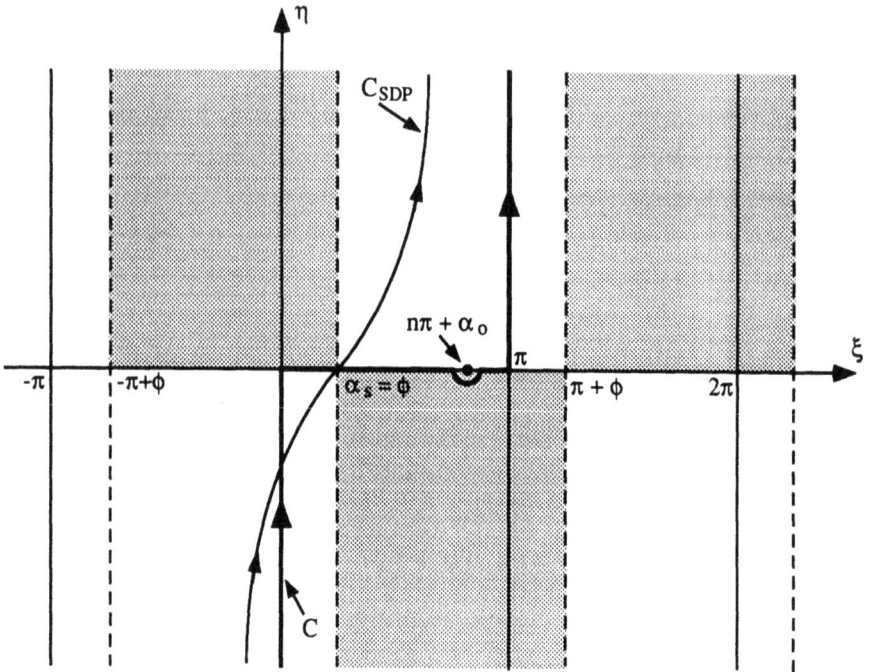

Figure C–4: *The complex α plane showing the path C in (C.30) and the steepest descent path C_{SDP}. The open regions are the half strips satisfying (C.32) and (C.35) where C_{SDP} may lie*

References

Abramowitz, M. and Stegun, I. A. (1964), *Handbook of Mathematical Functions*, National Bureau of Standards, Washington DC, p. 255.

Churchill, R. V. (1960), *Complex Variables and Applications*, McGraw-Hill Book Co., New York, pp. 34–38.

Clemmow, P. C. (1966), *The Plane Wave Spectrum Representation of Electromagnetic Fields*, Pergamon Press, New York.

Felsen, L. B. and Marcuvitz, N. (1973), *Radiation and Scattering of Waves*, Prentice-Hall, Inc., Englewood Cliffs NJ, Chapter 4.

Gennarelli, C. and Palumbo, L. (1984), "A uniform asymptotic expansion of a typical diffraction integral with many coalescing simple pole singularities and a first-order saddle point", *IEEE Trans. Antennas Propagat.*, **AP-32**, pp. 1122–1124.

Gradshteyn, I. S. and Ryzhik, I. M.(1965), *Table of Integrals, Series, and Products*, Academic Press, New York, p. 338.

Rojas, R. G. (1987), "Comparison between two asymptotic methods", *IEEE Trans. Antennas Propagat.*, **AP-35**, pp. 1489–1492.

Shafer, R. H. and Kouyoumjian, R. G. (1967), "Higher order terms in the saddle point approximation", *Proc. IEEE*, **55**, pp. 1496–1497.

Van der Waerden, B. L. (1951), "On the method of saddle points", *Appl. Sci. Res.*, **B2**, pp. 33–45.

Volakis, J. L. and Herman, M. I. (1986), "A uniform asymptotic evaluation of integrals", *Proc. IEEE*, **74**, pp. 1043–1044.

Yip, E. L. and Chiavetta, R. J. (1987), "Comparison of uniform asymptotic expansions of diffraction integrals", *IEEE Trans. Antennas Propagat.*, **AP-35**, pp. 1179–1180.

Appendix D

PO diffraction coefficient for impedance wedges

The physical optics (PO) currents are discontinuous at any edge or other line discontinuity in the surface slope, and, since they are non-zero only over the illuminated portion of a body, they may also be discontinuous at a shadow boundary. The discontinuity at an edge yields the PO diffracted field. This is part of the "true" edge-diffracted field, but the PTD formulation (see Section 4.4) requires that we separate it out. As an example, for the metallic half-plane in Fig. 3–11, the PO diffracted field is

$$E_z^{\mathrm{d_{po}}} = \frac{e^{-jk_0\rho}}{\sqrt{\rho}} \, D_{\mathrm{E}}^{\mathrm{po}}(\phi, \phi_0) \tag{D.1}$$

with

$$D_{\mathrm{E}}^{\mathrm{po}}(\phi, \phi_0) = \frac{e^{-j\pi/4}}{2\sqrt{2\pi k_0}} \left(\tan\frac{\phi + \phi_0}{2} - \tan\frac{\phi - \phi_0}{2} \right) \tag{D.2}$$

In accordance with PTD, the fringe wave field is then

$$E_z^{\mathrm{f}} = \frac{e^{-jk_0\rho}}{\sqrt{\rho}} \, D_{\mathrm{E}}^{\mathrm{f}}(\phi, \phi_0) \tag{D.3}$$

where

$$D_{\mathrm{E}}^{\mathrm{f}}(\phi, \phi_0) = D_{\mathrm{E}}^{\mathrm{nu}}(\phi, \phi_0) - D_{\mathrm{E}}^{\mathrm{po}}(\phi, \phi_0) \tag{D.4}$$

and

$$D_{\mathrm{E}}^{\mathrm{nu}}(\phi, \phi_0) = \frac{e^{-j\pi/4}}{2\sqrt{2\pi k_0}} \left(\sec\frac{\phi + \phi_0}{2} - \sec\frac{\phi - \phi_0}{2} \right) \tag{D.5}$$

obtained from (3.46) by putting $\bar{\eta} = 0$. In contrast to (D.2) and (D.5) which are infinite at $\phi = \pi \pm \phi_0$, the fringe wave diffraction coefficient (D.4) is continuous there.

We now derive the PO diffraction coefficient for an impedance wedge illuminated by the plane wave at skew incidence. The incident field is written as

$$\mathbf{E}^{\mathrm{i}} = \hat{e}e^{-jk_0 \mathrm{i}.\mathbf{r}}, \qquad Z_0\mathbf{H}^{\mathrm{i}} = \hat{h}e^{-jk_0 \mathrm{i}.\mathbf{r}} \tag{D.6}$$

where $\hat{\imath}.\hat{e} = 0$, $\hat{\imath} \times \hat{e} = \hat{h}$ and the direction of incidence is

$$\hat{\imath} = -\hat{x}\sin\beta^i\cos\phi_0 - \hat{y}\sin\beta^i\sin\phi_0 + \hat{z}\cos\beta^i \qquad (D.7)$$

The angles ϕ_0 and β^i are shown in Fig. D–1, with ϕ_0 measured from the upper

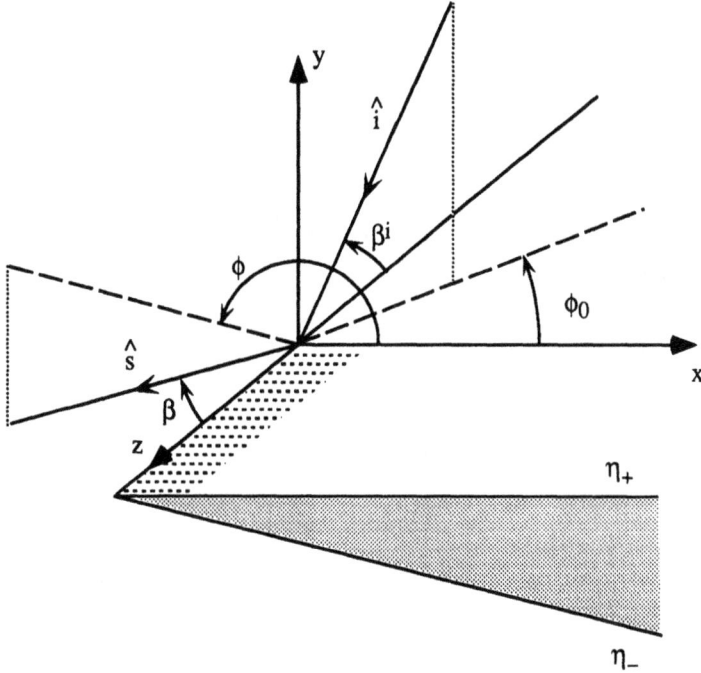

Figure D–1: *Edge geometry and angle definitions*

(+) face of the wedge and, in the special case of normal incidence, $\beta^i = \pi/2$. In carrying out the analysis, it is convenient to work with the z components of $\hat{e}$ and $\hat{h}$, and from (4.46) the other components are

$$
\begin{aligned}
e_x &= (\cos\beta^i\cos\phi_0\, e_z + \sin\phi_0\, h_z)/\sin\beta^i \\[4pt]
e_y &= (\cos\beta^i\sin\phi_0\, e_z - \cos\phi_0\, h_z)/\sin\beta^i \\[4pt]
h_x &= (-\sin\phi_0\, e_z + \cos\beta^i\cos\phi_0\, h_z)/\sin\beta^i \\[4pt]
h_y &= (\cos\phi_0\, e_z + \cos\beta^i\sin\phi_0\, h_z)/\sin\beta^i
\end{aligned}
\qquad (D.8)
$$

The PO surface field is the sum of the incident and reflected fields, *viz.*

$$\mathbf{E}^{po} = \mathbf{E}^i + \mathbf{E}^r, \qquad \mathbf{H}^{po} = \mathbf{H}^i + \mathbf{H}^r$$

and the associated electric and magnetic currents are

$$\mathbf{J}_e^{po} = \hat{n} \times \mathbf{H}^{po}, \qquad \mathbf{J}_m^{po} = -\hat{n} \times \mathbf{E}^{po} \tag{D.9}$$

respectively, where $\hat{n}$ is the outward unit vector normal. With a planar surface the reflected field is obtained by enforcing the boundary condition (2.36) as if the surface were infinite in extent. For the upper face of the wedge, application of the conditions

$$E_x = \bar{\eta}_+ Z_0 H_z, \qquad E_z = -\bar{\eta}_+ Z_0 H_x$$

gives

$$E_z^r = R_E e^{-jk_0 \hat{s}^r.\mathbf{r}}, \qquad Z_0 H_z^r = R_H e^{-jk_0 \hat{s}^r.\mathbf{r}}$$

where

$$\hat{s}^r = -\hat{x} \sin \beta^i \cos \phi_0 + \hat{y} \sin \beta^i \sin \phi_0 + \hat{z} \cos \beta^i$$

is a unit vector in the direction of the reflected wave, and

$$
\begin{aligned}
R_E &= \gamma \Bigg[\left\{ \left(\sin \phi_0 + \bar{\eta}_+ \sin \beta^i \right) \left(\sin \phi_0 - \frac{\sin \beta^i}{\bar{\eta}_+} \right) - \cos^2 \beta^i \cos^2 \phi_0 \right\} e_z \\
&\qquad - 2 \cos \beta^i \sin \phi_0 \cos \phi_0 \, h_z \Bigg]
\end{aligned}
$$

$$
\begin{aligned}
R_H &= \gamma \Bigg[2 \cos \beta^i \sin \phi_0 \cos \phi_0 \, e_z \\
&\qquad + \left\{ \left(\sin \phi_0 - \bar{\eta}_+ \sin \beta^i \right) \left(\sin \phi_0 + \frac{\sin \beta^i}{\bar{\eta}_+} \right) - \cos^2 \beta^i \cos^2 \phi_0 \right\} h_z \Bigg]
\end{aligned}
$$

with

$$\gamma = \left(1 + \bar{\eta}_+ \sin \beta^i \sin \phi_0 \right)^{-1} \left(1 + \frac{1}{\bar{\eta}_+} \sin \beta^i \sin \phi_0 \right)^{-1}$$

Hence

$$
\begin{aligned}
Z_0 \mathbf{J}_e^{po}(\mathbf{r}) &= \mathbf{j} e^{jk_0 (x \sin \beta^i \cos \phi_0 - z \cos \beta^i)} \\
\mathbf{J}_m^{po}(\mathbf{r}) &= \mathbf{m} e^{jk_0 (x \sin \beta^i \cos \phi_0 - z \cos \beta^i)}
\end{aligned}
\tag{D.10}
$$

where $\mathbf{j}$ and $\mathbf{m}$ are

$$
\begin{aligned}
\mathbf{j} &= 2\gamma \sin \phi_0 \Bigg[\left\{ \cos \beta^i \cos \phi_0 \, e_z + \left(\sin \phi_0 + \frac{\sin \beta^i}{\bar{\eta}_+} \right) h_z \right\} \hat{x} \\
&\qquad + \frac{1}{\bar{\eta}_+} \left\{ \left(\sin \phi_0 + \bar{\eta}_+ \sin \beta^i \right) e_z - \cos \beta^i \cos \phi_0 \, h_z \right\} \hat{z} \Bigg]
\end{aligned}
\tag{D.11}
$$

$$
\begin{aligned}
\mathbf{m} &= 2\gamma \sin \phi_0 \Bigg[\left\{ - \left(\sin \phi_0 + \bar{\eta}_+ \sin \beta^i \right) e_z + \cos \beta^i \cos \phi_0 \, h_z \right\} \hat{x} \\
&\qquad + \bar{\eta}_+ \left\{ \cos \beta^i \cos \phi_0 \, e_z + \left(\sin \phi_0 + \frac{\sin \beta^i}{\bar{\eta}_+} \right) h_z \right\} \hat{z} \Bigg]
\end{aligned}
$$

As expected, these are dual quantities which depend only on the incident field and the surface impedance.

The field generated by these currents is

$$\mathbf{E}^{po} = \iint_{S_{lit}} \{Z_0 \hat{s} \times \hat{s} \times \mathbf{J}_e^{po}(\mathbf{r}') + \hat{s} \times \mathbf{J}_m^{po}(\mathbf{r}')\} \frac{e^{-jk_0|\mathbf{r}-\mathbf{r}'|}}{|\mathbf{r} - \mathbf{r}'|} \, dS' \qquad (D.12)$$

where the illuminated surface S_{lit} is the upper face of the wedge and $\hat{s}$ is a unit vector in the direction of the observation point $\mathbf{r}$. On inserting the expressions (D.10) and (D.11), we obtain

$$\mathbf{E}^{po} = \frac{jk_0}{4\pi} \{\hat{s} \times (\hat{s} \times \mathbf{j}) + \hat{s} \times \mathbf{m}\} \int_0^\infty \int_{-\infty}^\infty e^{jk_0(x' \sin \beta^i \cos \phi_0 - z' \cos \beta^i)}$$

$$\cdot \frac{e^{-jk_0\sqrt{(x-x')^2+y^2+(z-z')^2}}}{\sqrt{(x - x')^2 + y^2 + (z - z')^2}} \, dz' \, dx'$$

The z' integration is easy to perform. If $u = \sqrt{(x - x')^2 + y^2}$, the substitution $z' - z = u \sinh \alpha$ gives

$$\int_{-\infty}^\infty \frac{e^{-jk_0\sqrt{(x-x')^2+y^2+(z-z')^2}}}{\sqrt{(x - x')^2 + y^2 + (z - z')^2}} e^{-jk_0 z' \cos \beta^i} \, dz'$$

$$= e^{-jk_0 z \cos \beta^i} \int_{-\infty}^\infty e^{-jk_0 u(\cosh \alpha + \cos \beta^i \sinh \alpha)} \, d\alpha$$

$$= e^{-jk_0 z \cos \beta^i} \int_{-\infty}^\infty e^{-jk_0 u \sin \beta^i \cosh \alpha} \, d\alpha$$

$$= -j\pi e^{-jk_0 z \cos \beta^i} H_0^{(2)}(k_0 u \sin \beta^i)$$

where $H_0^{(2)}$ is the zeroth order Hankel function of the second kind, and therefore

$$\mathbf{E}^{po} = \frac{k_0}{4} \{\hat{s} \times (\hat{s} \times \mathbf{j}) + \hat{s} \times \mathbf{m}\} e^{-jk_0 \cos \beta^i}$$

$$\cdot \int_0^\infty H_0^{(2)} \left(k_0 \sin \beta^i \sqrt{(x - x')^2 + y^2}\right) e^{jk_0 x' \sin \beta^i \cos \phi_0} \, dx' \qquad (D.13)$$

The PO edge-diffracted field is provided by the lower limit in the integral (MICHAELI (1986)) and, since x' is small close to the edge, we can replace the Hankel function by its large argument expansion to obtain

$$\mathbf{E}^{d_{po}} \simeq \frac{k_0}{4} \{\hat{s} \times (\hat{s} \times \mathbf{j}) + \hat{s} \times \mathbf{m}\} \sqrt{\frac{2j}{\pi k_0 \rho \sin \beta^i}} e^{-jk_0 \hat{s}.\mathbf{r}}$$

$$\cdot \int_0^\infty e^{jk_0 x' \sin \beta^i \cos \phi_0} \, dx' \qquad (D.14)$$

valid if $k_0 \rho \sin \beta^i \gg 1$ where $\rho = \sqrt{x^2 + y^2}$. Because the edge is infinite in extent, diffraction is confined to the Keller cone for which $\beta = \beta^i$ in Fig. D-1. Then

$$\hat{s} = \hat{y} \sin \beta^i \cos \phi + \hat{x} \sin \beta^i \sin \phi + \hat{z} \cos \beta^i$$

and in (D.14) we may set $\hat{s}.\mathbf{r} = s$ so that $\rho = s \sin \beta^i$. On carrying out the integration and retaining only the contribution from the lower limit, we find

$$\mathbf{E}^{\mathrm{dpo}} = -\frac{e^{-jk_0 s}}{\sqrt{s}} \frac{e^{-j\pi/4}}{2\sqrt{2\pi k_0}} \frac{1}{\sin \beta^i (\cos \phi + \cos \phi_0)} \{\hat{s} \times (\hat{s} \times \mathbf{j}) + \hat{s} \times \mathbf{m}\} \quad (\mathrm{D}.15)$$

with $\mathbf{j}$ and $\mathbf{m}$ defined in (D.11).

For use in (4.106) we must first express (D.15) in ray-fixed coordinates and, since its form is analogous to (3.179), the matrix representation (3.183) is applicable. Thus, for the upper face, the PO diffracted field can be obtained from (3.183) by making the substitutions

$$U(\bar{\eta}_{e1}, \bar{\eta}_{e2}) \quad \to \quad U^{\mathrm{po}}(\bar{\eta}_+) = \frac{\sin \phi_0}{\cos \phi + \cos \phi_0} \frac{1 - \bar{\eta}_+ \sin \beta^i \sin \phi}{1 + \bar{\eta}_+ \sin \beta^i \sin \phi_0}$$
$$\cdot \left(\cos^2 \beta^i - \sin^2 \beta^i \cos \phi \cos \phi_0\right)$$

$$V(\bar{\eta}_{e1}, \bar{\eta}_{e2}) \quad \to \quad V^{\mathrm{po}}(\bar{\eta}_+) = -\sin \beta^i \cos \beta^i \sin \phi_0 \frac{1 - \bar{\eta}_+ \sin \beta^i \sin \phi}{\bar{\eta}_+ + \sin \beta^i \sin \phi_0}$$

For the lower face, the required substitutions are $\bar{\eta}_+ \to \bar{\eta}_-$, $\phi_0 \to n\pi - \phi_0$, $\phi \to n\pi - \phi$ and $\beta^i \to \pi - \beta^i$, and, if both faces of the wedge are illuminated, the PO diffracted field is the sum of the contributions from each.

References

Michaeli, A. (1986), "Elimination of infinities in equivalent edge currents, part I: fringe current components", *IEEE Trans. Antennas Propagat.*, **AP-34**, pp. 912–918.

Appendix E

Determination of constants

The constants a_1, a_2, a_3, A_0 and A_1 in Section 5.6.2 are related by (5.213) with $\alpha = \frac{3\pi}{4} - \theta_m$, $\frac{\pi}{4} = \theta_m$ $(m = 1, 2)$, and, since there are only four equations, it is clear that one constant is undetermined.

When $\alpha = \frac{3\pi}{4} - \theta_1$, (5.200) becomes

$$a_1 \cos \frac{2\theta_1}{3} - a_2 \cos 2\theta_1 + a_3 \cos \frac{10\theta_1}{3} - \frac{1}{6} \cos \frac{14\theta_1}{3}$$

$$= \tfrac{1}{2} \sin \theta_1 (\sin \theta_1 + \cos \theta_1)\Big\{ (\sin \theta_2 + \cos \theta_1) K A_0 + \tfrac{1}{2} \cos \theta_1 (\sin \theta_1$$

$$+ \cos \theta_1 + 2 \sin \theta_2) K A_1 \Big\} + C \left(\frac{1}{2} - \frac{2\theta_1}{3\pi} \right) \sin \theta_1 \cos \theta_1 \qquad \text{(E.1)}$$

and for $\alpha = \frac{\pi}{4} - \theta_1$ we have

$$\frac{1}{2} a_1 \left(\cos \frac{2\theta_1}{3} - \sqrt{3} \sin \frac{2\theta_1}{3} \right) + a_2 \cos \theta_1$$

$$+ \frac{1}{2} a_3 \left(\cos \frac{10\theta_1}{3} + \sqrt{3} \sin \frac{10\theta_1}{3} \right) - \frac{1}{12} \left(\cos \frac{14\theta_1}{3} - \sqrt{3} \sin \frac{14\theta_1}{3} \right)$$

$$= \tfrac{1}{2} \sin \theta_1 (\sin \theta_1 - \cos \theta_1)\Big\{ -(\sin \theta_2 - \cos \theta_1) K A_0 + \tfrac{1}{2} \cos \theta_1 (\sin \theta_1$$

$$- \cos \theta_1 + 2 \sin \theta_2) K A_1 \Big\} - C \left(\frac{1}{6} - \frac{2\theta_1}{3\pi} \right) \sin \theta_1 \cos \theta_1 \qquad \text{(E.2)}$$

The addition of these gives

$$a_1 \sin \beta_1 - a_3 \sin 5\beta_1 = \frac{1}{6} \sin 7\beta_1 + \frac{2}{3\sqrt{3}} C \sin 3\beta_1 \qquad \text{(E.3)}$$

where

$$\beta_m = \frac{2\theta_m}{3} - \frac{\pi}{3} \qquad (m = 1, 2) \qquad \text{(E.4)}$$

and there is an equation similar to (E.3) with β_1 replaced by β_2. These are sufficient to specify a_1 and a_3 in terms of C. By expanding $(\sin \beta_m)^{-1} \sin 3\beta_m$ etc. in powers of $\cos^2 \beta_m$ and using the fact that

$$\cos 2\beta_m = \gamma_m \qquad \text{(E.5)}$$

345

it can be shown that

$$a_1 \;=\; -\tfrac{2}{3}\left(\gamma_1\gamma_2 + \tfrac{1}{4}\right) - \tfrac{4}{3}\left(2\gamma_1\gamma_2 + \gamma_1 + \gamma_2 + 1\right)\Gamma \tag{E.6}$$

$$a_3 \;=\; -\tfrac{1}{3}\left(\gamma_1 + \gamma_2 - \tfrac{1}{2}\right) + \tfrac{2}{3}\Gamma \tag{E.7}$$

$$C \;=\; \tfrac{\sqrt{3}}{2}\left(2\gamma_1\gamma_2 - \gamma_1 - \gamma_2 + \tfrac{1}{2}\right) - 2\sqrt{3}\left(\gamma_1 + \gamma_2 + \tfrac{1}{2}\right)\Gamma \tag{E.8}$$

for any Γ.

To complete the specification we need another equation connecting C and Γ. Subtraction of (E.2) from (E.1) gives

$$a_1 \cos\beta_1 + 2a_2\cos 3\beta_1 + a_3\cos 5\beta_1 = \tfrac{1}{6}\cos 7\beta_1$$
$$+\; \frac{\sin\theta_1\cos 2\theta_1}{1 + 2\sin\theta_1\sin\theta_2}\left\{\cos 2\theta_2 - \sin\theta_1(\sin\theta_1 + \sin\theta_2)\right\}KA_0$$
$$+\; \sin 2\theta_1\left\{\frac{\cos\theta_1(\sin\theta_1 + \sin\theta_2)}{1 + 2\sin\theta_1\sin\theta_2} + \frac{1}{3} - \frac{2\theta_1}{3\pi}\right\}C$$

and there is again a similar equation with β_1, θ_1 and θ_2 replaced by β_2, θ_2 and θ_1, respectively. On eliminating a_2 from the two equations, we obtain

$$a_1(\cos\beta_1\cos 3\beta_2 - \cos\beta_2\cos 3\beta_1) + a_3(\cos 5\beta_1\cos 3\beta_2 - \cos 5\beta_2\cos 3\beta_1)$$

$$=\; \tfrac{1}{6}(\cos 7\beta_1\cos 3\beta_2 - \cos 7\beta_2\cos 3\beta_1)$$

$$-\;\frac{1}{2}\frac{(\sin\theta_1 - \sin\theta_2)\cos 2\theta_1\cos 2\theta_2(\cos 2\theta_1 + \cos 2\theta_2)}{1 + 2\sin\theta_1\sin\theta_2}KA_0$$

$$-\;\frac{\sin^2\theta_1 - \sin^2\theta_2}{1 + 2\sin\theta_1\sin\theta_2}\left\{\cos 2\theta_1 + \cos 2\theta_2\right.$$

$$\left. +\; 2\sin\theta_1\sin\theta_2(1 + 2\sin\theta_1\sin\theta_2)\right\}C$$

$$-\;\left\{\sin 2\theta_1\cos 2\theta_2\left(\frac{1}{3} - \frac{2\theta_1}{3\pi}\right) - \sin 2\theta_2\cos 2\theta_1\left(\frac{1}{3} - \frac{2\theta_2}{3\pi}\right)\right\}C \tag{E.9}$$

where we have used the fact that

$$\cos 3\beta_m = -\cos 2\theta_m$$

and by a similar process to that used before, (E.9) can be written as

$$a_1 - 2(2\gamma_1\gamma_2 - \gamma_1 - \gamma_2 + 1)a_3 = -\frac{1}{6}\left\{4\gamma_1^2\gamma_2^2 + 8\gamma_1\gamma_2(\gamma_1 + \gamma_2)\right.$$

$$\left. -\; 8(\gamma_1^2 + \gamma_2^2) - 24\gamma_1\gamma_2 - 14(\gamma_1 + \gamma_2) - 1\right\}$$

$$-\;\frac{1}{8}\frac{\cos 2\theta_1\cos 2\theta_2(\cos 2\theta_1 + \cos 2\theta_2)}{\cos\theta_1\cos\theta_2(\sin\theta_1 + \sin\theta_2)(1 + 2\sin\theta_1\sin\theta_2)}KA_0$$

$$-\frac{1}{4\cos\theta_1\cos\theta_2}\left\{\frac{\cos 2\theta_1+\cos 2\theta_2}{1+2\sin\theta_1\sin\theta_2}+2\sin\theta_1\sin\theta_2\right\}C$$

$$+\frac{\sin 2\theta_1\cos 2\theta_2\left(\dfrac{1}{3}-\dfrac{2\theta_1}{3\pi}\right)-\sin 2\theta_2\cos 2\theta_1\left(\dfrac{1}{3}-\dfrac{2\theta_2}{3\pi}\right)}{2\cos\theta_1\cos\theta_2(\cos 2\theta_1-\cos 2\theta_2)}C$$

Finally, on inserting the expressions for a_1, a_3 and C in terms of Γ, we obtain

$$\{2\gamma_1\gamma_2+1+(\gamma_1+\gamma_2+\tfrac{1}{2})X(\theta_1,\theta_2)\}\Gamma =$$

$$\frac{1}{4}\Big\{\gamma_1^2\gamma_2^2+4\gamma_1\gamma_2(\gamma_1+\gamma_2)-3(\gamma_1+\gamma_2)^2$$

$$-10\gamma_1\gamma_2-2(\gamma_1+\gamma_2)-\tfrac{7}{2}+(2\gamma_1\gamma_2-\gamma_1-\gamma_2+\tfrac{1}{2})X(\theta_1,\theta_2)\Big\}$$

$$+\frac{3}{64}\frac{\cos 2\theta_1\cos 2\theta_2(\cos 2\theta_1+\cos 2\theta_2)}{\cos\theta_1\cos\theta_2(\sin\theta_1+\sin\theta_2)(1+2\sin\theta_1\sin\theta_2)}KA_0 \qquad (E.10)$$

where

$$X(\theta_1,\theta_2)=\frac{3\sqrt{3}}{16\cos\theta_1\cos\theta_2}\Bigg\{\frac{\cos 2\theta_1+\cos 2\theta_2}{1+2\sin\theta_1\sin\theta_2}+2\sin\theta_1\sin\theta_2$$

$$-\frac{2}{\cos 2\theta_1-\cos 2\theta_2}\Bigg[\sin 2\theta_1\cos 2\theta_2\left(\frac{1}{3}-\frac{2\theta_1}{3\pi}\right)$$

$$-\sin 2\theta_2\cos 2\theta_1\left(\frac{1}{3}-\frac{2\theta_2}{3\pi}\right)\Bigg]\Bigg\} \qquad (E.11)$$

Thus, if KA_0 is known, so is Γ, and, since the contact condition (5.193) specifies that $A_0 = 0$, Γ is given by (E.10).

Author index

Subject index

www.ingramcontent.com/pod-product-compliance
Lightning Source LLC
Chambersburg PA
CBHW050521190326
41458CB00005B/1621